MW01093821

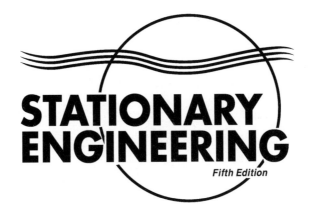

STATIONARY ENGINEERING

Fifth Edition

atp AMERICAN TECHNICAL PUBLISHERS
ORLAND PARK, ILLINOIS

Frederick M. Steingress
Harold J. Frost
Daryl R. Walker

American Technical Publishers, Inc., Editorial Staff

Editor in Chief:
  Jonathan F. Gosse
Vice President—Production:
  Peter A. Zurlis
Assistant Production Manager:
  Nicole D. Bigos
Copy Editor:
  Talia J. Lambarki
Editoiral Assistant:
  Alex C. Tulik

Cover Design:
  Jennifer M. Hines
Illustration/Layout:
  Bethany J. Fisher
Digital Resources:
  Robert E. Stickley

5 6 7 8 9 – 16 – 9 8 7 6 5 4 3 2 1

Printed in the United States of America

ISBN 978-0-8269-4329-3

 This book is printed on recycled paper.

ACKNOWLEDGMENTS

The authors and publisher are grateful to the following companies and organizations for providing photographs, technical information, and assistance:

ASI Robicon

Babcock & Wilcox Co.

Bacharach, Inc.

Cleaver-Brooks

Cooper Industries Bussmann Division

Copper Development Association

Cutler-Hammer

Dresser Industries, Inc.

Fireye, Inc.

Fluke Corporation

Furnas Electric Company

General Electric Company

Hach Company, USA

H&C Heat Transfer Solutions Inc.

Horlick Company, Inc.

ITT Bell & Gossett

Motoman, Inc.

Pollulert Systems/Mallory Components Group

Rockwell Automation, Allen-Bradley Company

Rockwell Automation, Inc.

Sarco Company, Inc.

Siemens

Teledyne Farris Engineering

Trico Corporation

U.S. Green Building Council

Wheelabrator Spokane Inc.

Woodward, Inc.

Worthington Pump

Zurn Constructors, Inc.

CONTENTS

LEARNER RESOURCES

- Quick Quizzes®
- Illustrated Glossary
- Reciprocating Steam Engines
- Sample Licensing Exams
- Flash Cards
- Master Math® Problems
- Media Library
- ATPeResources.com

INTRODUCTION

Stationary Engineering is a comprehensive reference for all aspects of boiler operation and auxiliary equipment. The textbook can be used for licensing examination preparation, industrial classes, or as a reference for studying boiler operation principles and upgrading skills. The sample licensing examination chapters include 1000 practice questions and are presented as multiple choice and essay questions.

New to this edition is coverage of the National Ambient Air Quality Standards (NAAQS), the Clean Air Act, Source Performance Standards, and Classification of Emissions Sources. Expanded topics include deaerators, safety valve springs, natural gas burners, and mechanical-hydraulic governors. Also incorporated are the latest methods of testing steam traps, colorimetric and instrumental tests, and other test methods used to test boiler water.

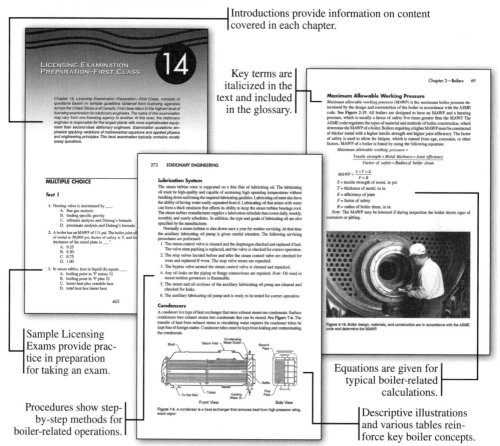

Introductions provide information on content covered in each chapter.

Key terms are italicized in the text and included in the glossary.

Sample Licensing Exams provide practice in preparation for taking an exam.

Procedures show step-by-step methods for boiler-related operations.

Equations are given for typical boiler-related calculations.

Descriptive illustrations and various tables reinforce key boiler concepts.

LEARNER RESOURCES

The Learner Resources include the following:

- **Quick Quizzes**® that provide interactive questions for each chapter, with embedded links to highlighted content within the textbook and to the Illustrated Glossary

- An **Illustrated Glossary** that defines common boiler operation and auxiliary equipment terms and provides links to illustrations for selected terms

- **Reciprocating Steam Engines** covers the theory and operating principles of reciprocating steam engines

- **Sample Licensing Exams** provide 150 additional practice problems for taking a licensing exam

- **Flash Cards** that provide a self-study/review tool to match terms and definitions found in the textbook

- **Master Math**® **Problems** provide 100 practice applications of trade-related math to complement the math in the textbook

- A **Media Library** that consists of videos and animations that reinforce and expand textbook content

- **ATPeResources.com**, which provides access to online reference materials for continued learning

STATIONARY ENGINEERING PRINCIPLES

Steam generated in boilers used for heating, generating electricity, and industrial processes must be carefully controlled and managed by a stationary engineer. Boilers and auxiliary equipment are subjected to stress and high temperatures during operation and must be designed and constructed of materials that withstand such forces.

A stationary engineer must know the basic principles that cause stress and high temperatures in a boiler and auxiliary equipment. Knowledge of these basic principles provides a stationary engineer with the reasons why and how things act in a plant. This knowledge, when combined with the necessary operating skills, provides the background necessary for successful plant management.

BOILER OPERATION THEORY

Steam boilers are used for heating buildings and for various industrial processes. A *boiler* is a closed vessel containing water. The water in a boiler is pressurized and turned into steam when heat is added. The steam is then directed to various locations for use. When a fuel is burned, the chemical energy in the fuel is transformed into heat. The heat, which is a form of energy, is released in the steam. Auxiliary equipment uses steam produced by the boiler to perform work.

Basic principles of physics and chemistry are involved in the operation of a boiler. A stationary engineer who understands these basic principles is better qualified to operate the boiler with maximum safety and efficiency. Basic principles that are involved in boiler operation include pressure, force, and area; work; and power.

Pressure, Force, and Area

In nature, forces are present at all times. For example, wind as a force acts on a sail or windmill and causes motion. Boiler room equipment controls and directs various forces to perform specific functions. For example, the force of steam when directed against a steam turbine rotor causes the rotor to rotate. Another example of force is steam pressure applied to the area of a safety valve disc. **See Figure 1-1.**

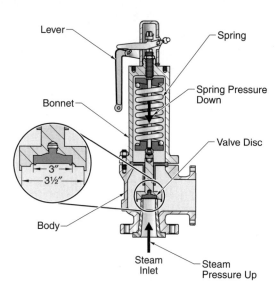

Figure 1-1. The force of steam pressure acts on the valve disc of the safety valve. When the force of steam pressure exceeds the force of the safety valve spring, the valve pops open.

Force is a pushing or pulling motion and is commonly measured in pounds. The amount of surface affected by the force is the area. *Area* is a surface measured in square inches or square feet. For a rectangle, area is equal to length times width. For example, the area of a square 2″ by 2″ is 4 sq in.

$Area = Length \times Width$

$A = L \times W$

$A = 2'' \times 2''$

$A = \textbf{4 sq in.}$

A surface that is 2′ by 2′ has an area of 4 sq ft.

$A = L \times W$

$A = 2' \times 2'$

$A = \textbf{4 sq ft}$

For a circle, area is equal to 0.7854 times the diameter squared. For example, the area of a circle 5″ in diameter is 19.635 sq in.

$A = 0.7854 \times D^2$

$A = 0.7854 \times 5 \times 5$

$A = \textbf{19.635 sq in.}$

Force, such as with steam and water pistons and safety valve discs, applied to areas of surfaces is commonly calculated by a stationary engineer. **See Figure 1-2.**

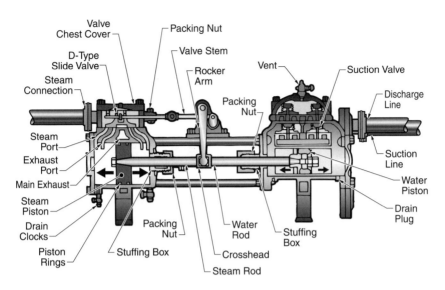

Figure 1-2. Steam engines follow the principle that the amount of force is equal to pressure times area.

Area and force are used to determine pressure. *Pressure* is force acting on a unit of area. Pressure is equal to force divided by area. Pressure readings from pressure gauges are used by a stationary engineer to assist in the safe and efficient operation of a plant.

$$Pressure = \frac{Force}{Area}$$

$$P = \frac{F}{A}$$

Pressure is measured in pounds per square inch (psi) or pounds per square foot (psf). A force of 200 lb applied to an area of 4 sq in. results in 50 psi of pressure exerted.

$$P = \frac{F}{A}$$

$$P = \frac{200 \text{ lb}}{4 \text{ sq in.}}$$

$$P = \textbf{50 psi}$$

A force of 200 lb applied to an area of 4 sq ft results in 50 psf of pressure exerted.

$$P = \frac{F}{A}$$

$$P = \frac{200 \text{ lb}}{4 \text{ sq ft}}$$

$$P = \textbf{50 psf}$$

If any two factors of an equation are known, the third factor can be found. Force is equal to pressure times area, and area is equal to force divided by pressure.

$Force = Pressure \times Area$

$F = P \times A$

$Area = \dfrac{Force}{Pressure}$

$A = \dfrac{F}{P}$

EXAMPLES

1. What is the pressure if a force of 1000 lb is applied to an area of 4 sq in.?

$P = \dfrac{F}{A}$

$P = \dfrac{1000 \text{ lb}}{4 \text{ sq in.}}$

$P = 250$ psi

2. What is the pressure on a surface if a force of 1500 lb is applied to a rectangular area 5" by 3"?

$P = \dfrac{F}{A}$

$P = \dfrac{1500 \text{ lb}}{5'' \times 3''}$

$P = \dfrac{1500 \text{ lb}}{15 \text{ sq in.}}$

$P = 100$ psi

3. What is the pressure on the surface if a force of 707 lb is applied to a disc 3" in diameter?

$P = \dfrac{F}{A}$

$P = \dfrac{707 \text{ lb}}{0.7854 \times 3'' \times 3''}$

$P = \dfrac{707 \text{ lb}}{7.0686 \text{ sq in.}}$

$P = 100$ psi

4. What is the area a 1500 lb force is acting on if it exerts a pressure of 100 psi?

$$A = \frac{F}{P}$$

$$A = \frac{1500 \text{ lb}}{100 \text{ psi}}$$

$A = $ **15 sq in.**

5. What is the area a 707 lb force is acting on if it exerts a pressure of 100 psi?

$$A = \frac{F}{P}$$

$$A = \frac{707 \text{ lb}}{100 \text{ psi}}$$

$A = $ **7.07 sq in.**

6. What is the force on the bottom of a tank that has base dimensions of 6′ by 12′ and contains a water depth of 7′? *Note:* Feet must be converted to inches. There is a pressure of 0.433 psi per vertical foot (depth) of water.

$F = P \times A$

$F = (0.433 \times 7) \text{ psi} \times [(6' \times 12) \times (12' \times 12)] \text{ sq in.}$

$F = 3.031 \text{ psi} \times (72 \times 144) \text{ sq in.}$

$F = 3.031 \text{ psi} \times 10{,}368 \text{ sq in.}$

$F = $ **31,425 lb**

7. What is the force on a flat head of a steam boiler drum (tube sheet) with a pressure of 100 psi and an area of 1000 sq in.?

$F = P \times A$

$F = 100 \text{ psi} \times 1000 \text{ sq in.}$

$F = $ **100,000 lb**

Work

Work is the movement of an object by a constant force to a specific distance. The units of distance are stated in linear measurement (feet). Force is a push or pull measured in pounds (weight). Before machines were invented, people did much of the physical work. The amount of physical work performed was measured by how tired a person was. When machines were designed and built, a measurement of work produced became necessary. A standard unit of work provided manufacturers throughout the world with consistent measurements.

The amount of work performed is the product of the force and the distance moved. The unit of measurement of work is the foot-pound (ft-lb).

$Work = Force \times Distance$

$W = F \text{ (in lb)} \times D \text{ (in ft)}$

$W = F \times D$

$W = ft\text{-}lb$

Again, if any two factors of an equation are known, the third factor can be found. Distance is equal to work divided by force, and force is equal to work divided by distance.

$Distance = \dfrac{Work}{Force}$

$D = \dfrac{W}{F}$

$Force = \dfrac{Work}{Distance}$

$F = \dfrac{W}{D}$

EXAMPLES

1. How many ft-lb of work are done if a force of 80 lb is applied and moves an object 100'?

   $W = F \times D$

   $W = 80 \text{ lb} \times 100'$

   $W = \textbf{8000 ft-lb}$

2. How high is a 2500 lb load lifted by a hoist performing 50,000 ft-lb of work?

   $D = \dfrac{W}{F}$

   $D = \dfrac{50{,}000 \text{ ft-lb}}{2500 \text{ lb}}$

   $D = \textbf{20}'$

3. What is the force applied to move an object if 450 ft-lb of work is done while moving this object 15'?

   $F = \dfrac{W}{D}$

   $F = \dfrac{450 \text{ ft-lb}}{15'}$

   $F = \textbf{30 lb}$

Power

Power is the rate at which work is performed. The unit of measurement for power is ft-lb per unit of time. For example, if minutes were the unit of time, power would be expressed in ft-lb/min.

$$Power = \frac{Work}{Time}$$

$$P = \frac{W}{T}$$

$$Work = Force \times Distance$$

$$W = F \times D$$

$$Power = \frac{Force \times Distance}{Time}$$

$$P = \frac{F \times D}{T}$$

$$Power = \frac{ft\text{-}lb}{Time}$$

$$P = \frac{ft\text{-}lb}{T}$$

EXAMPLE

1. What is the power output of an engine that performs 140,000 ft-lb of work in 4 minutes?

$$P = \frac{W}{T}$$

$$P = \frac{ft\text{-}lb}{T}$$

$$P = \frac{140,000 \text{ ft-lb}}{4 \text{ min}}$$

$$P = \textbf{35,000 ft-lb/min}$$

Before machines were invented, horses were used as a primary source of power for work. It was determined that a horse could move 33,000 lb a distance of 1′ in 1 minute. This amount was designated as 1 horsepower (HP). Horsepower was accepted as the standard measure of power. One mechanical HP is equal to 33,000 ft-lb of work per minute.

1 mechanical HP = 33,000 ft-lb of work/min

Note: Use the answer from the previous equation.

$$Horsepower = \frac{Power}{33,000 \text{ ft-lb/min}}$$

$$HP = \frac{P}{33,000 \text{ ft-lb/min}}$$

$$HP = \frac{35,000 \text{ ft-lb/min}}{33,000 \text{ ft-lb/min}}$$

$$HP = \textbf{1.06 HP}$$

EXAMPLES

1. How much horsepower will an engine develop if it does 168,000 ft-lb of work in 2½ minutes?

$$HP = \frac{W}{T \times 33,000}$$

$$HP = \frac{168,000 \text{ ft-lb}}{2.5 \text{ min} \times 33,000 \text{ ft-lb/min}}$$

$$HP = \frac{168,000 \text{ ft-lb}}{82,500 \text{ ft-lb}}$$

$$HP = \textbf{2.04 HP}$$

2. A pump delivers 2500 lb of water per minute and lifts it 280′. What is the horsepower of the pump? *Note:* Do not factor in friction losses.

$$HP = \frac{F \times D}{T \times 33,000 \text{ ft-lb/min}}$$

$$HP = \frac{2500 \text{ lb} \times 280′}{1 \text{ min} \times 33,000 \text{ ft-lb/min}}$$

$$HP = \frac{700,000 \text{ ft-lb}}{33,000 \text{ ft-lb}}$$

$$HP = \frac{700 \text{ ft-lb}}{33 \text{ ft-lb}}$$

$$HP = \textbf{21.21 HP}$$

HEAT AND ENERGY

Energy is the ability to do work. Two forms of energy are potential energy and kinetic energy. *Potential energy* is energy associated with position or shape. Examples of potential energy are a weight held above a floor or a slingshot in firing position. The weight above a floor has potential energy because of its location. The slingshot has potential energy because of its shape in the firing position. *Kinetic energy* is energy due to a body in motion. For example, a moving train is an example of kinetic energy. Energy is measured in ft-lb.

Heat is energy caused by molecules in motion within a substance. To be in motion, molecules must have kinetic energy. If heat is added to a substance, it causes molecules to increase in velocity. This results in an increase of energy. If heat is removed from a substance, molecules decrease in velocity and there is a decrease in energy.

A substance can be in a solid, liquid, or gas (vapor) state. For example, water when solid is ice, when liquid is water, and when gas is steam. The change to or from one state to another requires the addition or removal of heat. When enough heat is added to ice, it changes to water. When enough heat is added to water, it turns into steam. **See Figure 1-3.**

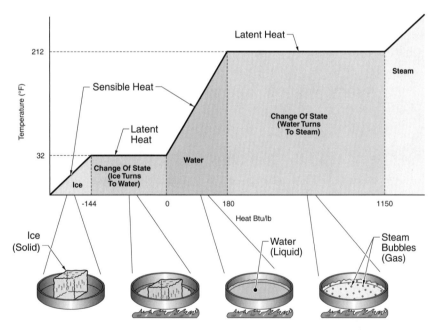

Figure 1-3. The addition or removal of heat causes a substance to change from one state to another.

Latent heat (hidden heat) is heat added to a substance that changes its state without a change of temperature. For example, when water starts to boil, as more heat is added, the water turns to steam with no change in temperature.

Latent heat of fusion is the heat required to change the state of 1 lb of a substance from a solid to a liquid at a constant temperature. For example, 144 Btu/lb is required to change 1 lb of ice to 1 lb of water at 32°F.

Sensible heat is heat that changes the temperature of a substance, but not its state. For example, heat applied to water, changing its temperature from 45°F to 55°F, is sensible heat.

Latent heat of evaporation is the heat necessary to change 1 lb of a substance from a liquid to a gas at a constant temperature. For example, 1 lb of water at 212°F converted to 1 lb of steam at 212°F takes 970.3 Btu.

Specific heat is the amount of heat required to raise the temperature of 1 lb of a substance 1°F. For example, the specific heat of ice is 0.5 Btu. For every 0.5 Btu added per pound of ice, the temperature is raised 1°F.

Temperature

Temperature is the measurement of the degree or intensity of heat. The temperature of matter is a comparison of the degree of hotness or coldness on a specific scale. Temperature is measured with a thermometer or an instrument in degrees (Fahrenheit or Celsius). The quantity of heat is the amount of heat measured in Btu from a given base. The base used for water is 32°F. One Btu is the amount of heat necessary to change the temperature of 1 lb of water 1°F at 59°F. The quantity of latent heat added or removed from a substance is found by using the following equation:

$$Quantity\ of\ heat = Weight\ of\ substance \times Latent\ heat$$
$$Q\ (in\ Btu) = W\ (in\ lb) \times Lh\ (in\ Btu/lb\ to\ change\ its\ state)$$
$$Q = W \times Lh$$

EXAMPLE

1. How much heat is required to change 1000 lb of water to steam at 212°F? *Note:* The latent heat of evaporation is 970.3 Btu/lb at atmospheric pressure.

   $Q = W \times Lh$

   $Q = 1000\ lb \times 970.3\ Btu/lb$

   $Q = 970,300\ Btu$

Heat Transfer

Heat flows or is transferred from one substance to another when a temperature difference exists. Heat is always transferred from a substance with a high temperature to a substance with a low temperature. The rate of heat transfer is proportional to the temperature difference between the two substances. Heat transfer rate increases with the temperature difference between two substances. Three methods of heat transfer are conduction, convection, and radiation.

Conduction is heat transfer that occurs when molecules come into direct contact with each other, and energy is passed from one to the other. For example, if one end of a metal rod is heated, heat is transferred by conduction to the other end. By heating the end of the metal rod, molecules are heated and move faster. The faster-moving molecules transfer energy from molecule to molecule across the metal rod.

Convection is heat transfer that occurs when heat is transferred by currents in a fluid. For example, in a pot of water with heat applied to the bottom, the warm water rises and is replaced by cooler water. The movement of water creates a current that continues as long as heat is applied. Heat transfer by convection occurs in air and hot-water heating systems. Air is heated as it passes around a radiator of the hot-water heating systems. The heated air rises and is replaced with cooler air at the bottom. The movement of air creates a current as long as heat is applied.

Radiation is heat transfer that occurs when heat is transferred without a material carrier. For example, light waves reaching the earth from the sun are a form of radiation. Similarly, when a furnace door is opened, the radiant heat of the fire is sensed immediately.

The quantity of sensible heat added to or removed from a substance is found by using the following equation:

Quantity of heat = Weight of substance × Specific heat ×
(Final temperature – Initial temperature)

Q (in Btu) = W (in lb) × Sph (in Btu/lb/°F) × [T_1 (in °F) – T_2 (in °F)]

$$Q = W \times Sph \times (T_1 - T_2)$$

EXAMPLE

1. How much heat transfer is required to raise the temperature of 1000 lb of water from 92°F to 212°F? *Note:* Specific heat of water is 1 Btu/lb/°F.

$Q = W \times Sph \times (T_1 - T_2)$
Q = 1000 lb × 1 Btu/lb/°F × (212 – 92)°F
Q = 1000 lb × 120 Btu/lb
Q = **120,000 Btu**

Temperature Conversion

Sometimes it is necessary to convert degrees Fahrenheit (°F) to degrees Celsius (°C) or degrees Celsius to degrees Fahrenheit. It also may be necessary to convert degrees Fahrenheit to absolute temperature in degrees Rankine (°R) and to convert degrees Celsius to absolute temperature in degrees Kelvin (K). These conversions are commonly required in problem solving.

There are 180° between the freezing point and boiling point on the Fahrenheit scale (212° − 32° = 180°) and 100° on the Celsius scale (100° − 0° = 100°). **See Figure 1-4.** This does not mean that 180°F equals 100°C. A difference in temperature readings of 180°F is equal to a difference in temperature readings of 100°C. This is a ratio of 180 to 100, or 1.8 to 1.0. If a number of degrees Celsius is multiplied by 1.8, the result is a corresponding number of degrees Fahrenheit. However, because 32°F is equal to 0°C, 32°F must be added so that the reading corresponds on both scales. The following equation is used for temperature conversion between Fahrenheit and Celsius:

$$°Fahrenheit = (1.8 \times °Celsius) + 32$$

$$°F = (1.8 \times °C) + 32$$

$$°Celsius = \frac{°Fahrenheit - 32}{1.8}$$

$$°C = \frac{°F - 32}{1.8}$$

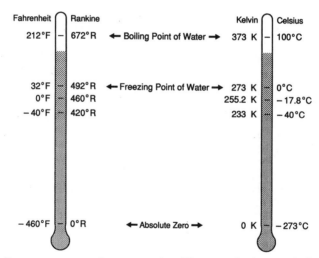

Figure 1-4. Temperatures must be converted to different scales in some boiler calculations.

To convert degrees Fahrenheit to absolute temperature on the Rankine scale, add 460. To convert degrees Celsius to absolute temperature on the Kelvin scale, add 273.15. This is typically rounded to 273. *Note:* The degree symbol (°) is not used with the Kelvin scale. The equations are as follows:

$$°Rankine = °Fahrenheit + 460$$

$$°R = °F + 460$$

$$Kelvin = °Celsius + 273$$

$$K = °C + 273$$

EXAMPLES

1. What is 60°C on the Fahrenheit scale?

   $°F = (1.8 \times °C) + 32$

   $°F = (1.8 \times 60°C) + 32$

   $°F = 108 + 32$

   $°F = \mathbf{140°F}$

2. What is 180°F on the Celsius scale?

   $°C = \dfrac{°F - 32}{1.8}$

   $°C = \dfrac{180 - 32}{1.8}$

   $°C = \dfrac{148}{1.8}$

   $°C = \mathbf{82.2°C}$

3. What is 140°F on the Rankine scale?

   $°R = °F + 460$

   $°R = 140 + 460$

   $°R = \mathbf{600°R}$

4. What is 82°C on the Kelvin scale?

   $K = °C + 273$

   $K = 82 + 273$

   $K = \mathbf{355\ K}$

Heat and Mechanical Energy Conversion

Heat is a form of energy that can be converted into other forms of energy. Mechanical energy is measured in ft-lb, and heat is measured in British thermal units (Btu). One Btu is equal to 778 ft-lb of mechanical energy. To convert Btu to ft-lb, the number of Btu is multiplied by 778.

$$ft\text{-}lb = Btu \times 778$$

To convert ft-lb to Btu, divide ft-lb by 778.

$$Btu = \frac{ft\text{-}lb}{778}$$

EXAMPLES

1. The total amount of heat 1 lb of steam has as it leaves the boiler is 1200 Btu. What is the equivalent in ft-lb?

   ft-lb = Btu × 778

   ft-lb = 1200 Btu × 778

   ft-lb = **933,600 ft-lb**

2. There are 33,000 ft-lb/min in a mechanical horsepower. What is the equivalent in Btu/min?

   $$Btu/min = \frac{ft\text{-}lb/min}{778}$$

   $$Btu/min = \frac{33,000 \ ft\text{-}lb/min}{778}$$

   Btu/min = **42.4 Btu/min**

Gas Laws

The relationship between the volume, temperature, and pressure of gases follows specific laws. To use the gas law formulas correctly, the temperature must be measured or corrected to absolute degrees (°R) and pressure must be measured or corrected to pounds per square inch absolute (psia). *Charles' first law* states that, for a constant pressure process, gas volume increases as temperature increases. As temperature decreases, volume decreases. This relationship is expressed in the following equation:

$$\frac{Volume_1}{Temperature_1} = \frac{Volume_2}{Temperature_2}$$

$$\frac{V_1}{T_1} = \frac{V_2}{T_2}$$

V_1 and T_1 are the volume and temperature of the gas before it is heated or cooled. V_2 and T_2 are the volume and temperature of the gas after it is heated or cooled.

Charles' second law states that, for a constant volume process, pressure increases as temperature increases. As temperature decreases, pressure decreases. This relationship is expressed in the following equation:

$$\frac{Pressure_1}{Temperature_1} = \frac{Pressure_2}{Temperature_2}$$

$$\frac{P_1}{T_1} = \frac{P_2}{T_2}$$

P_1 and T_1 are the pressure and temperature of the gas before it is heated or cooled. P_2 and T_2 are the pressure and temperature of the gas after it is heated or cooled. Volume is measured in cubic feet (cu ft).

EXAMPLES

1. What is the final volume of 10 cu ft of gas at 90°F when heated to 200°F at a constant pressure?

$$\frac{V_1}{T_1} = \frac{V_2}{T_2}$$

$$T_1 V_2 = V_1 T_2$$

$$\frac{T_1 V_2}{T_1} = \frac{V_1 T_2}{T_1}$$

$$V_2 = \frac{V_1 T_2}{T_1}$$

Note: °R = °F + 460

$$V_2 = \frac{10 \text{ cu ft} \times (200 + 460)°R}{(90 + 460)°R}$$

$$V_2 = \frac{10 \text{ cu ft} \times 660°R}{550°R}$$

$$V_2 = \frac{6600}{550}$$

V_2 = **12 cu ft**

2. A gas with an initial temperature of 60°F has a volume of 5 cu ft. It is heated at a constant pressure to 580°F. What is its final volume?

$$\frac{V_1}{T_1} = \frac{V_2}{T_2}$$

$$V_2 = \frac{V_1 T_2}{T_1}$$

$$V_2 = \frac{5 \text{ cu ft} \times (580 + 460)°R}{(60 + 460)°R}$$

$$V_2 = \frac{5 \text{ cu ft} \times 1040°R}{520°R}$$

$$V_2 = \frac{5200}{520}$$

V_2 = **10 cu ft**

3. What is the pressure on a gas tank with an initial temperature of 90°F and a pressure of 100 psia (85.3 psig) when heated to 200°F at a constant volume? *Note:* Absolute pressure is equal to gauge pressure plus atmospheric pressure.

$$\frac{P_1}{T_1} = \frac{P_2}{T_2}$$

$$P_2 T_1 = P_1 T_2$$

$$\frac{P_2 T_1}{T_1} = \frac{P_1 T_2}{T_1}$$

$$P_2 = \frac{P_1 T_2}{T_1}$$

$$P_2 = \frac{100 \text{ psia} \times (200 + 460)°R}{(90 + 460)°R}$$

$$P_2 = \frac{100 \text{ psia} \times 660°R}{550°R}$$

$$P_2 = \frac{66,000}{550}$$

$$P_2 = \textbf{120 psia (105.3 psig)}$$

4. What is the pressure of a gas tank with a pressure of 15 psia (0.3 psig) and a temperature of 60°F if the tank is heated to 580°F at a constant volume?

$$\frac{P_1}{T_1} = \frac{P_2}{T_2}$$

$$P_2 = \frac{P_1 T_2}{T_1}$$

$$P_2 = \frac{15 \text{ psia} \times (580 + 460)°R}{(60 + 460)°R}$$

$$P_2 = \frac{15 \text{ psia} \times 1040°R}{520°R}$$

$$P_2 = \frac{15,600}{520}$$

$$P_2 = \textbf{30 psia (15.3 psig)}$$

Boyle's law states that there is a relationship between the volume, temperature, and pressure of gases. With a constant temperature process for each change of pressure, there is an inverse (opposite) change of volume. When pressure increases, volume decreases. When pressure decreases, volume increases. This relationship is expressed in the following equation:

$$Pressure_1 \times Volume_1 = Pressure_2 \times Volume_2$$

$$P_1 V_1 = P_2 V_2$$

P_1 and V_1 are the pressure and volume of the gas before change. P_2 and V_2 are the pressure and volume of the gas after change. Temperature is constant. Pressure is measured in psia and volume is measured in cu ft.

EXAMPLES

1. What is the final pressure of 1 cu ft of air with an initial pressure of 100 psia (85.3 psig) that is expanded at a constant temperature to 2 cu ft?

$$P_1V_1 = P_2V_2$$

$$\frac{P_1V_1}{V_2} = \frac{P_2V_2}{V_2}$$

$$P_2 = \frac{P_1V_1}{V_2}$$

$$P_2 = \frac{100 \text{ psia} \times 1 \text{ cu ft}}{2 \text{ cu ft}}$$

$$P_2 = \frac{100}{2}$$

$P_2 =$ **50 psia (35.3 psig)**

2. Five pounds of air have an initial volume of 5 cu ft and an initial pressure of 20 psia (5.3 psig). What is the final pressure if the air is expanded at a constant temperature to a final volume of 10 cu ft?

$$P_1V_1 = P_2V_2$$

$$P_2 = \frac{P_1V_1}{V_2}$$

$$P_2 = \frac{20 \text{ psia} \times 5 \text{ cu ft}}{10 \text{ cu ft}}$$

$$P_2 = \frac{100}{10}$$

$P_2 =$ **10 psia (−4.7 psig)**

Combining Charles' and Boyle's laws provides the basis for the general gas law.

Charles' first law: $\dfrac{V_1}{T_1} = \dfrac{V_2}{T_2}$

Charles' second law: $\dfrac{P_1}{T_1} = \dfrac{P_2}{T_2}$

Boyle's law: $P_1V_1 = P_2V_2$

The *general gas law* states that the product of the pressure and volume divided by the temperature before change is equal to the product of the pressure and volume divided by the temperature after change. This relationship is expressed in the following equation:

$$\frac{Pressure_1 \times Volume_1}{Temperature_1} = \frac{Pressure_2 \times Volume_2}{Temperature_2}$$

$$\frac{P_1 V_1}{T_1} = \frac{P_2 V_2}{T_2}$$

Pressure is measured in psia, volume in cu ft, and temperature in °R.

EXAMPLES

1. Use the following data to determine the final gas volume.

P_1 = 15 psia

P_2 = 30 psia

T_1 = 70°F

T_2 = 600°F

V_1 = 1 cu ft

$$\frac{P_1 V_1}{T_1} = \frac{P_2 V_2}{T_2}$$

$$P_2 V_2 T_1 = P_1 V_1 T_2$$

$$V_2 = \frac{P_1 V_1 T_2}{P_2 T_1}$$

$$V_2 = \frac{15 \text{ psia} \times 1 \text{ cu ft} \times (600 + 460)°R}{30 \text{ psia} \times (70 + 460)°R}$$

$$V_2 = \frac{15 \text{ psia} \times 1 \text{ cu ft} \times 1060°R}{30 \text{ psia} \times 530°R}$$

$$V_2 = \frac{15,900}{15,900}$$

V_2 = **1 cu ft**

2. Use the following data to determine the final gas pressure.

P_1 = 30 psia

T_1 = 700°R

T_2 = 1400°R

V_1 = 25 cu ft

V_2 = 15 cu ft

$$\frac{P_1 V_1}{T_1} = \frac{P_2 V_2}{T_2}$$

$$P_2 = \frac{P_1 V_1 T_2}{V_2 T_1}$$

$$P_2 = \frac{30 \text{ psia} \times 25 \text{ cu ft} \times 1400°R}{15 \text{ cu ft} \times 700°R}$$

$$P_2 = \frac{1,050,000}{10,500}$$

P_2 = **100 psia**

SIMPLE MACHINES

Simple machines are tools used to convert energy to useful work in the boiler and auxiliary equipment. Force is used by simple machines to perform specific tasks. Force is a pushing or pulling motion acting on another object. *Stress* is the result of forces acting on an object. Different types of stress include compression stress, shear stress, and tensile stress. The type of stress depends on how the force is applied. **See Figure 1-5.**

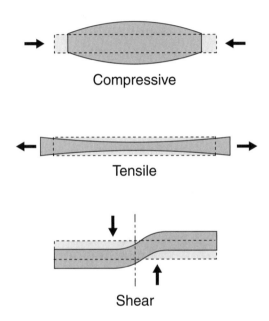

Compressive

Tensile

Shear

Figure 1-5. Boilers in operation are subjected to many stresses. As pressure in the boiler increases, stress also increases.

Compression stress is stress that is the result of force applied to an object that presses or squeezes the object. For example, tubes in a firetube boiler have force applied on them from steam pressure in the boiler. *Tensile stress* is stress that is the result of force applied to an object that stretches an object. For example, tube sheets in a firetube boiler could be stretched by stress from steam pressure. *Shear stress* is stress that is the result of force applied to an object that shears or cuts through an object. For example, a stay applies stress in a direction opposite the stress applied by the steam pressure in the boiler.

Forces can act on an object in several ways. **See Figure 1-6.** When forces are acting in the same direction, the resultant force is the sum of the forces. When forces are acting in opposite directions, the resultant force is the difference between the forces. When forces are acting at an angle to each other, the resultant force is determined by the resolution of all forces acting on the object.

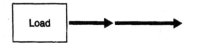

Forces Acting in Same Direction

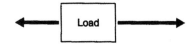

Forces Acting in Opposite Directions

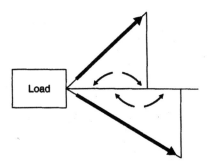

Resolution of All Forces Acting on an Object

Figure 1-6. Force is a pushing or pulling motion and can act on an object in several ways.

For example, two forces are acting on a load at an angle to each other. This can happen when a load is suspended from a cable or wire at an angle and another cable or wire is used from below to keep the load from moving. One force is the upward and sideways pull of the first cable. The other force is the downward and sideways pull of the first cable. The first force can be resolved into F_1 in the horizontal direction and F_1' in the vertical direction. The second force can be resolved into F_2 in the horizontal direction and F_2' in the vertical direction. The sum of F_1 and F_2 is the total force in the horizontal direction. The sum of F_1' and F_2' is the total force in the vertical direction. The values of the forces may be positive or negative depending on the direction of the forces relative to the directions of positive on the horizontal and vertical axes. Up and to the right are positive and down and to the left are negative.

Resultant force of forces acting in the same direction is determined by the sum of the forces.

$Resultant\ force = Force_1 + Force_2$

$Resultant\ force = F_1 + F_2$

Resultant force of forces acting in opposite directions is determined by the sum of the forces, with one of the forces being negative because they are in opposite direction. If the sum of the positive and negative forces is positive, the net force is in the direction of the force in the positive direction. If the sum of the positive and negative forces is negative, the net force is in the direction of the force in the negative direction.

$Resultant\ force = Force_1 + Force_2$

$Resultant\ force = F_1 + F_2$

Resultant force of forces acting at angles to each other is determined by the sum of the forces in the horizontal direction and the forces in the vertical direction.

$Resultant\ force\ (horizontal) = Force_1 + Force_2$

$Resultant\ force\ (horizontal) = F_1 + F_2$

$Resultant\ force\ (vertical) = Force_1' + Force_2'$

$Resultant\ force\ (vertical) = F_1' + F_2'$

EXAMPLES

1. What is the resultant force of forces acting in the same direction if F_1 is 600 lb and F_2 is 900 lb?

   $Resultant\ force = F_1 + F_2$

   $Resultant\ force = 600\ lb + 900\ lb$

   $Resultant\ force =$ **1500 lb**

2. What is the resultant force of forces acting in opposite directions if F_1 is 800 lb and F_2 is 600 lb?

   $Resultant\ force = F_1 + F_2$

   $Resultant\ force = 800\ lb + (-600)\ lb$

   $Resultant\ force =$ **200 lb**

   This resultant force is in the same direction as force F_1.

3. What is the resultant horizontal force of forces acting at an angle to each other if the horizontal force component F_1 is 340 lb and F_2 is 540 lb? Both horizontal forces are in the same direction.

   $Resultant\ force = F_1 + F_2$

   $Resultant\ force = 340\ lb + 540\ lb$

   $Resultant\ force =$ **880 lb**

Levers

A *lever* is a simple machine formed by a rigid bar that pivots on a fixed body (fulcrum) with both resistance and effort (force) applied. **See Figure 1-7.** There are three classes of levers, as follows:

• Class 1—Basic lever-resistance or effort is applied on either side of the fulcrum

• Class 2—Resistance and effort are applied on the same side of the fulcrum, with the effort at the greater distance from the fulcrum

• Class 3—Similar to Class 2 except the effort is applied between the fulcrum and the resistance

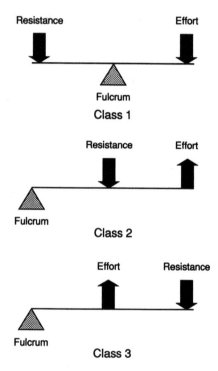

Figure 1-7. A lever is used to obtain a mechanical advantage to overcome a large resistance with a smaller force.

The purpose of a lever is to obtain a mechanical advantage to overcome a large resistance with a small effort. The lever principle is used in tools such as crowbars, pliers, and wrenches. To solve problems involving levers, the law of moments is used. The *law of moments* states that if the lever is to be in equilibrium (balanced with no motion), the sum of clockwise moments must equal the sum of counterclockwise moments. A *moment* is the product of an effort or resistance (force) times its distance from the fulcrum. **See Figure 1-8.**

Clockwise moments = Counterclockwise moments

Resistance × *Distance$_R$* = *Effort* × *Distance$_E$*

$R \times D_R = E \times D_E$

Resistance is measured in lb, distance in ft, and effort in lb.

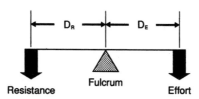

Figure 1-8. The law of moments states that for a lever to be balanced, the sum of clockwise moments must equal the sum of counterclockwise moments.

EXAMPLES

1. What effort must be applied to hold the lever in equilibrium?

$D_E = 125'$

$D_R = 25'$

$R = 100 \text{ lb}$

$R \times D_R = E \times D_E$

$\dfrac{R \times D_R}{D_E} = \dfrac{E \times D_E}{D_E}$

$E = \dfrac{R \times D_R}{D_E}$

$E = \dfrac{100 \text{ lb} \times 25'}{125'}$

$E = \dfrac{2500}{125}$

$E = \mathbf{20 \ lb}$

2. What resistance must be applied to hold the lever in equilibrium?

$D_E = 150'$

$D_R = 50'$

$E = 50 \text{ lb}$

$R \times D_R = E \times D_E$

$R = \dfrac{E \times D_E}{D_R}$

$R = \dfrac{50 \text{ lb} \times 150'}{50'}$

$R = \dfrac{7500}{50}$

$R = \mathbf{150 \ lb}$

3. What is the missing value?

$D_E = 20'$

$E = 8$ lb

$R = 40$ lb

$R \times D_R = E \times D_E$

$D_R = \dfrac{E \times D_E}{R}$

$D_R = \dfrac{8 \text{ lb} \times 20'}{40 \text{ lb}}$

$D_R = \dfrac{160}{40}$

$D_R = \mathbf{4'}$

4. What is the missing value?

$D_E = 15'$

$E = 15$ lb

$R = 45$ lb

$R \times D_R = E \times D_E$

$D_R = \dfrac{E \times D_E}{R}$

$D_R = \dfrac{15 \text{ lb} \times 15'}{45 \text{ lb}}$

$D_R = \dfrac{225}{45}$

$D_R = \mathbf{5'}$

5. What is the missing value?

$D_E = 12'$

$D_R = 6'$

$R = 20$ lb

$R \times D_R = E \times D_E$

$E = \dfrac{R \times D_R}{D_E}$

$E = \dfrac{20 \text{ lb} \times 6'}{12'}$

$E = \dfrac{120}{12}$

$E = \mathbf{10 \text{ lb}}$

6. What is the missing value?

$D_E = 8'$

$D_R = 3'$

$R = 12$ lb

$R \times D_R = E \times D_E$

$E = \dfrac{R \times D_R}{D_E}$

$E = \dfrac{12 \text{ lb} \times 3'}{8'}$

$E = \dfrac{36}{8}$

$E = \textbf{4.5 lb}$

7. What is the missing value?

$D_{E1} = 5'$

$D_{E2} = 8'$

$D_R = 12'$

$E_1 = 12$ lb

$E_2 = 5$ lb

$R \times D_R = (E_1 \times D_{E1}) + (E_2 \times D_{E2})$

$\dfrac{R \times D_R}{D_R} = \dfrac{(E_1 \times D_{E1}) + (E_2 \times D_{E2})}{D_R}$

$R = \dfrac{(E_1 \times D_{E1}) + (E_2 \times D_{E2})}{D_R}$

$R = \dfrac{(12 \text{ lb} \times 5') + (5 \text{ lb} \times 8')}{12'}$

$R = \dfrac{60 + 40}{12}$

$R = \dfrac{100}{12}$

$R = \textbf{8.33 lb}$

8. What is the missing value?

$D_{E1} = 4'$

$D_{E2} = 6'$

$D_R = 15'$

$E_1 = 7$ lb

$E_2 = 5$ lb

$$R \times D_R = (E_1 \times D_{E1}) + (E_2 \times D_{E2})$$

$$R = \frac{(E_1 \times D_{E1}) + (E_2 \times D_{E2})}{D_R}$$

$$R = \frac{(7\text{ lb} \times 4') + (5\text{ lb} \times 6')}{15'}$$

$$R = \frac{28 + 30}{15}$$

$$R = \frac{58}{15}$$

$$R = \textbf{3.87 lb}$$

Inclined Plane

An *inclined plane* is a simple machine formed by two surfaces at an acute angle to each other. Inclined planes follow the principle that when a load is moved from the bottom to the top of the inclined plane, the load is moved a greater distance using a smaller force than would be required if it were lifted vertically. **See Figure 1-9.** The ability to move a load using a smaller force is called mechanical advantage. Mechanical advantage is expressed in the following equation:

Force × Length = Resistance × Height

$F \times L = R \times H$

F = force needed to move the load

L = length of incline

R = resistance due to load

H = height

Note: This equation does not include extra force needed to overcome friction.

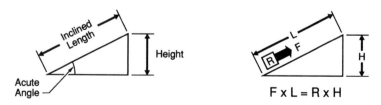

$$F \times L = R \times H$$

Figure 1-9. The load moves a greater distance on an inclined plane than if lifted vertically.

EXAMPLES

1. What is the force needed to move a 100 lb weight 12′ along an inclined plane to a point 5′ off the ground?

$$F \times L = R \times H$$

$$F = \frac{R \times H}{L}$$

$$F = \frac{100 \text{ lb} \times 5'}{12'}$$

$$F = \frac{500}{12}$$

$$F = \textbf{41.67 lb}$$

2. What is the vertical height of an inclined plane?

$$F = 25 \text{ lb}$$

$$L = 40'$$

$$R = 200 \text{ lb}$$

$$F \times L = R \times H$$

$$H = \frac{F \times L}{R}$$

$$H = \frac{25 \text{ lb} \times 40'}{200 \text{ lb}}$$

$$H = \frac{1000}{200}$$

$$H = \textbf{5}'$$

3. What is the missing value?

$$F = 15 \text{ lb}$$

$$H = 5'$$

$$R = 150 \text{ lb}$$

$$F \times L = R \times H$$

$$L = \frac{150 \text{ lb} \times 5'}{15 \text{ lb}}$$

$$L = \frac{750}{15}$$

$$L = \textbf{50}'$$

Wheel and Axle

A *wheel and axle* is a simple machine formed by an outer circular rim (wheel) fastened to an inner rim (axle). **See Figure 1-10.** When a small force is applied to the outer rim, it can lift a larger load on the inner rim. For example, when a water bucket is lifted out of a well, the rotating crank handle forms the outer rim where force is applied. The axle is the shaft the rope winds on. The center of the axle is the fulcrum. Force is measured in lb, distance (radius) in inches (in.), and resistance in lb. This relationship is expressed in the following equation:

> *Moments of force = Moments of resistance*
>
> *Force × Distance$_F$ = Resistance × Distance$_R$*
>
> $F \times D_F = R \times D_R$

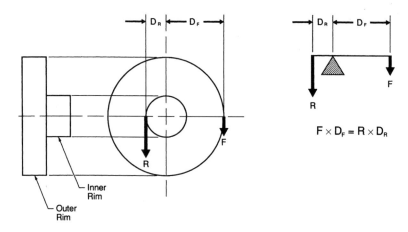

Figure 1-10. Using a wheel and axle, a small force on the outer rim can lift a larger load on the inner rim.

EXAMPLES

1. What force is required to lift a 40 lb bucket of water from a well that has an 18″ long handle and a 6″ diameter shaft?

Radius = ½ Diameter

Radius = ½ × 6″

Radius = 3″

$F \times D_F = R \times D_R$

$$\dfrac{F \times D_F}{D_F} = \dfrac{R \times D_R}{D_F}$$

$$F = \dfrac{R \times D_R}{D_F}$$

$$F = \frac{40\ lb \times 3''}{18''}$$

$$F = \frac{120}{18}$$

$F = \textbf{6.67 lb}$

2. How much of a load can be lifted with a 20 lb force using a 3' (36") diameter wheel and a 6" diameter axle?

Radius = ½ Diameter

Radius = ½ × 36"

Radius = **18"**

Radius = ½ Diameter

Radius = ½ × 6"

Radius = **3"**

$F \times D_F = R \times D_R$

$$R = \frac{F \times D_F}{D_R}$$

$$R = \frac{20\ lb \times 18''}{3''}$$

$R = \textbf{120 lb}$

Torque

Torque is force on a body that causes the body to turn. Torque produces rotary motion. It is measured in in-lb or ft-lb. There are 12 in-lb in 1 ft-lb. For example, 4500 in-lb is exerted by a force of 500 lb acting at a radius of 9", and 1 ft-lb of torque is exerted by a force of 1 lb acting at a radius of 1'. **See Figure 1-11.**

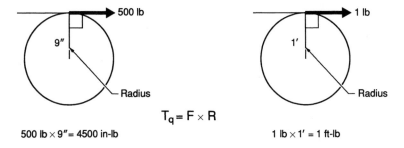

$$T_q = F \times R$$

500 lb × 9"= 4500 in-lb   1 lb × 1' = 1 ft-lb

Figure 1-11. The torque of a rotating element is its turning moment, measured in in-lb or ft-lb.

The torque of any rotating element is its turning moment. Turning moment is the product of the force and the perpendicular distance from the axis of rotation to the line of action of the force.

Torque = Force × Radius

$T_q = F \times R$

EXAMPLES

1. What is the torque developed if a 500 lb force is exerted on an 18″ diameter flywheel?

   $T_q = F \times R$

   $T_q = 500 \text{ lb} \times 6″$

   $T_q = \textbf{4500 in-lb}$

2. What is the torque developed if a 2000 lb force acts on a 2′ diameter shaft?

   Radius = ½ Diameter

   Radius = ½ × 2′

   Radius = 1′

   $T_q = F \times R$

   $T_q = 2000 \text{ lb} \times 1′$

   $T_q = \textbf{2000 ft-lb}$

Fluid Pressure

A *fluid* is any substance that flows, such as a liquid or a gas. A fluid takes the shape of the container it is in, regardless of the container shape. When pressure is applied to a fluid, pressure is transmitted in all directions. **See Figure 1-12.** This principle is used in a hydraulic press and in a deadweight pressure gauge tester. Force on the small piston times the distance traveled by the small piston equals force on the large piston times the distance traveled by the large piston. This relationship is expressed in the following equation:

Force on small piston × Distance traveled by small piston =
    Force on large piston × Distance traveled by large piston

$F \times D = F' \times D'$

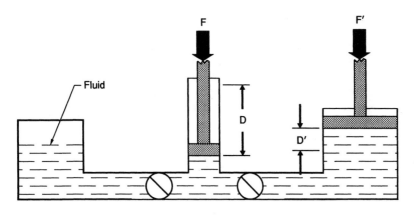

$$F \times D = F' \times D'$$

Figure 1-12. Pressure applied to a fluid is transmitted in all directions.

EXAMPLES

1. A 25 lb force is applied to a piston that moves 12". The second piston connected to the same hydraulic system only moves ½". What is the force exerted on the second piston?

$$F \times D = F' \times D'$$

$$F' = \frac{F \times D}{D'}$$

$$F' = \frac{25 \text{ lb} \times 12''}{0.5''}$$

$$F' = \frac{300}{0.5}$$

$$F' = \textbf{600 lb}$$

2. A 50 lb force is applied on the small piston and a 1500 lb force is applied on the large piston, moving it a distance of 0.1". How far does the small piston travel?

$$F \times D = F' \times D'$$

$$D = \frac{F' \times D'}{F}$$

$$D = \frac{1500 \text{ lb} \times 0.1''}{50 \text{ lb}}$$

$$D = \frac{150}{50}$$

$$D = \textbf{3}''$$

CHEMISTRY AND pH

Chemistry is the study of the composition and chemical properties of substances. A molecule is the smallest unit of a chemical compound. Chemical compounds are typically made up of many identical molecules. For example, a molecule of sodium and a molecule of chlorine chemically combine to form a molecule of table salt (sodium chloride). Many molecules of table salt come together to form salt crystals. Water is a compound composed of hydrogen and oxygen. Carbon dioxide is a compound composed of carbon and oxygen. In the boiler room, chemical elements such as carbon, sulfur, hydrogen, and oxygen form compounds. For example,

carbon + oxygen → carbon dioxide

$$C + O_2 \rightarrow CO_2$$

sulfur + oxygen → sulfur dioxide

$$S + O_2 \rightarrow SO_2$$

hydrogen + oxygen → water

$$2H_2 + O_2 \rightarrow 2H_2O$$

An *acid* is a compound that contains hydrogen (H) and is capable of reacting with a base, such as sulfuric acid (H_2SO_4) or hydrochloric acid (HCl). An *alkali,* or *base,* is a compound that contains hydroxide (OH) and is capable of reacting with acid, such as sodium hydroxide (NaOH) and calcium hydroxide [$Ca(OH)_2$]. A pH scale is used to classify a solution as acidic, alkaline, or neutral. **See Figure 1-13.** A solution with a pH less than 7 is acidic. A solution with a pH more than 7 is alkaline. A solution with a pH of 7 is neutral. When an acid is combined with an alkali in solution, the result is a salt and water. For example,

sulfuric acid + sodium hydroxide → sodium sulfate (a salt) + water

$$H_2SO_4 + 2NaOH \rightarrow Na_2SO_4 + 2H_2O$$

hydrochloric acid + calcium hydroxide → calcium chloride (a salt) + water

$$2HCl + Ca(OH)_2 \rightarrow CaCl_2 + 2H_2O$$

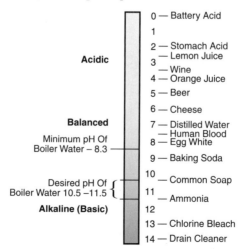

	0 — Battery Acid
	1
Acidic	2 — Stomach Acid
	— Lemon Juice
	3
	— Wine
	4 — Orange Juice
	5 — Beer
	6 — Cheese
Balanced	7 — Distilled Water
Minimum pH Of	— Human Blood
Boiler Water – 8.3	8 — Egg White
	9 — Baking Soda
	10
Desired pH Of	— Common Soap
Boiler Water 10.5 –11.5	11
	— Ammonia
Alkaline (Basic)	12
	13 — Chlorine Bleach
	14 — Drain Cleaner

Figure 1-13. The pH of a solution indicates whether the solution is acidic, neutral, or alkaline.

STEAM PROPERTIES

Steam is water in a vapor condition. Steam often does not follow the gas laws because when the temperature and pressure change, the steam may condense out as water. Superheated steam usually follows the gas laws as long as it is kept away from condensing conditions. The change from water to steam takes place at a temperature that is dependent on the pressure within the vessel. In other words, there is an equilibrium between steam pressure and its corresponding temperature. At atmospheric pressure (0 psig, 14.7 psia), water boils at 212°F. Water is releasing steam at 212°F while the water also is at 212°F. When pressure is increased to 100 psig, the water boils at 337.9°F and the steam is also at 337.9°F.

Saturated steam is a vapor at a temperature that corresponds with its pressure. Any removal of heat from the steam causes a portion of the steam to turn back into a liquid state. For example, water is heated and starts to boil at 212°F and continues to boil unless the pressure changes.

The boiling point of water at various pressures can be found in Dry Saturated Steam tables. The pressures are listed in psia and temperatures are in °F. When steam is removed from its liquid and heated to a higher temperature, it is superheated a given number of degrees. *Superheated steam* is a vapor at a temperature above its corresponding pressure.

Heat is added to the steam in the superheater after it is removed from its liquid. For example, if saturated steam is passed through a superheater to add heat, its temperature would increase, but not its pressure. The number of degrees of superheat is the difference between saturated steam temperature at the given pressure and actual steam temperature. Superheated Steam tables can be used to determine the amount of heat in superheated steam. **See Figure 1-14.** As steam pressure increases, steam temperature increases. Steam pressure leaving the superheater remains the same. Only its temperature increases.

The boiling point of water increases with increases in pressure until the critical pressure-temperature point is reached. *Critical pressure* is the pressure where water and saturated steam are indistinguishable. Critical pressure is 3206 psia and its corresponding temperature is 705°F. There is no change of state when the pressure is increased or if heat is added and the fluid has properties of both steam and water. Above this temperature, water cannot be liquefied, no matter the pressure.

Terms commonly found in Dry Saturated Steam tables include the following:

- *Gauge pressure*—Pressure recorded on a steam pressure gauge
- *Absolute pressure*—Gauge pressure plus atmospheric pressure
- *Latent heat*—Hidden heat; changes a substance's state but not its temperature
- *Latent heat of fusion*—Changes ice from 32°F to water at 32°F; requires 144 Btu/lb
- *Sensible heat*—Changes a substance's temperature but not its state; water at 32°F to water at 212°F requires 180 Btu/lb
- *Latent heat of evaporation*—Changes water at 212°F to steam at 212°F; requires 970.3 Btu/lb
- *Enthalpy*—Total heat of steam; the sum of sensible and latent heat above the base of 32°F liquid

  Note: Sensible heat becomes the enthalpy of liquid, latent heat becomes the enthalpy of evaporation, and total heat becomes the enthalpy of steam.

Properties of Superheated Steam

(v = specific volume, cu ft/lb; h = enthalpy, Btu/lb; s = entropy)

Abs press., psi (sat. temp.)		Temperature, °F									
		400	500	600	700	800	900	1000	1100	1200	1400
1 (101.74)	v	512.0	571.6	631.2	690.8	750.4	809.9	869.5	929.1	988.7	1107.8
	h	1241.7	1288.3	1335.8	1383.8	1432.8	1482.7	1533.5	1585.2	1637.7	1745.7
	s	2.1720	2.2233	2.2702	2.3137	2.3542	2.3932	2.4283	2.4625	2.4952	2.5566
5 (162.24)	v	102.26	114.22	126.16	138.10	150.03	161.95	173.87	185.79	197.71	221.6
	h	1241.2	1288.0	1335.4	1383.6	1432.7	1482.6	1533.4	1585.1	1637.7	1745.7
	s	1.9942	2.0456	2.0927	2.1361	2.1767	2.2148	2.2509	2.2851	2.3178	2.3792
10 (193.21)	v	51.04	57.05	63.03	69.01	74.98	80.95	86.92	92.88	98.84	110.77
	h	1240.6	1287.5	1335.1	1383.4	1432.5	1482.4	1533.2	1585.0	1637.6	1745.6
	s	1.9172	1.9689	2.0160	2.0596	2.1002	2.1383	2.1744	2.2086	2.2413	2.3028
14.696 (212.00)	v	34.68	38.78	42.86	46.94	51.00	55.07	59.13	63.19	67.25	75.37
	h	1239.9	1287.1	1334.8	1383.2	1432.3	1482.3	1533.1	1584.8	1637.5	1745.5
	s	1.8743	1.9261	1.9734	2.0170	2.0576	2.0958	213.19	2.1662	2.1989	2.2603
20 (227.96)	v	25.43	28.46	31.47	34.47	37.46	40.45	43.44	46.42	49.41	55.37
	h	1239.2	1286.6	1334.4	1382.9	1432.1	1482.1	1533.0	1584.7	1637.4	1745.4
	s	1.8396	1.8918	1.9392	1.9829	2.0235	2.0618	2.0978	2.1321	2.1648	2.2263
40 (267.25)	v	12.628	14.168	15.688	17.198	18.702	20.20	21.70	23.20	24.69	27.68
	h	1236.5	1284.8	1333.1	1381.9	1431.3	1481.4	1532.4	1584.3	1637.0	1745.1
	s	1.7608	1.8140	1.8619	1.9058	1.9467	1.9850	2.0212	2.0555	2.0883	2.1498
60 (292.71)	v	8.357	9.403	10.427	11.441	12.449	13.452	14.454	15.453	16.451	18.446
	h	1233.6	1283.0	1331.8	1380.9	1430.5	1480.8	1531.9	1583.8	1636.6	1744.8
	s	1.7135	1.7678	1.8162	1.8605	1.9015	1.9400	1.9762	2.0106	2.0434	2.1049
80 (312.03)	v	6.220	7.020	7.797	8.562	9.322	10.077	10.830	11.582	12.332	13.830
	h	1230.7	1281.1	1330.5	1379.9	1429.7	1480.1	1531.3	1583.4	1636.2	1744.5
	s	1.6791	1.7346	1.7836	1.8281	1.8694	1.9079	1.9442	1.9787	2.0115	2.0731

Figure 1-14. Superheated steam is steam at a higher temperature than its corresponding pressure.

Dry Saturated Steam tables indicate that saturated steam has a corresponding temperature and pressure, and, as the pressure increases, the following changes take place:
- Boiling point increases
- Sensible heat increases
- Latent heat decreases
- Total heat increases slowly until the pressure is approximately 450 psia, then decreases slowly until pressure approaches the critical pressure of 3206 psia

Dry Saturated Steam tables are divided into columns. The first two columns are temperature and pressure, which indicate the pressure-temperature relationship of dry saturated steam. For each pressure listed, there is a corresponding temperature. The columns are headed by the letters t for temperature in °F, and p for pressure in psia. Other tables have the pressure in psig. **See Figure 1-15.**

Dry Saturated Steam Table								
		Specific volume, cu ft/lb		Enthalpy, Btu/lb			Entropy		
Abs press., psi p	Temp., °F t	Sat. liquid v_f	Sat. vapor v_g	Sat. liquid h_f	Evap. h_{fg}	Sat. vapor h_g	Sat. liquid s_f	Evap. s_{fg}	Sat. vapor s_g
1.0	101.74	0.01614	333.6	69.70	1036.3	1106.0	0.1326	1.8456	1.9782
2.0	126.08	0.01623	173.73	93.99	1022.2	1116.2	0.1749	1.7451	1.9200
3.0	141.48	0.01630	118.71	109.37	1013.2	1122.6	0.2008	1.6855	1.8863
4.0	152.97	0.01636	90.63	120.86	1006.4	1127.3	0.2198	1.6427	1.8625
5.0	164.24	0.01640	73.52	130.13	1001.0	1131.1	0.2347	1.6094	1.8441
6.0	170.06	0.01645	61.98	137.96	996.2	1134.2	0.2472	1.5820	1.8292
7.0	176.85	0.01649	53.64	144.76	992.1	1136.9	0.2581	1.5586	1.8167
8.0	182.86	0.01653	47.34	150.79	998.5	1139.3	0.2674	1.5383	1.8057
9.0	188.28	0.01656	42.40	156.22	985.2	1141.4	0.2759	1.5203	1.7962
10	193.21	0.01659	38.42	161.17	982.1	1143.3	0.2835	1.5041	1.7876
14.696	212.00	0.01672	26.80	180.07	970.3	1150.4	0.3120	1.4446	1.7566
15	213.03	0.01672	26.29	181.11	969.7	1150.8	0.3135	1.4415	1.7549
20	227.96	0.01683	20.089	196.16	960.1	1156.3	0.3356	1.3962	1.7319
25	240.07	0.01692	16.303	208.42	952.1	1160.6	0.3533	1.3606	1.7139
30	250.33	0.01701	13.746	218.82	945.3	1164.1	0.3680	1.3313	1.6993
35	259.28	0.01708	11.898	227.91	939.2	1167.1	0.3807	1.3063	1.6870
40	267.25	0.01715	10.498	236.03	933.7	1169.7	0.3919	1.2844	1.6763

Figure 1-15. As saturated steam pressure is increased, boiling point of water, latent heat, and total heat are affected.

The three columns listed under enthalpy list heat in saturated liquid, heat of evaporation, and heat in saturated vapor. The sum of heat in saturated liquid and heat of evaporation is equal to heat in saturated vapor. When using Dry Saturated Steam tables, it may be necessary to interpolate values listed to obtain the required information.

As pressure in a boiler increases, the temperature of the water and steam increases, and the boiling point increases. The enthalpy of the saturated liquid also increases, but the enthalpy of evaporation decreases. The enthalpy of the saturated vapor increases up to 466.9 psia and 460°F. It then slowly decreases. Basic principles are shown by the data listed in Dry Saturated Steam tables, including the following:

- Boiling point changes with pressure
- Heat in liquid increases with pressure (sensible heat)
- Heat of evaporation decreases with pressure (latent heat)
- Total heat in vapor increases, then decreases
- Total heat in normal dry saturated steam is within boiler operating range (approximately 1200 Btu/lb)

EXAMPLES

1. Use the Dry Saturated Steam table in the Appendix to interpolate and find the following at a steam pressure of 24 psia:

   A. Enthalpy of saturated liquid

   B. Latent heat of evaporation

   C. Enthalpy of saturated vapor

   D. Saturated steam temperature

   Note: In Dry Saturated Steam tables, 24 psia is not recorded, but 20 psia and 25 psia are recorded. Use the table values at 20 and 25 and interpolate between them.

   A. *Enthalpy of saturated liquid at 24 psia:*

      24 − 20 = 4

      25 − 20 = 5

      208.42 − 196.16 = 12.26

      ⁴⁄₅ × 12.26 = 9.808

      196.16 + 9.808 = 205.968

      Enthalpy of saturated liquid at 24 psia = **205.968 Btu/lb**

   B. *Latent heat of evaporation at 24 psia:*

      ⁴⁄₅ × 8.0 = 6.4

      Because the latent heat of evaporation decreases as the pressure increases, subtract 6.4 from the 960.1 to give the latent heat of evaporation at 24 psia.

      960.1 − 6.4 = 953.7

      Latent heat of evaporation at 24 psia = **953.7 Btu/lb**

C. *Enthalpy of saturated vapor at 24 psia:*

$\frac{4}{5} \times 4.3 = 3.44$

$1156.3 + 3.44 = 1159.74$

Enthalpy of saturated vapor at 24 psia = **1159.74 Btu/lb**

D. *Saturated steam temperature at 24 psia:*

$\frac{4}{5} \times 12.11 = 9.688$

$227.96 + 9.688 = 237.648$

Saturated steam temperature at 24 psia = **237.648°F**

2. Use the Dry Saturated Steam table in the Appendix to interpolate and find the following at a steam pressure of 264 psia:

A. Enthalpy of saturated liquid

B. Latent heat of evaporation

C. Enthalpy of saturated vapor

D. Saturated steam temperature

Note: In Dry Saturated Steam tables, 264 psia is not recorded, but 250 psia and 300 psia are recorded.

A. *Enthalpy of saturated liquid at 264 psia:*

$264 - 250 = 14$

$300 - 250 = 50$

$393.84 - 376.00 = 17.84$

$\frac{14}{50} \times 17.84 = 4.9952$

$376.00 + 4.9952 = 380.9952$

Enthalpy of saturated liquid at 264 psia = **380.9952 Btu/lb**

B. *Latent heat of evaporation at 264 psia:*

$264 - 250 = 14$

$300 - 250 = 50$

$825.1 - 809.0 = 16.1$

$\frac{14}{50} \times 16.1 = 4.508$

Because the latent heat of evaporation decreases as the pressure increases, subtract the following:

$825.100 - 4.508 = 820.592$

Latent heat of evaporation at 264 psia = **820.592 Btu/lb**

C. *Enthalpy of saturated vapor at 264 psia:*

$264 - 250 = 14$

$300 - 250 = 50$

$1202.8 - 1201.1 = 1.7$

$^{14}/_{50} \times 1.7 = 0.476$

$1201.1 + 0.476 = 1201.576$

Enthalpy of saturated vapor at 264 psia = **1201.576 Btu/lb**

D. *Saturated steam temperature at 264 psia:*

$264 - 250 = 14$

$300 - 250 = 50$

$417.33 - 400.95 = 16.38$

$^{14}/_{50} \times 16.38 = 4.5864$

$400.95 + 4.5864 = 405.5364$

Saturated steam temperature at 264 psia = **405.5364°F**

In Dry Saturated Steam tables, the third and fourth columns pertain to specific volume of saturated steam. For each pressure, there is a specific volume for the saturated liquid and saturated vapor. Specific volume is measured in cu ft/lb.

As pressure in a boiler increases, the specific volume of the saturated liquid increases. **See Figure 1-16.** This results from the corresponding increase in temperature, not from the pressure directly. The specific volume of saturated vapor decreases with an increase in pressure, as indicated in Dry Saturated Steam tables. With an increase in pressure, specific volume of the saturated liquid increases and specific volume of the saturated vapor decreases.

Specific volume of dry saturated steam at pressures other than those given in Dry Saturated Steam tables can be found by interpolating values.

Figure 1-16. As pressure in the boiler increases, the specific volume of saturated liquid increases and the specific volume of saturated vapor decreases.

EXAMPLES

1. Use the Dry Saturated Steam table in the Appendix to interpolate and find the following at 102 psia:

   A. Specific volume of liquid

   B. Specific volume of vapor

   A. *Specific volume of liquid at 102 psia:*

      $\frac{2}{10} \times 0.00008 = 0.000016$

      $0.01774 + 0.000016 = 0.017756$ cu ft/lb

      Specific volume of liquid at 102 psia = **0.017756 cu ft/lb**

   B. *Specific volume of vapor at 102 psia:*

      $\frac{2}{10} \times 0.383 = 0.0766$

      $4.432 - 0.0766 = 4.3554$

      Specific volume of vapor at 102 psia = **4.3554 cu ft/lb**

CHAPTER 1 LEARNER RESOURCES

ATPeResources.com/QuickLinks
Access Code: 735728

*The boiler and accessories are key components of industrial plants requiring steam for plant processes or the generation of electricity. Different types of boilers are used for specific tasks based on the pressure and volume required. The design and construction of the boiler and fittings must be in accordance with the American Society of Mechanical Engineers (ASME) code. The ASME also classifies boilers for specific applications. **See Appendix.** In normal operation, the boiler is subjected to high temperatures and pressures. Knowledge of how the boiler operates and the stresses that occur in the boiler will improve the ability of a stationary engineer to operate the plant safely.*

BOILERS

A boiler is a closed pressure vessel containing water. The water in the boiler is heated to produce steam or hot water. Steam or hot water is directed to different locations through the supply piping for use to produce work. Boilers were first used for warming water and are of Roman and Greek origin. Early boilers were recovered from the ruins of Pompeii. In 1698, Thomas Savery developed a steam-driven water pump. As the steam condensed, a vacuum was created, causing water to be drawn into the cylinder. The boiler continued to be refined and developed for industrial use. Boilers are either firetube or watertube boilers. In addition, boilers can be package boilers and special-purpose boilers.

Firetube Boilers

In firetube boilers, gases of combustion pass through tubes that are surrounded by water. Firetube boilers are used in plants that require moderate pressures and moderate demands for quantities of steam per hour. Firetube boilers are classified as horizontal return tubular (HRT), scotch marine, and vertical firetube boilers. Vertical firetube boilers are not commonly used in plants generating steam for plant process and electricity.

   Horizontal return tubular (HRT) boilers are firetube boilers that consist of a drum fitted with tubes suspended over a furnace. The tubes are rolled, expanded, and beaded over in front and rear tube sheets. HRT boilers have a large furnace volume. In addition, because the bottom of the drum is exposed to the radiant heat from the furnace, in the past, it was prone to develop leaks. **See Figure 2-1.**

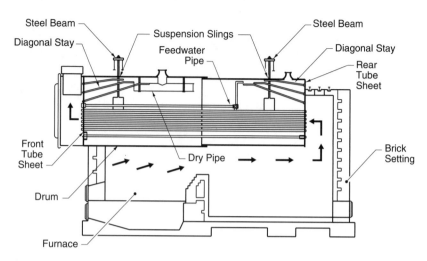

Figure 2-1. HRT boilers are externally fired and are suspended over a furnace.

Scotch marine boilers are firetube boilers equipped with an internal furnace that is completely surrounded by water, which increases the boiler heating surface and in turn increases boiler efficiency. The internal furnace is sometimes corrugated to increase strength, increase heating surface, and allow for expansion and contraction. The scotch marine boiler was used for many years on ships because of their compactness. Scotch marine boilers used in industry today are modified to meet stationary plant needs. **See Figure 2-2.**

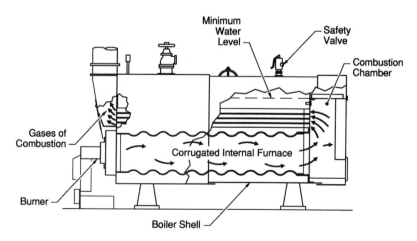

Figure 2-2. Scotch marine boilers have an internal corrugated furnace that allows for expansion and contraction.

Watertube Boilers

Watertube boilers are boilers in which water passes through tubes that are surrounded by gases of combustion. Watertube boilers were developed because industry needed steam at higher pressures and boilers with larger steam capacities. **See Figure 2-3.**

Cleaver-Brooks

Figure 2-3. Watertube boilers generate steam rapidly and can handle fluctuating steam loads.

Firetube boilers required such large diameters and plate thicknesses to meet these demands that they were no longer practical or safe. Watertube boilers require less water than firetube boilers and can respond to changes in steam demand more rapidly. Watertube boilers have straight or bent tubes, are rapid steamers, and carry considerably less water than firetube boilers for a given output. In addition, watertube boilers can carry higher steam pressures, respond to changes in steam demand more rapidly, and are less susceptible to boiler explosion. A *flex-tube watertube boiler* is a bent-tube boiler that has shaped tubes surrounded by gases of combustion. The shaped tubes provide more heating surface for greater efficiency than straight-tube watertube boilers. They are used for low and high pressure steam or hot water applications. **See Figure 2-4.** The development of watertube boilers has not eliminated the need for firetube boilers in industry. Firetube boilers are appropriate where moderate pressures and quantities of steam are required.

Package Boilers

Package boilers are assembled at the factory and shipped and installed as a unit. Package boilers are self-contained units and require minimal installation work in the field. They usually only need electrical, water, fuel, and chimney connections to be operative. **See Figure 2-5.** Package boilers are test-fired before they are shipped. Efficiency reports are supplied to the buyer. Production cost is less expensive for package boilers than for field-erected boilers. In addition, production quality of package boilers manufactured at the factory can be more closely monitored than field-erected boilers. Package boilers may be firetube or watertube boilers.

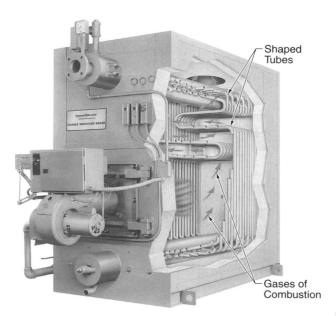

Cleaver-Brooks

Figure 2-4. A flex-tube watertube boiler has shaped tubes to provide more heating surface and greater efficiency than a straight-tube watertube boiler.

Cleaver-Brooks

Figure 2-5. Package boilers are supplied complete with controls, combustion equipment, and other appliances attached.

Special-Purpose Boilers

Special-purpose boilers are designed for specific applications, and they are used to dispose of industrial waste. Special-purpose boilers use equipment that may require specialized training by manufacturers of the equipment. The method of generating steam varies among special-purpose boilers, but the end product is steam. Special-purpose boilers include the following:

Public utility boilers are boilers designed for high steam output and are built on a large scale.

Electric boilers are boilers that are very clean and do not require fuel storage or ash removal, but have high operating costs. Electric boilers have been in use in Europe since the early 1900s.

Membrane boilers are boilers that have water directed through formed metal membranes connected to chambers that serve the same purpose as the upper and lower drums. The boiler vessel is gas tight, insulated on the outside, and covered by a casing constructed of removable, formed steel. Membrane watertube boilers employ a recent design and are used where space is limited for hot water or low- or high-pressure steam requirements. A membrane watertube boiler, when compared to other boiler designs, offers better efficiency and can result in savings up to 50% in floor space and up to 40% in overall weight. **See Figure 2-6.**

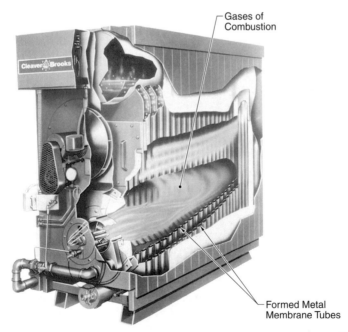

Gases of Combustion

Formed Metal Membrane Tubes

Cleaver-Brooks

Figure 2-6. A membrane watertube boiler uses formed metal membrane tubes to direct water in the boiler.

High-temperature, high-pressure hot water boilers are special-purpose boilers used where high-temperature, high-pressure water is circulated to various buildings. Heat is transferred by heat exchangers where temperature is reduced to safe levels. Some airports use a high-temperature, high-pressure hot water system for melting snow on runways without heat exchangers. However, it is unsafe to circulate high-temperature, high-pressure water in or around the public without the use of heat exchangers. Any leak would cause the water to flash into steam. Hot water is circulated through the boiler and system using circulating pumps.

Natural circulation boilers use the movement of heated and cooled water for circulation. As heated water rises, cooler water drops down to take its place. Steam bubbles form and move to the steam and water drum and increase water circulation.

Forced circulation boilers are special-purpose boilers that use pumps to increase water circulation and was developed because natural circulation was not sufficient in some boilers. Pumps were used to increase water circulation to prevent tubes and heating surfaces from overheating and causing blisters or bags. Older scotch marine boilers had a hydrokineter valve that circulated water from the bottom of the boiler when the boiler was first started up. Cold water from the bottom of the boiler was then forced up and circulated.

Industrial waste boilers are special-purpose boilers designed to dispose of and utilize industrial waste and/or by-products for the production of steam. Design engineers developed special-purpose boilers and methods of combustion to handle specific industrial wastes. For example, the sugar refining process using sugarcane produces large quantities of bagasse, an industrial by-product. The coffee industry produces large amounts of residue from making instant coffee. The paper industry produces industrial waste in the form of black liquor residue consisting primarily of sulfuric acid that must be disposed of. Sewage disposal plants have quantities of methane gas that are burned off in the atmosphere without utilizing the Btu content. Special stokers to burn bagasse, pulverizers to burn the coffee and paper residue, and special burners to burn the methane gas from sewage plants were developed.

Condensing boilers are hot-water heating boilers designed to recover heat that is normally discharged up the stack. **See Figure 2-7.** When a condensing boiler is working at peak efficiency, the water vapor produced by burning the fuel in the boiler condenses back into condensate. Condensing the exhaust gases releases the latent heat of vaporization of the water. This is a more significant source of energy than the transfer of heat by cooling the vapor. A condensing boiler regains about 970 Btu/pound of condensate by recovering the latent heat.

Condensing boilers need a drain pipe for the condensate collected during operation. The condensate is usually mildly acidic because of sulfur and nitrogen impurities in the fuel. In some applications, no special treatment of the condensate is necessary, but most manufacturers recommend some means of neutralizing this condensate. The final exhaust from a condensing boiler has a lower temperature than the exhaust from a conventional boiler, so natural convection does not work and a fan is required to expel the final exhaust. This has the added benefit of using low-temperature exhaust piping and does not require insulation or a chimney or stack. Typical condensing boiler efficiencies are around 90%. Many condensing boilers have earned an Energy Star rating.

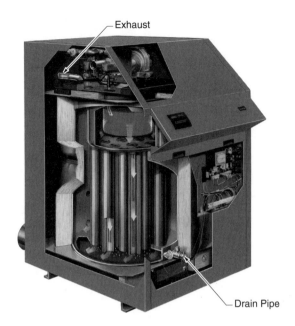

Exhaust

Drain Pipe

Cleaver-Brooks

Figure 2-7. Condensing boilers are hot-water heating boilers designed to recover heat that is normally discharged up the stack.

BOILER CONSTRUCTION

Boiler drums and shells, braces, stays, and tubes are subjected to continual stress and high temperatures when the boiler is operating. The ASME code establishes standards for the design and construction of boilers, fittings, and accessories. The ASME code ensures that the boiler construction and fittings are made of suitable materials to withstand forces and temperatures. Stresses occurring in a boiler that must be considered include compression, shear, and tensile stress. Boiler design and construction must allow for expansion and contraction according to the ASME code for the specified operating temperatures and pressures.

Modern boilers are of welded construction. Welded seams provide a 100% efficiency rating with ideal conditions. Metal used in boiler construction must be stress-relieved, joints ground smooth, and welds X-rayed for defects. Welds performed in the field are usually given an efficiency rating of 90%. All welding on a steam boiler must be done by certified welders and conform to the specifications of the ASME code. In addition, welded construction eliminates the possibility of caustic embrittlement around rivets joining boiler parts. *Caustic embrittlement* is the collection of high alkaline material that leads to breakdown and weakening of boiler metal.

Firetube Boiler Construction

With newer construction techniques, firetube boilers are rated up to 750 boiler horse-power (BHP). Welded construction in new firetube boilers reduces the possibility of caustic embrittlement and allows a higher maximum allowable working pressure (MAWP) than a firetube boiler with riveted construction.

There are still a few older firetube boilers in the field that are of riveted construction. Riveted construction of the longitudinal and circumferential seams in boilers requires lap or butt-strap seams. Lap and butt-strap seams can be single-, double-, or triple-riveted, depending upon the strength required. The joint design and number of rivets used determine the efficiency of the joint. **See Figure 2-8.** All lap seams are limited to boiler drums up to 36″ in diameter and a maximum of 100 psi steam-carrying capacity.

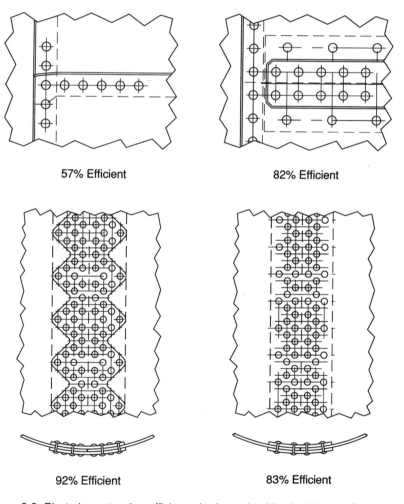

57% Efficient   82% Efficient

92% Efficient   83% Efficient

Figure 2-8. Riveted construction efficiency is determined by the joint configuration and number of rivets.

HRT boilers with riveted lap-seam construction have a maximum boiler life of 20 years per ASME code. HRT boilers with riveted butt-strap construction have a boiler life based on the strength of the joint. Boiler life does not consider corrosion or caustic embrittlement present along the riveted seam.

Tube sheets on firetube boilers are flat and require stays for support. *Stays* are devices used to reinforce flat surfaces that are subjected to boiler pressure. For example, stays are used for reinforcing tube sheets, combustion chambers, and water legs found on firetube boilers. Different stays are required to control different stresses and forces present in the boiler, including solid through stays, solid staybolts, hollow staybolts, girder stays, and dog stays. **See Figure 2-9.**

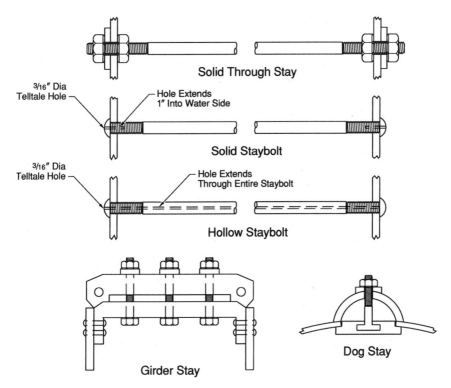

Figure 2-9. Firetube boilers require different stays and staybolts to control the various forces present in the boiler.

Solid through stays are used to hold front and rear tube sheets together and have nuts inside and outside the boiler drum. Solid and hollow staybolts hold inner and outer wrapper sheets together and have ³⁄₁₆″ diameter telltale holes. If staybolts fail, water will leak out of the telltale hole, warning the stationary engineer. Solid staybolts have telltale holes extending 1″ into the water side of the boiler. Hollow staybolts have a hole extending through the entire staybolt. Girder stays are used to strengthen the flat surface of the crown sheet on a Scotch marine boiler. Dog stays are used for manhole and handhole covers.

Stresses develop in the boiler when it is fired. The boiler construction determines how stress is controlled. The primary stresses found in boilers are compression, shear, and tensile stress. Compression stress is the result of force applied to an object that tends to press on or squeeze the object. Shear stress is the result of force that is applied to an object that tends to shear or cut through the object. Tensile stress is the result of force that is applied to an object that tends to stretch the object. Compression stress occurs in the boiler drum when a vacuum is formed. Shear stress occurs in rivets used to join boiler parts. Tensile stress occurs in boiler plates and staybolts.

The steel selected for use in boiler construction must be able to withstand temperature and pressure in the boiler. High temperatures affect the strength of steel more than pressure. The tensile strength of steel used in boiler construction varies from 50,000 to 65,000 psi.

The safe working pressure (SWP) for a boiler with flat surfaces that are held together by staybolts is found by using the following equation:

$$Safe\ working\ pressure = \frac{Area\ of\ staybolt \times Stress\ constant\ of\ staybolt}{Area\ covered\ by\ staybolt}$$

$$SWP = \frac{a \times C}{A}$$

SWP = safe working pressure, in psi

a = area of the staybolt, in sq in. *Note:* The area of the staybolt is found by using the equation, $a = 0.7854D^2$.

C = stress constant of staybolt (as specified in ASME code)

A = pitch × pitch′, or P × P′ (area covered by staybolt in sq in.) *Note:* This area refers to the strength of the staybolt, not the strength of the plate.

To determine the area covered by the staybolts, a rectangle using the centers of the staybolts as connecting points is formed. **See Figure 2-10.**

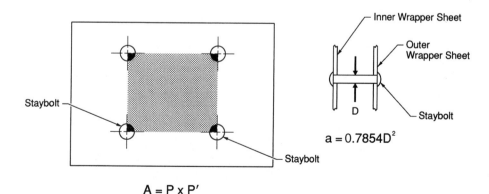

Figure 2-10. Staybolts hold the inner and outer wrapper sheets together on firetube boilers.

EXAMPLES

1. A staybolt has a diameter of 1″ and a pitch of 7″ × 8″. What is the SWP if the stress constant is 12,000 psi?

$$SWP = \frac{a \times C}{A}$$

$$SWP = \frac{0.7854D^2 \times C}{A}$$

$$SWP = \frac{0.7854 \times 1 \times 1 \times 12,000}{7 \times 8}$$

$SWP = \mathbf{168.3\ psi}$

2. A staybolt has a diameter of 1.5″ and a pitch of 7″ × 8″. What is the SWP if the stress constant is 12,500 psi?

$$SWP = \frac{a \times C}{A}$$

$$SWP = \frac{0.7854D^2 \times C}{A}$$

$$SWP = \frac{0.7854 \times 1.5 \times 1.5 \times 12,500}{7 \times 8}$$

$SWP = \mathbf{394.5\ psi}$

The minimum diameter of the staybolt needed for a given boiler pressure is found by using the following equation:

$$Safe\ working\ pressure = \frac{Area\ of\ staybolt \times Stress\ constant\ of\ staybolt}{Area\ covered\ by\ staybolt}$$

$$SWP = \frac{a \times C}{A}$$

$$SWP \times A = a \times C$$

$$A = P \times P'$$

$$a = 0.7854D^2$$

$$SWP \times A = 0.7854D^2 \times C$$

$$\frac{SWP \times A}{0.7854 \times C} = D^2$$

$$D^2 = \frac{SWP \times A}{0.7854 \times C}$$

$$D = \sqrt{\frac{SWP \times A}{0.7854 \times C}}$$

D = diameter of staybolt, in in.

SWP = safe working pressure, in psi

A = pitch × pitch′, or P × P′, in sq in.

C = stress constant of staybolt

EXAMPLES

1. What is the minimum diameter of the staybolts needed if the SWP is 150 psi, the pitch is 7″ × 8″, and the stress constant is 13,000 psi?

$$D = \sqrt{\frac{SWP \times A}{0.7854 \times C}}$$

$$D = \sqrt{\frac{150 \times 7 \times 8}{0.7854 \times 13,000}}$$

$$D = \sqrt{\frac{8400}{10,210.2}}$$

$$D = \sqrt{0.8227067}$$

$$D = 0.91″$$

2. What is the minimum diameter of the staybolts needed if the SWP is 160 psi, the pitch is 5″ × 5″, and the stress constant is 12,000 psi?

$$D = \sqrt{\frac{SWP \times A}{0.7854 \times C}}$$

$$D = \sqrt{\frac{160 \times 5 \times 5}{0.7854 \times 12,000}}$$

$$D = \sqrt{\frac{4000}{9424.8}}$$

$$D = \sqrt{0.424412}$$

$$D = 0.65″$$

Watertube Boiler Construction

Watertube boilers have water passing through tubes that are surrounded by the heat and gases of combustion. Watertube boilers can be single-drum or multidrum, straight-tube or bent-tube, and single-pass or multipass.

Watertube boilers carry a smaller volume of water than firetube boilers and can re-spond quickly to fluctuating steam loads. For strength in watertube boilers, flat surfaces were replaced with curved surfaces that do not require stays. The increased strength allowed watertube boilers to carry high pressures with high superheat temperatures.

Watertube boilers are externally fired and require refractory material within the furnace area. Because of the spalling and erosion of the refractory, waterwalls became necessary to reduce maintenance costs. Waterwalls are constructed of closely spaced vertical or horizontally placed tubes in the furnace walls. **See Figure 2-11.**

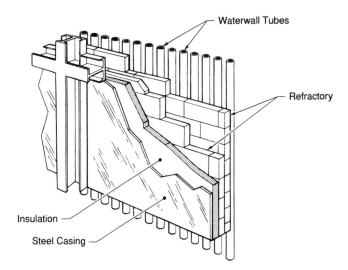

Figure 2-11. Waterwalls increase capacity of the pounds of steam generated per hour.

The development of waterwalls also allowed boilers to be built with smaller furnace volume per boiler horsepower. This produced higher heat release and reduced excess air. Without waterwalls, stokers and pulverized coal burners could never have been developed. **See Figure 2-12.** According to Section I, Power Boilers, of the ASME code, the minimum pounds of steam per hour per square foot of waterwall heating surface generated is:

• hand-fired coal boilers—8 pounds of steam per hour per square foot of waterwall heating surface

• stoker-fired coal boilers—12 pounds of steam per hour per square foot of water wall heating surface

• pulverized coal, fuel oil, or gas boilers—16 pounds of steam per hour per square foot of waterwall heating surface

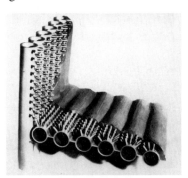

Babcock & Wilcox Co.

Figure 2-12. Studded waterwalls hold refractory that protects the tubes from extreme heat and abrasion.

The number of pounds of steam per hour generated by the waterwall heating surface is found by using the following equation:

Pounds of steam = Waterwall heating surface × 16

$W = HS_{ww} \times 16$

W = pounds of steam generated per hour

HS_{ww} = waterwall heating surface

16 = pounds of steam generated per hour per square foot of waterwall heating surface

EXAMPLES

1. A watertube boiler using pulverized coal has 2000 sq ft of waterwall heating surface. How many lb of steam/hr are generated by the waterwall heating surface?

$W = HS_{ww} \times 16$

$W = 2000 \times 16$

W = 32,000 lb of steam/hr

2. A watertube boiler using fuel oil has 1500 sq ft of waterwall heating surface. How many lb of steam/hr are generated by the waterwall heating surface?

$W = HS_{ww} \times 16$

$W = 1500 \times 16$

W = 24,000 lb of steam/hr

BOILER OPERATION CALCULATIONS

A stationary engineer is not responsible for the design or construction of boilers. However, a stationary engineer must understand the calculations used in principles of boiler operation. Common calculations a stationary engineer must be familiar with include boiler heating surface, boiler horsepower, factor of evaporation, developed boiler horsepower, thermal efficiency, total force, and maximum allowable working pressure.

Stress is the result of force acting on an object. The material of the object resists this force. Stress is found by using the following equation:

$$Stress = \frac{Force \text{ (acting on object)}}{Area \text{ (material resisting the force)}}$$

This equation is used when analyzing the strength of longitudinal seams of a boiler. The force tending to cause a rupture along a longitudinal seam is equal to the pressure in the drum multiplied by the area it is acting on.

Total force = Pressure × Area

If the longitudinal seam has an efficiency of less than 100%, the break will most likely occur at the longitudinal seam. **See Figure 2-13.**

The force acting on the drum that would cause it to rupture is the pressure in the drum times the projected area. The projected area of the drum is the diameter times the length of the drum. **See Figure 2-14.**

Area (A-B-C-D) = *Diameter* × *Length of drum*

Total force = *Pressure* × *Area* (A-B-C-D)

TF = *P* × *A*

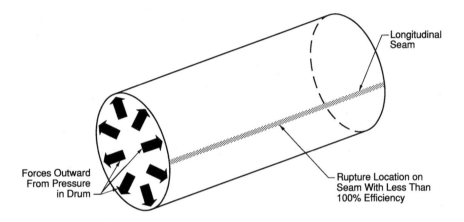

Figure 2-13. The force that can cause a rupture on the longitudinal seam is equal to the pressure in the drum times the area being acted on.

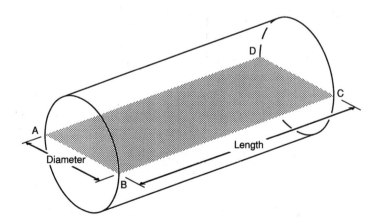

Figure 2-14. The projected area (A-B-C-D) is the area that the pressure is acting on in the boiler drum.

The area that is resisting a break consists of two wall thicknesses of the boiler drum. Therefore, the area is two times the product of the thickness and the length of the drum. **See Figure 2-15.**

$Area = 2 \times Thickness \times Length$

$A = 2 \times T \times L$

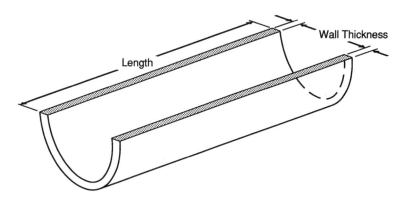

Figure 2-15. The area resisting the rupture is two times the product of the thickness and the length of the drum.

Using this information, the stress on the longitudinal seams is found by using the following equation:

$$Longitudinal\ stress = \frac{Total\ force}{Area}$$

$$Longitudinal\ stress = \frac{Pressure \times Diameter \times Length}{2 \times Thickness \times Length}$$

$$S_l = \frac{P \times D \times L}{2 \times T \times L}$$

$$S_l = \frac{P \times D}{2T}$$

Note: Drum length does not affect the stress on the longitudinal seam.

The stress on the circumferential seam can also be found from the same basic equation:

$$Stress = \frac{Force}{Area}$$

The force tending to cause a rupture along the circumferential seam is equal to the pressure in the drum multiplied by the area it is acting on. **See Figure 2-16.** The break would occur across section A-B if the circumferential seam has an efficiency of less than 100%.

$Total\ force = Pressure \times Area$

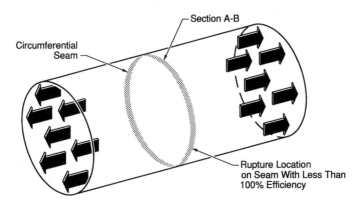

Figure 2-16. The force that can cause a rupture on the circumferential seam is equal to the pressure in the drum times the area being acted on.

The area on which the pressure is acting is the drum head. The area of the drum head is found by using the following equation:

$Area = 0.7854D^2$

$Total\ force = Pressure \times Area\ of\ drum\ head$

$TF = P \times A$

$TF = P \times 0.7854D^2$

The area resisting the break is the product of the circumference of the drum and the thickness of the material. **See Figure 2-17.**

$Area = Circumference \times Thickness$

$Circumference = \pi \times D$, or $3.1416 \times D$

$A = 3.1416 \times D \times T$

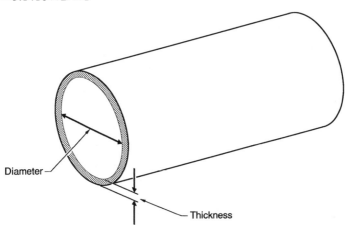

Figure 2-17. The area of the boiler resisting force is determined by the product of the circumference and wall thickness.

Using this information, the stress on a circumferential seam is found by using the following equation:

$$Circumferential\ stress = \frac{Total\ force}{Area}$$

$$S_c = \frac{P \times 0.7854D^2}{3.1416 \times D \times T}$$

$$S_c = \frac{P \times D}{4T}$$

EXAMPLES

1. What are the longitudinal and circumferential stresses on a boiler with a steam pressure of 175 psi, a 36″ diameter drum that is 12′ long, and a plate thickness of ½″?

$$S_l = \frac{P \times D}{2T}$$

$$S_l = \frac{175 \times 36}{2 \times 0.5}$$

S_l = **6300 psi**

$$S_c = \frac{P \times D}{4T}$$

$$S_c = \frac{175 \times 36}{4 \times 0.5}$$

S_c = **3150 psi**

2. What are the longitudinal and circumferential stresses on a boiler with a steam pressure of 200 psi, a 40″ diameter drum that is 12′ long, and a plate thickness of ½″?

$$S_l = \frac{P \times D}{2T}$$

$$S_l = \frac{200 \times 40}{2 \times 0.5}$$

S_l = **8000 psi**

$$S_c = \frac{P \times D}{4T}$$

$$S_c = \frac{200 \times 40}{4 \times 0.5}$$

S_c = **4000 psi**

Longitudinal stress is twice as much as circumferential stress. Because of the additional stress on longitudinal seams, they must be constructed to resist greater stress than circumferential seams. A stationary engineer must monitor longitudinal seams more closely than circumferential seams.

Boiler Heating Surface

Boiler heating surface is the part of the boiler that has fire and gases of combustion on one side and water on the other side. In a firetube boiler, heat from gases of combustion passes through the tubes that are surrounded by water. Boiler heating surface is found by using the following equation:

$Heating\ surface = Circumference\ of\ tubes \times Length\ of\ tubes \times Number\ of\ tubes$

$HS = C \times L \times N$

$Circumference = \pi \times Diameter\ of\ tube$

$C = 3.1416 \times D$

$HS = 3.1416 \times D \times L \times N$

Heating surface is measured in sq ft and the diameter of the tubes is measured in ft.

Note: For firetube boilers, use the inside diameter (ID) of the tube to determine the circumference of the tube. The ID is determined by taking the outside diameter (OD) minus two times the tube wall thickness. For watertube boilers, use the OD to determine the circumference of the tube. **See Figure 2-18.**

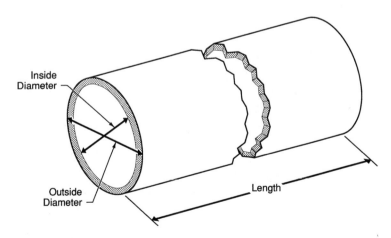

Figure 2-18. The heating surface in a firetube boiler is calculated using the inside diameter of the tube. The heating surface in a watertube boiler is calculated using the outside diameter of the tube.

EXAMPLES

1. What is the heating surface of a firetube boiler with a ⅛" thick wall and 300 tubes that are 15′ long and 3" in diameter?

$HS = 3.1416 \times D \times L \times N$

$D = 3 - (⅛ + ⅛) = 2¾"$

Note: Change D to feet: $\dfrac{2.75}{12}$

$HS = 3.1416 \times \dfrac{2.75}{12} \times 15 \times 300$

$HS = $ **3239.8 sq ft**

2. What is the heating surface of a watertube boiler with a ⅛" thick wall and 300 tubes that are 15′ long and 3" in diameter?

$HS = 3.1416 \times D \times L \times N$

Note: Change D to feet: $\dfrac{3}{12}$

$HS = 3.1416 \times \dfrac{3}{12} \times 15 \times 300$

$HS = $ **3534.3 sq ft**

Boiler Horsepower

Boilers are rated in boiler horsepower (BHP). In the past, BHP was roughly estimated by the number of square feet of heating surface. For example, in watertube boilers, 10 sq ft of heating surface was equal to 1 BHP. In horizontal firetube boilers, 12 sq ft of heating surface was equal to 1 BHP. In vertical firetube boilers, 14 sq ft of heating surface was equal to 1 BHP. This method of determining BHP was not very accurate and was replaced by an equation using the factor of evaporation.

Factor of Evaporation

Some boiler manufacturers claim that their package boilers produce 1 BHP for every 5 sq ft of heating surface. To standardize BHP ratings, 1 BHP is equal to the evaporation (generation) of 34.5 lb of water per hour from and at 212°F as adopted by the ASME. This method of determining BHP is accurate at 212°F. However, boilers do not operate with steam-water temperatures of 212°F. A *factor of evaporation* is a correction factor used to compensate for the difference between evaporation at 212°F and evaporation at the corresponding temperature in a boiler. By using the factor of evaporation, developed boiler horsepower (DBHP) can be determined at any time. This allows BHP to be measured precisely throughout the industry.

$$Developed\,boiler\,horsepower = \frac{Pounds\,of\,steam\,per\,hour \times Factor\,of\,evaporation}{34.5}$$

$$DBHP = \frac{W_s \times FE}{34.5}$$

One BHP is also equivalent to 33,475 Btu/hr. This results from multiplying 970.3 Btu (the Btu needed for latent heat of evaporation) by 34.5 lb (pounds of water evaporated per hour from and at 212°F for 1 HP).

$$BHP = 970.3 \text{ Btu/lb} \times 34.5 \text{ lb}$$

$$BHP = 33,475 \text{ Btu/hr}$$

The pounds of steam evaporated per hour can also be obtained from a steam flow meter.

To solve a problem including DBHP, first find the factor of evaporation of a given boiler by using the following equation:

$$Factor\,of\,evaporation = \frac{Enthalpy\,of\,steam - Enthalpy\,of\,feedwater}{970.3}$$

$$FE = \frac{H_s - \left(T_{fw} - 32\right)}{970.3}$$

H_s = enthalpy of steam

T_{fw} = feedwater temperature

32 = base temperature for determining enthalpy (heat in substance)

970.3 = latent heat of vaporization of water at 212°F

Note: $T_{fw} - 32$ is the enthalpy of the feedwater entering the boiler drum.

EXAMPLES

1. What is the factor of evaporation of a boiler generating steam at 400 psia that contains 1204.5 Btu/lb? The feedwater temperature is 190°F.

$$FE = \frac{H_s - \left(T_{fw} - 32\right)}{970.3}$$

$$FE = \frac{1204.5 - \left(190 - 32\right)}{970.3}$$

$$FE = \frac{1204.5 - 158}{970.3}$$

$$FE = \mathbf{1.079}$$

2. What is the factor of evaporation of a boiler generating steam at 120 psia that contains 1190.4 Btu/lb? The feedwater temperature is 210°F.

$$FE = \frac{H_s - (T_{fw} - 32)}{970.3}$$

$$FE = \frac{1190.4 - (210 - 32)}{970.3}$$

$$FE = \frac{1190.4 - 178}{970.3}$$

$$FE = \textbf{1.043}$$

To find the DBHP, the factor of evaporation must be included.

$$Developed\,boiler\,horsepower = \frac{Pounds\,of\,steam\,per\,hour \times Factor\,of\,evaporation}{34.5}$$

$$DBHP = \frac{W_s \times FE}{34.5}$$

EXAMPLES

1. A boiler generates 10,000 lb of steam/hr and has a factor of evaporation of 1.24. What is the developed boiler horsepower?

$$DBHP = \frac{W_s \times FE}{34.5}$$

$$DBHP = \frac{10,000 \times 1.25}{34.5}$$

$$DBHP = \textbf{359.42}$$

2. A boiler generates 5500 lb of steam/hr and has a factor of evaporation of 1.078. What is the developed boiler horsepower?

$$DBHP = \frac{W_s \times FE}{34.5}$$

$$DBHP = \frac{5500 \times 1.078}{34.5}$$

$$DBHP = \textbf{171.86}$$

If the factor of evaporation is unknown, the DBHP may be solved in one step instead of two.

$$DBHP = \frac{W_s \times FE}{34.5}$$

$$FE = \frac{H_s - (T_{fw} - 32)}{970.3}$$

$$DBHP = \frac{W_s\left[H_s - \left(T_{fw} - 32\right)\right]}{34.5 \times 970.3}$$

$$DBHP = \frac{W_s\left[H_s - \left(T_{fw} - 32\right)\right]}{33,475}$$

$DBHP$ = developed boiler horsepower

W_s = pounds of steam per hour

H_s = enthalpy of steam

T_{fw} = feedwater temperature

32 = base temperature for determining enthalpy (heat in substance)

33,475 = number of Btu in 1 BHP

Note: Total heat output (in Btu) by the boiler per hour is determined. This total is divided by 33,475, which is the number of Btu in 1 BHP.

EXAMPLES

1. A boiler generates 30,000 lb of steam/hr. The steam pressure is 130 psia and the steam contains 1191.7 Btu/lb. The feedwater temperature is 210°F. What is the developed boiler horsepower?

$$DBHP = \frac{W_s\left[H_s - \left(T_{fw} - 32\right)\right]}{33,475}$$

$$DBHP = \frac{30,000\left[1191.7 - \left(210 - 32\right)\right]}{33,475}$$

$$DBHP = \frac{30,000 \times 1013.7}{33,475}$$

$DBHP$ = **908.469**

2. A boiler generates 9100 lb of steam/hr. The steam pressure is 150 psia and the steam contains 1194.1 Btu/lb. The feedwater temperature is 172°F. What is the developed boiler horsepower?

$$DBHP = \frac{W_s\left[H_s - \left(T_{fw} - 32\right)\right]}{33,475}$$

$$DBHP = \frac{9100\left[1194.1 - \left(172 - 32\right)\right]}{33,475}$$

$$DBHP = \frac{9100\left(1194.1 - 140\right)}{33,475}$$

$$DBHP = \frac{9100 \times 1054.1}{33,475}$$

$DBHP$ = **286.55**

Equivalent evaporation is the amount of steam that a boiler would produce if converted back to standard conditions at 212°F.

Note: $\dfrac{H_s - H_{fw}}{970.3} = FE$

$W_e = W_s \times FE$

Note: W_e is equal to equivalent evaporation.

Pounds of steam/hr from and at 212°F =

$$\frac{Pounds\ of\ steam/hr\ \left(Enthalpy\ of\ steam - Enthalpy\ of\ feedwater\right)}{970.3}$$

$$W_e = \frac{W_s\left(H_s - H_{fw}\right)}{970.3}$$

W_e = pounds of steam per hour from and at 212°F

W_s = pounds of steam per hour as produced

H_s = enthalpy of steam, in Btu/lb

H_{fw} = enthalpy of feedwater

EXAMPLES

1. A boiler generates 20,000 lb of steam/hr. The steam pressure is 150 psia and the steam contains 1195 Btu/lb. The feedwater temperature is 175°F. What is the equivalent evaporation produced?

$$W_e = \frac{W_s\left[H_s - \left(T_{fw} - 32\right)\right]}{970.3}$$

Note: $T_{fw} - 32$ is the enthalpy of the feedwater.

$$W_e = \frac{20,000\left[1195 - \left(175 - 32\right)\right]}{970.3}$$

$$W_e = \frac{20,000\left(1195 - 143\right)}{970.3}$$

$$W_e = \frac{20,000 \times 1052}{970.3}$$

$$W_e = \frac{21,040,000}{970.3}$$

W_e = **21,684 lb/hr**

2. A boiler generates 50,000 lb of steam/hr. The steam contains 1200 Btu/lb. The feedwater temperature is 210°F. What is the equivalent evaporation produced?

$$W_e = \frac{W_s\left[H_s - (T_{fw} - 32)\right]}{970.3}$$

$$W_e = \frac{50,000\left[1200 - (210 - 32)\right]}{970.3}$$

$$W_e = \frac{50,000(1200 - 178)}{970.3}$$

$$W_e = \frac{50,000 \times 1022}{970.3}$$

$$W_e = \frac{51,100,000}{970.3}$$

$$W_e = \textbf{52,664.124 lb/hr}$$

Thermal Efficiency

Thermal efficiency, or *boiler efficiency,* is the ratio of heat output to heat input of a boiler. Thermal efficiency can be used to determine the amount of fuel required to generate a given quantity of steam. Thermal efficiency is found by using the following equation:

Thermal efficiency =

$$\frac{Pounds\ of\ steam/hr\ \left[\ Btu\ content\ of\ steam - (Feedwater\ temperature - 32)\right]}{Units\ of\ fuel/hr\ \times Btu\ content\ per\ unit\ of\ fuel}$$

$$TE = \frac{W_s\left[H_s - (T_{fw} - 32)\right]}{W_f \times C}$$

TE = thermal efficiency

W_s = pounds of steam per hour

H_s = enthalpy of steam

T_{fw} = feedwater temperature

32 = base temperature for determining enthalpy (heat in substance)

W_f = units of fuel per hour

C = Btu content per unit of fuel

EXAMPLES

1. A boiler generates 50,000 lb of steam/hr at a pressure of 350 psi. The steam contains 1203.9 Btu/lb. The feedwater temperature is 210°F. The boiler consumed 450 gallons (gal.) of fuel oil containing 142,000 Btu/gal. What is the thermal efficiency?

$$TE = \frac{W_s\left[H_s - (T_{fw} - 32)\right]}{W_f \times C}$$

$$TE = \frac{50,000\left[1203.9 - (210 - 32)\right]}{450 \times 142,000}$$

$$TE = \frac{50,000 \times 1025.9}{450 \times 142,000}$$

$$TE = \frac{51,295,000}{63,900,000}$$

$TE = \mathbf{0.8027}$ or $\mathbf{80.27\%}$

2. A boiler uses 1200 therms/hr of natural gas while generating 100,500 lb of steam. The steam contains 1300 Btu/lb. The feedwater temperature is 300°F. What is the thermal efficiency? *Note:* 1 therm = 100,000 Btu.

$$TE = \frac{W_s\left[H_s - (T_{fw} - 32)\right]}{W_f \times C}$$

$$TE = \frac{100,500\left[1300 - (300 - 32)\right]}{1200 \times 100,000}$$

$$TE = \frac{100,500 \times 1032}{1200 \times 100,000}$$

$$TE = \frac{103,716,000}{120,000,000}$$

$TE = \mathbf{0.8643}$ or $\mathbf{86.43\%}$

In some plants, it is more convenient to determine a daily evaporation rate rather than determining thermal efficiency. In a given plant, the feedwater temperature is usually constant, and the heat content of the steam remains constant along with the Btu content of the fuel. The only variables are the lb of steam/hr generated and the amount of fuel being consumed. The lb of steam generated per unit of fuel is found by using the following equation:

$$Evaporation\ rate = \frac{Pounds\ of\ steam\ per\ hour}{Units\ of\ fuel\ per\ hour}$$

$$ER = \frac{W_s}{W_f}$$

ER = evaporation rate, in pounds of steam per hour

W_s = pounds of steam per hour

W_f = units of fuel per hour

Note: Pounds of steam per hour can also be determined using a steam flow meter and fuel consumption data is found from fuel flow meters or coal scales.

EXAMPLES

1. A boiler generates 50,000 lb of steam/hr while using 450 gal. of fuel oil. What is the evaporation rate?

$$ER = \frac{W_s}{W_f}$$

$$ER = \frac{50,000}{450}$$

ER = **111.11 lb of steam/gal. of fuel oil**

2. A boiler generates 100,500 lb of steam/hr while using 1200 therms of natural gas. What is the evaporation rate?

$$ER = \frac{W_s}{W_f}$$

$$ER = \frac{100,500}{1200}$$

ER = **83.75 lb of steam/therm of natural gas**

Total Force

Total force is equal to area in square inches times the steam pressure. The total force acting on a boiler drum surface is the total area in square inches of the drum times the steam pressure in the boiler drum. To find the total area of the boiler drum, the area of the front head, the area of the shell, and the area of the rear head must be known.

Area of front head = $0.7854 \times Diameter^2$

$A_{fh} = 0.7854D^2$

Area of boiler shell = $Circumference \times Length$

$A_s = C \times L$

Note: $C = \pi \times D$

Area of rear head = $0.7854 \times Diameter^2$

$A_{rh} = 0.7854D^2$

$Total\ force = Pressure \times Area$

$TF = P(A_h + A_s)$

TF = total force, in lb

P = steam pressure, in psi

A_h = area of front and rear heads, in sq in.

A_s = area of boiler shell, in sq in.

EXAMPLES

1. A boiler operates with a pressure of 300 psi. The boiler drum is 9′ long and has a diameter of 4′. What is the total force acting on the boiler drum surface?

$TF = P(A_h + A_s)$

$A_h = 2 \times 0.7854D^2$

$A_h = 2 \times 0.7854(4 \times 12)(4 \times 12)$

$A_h =$ **3619.1 sq in.**

Note: This is area of both front and rear heads.

$A_s = C \times L$

$A_s = \pi \times D \times L$

$A_s = 3.1416(4 \times 12)(9 \times 12)$

$A_s =$ **16,286 sq in.**

$TF = 300(3619.1 + 16,286)$

$TF = 300 \times 19,905.1$

$TF =$ **5,971,530 lb**

2. A boiler operates with a pressure of 250 psi. The boiler drum is 12′ long and has a diameter of 3.5′. What is the total force acting on the boiler drum surface?

$TF = P(A_h + A_s)$

$A_h = 2 \times 0.7854D^2$

$A_h = 2 \times 0.7854(3.5 \times 12)(3.5 \times 12)$

$A_h =$ **2770.9 sq in.**

$A_s = \pi \times D \times L$

$A_s = 3.1416(3.5 \times 12)(12 \times 12)$

$A_s =$ **19,000.4 sq in.**

$TF = 250(2770.9 + 19,000.4)$

$TF = 250(21,771.3)$

$TF =$ **5,442,825 lb**

Maximum Allowable Working Pressure

Maximum allowable working pressure (MAWP) is the maximum boiler pressure determined by the design and construction of the boiler in accordance with the ASME code. **See Figure 2-19.** All boilers are designed to have an MAWP and a bursting pressure, which is usually a factor of safety five times greater than the MAWP. The ASME code regulates the types of material and methods of boiler construction, which determine the MAWP of a boiler. Boilers requiring a higher MAWP must be constructed of thicker metal with a higher tensile strength and higher joint efficiency. The factor of safety is used to allow for fatigue, which is caused from age, corrosion, or other factors. MAWP of a boiler is found by using the following equation:

Maximum allowable working pressure =

$$\frac{Tensile\ strength \times Metal\ thickness \times Joint\ efficiency}{Factor\ of\ safety \times Radius\ of\ boiler\ drum}$$

$$MAWP = \frac{S \times T \times E}{F \times R}$$

S = tensile strength of metal, in psi

T = thickness of metal, in in.

E = efficiency of joint

F = factor of safety

R = radius of boiler drum, in in.

Note: The MAWP may be lowered if during inspection the boiler shows signs of corrosion or pitting.

Figure 2-19. Boiler design, materials, and construction are in accordance with the ASME code and determine the MAWP.

Bursting pressure is found by using the following equation:

Bursting pressure = MAWP × Factor of safety

$BP = MAWP \times F$

EXAMPLES

1. A 72″ boiler drum is ½″ thick, and its metal has a tensile strength of 65,000 psi. The joints are 82% efficient and the factor of safety is 5. What are the MAWP and bursting pressure? *Note:* The factor of safety is the ratio of bursting pressure to MAWP.

    $$MAWP = \frac{S \times T \times E}{F \times R}$$

    Note: Radius = ½D

    $$MAWP = \frac{65,000 \times 0.5 \times 0.82}{5 \times 36}$$

    $$MAWP = \frac{26,650}{180}$$

    $MAWP =$ **148 psi**

    $BP = MAWP \times F$

    $BP = 148 \times 5$

    $BP =$ **740 psi**

2. A 36″ boiler drum is ¾″ thick and its metal has a tensile strength of 55,000 psi. The joints are 87% efficient and the safety factor is 5. What are the MAWP and bursting pressure?

    $$MAWP = \frac{S \times T \times E}{F \times R}$$

    $$MAWP = \frac{55,000 \times 0.75 \times 0.87}{5 \times 18}$$

    $$MAWP = \frac{35,887.5}{90}$$

    $MAWP =$ **398.75 psi**

    $BP = MAWP \times F$

    $BP = 398.75 \times 5$

    $BP =$ **1994 psi**

CHAPTER 2 LEARNER RESOURCES

ATPeResources.com/QuickLinks
Access Code: 735728

Steam and water fittings and accessories are required for safety and efficiency in the operation of a boiler. Steam and water fittings and accessories must be constructed in accordance with the ASME code. In addition, materials used in the construction of steam and water fittings and accessories must conform to the ASME code.

Fittings used on firetube boilers and watertube boilers serve the same function. The frequency of testing and testing procedures of fittings vary, depending on the operating temperatures and pressures of the boiler.

STEAM FITTINGS AND ACCESSORIES

Steam fittings (trim) are attached to the boiler and are necessary for its safe and efficient operation. Steam accessories are pieces of equipment not directly attached to the boiler and are also required for the operation of the boiler. Steam fittings must be constructed in accordance with the ASME code. The ASME code also specifies how these fittings are to be attached to the boiler as determined by the temperatures and pressures present in the boiler.

Safety Valves

A *safety valve* is a valve used to release pressure and protect the boiler from exceeding its MAWP and is the most important valve on the boiler. If the MAWP is exceeded, a failure on the pressure side of the boiler could occur. Safety valves must be located at the highest part of the steam side of the boiler. They must be connected to the boiler shell according to ASME code, with no intervening valves between the safety valve and the boiler. The only type of safety valve allowed is the spring-loaded pop-type safety valve. **See Figure 3-1.** Lever-type and deadweight safety valves can be easily tampered with and are not permitted.

The ASME code requires that every boiler have at least one safety valve. Boilers with more than 500 sq ft of heating surface shall have two or more safety valves. Safety valves over 3″ in diameter for pressures over 15 psi shall have a flanged or a weld-end inlet. Safety valve discharge piping must be installed and connected to the safety valve, which allows proper discharge of steam, but does not place strain on the safety valve. **See Figure 3-2.**

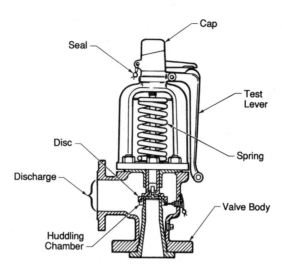

Figure 3-1. Spring-loaded pop-type safety valves are used to protect the boiler from excessive pressure.

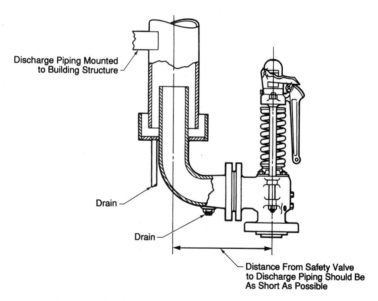

Figure 3-2. Discharge piping from the safety valve must not interrupt the flow of steam or place a strain on the safety valve.

Safety valve capacity must be enough to discharge all steam the boiler can generate without allowing the pressure to rise more than 6% above the MAWP. Safety valves must be designed to open fully at a predetermined pressure. Safety valves must remain open until there is a definite drop in pressure. *Blowdown*, or *blowback*, is the drop in pressure that allows the safety valve to close. Blowdown is normally 2 lb to 8 lb below popping pressure, but not more than 4% of set pressure. Safety valves must close tight without chattering. Once the safety valve has reseated, it must remain tightly closed. Setting or adjusting safety valves must be done by qualified personnel who are familiar with safety valve construction, operation, and maintenance. Repairs must only be done by the manufacturer or an authorized representative.

Safety valve springs are calibrated for the correct popping pressure. For pressures up to 250 psi, the safety valve can be set 10% above or 10% below this pressure. For pressures over 250 psi, safety valves can only be set 5% above or 5% below the set pressure without changing the safety valve spring. One or more boiler safety valves shall be set at or below MAWP. If additional safety valves are used, the highest pressure setting shall not exceed the MAWP by more than 3%.

The complete range of pressure settings of all safety valves on a boiler shall not exceed 10% of the highest pressure to which any safety valve is set. For example, if the safety valve with the highest setting is at 100 psi, the safety valve with the lowest setting cannot be lower than 90 psi. Any change in the settings of a safety valve blowdown or pressure requires a new safety valve data plate. **See Figure 3-3.** The safety valve data plate should include the following:

- manufacturer's name or trademark
- manufacturer's design or type number
- size of safety valve (in inches), seat diameter
- popping pressure setting (in psig)
- blowdown (in psi)
- capacity (in lb/hr)
- lift of the valve (in inches)
- year built or code mark
- ASME symbol
- serial number

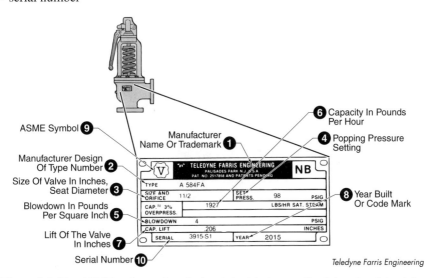

Teledyne Farris Engineering

Figure 3-3. The ASME code requires that certain data be on all safety valve data plates.

Popping-pressure tolerance is the difference between the set pressure and the popping pressure of a safety valve. Popping-pressure tolerance, plus or minus, is:

- 2 psi for pressures up to 70 psi
- 3% for pressures from over 70 psi to 300 psi
- 10 psi for pressures over 300 psi to 1000 psi
- 1% for pressures over 1000 psi

For example, at 70 psi, safety valves pop open between 68 psi and 72 psi. At 200 psi, safety valves pop open between 194 psi and 206 psi. At 500 psi, safety valves pop open between 490 psi and 510 psi. At 1200 psi, safety valves pop open between 1188 psi and 1212 psi and are still within popping-pressure tolerances.

Normally, valves under pressure are opened slowly. However, safety valves are designed to pop open. A spring on the safety valve exerts a downward force, which keeps the valve disc in contact with the valve seat. **See Figure 3-4.** Steam pressure acts on the safety valve disc and exerts an upward force that tries to force the valve open. The total force of steam trying to overcome the spring force is equal to the area of the safety valve disc times the steam pressure.

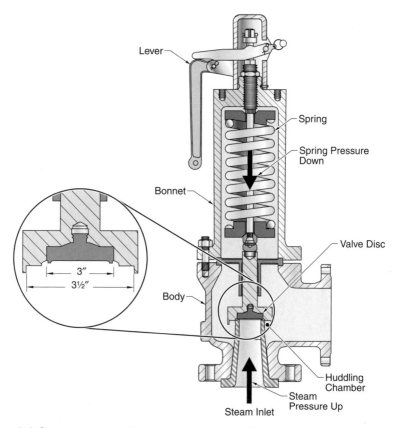

Figure 3-4. Steam pressure acting on the safety valve disc must develop a force that overcomes the force of the safety valve spring before the valve will open.

The safety valve spring must exert a downward force equal to the steam pressure force to keep the valve closed. When steam pressure starts to overcome the force of the safety valve spring, the safety valve will start to open (feather). The steam then enters the huddling chamber to expose steam to a larger area of the valve disc. This increases the force, which causes the safety valve to open quickly, or pop open. **See Figure 3-5.** To find the total force, first calculate the area of the valve disc of the safety valve by applying the following formula:

$$A = 0.7854 \times D^2$$

where

A = area (in in.)

D = diameter (in in.)

Next, multiply the area times the pressure.

$$TF = P \times A$$

where

TF = total force (in lb)

P = pressure (in psi)

A = area (in in.)

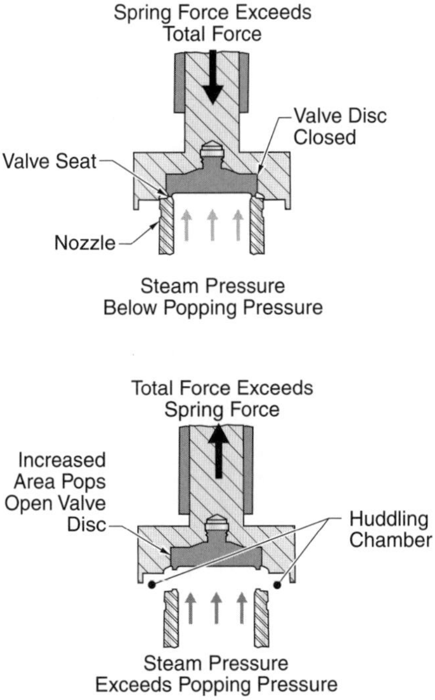

Spring Force Exceeds Total Force

Valve Disc Closed

Valve Seat

Nozzle

Steam Pressure Below Popping Pressure

Total Force Exceeds Spring Force

Valve Disc Lifted Off Valve Seat

Huddling Chamber

Steam Pressure Exceeds Popping Pressure

Total Force Exceeds Spring Force

Increased Area Pops Open Valve Disc

Huddling Chamber

Steam Pressure Exceeds Popping Pressure

Spring Force Exceeds Total Force

Valve Disc On Valve Seat

Steam Pressure Reduces Below Set Pressure

Figure 3-5. As soon as the steam pressure starts to overcome the force of the spring, the valve will slowly start to open. This allows the steam to enter the huddling chamber and force the valve to pop open.

For example, a steam pressure of 100 psi and a valve disc diameter of 3″ has a total force of 706.86 lb.

$TF = P \times A$

$TF = P \times 0.7854 \times D^2$

$TF = 100 \times 0.7854 \times 3 \times 3$

$TF = \textbf{706.86 lb}$

The spring must exert a downward force equal to the upward force to keep the valve closed. As soon as the steam pressure starts to overcome the force of the spring, the valve will slowly start to open (feather). This allows the steam to enter a huddling chamber, which exposes a larger area for the steam pressure to act on.

This increases the total upward force, causing the safety valve to open quickly, or pop open. Using the total dimensions, the diameter of the valve disc and the huddling chamber exposed to steam pressure is now 3.5″. The total force increases as follows:

$TF = P \times A$

$TF = P \times 0.7854 \times D^2$

$TF = 100 \times 0.7854 \times 3.5 \times 3.5$

$TF = \textbf{962.11 lb}$

The total force jumps from 706.86 lb to 962.11 lb, an increase of 255.25 lb. This greater total force overcomes the spring force, causing the valve to pop open.

EXAMPLES

1. A boiler has a pressure of 100 psi and a 2″ diameter safety valve. What is the total force acting on the safety valve? *Note:* Safety valve diameter includes the huddling chamber.

$TF = P \times A$

$TF = 100 \times 0.7854 D^2$

$TF = 100 \times 0.7854 \times 2 \times 2$

$TF = 100 \times 3.1416$

$TF = \textbf{314.16 lb}$

2. A boiler has a pressure of 100 psi and a 2½″ diameter safety valve. What is the total force acting on the safety valve?

$TF = P \times A$

$TF = 100 \times 0.7854 D^2$

$TF = 100 \times 0.7854 \times 2.5 \times 2.5$

$TF = 100 \times 4.90875$

$TF = \textbf{490.88 lb}$

3. A boiler has a pressure of 150 psi and a 3″ diameter safety valve. What is the total force acting on the safety valve?

$TF = P \times A$

$TF = 150 \times 0.7854D^2$

$TF = 150 \times 0.7854 \times 3 \times 3$

$TF = 150 \times 7.0686$

$TF = \mathbf{1060.29\ lb}$

4. A 3″ diameter safety valve has a spring that exerts a total force of 1060.29 lb. What is the popping pressure?

$TF = P \times A$

$\dfrac{TF}{A} = P$

$P = \dfrac{TF}{A}$

$P = \dfrac{1060.29}{0.7854 \times 3 \times 3}$

$P = \dfrac{1060.29}{7.0686}$

$P = \mathbf{150\ psi}$

5. A 2″ diameter safety valve has a total force of 650 lb acting on its disc and is at the point of popping. What is the pressure on the boiler?

$TF = P \times A$

$P = \dfrac{TF}{A}$

$P = \dfrac{650}{0.7854 \times 2 \times 2}$

$P = \dfrac{650}{3.1416}$

$P = \mathbf{206.9\ psi}$

6. A boiler has a pressure of 150 psi and a safety valve spring that exerts a 1060.29 lb force. What diameter safety valve is required if the valve is to pop at 150 psi?

$TF = P \times A$

$TF = P \times 0.7854D^2$

$\dfrac{TF}{0.7854 \times P} = \dfrac{0.7854D^2 \times P}{0.7854 \times P}$

$\dfrac{TF}{0.7854 \times P} = D^2$

$$D^2 = \frac{TF}{0.7854P}$$

$$\sqrt{D^2} = \sqrt{\frac{TF}{0.7854P}}$$

$$D = \sqrt{\frac{TF}{0.7854P}}$$

$$D = \sqrt{\frac{1060.29}{0.7854 \times 150}}$$

$$D = \sqrt{\frac{1060.29}{117.81}}$$

$$D = \sqrt{9}$$

$$D = 3''$$

7. A boiler has a pressure of 200 psi and a safety valve spring that exerts a downward force of 880 lb. What is the safety valve diameter required to open the valve at 200 psi?

$$D = \sqrt{\frac{TF}{0.7854P}}$$

$$D = \sqrt{\frac{880}{0.7854 \times 200}}$$

$$D = \sqrt{\frac{880}{157.08}}$$

$$D = 2.367''$$

Safety Valve Size. To determine the safety valve size required for a boiler, the boiler capacity must be known. Boiler capacity is rated in pounds of steam generated per hour. According to ASME code, boiler capacity in pounds of steam per hour is determined by the maximum amount of fuel consumed by the burners of the boiler. Pounds of steam per hour is found by using the following equation:

Pounds of steam/hour =

$$\frac{Quantity\ of\ fuel\ burned/hour \times Heating\ value/unit\ of\ fuel \times 0.75}{1100}$$

$$W = \frac{c \times H \times 0.75}{1100}$$

Note: 0.75 is the assumed boiler efficiency; 1100 is the assumed Btu value of 1 lb of steam.

According to ASME code,

$$W = \frac{c \times H \times 0.75}{h}$$

W = pounds of steam generated per hour

c = quantity of fuel burned per hour

H = heating value per unit of fuel

h = constant, the latent heat of the steam (in Btu/lb)

Boiler capacity is expressed in pounds of steam produced per hour. The safety valve manufacturer must produce a valve or valves capable of relieving that capacity, within the tolerances set by ASME, at the MAWP for the specific boiler.

EXAMPLES

1. A burner has a maximum firing rate of 900 lb of fuel/hr, and the fuel has a heat content of 18,000 Btu/lb. How many lb of steam/hr will this boiler generate?

$$W = \frac{c \times H \times 0.75}{h}$$

$$W = \frac{900 \times 18,000 \times 0.75}{1100}$$

$$W = \frac{12,150,000}{1100}$$

W = **11,045 lb/hr**

2. A burner has a maximum firing rate of 50,000 cu ft of gas/hr, and the gas has a heat content of 1050 Btu/cu ft. What is the boiler capacity in lb of steam/hr?

$$W = \frac{c \times H \times 0.75}{h}$$

$$W = \frac{50,000 \times 1050 \times 0.75}{1100}$$

$$W = \frac{39,375,000}{1100}$$

W = **35,795 lb/hr**

When boiler capacity in lb/hr of a boiler is unknown, it is found by the number of sq ft of heating surface in the boiler and by using the following equations:
For firetube boilers:

$$W = (8 \times HS_b) + (14 \times HS_{ww})$$

For watertube boilers:

$$W = (10 \times HS_b) + (16 \times HS_{ww})$$

W = pounds of steam generated per hour

HS_b = boiler heating surface (in sq ft)

HS_{ww} = waterwall heating surface in furnace (in sq ft)

EXAMPLES

1. A firetube boiler has 400 sq ft of boiler heating surface and 80 sq ft of waterwalls. What is the boiler capacity?

$W = (8 \times HS_b) + (14 \times HS_{ww})$

$W = (8 \times 400) + (14 \times 80)$

$W = 3200 + 1120$

$W = $ **4320 lb/hr**

2. A firetube boiler has 2200 sq ft of boiler heating surface but is not equipped with waterwalls. What is the boiler capacity?

$W = (8 \times HS_b) + (14 \times HS_{ww})$

$W = (8 \times 2200) + (14 \times 0)$

$W = 17,600 + 0$

$W = $ **17,600 lb/hr**

3. A watertube boiler has a total of 5000 sq ft of boiler heating surface, 800 sq ft of which are waterwalls. What is the boiler capacity?

$W = (10 \times HS_b) + (16 \times HS_{ww})$

$W = (10 \times 4200) + (16 \times 800)$

$W = 42,000 + 12,800$

$W = $ **54,800 lb/hr**

4. A watertube boiler has a total of 5000 sq ft of boiler heating surface, 750 sq ft of which are waterwalls. What is the boiler capacity?

$W = (10 \times HS_b) + (16 \times HS_{ww})$

$W = (10 \times 4250) + (16 \times 750)$

$W = 42,500 + 12,000$

$W = $ **54,500 lb/hr**

Boiler capacity is based on the amount and type of fuel used, or the amount of heating surface of the boiler. The boiler capacity can then be used to determine the approximate area of the safety valves needed on the boiler. *Note:* The boiler manufacturer and safety valve manufacturer will specify the proper size safety valve(s) required per ASME code.

The safety valve area required is equal to 0.2 times the number of lb of steam/hr divided by the MAWP of the boiler plus 10.

$$Area = \frac{0.2 \times Pounds \ of \ steam/hr}{MAWP + 10}$$

$$A = \frac{0.2W}{MAWP + 10}$$

A = safety valve area (in sq in.)

W = pounds of steam generated per hour

$MAWP$ = maximum allowable working pressure (in psi)

EXAMPLES

1. A boiler generates 11,045 lb of steam/hr at 200 psi. What is the safety valve area required?

$$A = \frac{0.2W}{MAWP + 10}$$

$$A = \frac{0.2 \times 11,045}{200 + 10}$$

$$A = \frac{2209}{210}$$

A = **10.52 sq in.**

2. A watertube boiler generates a maximum of 35,795 lb of steam/hr at 300 psi. What is the safety valve area required to relieve that quantity of steam?

$$A = \frac{0.2W}{MAWP + 10}$$

$$A = \frac{0.2 \times 35,795}{300 + 10}$$

$$A = \frac{7159}{310}$$

A = **23.1 sq in.**

After the safety valve area has been found, the diameter of the valve(s) required can be found.

$$Area = 0.7854D^2$$

$$\frac{A}{0.7854} = \frac{0.7854D^2}{0.7854}$$

$$D^2 = \frac{A}{0.7854}$$

$$D = \sqrt{\frac{A}{0.7854}}$$

EXAMPLES

1. A boiler generates 11,045 lb of steam/hr at 200 psi. The area of its safety valve is 10.52 sq in. What safety valve diameter is required?

$$D = \sqrt{\frac{A}{0.7854}}$$

$$D = \sqrt{\frac{10.52}{0.7854}}$$

$$D = \sqrt{13.39}$$

$$D = \mathbf{3.66''}$$

Note: The boiler requires a 3.66″ diameter safety valve or two 2.6″ diameter safety valves to conform to the ASME code.

2. A watertube boiler generates a maximum of 35,795 lb of steam/hr at 300 psi. The area of its safety valve is 23.1 sq in. What safety valve diameter is required?

$$D = \sqrt{\frac{A}{0.7854}}$$

$$D = \sqrt{\frac{23.1}{0.7854}}$$

$$D = \sqrt{29.41}$$

$$D = \mathbf{5.42''}$$

Note: The boiler requires a 5.42″ diameter safety valve or three 3.13″ diameter safety valves to conform to the ASME code.

Main Steam Stop Valves

A *main steam stop valve* is a valve on the main steam line leaving the boiler that is used to cut the boiler in on-line or take the boiler off-line. The main steam line requires two main steam stop valves when boilers are in battery. The main steam line leaves the boiler and connects to the main steam header. The main steam header is the distribution line. From the main steam header, branch lines supply steam to working stations. The main steam line is constructed using steam bends to allow for expansion and contraction. **See Figure 3-6.** The main steam header must also have steam traps to remove condensate built up in the lines. Any condensate in the steam lines could lead to water hammer and a possible pipe rupture.

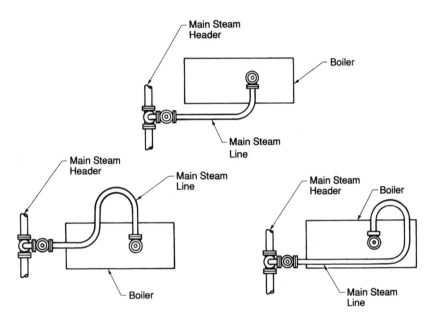

Figure 3-6. The main steam line from the boiler to the main steam header must be constructed with bends that allow for expansion and contraction.

The ASME code states that high-pressure boilers in batteries equipped with manhole openings must have two main steam stop valves with an ample free-blowing drain between them. **See Figure 3-7.** In addition, the ASME code also states that main steam stop valves must be of the outside stem and yoke (os&y) type that show if the valves are open or closed by the position of their stem. The main steam stop valves should be gate valves. Gate valves offer no resistance to the flow of steam and should always be wide open or completely closed. An automatic nonreturn valve may be used in place of one stop valve, but must be located as close to the boiler shell as practical.

The main steam stop valve is used to place a boiler in service or isolate it from the system for cleaning, inspection, or boiler repairs. An automatic nonreturn valve operates like a check valve when its stem is in the open postition to improve the safety and efficiency of the plant by cutting a boiler in on the line automatically or taking it off the line automatically. Automatic nonreturn valves also protect the system in case of a failure on the pressure side of any boiler on the line. For example, if a boiler dropped some tubes and the pressure in the boiler dropped below header pressure, the automatic nonreturn valve would close, taking the boiler off the line and preventing steam from backing out of the other boilers. The main steam stop valve and the automatic nonreturn valve should be dismantled, inspected, and overhauled annually.

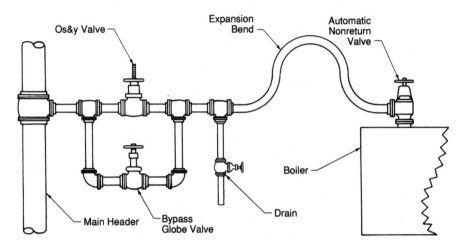

Figure 3-7. The main steam line must be equipped with two main steam stop valves and a free-blowing drain located between them.

The automatic nonreturn valve consists of a valve seat and disc connected to a piston in a cylinder, all enclosed in a valve body. **See Figure 3-8.** A stem attached to a hand wheel can be used to close the valve by hand by forcing the piston and disc down. When the stem is up, the valve is opened slowly by steam pressure under the disc of the valve.

Blowdown Valves and Lines

All boilers must have a blowdown valve with a blowdown pipe located at the lowest part of the water side of the firetube boiler, or on the mud drum on a watertube boiler. The blowdown pipe must run full size, with no reducers or bushings. All blowdown piping outside the boiler should be exposed for inspection.

Boilers with an MAWP exceeding 125 psi require extra-heavy blowdown pipe(s) from the boiler to the valve or valves per ASME code. All fittings between the boiler and valve shall be steel or extra-heavy bronze or brass, malleable iron, or cast iron suitable for the temperatures and pressures involved. Boilers exceeding an MAWP of 125 psi must have each bottom blowdown pipe fitted with at least a 250 lb standard valve or cock. Two extra-heavy valves, or an extra-heavy valve and cock are preferred.

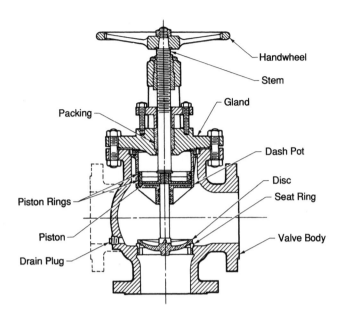

Figure 3-8. The automatic nonreturn valve cuts the boiler on-line and off-line automatically.

When the bottom blowdown pipe is exposed to direct furnace heat, it must be protected by firebrick and arranged so the pipe can be inspected. The opening in the boiler setting for the blowdown pipe should be arranged to allow for free expansion and contraction. Blowdown piping must be protected from freezing, heat, or corrosive conditions. The minimum size for blowdown pipes and fittings is 1″ and the maximum is 2½″. Surface blowdown pipes have no minimum size but have a maximum size of 2½″.

Waterwall blowdown valves shall be locked or sealed closed to prevent being opened by mistake when the boiler is firing. If the waterwall blowdown valves are opened, the loss of water or circulation will cause the waterwall tubes to overheat. Waterwall blowdown valves should only be blown down when the furnace has cooled properly. Waterwalls are usually blown down when the boiler is being taken off the line.

The frequency of blowing down the boiler is determined through boiler water analysis. However, the boiler should be blown down at least once every 24 hours. Makeup water contains a certain amount of scale-forming salts. Scale-forming salts start to settle on the boiler heating surfaces when the boiler water temperature reaches about 150°F. The settling of these salts on the boiler heating surfaces reduces the required heat transfer by insulating the surface. As a result, the boiler heating surfaces overheat and burn out. Chemicals added to the boiler water turn scale-forming salts into a non-adhering sludge. The sludge stays in suspension and settles to the lowest part of the water side of the boiler when the boiler is at a light load. Bottom blowdown valves and lines are located at the lowest part of the water side of the boiler to remove the sludge.

Boilers operating at 100 psi or higher must have two approved bottom blowdown valves. The two valves may be two slow-opening valves (screw-type), or one quick-opening valve and one slow-opening valve. The quick-opening valve must be located closest to the boiler shell. In blowing down, the quick-opening valve should be opened first and closed last. The quick-opening valve functions as a sealing valve. The slow-opening valve is located farthest from the boiler shell. The slow-opening valve is the blowing valve and takes all the wear and tear of blowing down. **See Figure 3-9.**

Bottom blowdown is used to remove sludge and sediment, control high water, control chemical concentration, and dump the boiler for cleaning and inspection.

There must be pressure on the boiler when blowing down. The only time there is no pressure is when the boiler is dumped. Before a boiler can be dumped, it must be cool with the air cock open to allow for the contraction of the cooling water and prevent a vacuum from forming.

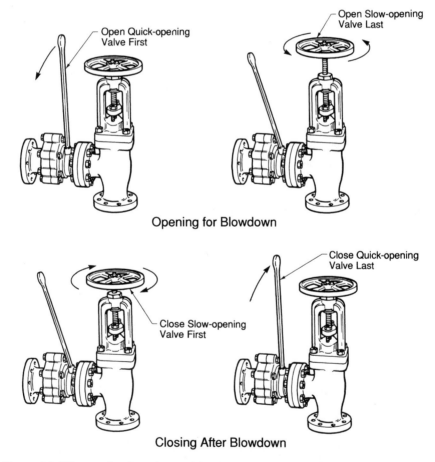

Open Quick-opening Valve First

Open Slow-opening Valve Last

Opening for Blowdown

Close Slow-opening Valve First

Close Quick-opening Valve Last

Closing After Blowdown

Figure 3-9. When performing a bottom blowdown, the quick-opening valve located closest to the boiler shell is opened first and closed last.

The best time to blow down a boiler is at its lightest load because sludge and sediment have had a chance to settle to the bottom of the water side of the boiler. When blowing down a boiler, blowdown lines are subjected to sudden, full boiler pressure and temperature without warm-up. The blowdown valves and piping must be designed for maximum capacity and allow for expansion and contraction. Straight-run globe valves or other valves that may impede boiler water flow or collect sediment must not be used. Any leak on the blowdown line is a serious condition and the boiler should be removed from service. An inspector should be notified and recommended procedures followed.

To prevent hot water and steam from entering the sewer system, a blowdown or flash tank is used between the blowdown line and the sewer. **See Figure 3-10.** Hot water and steam enter the top of the tank. Flash steam leaves through the vent and the hot water stays in the tank. As the level in the tank rises, cooler water from the bottom of the tank flows into the sewer. The blowdown tank equipment should be of sufficent capacity to prevent discharging water over 150°F and/or 5 psi into the city sewer. After blowing down a boiler, water remains in the blowdown tank to cool until another boiler is blown down. A siphon breaker is installed to prevent possible siphoning of the water from the tank.

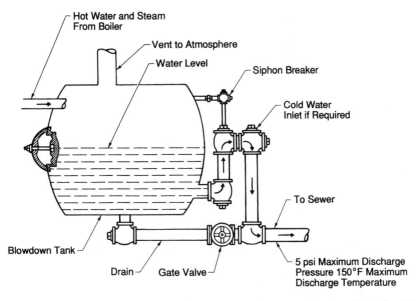

Figure 3-10. A flash tank recovers heat from steam and water coming from the boiler. Water is then discharged from the flash tank to the sewer.

A blowdown centrifugal separator can be used between the blowdown line and blowdown tank. **See Figure 3-11.** A *blowdown centrifugal separator* is a separator that reduces boiler pressure and temperature before the boiler water enters the blowdown tank. Whenever a blowdown centrifugal separator is used without a blowdown tank, it must allow for safe discharge.

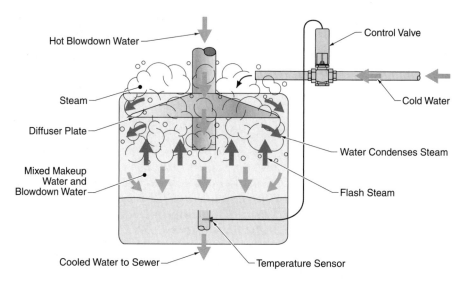

Hot Blowdown Water

Control Valve

Steam

Cold Water

Diffuser Plate

Water Condenses Steam

Mixed Makeup
Water and
Blowdown Water

Flash Steam

Cooled Water to Sewer

Temperature Sensor

Figure 3-11. A blowdown separator cools blowdown water by mixing it with cold water.

Some boilers are equipped with surface blowdown valves. Surface blowdown valves are located at the normal operating water level (NOWL) of the boiler and are used to remove impurities from the surface of boiler water. This reduces surface tension and prevents carryover of water and impurities. The surface blowdown line discharges to the blowdown tank.

Large boilers carrying heavy steam loads require closer control of the total dissolved solids in the boiler water. This is accomplished by using continuous proportioning of feedwater treatment in conjunction with a continuous blowdown line. The continuous blowdown line is located below the NOWL and discharges to a flash tank or continuous blowdown tank. The flash tank can be used to reclaim heat from boiler water from the continuous blowdown line. **See Figure 3-12.**

Water from the continuous blowdown line is piped to the flash tank. As the water pressure drops, some of the water flashes into steam. This steam flows to the open feedwater heater. Makeup water passes through coils below the water level in the flash tank and is heated. An internal overflow line is used to prevent an excessive water level in the flash tank. *Note:* For every 10°F rise in feedwater temperature, there is approximately a 1% savings in fuel.

Pressure Gauges

A *pressure gauge* is an instrument used to indicate suction and discharge pressures of fuel and water lines, air pressure, gas pressure, and pressures on steam engines and steam turbines. Steam plants have pressure gauges to indicate all the pressures required to ensure safe and efficient plant operation.

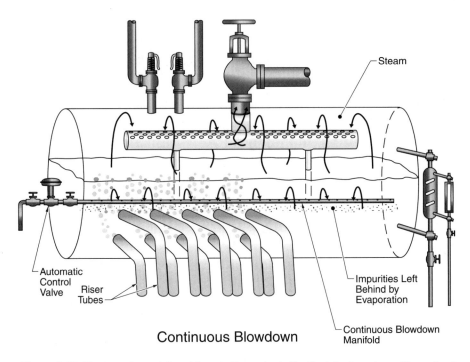

Continuous Blowdown

Automatic Control Valve
Riser Tubes
Steam
Impurities Left Behind by Evaporation
Continuous Blowdown Manifold

Figure 3-12. Heat can be reclaimed from boiler water in the flash tank and used to preheat make-up water supplied to the feedwater heater.

Pressure gauges are calibrated in psi. *Absolute pressure* is gauge pressure plus atmospheric pressure. Atmospheric pressure is equal to 14.7.

Absolute pressure = Gauge pressure + 14.7

A *vacuum gauge* is a pressure gauge used to measure pressure below atmospheric pressure in inches of mercury (Hg). For example, 4″ of mercury is expressed as 4″ Hg.

A *steam pressure gauge* is a pressure gauge used to indicate the steam pressure in the boiler. The ASME code requires that each boiler shall have its own steam pressure gauge. Steam pressure gauges must connect to the highest part of the steam side of the boiler. In addition, steam pressure gauges must be clearly visible from the operating platform and kept clean, well-lighted, and protected from extreme heat or cold. A siphon or similar device must be installed to protect the Bourdon tube in the steam pressure gauge from live steam. One-quarter inch brass or copper pipe may be used for temperatures up to 405°F. Steel or wrought iron pipe, when used, must be at least ½″ and the minimum size of a pipe siphon must be ¼″. The steam pressure gauge dial should have a range of approximately two times the safety valve setting, and not less than 1½ times the safety valve setting. This range is required when a hydrostatic test performed on a boiler raises the pressure to 1½ times the MAWP or safety valve setting. When ordering a new steam pressure gauge, the safety valve setting must be known.

Steam pressure gauges must also be graduated so that the pointer is in a nearly vertical position when at its normal operating pressure. Shutoff valves to steam pressure gauges should be lever-type cocks that show by their position whether they are open or closed. Some municipalities do not allow any valves or cocks between the boiler and the steam gauge. The ASME code also states that an additional valve or cock on the steam pressure gauge line must be locked or sealed open. **See Figure 3-13.**

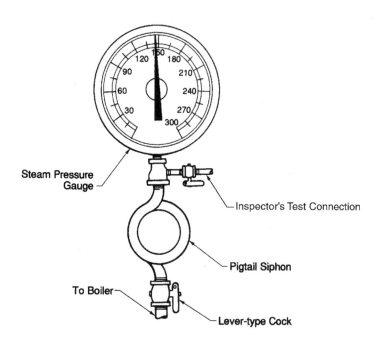

Figure 3-13. A shutoff valve to the steam pressure gauge should be a lever-type cock that indicates whether it is open or closed.

The steam pressure gauge must be connected to the highest part of the steam space of a boiler. It must also be visible to the operator. This presents a problem on boilers that have steam drums 40′ to 50′ or more above the floor level. A line must be run from the top of the steam drum down to where the steam pressure gauge is visible to the operator. For every vertical foot of piping filled with water, there is a pressure of 0.433 psi at the base. **See Figure 3-14.** This amount, 0.433 psi/vertical foot, is based on 1 cu ft of water weighing 62.4 lb at 60°F. Therefore, a column of water 1′ high with a cross section of 1 sq in. at 60°F exerts a pressure of 62.4 ÷ 144, or 0.433 psi. (There are 144 sq in. in 1 sq ft.) For example, a 100′ high pipe filled with water with a steam pressure gauge at the bottom reads 43.3 psi (100 × 0.433).

Pressure = psig + (Number of feet below boiler drum × 0.433)

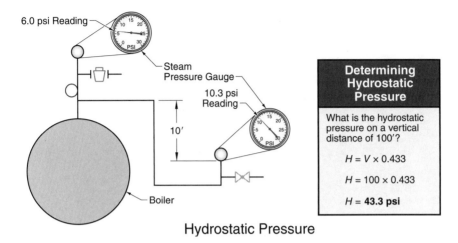

Hydrostatic Pressure

Figure 3-14. Steam pressure gauge readings can vary in hydrostatic pressure if there is a difference in height on the steam line.

EXAMPLES

1. A steam pressure gauge is located 20′ below the steam drum. The boiler has pressure of 100 psi. What would the steam pressure gauge read if it had not been corrected?

   $P = 100$ psig $+ (20 \times 0.433)$

   $P = 100 + 8.66$

   $P = \mathbf{109\ psig}$

2. A steam pressure gauge is located 35′ below the steam drum. The boiler has a pressure of 135 psi. What would the steam pressure gauge read if it had not been corrected?

   $P = 135$ psig $+ (35′ \times 0.433)$

   $P = 135 + 15.155$

   $P = \mathbf{150\ psig}$

Low Water Fuel Cutoff

A *low water fuel cutoff* is a safety device located slightly below the normal operating water level (NOWL) that shuts off fuel going to the burner if a low water condition occurs. A low water fuel cutoff prevents burning out of the tubes and a possible boiler explosion. The ASME code requires that all boilers have a low water fuel cutoff even though the boiler may be equipped with an automatic feedwater makeup. An automatic feedwater makeup can malfunction or there could be an interruption in the water supply. Depending on the application, the low water fuel cutoff may be integrated into the water column on the boiler. If only one low water fuel cutoff is used, it must have a manual reset. **See Figure 3-15.**

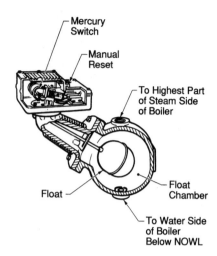

Figure 3-15. The low water fuel cutoff is located slightly below the NOWL.

The low water fuel cutoff is located slightly below the NOWL. The top line connects to the highest part of the steam side of the boiler. The bottom line connects to the water side well below the NOWL. A blowdown line and valve are used to keep the float chamber free of sludge and sediment. The primary function of a low water fuel cutoff is to de-energize the burner limit circuit and shut down the burner if the water level in the boiler drops below the safe operating level.

Other functions may be included on a low water fuel cutoff. For example, an alarm may sound at the same time the burner is shut down or it can start and stop the boiler feedwater pump to maintain water at a safe operating level in the boiler. A high water alarm may sound when the water level in the boiler exceeds a predetermined level. The boiler manufacturer specifies the type(s) of low water fuel cutoff required. In general, the larger the boiler, the more sophisticated the low water fuel cutoff controls.

The low water fuel cutoff should be blown down every shift on all operating boilers to ensure the test chamber is clear of sludge and sediment. Blowing down the low water fuel cutoff tests the low water cutoff circuit to verify that it is working. However, this is not a good operational test. The low water fuel cutoff should be operationally tested at least once a month with an evaporation test. The evaporation test involves cutting off the feedwater, which allows the water level in the boiler to drop by evaporation until the fuel is secured to the burner. The boiler operator must be present during the evaporation test to monitor the boiler. The low water fuel cutoff should be opened once a year for cleaning and inspection.

Water Column

A *water column* is a fitting used to slow down the turbulence of boiler water so an accurate reading of the water level in the gauge glass can be obtained. The water column must be located at the NOWL so that the lowest visible part of the gauge glass is 2″ to 3″ above the highest heating surface on a firetube boiler. The highest heating surface varies among firetube boilers. For example, in an HRT boiler, the highest heating surface is the top row of tubes. The firetube boiler manufacturer should be consulted to determine the highest heating surface. For watertube boilers, the water column must be located so that the lowest visible part of the gauge glass is 3″ above the lowest safe working water level per the boiler manufacturer specifications.

The water column and gauge glass should be checked for proper location. The proper location is checked by using measurements on the outside of the boiler. An alternate method is to add water until it is visible in the gauge glass when the boiler is off the line. A rule is then inserted through the manhole until it reaches the bottom of the drum on a watertube boiler or the highest heating surface on a firetube boiler to measure the amount of water in the boiler.

The water column is connected to the boiler with a minimum of 1″ pipe. The top line goes to the highest part of the steam side of the boiler. The bottom line connects below the NOWL. **See Figure 3-16.**

Valves used between the boiler and water column must be os&y valves or lever-type valves that show by their position whether they are open or closed. Some municipalities do not allow any valves between the boiler and the water column. The valves should be locked in the open position. Lines going to the water column must be fitted with a cross T for inspection and cleaning. The water column must be fitted with a minimum ¾″ blowdown valve. For pressures of 400 psi or higher, the lower water column connection to the boiler drum must be fitted with a shield or sleeve to reduce the effect of temperature differentials per ASME code.

The gauge glass is mounted on the water column. Gauge glasses are tubular (round) or flat, depending on the pressure carried in the boiler. Tubular gauge glasses are used with pressures up to 250 psi. Flat gauge glasses are used for pressures exceeding 250 psi. **See Figure 3-17.** The gauge glass must be fitted with a minimum ¼″ blowdown valve. Gauge glass connections to the water column must be at least ½″. Gauge glasses must have shields to protect the operator in case of gauge glass breakage. Gauge glasses must also be fitted with quick-closing valves so they may be taken out of service.

All boilers must have two means of determining boiler water level. There may be two gauge glasses, or one gauge glass and one remote level indicator, or one gauge glass and try cocks. Try cocks are mounted on the water column. **See Figure 3-18.** The three try cocks commonly found on the water column are the top, middle, and bottom try cocks. With water at the NOWL, the top try cock, when open, discharges steam. The middle try cock, when open, discharges water, steam, or a mixture of water and steam, depending on the actual water level. The bottom try cock, when open, discharges water. Water coming out of the top try cock indicates a high water level. The water level must be lowered by using the boiler bottom blowdown valves. Steam discharging from the bottom try cock indicates a low water condition. Try cocks are used as a secondary means to determine water level in boilers operating at pressures up to 250 psi. A low water condition requires removing the boiler from service and inspecting the boiler heating surface for signs of overheating.

If the top line going to the gauge glass is closed or clogged, the gauge glass will show a full glass. If the bottom line going to the gauge glass is closed or clogged, the water level in the gauge glass will be stationary (no movement), and the gauge glass will slowly start to fill up as a result of condensation of steam.

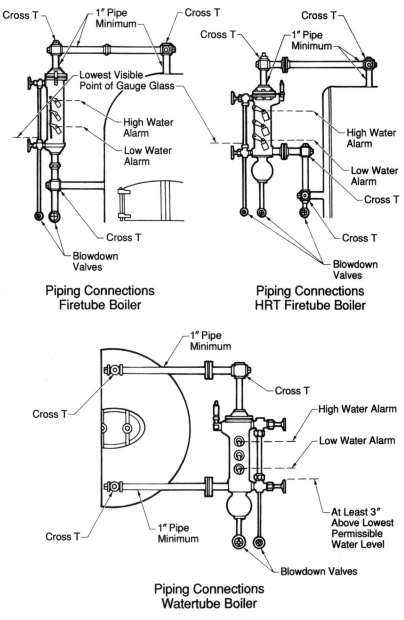

**Piping Connections
Firetube Boiler**

**Piping Connections
HRT Firetube Boiler**

**Piping Connections
Watertube Boiler**

Figure 3-16. The location of the water column is determined by the NOWL of the boiler.

Tubular Gauge Glass   Flat Gauge Glass

Figure 3-17. Gauge glasses for pressures up to 250 psi are round and for pressures exceeding 250 psi are flat.

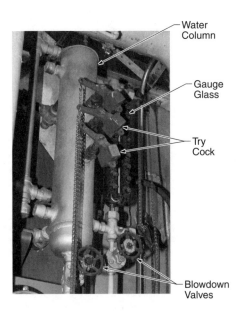

Figure 3-18. Gauge glasses and water columns use blowdown valves to remove sludge and sediment.

The materials used in water column construction vary with the boiler pressure. Cast iron can be used for pressures up to 250 psi and malleable iron can be used for pressures up to 350 psi. For pressures above 350 psi, steel construction is required per ASME code.

Internal Steam and Water Drum Fittings

Internal steam and water drum fittings are used to control direction and flow of steam and water inside the boiler. **See Figure 3-19.** Feedwater must be introduced into a boiler so that it does not come in contact with the boiler heating surfaces exposed to direct radiation or hot gases of combustion at any time. When pressures are 400 psi or higher, ASME code requires that the feedwater inlet through the drum be fitted with a shield, sleeve, or suitable method to reduce the effects of temperature differential in the boiler shell. If necessary, the discharge end of a feed pipe should be fitted with a baffle to divert the flow.

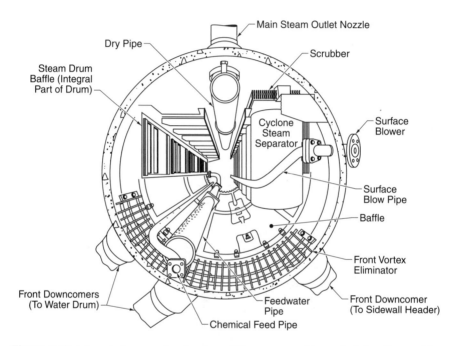

Figure 3-19. Internal steam and water drum fittings are used to control direction and flow of steam and water inside the boiler.

Feedwater supplied is cold compared to the temperatures on the combustion side of the boiler. If feedwater is allowed to impinge directly on the boiler heating surface, it would subject the boiler to thermal shock. The ASME code also requires that seamless steel pipe equal to Schedule 80 (extra-heavy) be used for pressures over 100 psi and temperatures over 220°F.

The internal feedwater line in steam drums of watertube boilers extends about 80% of the drum length. It is installed so that incoming feedwater is discharged below the surface of the water and over the entire length. This is accomplished using a perforated pipe or water trough.

In HRT boilers over 40″ in diameter, the feedwater line enters above the central row of tubes. The line discharges at approximately three-fifths the length from the end of the boiler subjected to the hottest gases of combustion. A perforated line or a trough must be used. When the boiler is off-line for inspection, the internal feedwater line should be carefully examined to make sure it is secure and there are no plugged holes in the perforated line. If a trough is used, it should be checked for cleanliness and properly secured.

When a boiler is operated at high loads or with a high water level, there is a tendency to carry over particles of water with the steam. A *dry pipe separator* is a separator used to control carryover, consisting of a closed pipe perforated at the top with drain holes on the bottom that remove moisture from steam. The top half is drilled with many small holes and connected at the top center to the main steam outlet from the boiler. Dry pipe separators have drain holes on the bottom that allow entrained moisture (water particles) to return to the boiler drum.

Dry pipe separators operate on the same principle as a steam separator. Steam enters the small holes on top. The steam has to change direction to leave through the steam outlet. The changing of direction causes any entrained moisture to be separated and returned to the boiler through the drain.

Dry pipe separators are very effective for small boilers or boilers with light steam loads. As the capacity of a boiler increases, the efficiency of the dry pipe separator decreases. Cyclone separators were developed to overcome the shortcomings of dry pipe separators. A *cyclone separator* is an internal fitting consisting of a number of cylinders (cyclones) set side by side along the length of the drum, and a baffle arrangement to direct the steam into a cyclone. Moisture is removed by centrifugal force as the steam is forced to rotate when it passes through the cyclones. For improved separation of steam and moisture, baffles and secondary scrubbers are installed. In addition, cyclone separators are required on boilers equipped with superheaters to keep carryover to a minimum.

Steam, saturated or superheated, is directed to points of use by steam lines. All steam lines must be properly insulated to prevent radiant heat loss from the piping and condensation of the steam in the piping. Regardless of the insulation quality, some condensation will occur. Condensation is removed before a steam engine by a receiver separator or before a steam turbine by a steam strainer separator. A *receiver separator* is a large piece of equipment that stores steam in large volumes. **See Figure 3-20.** Receiver separators remove condensate and absorb the shock of cutoff of the steam during the operation of reciprocating steam engines.

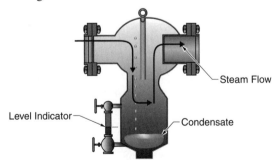

Receiver Separators

Figure 3-20. Receiver separators remove condensate and act as a buffer to absorb shock during steam cutoff to steam engines.

A *steam strainer separator* is an internal fitting used to remove any impurities in addition to condensate before the steam enters a steam turbine. **See Figure 3-21.** Condensate that impinges (hits) steam turbine blading causes severe damage to the blading by erosion or pitting, similar to the effect of sandblasting. Steam strainer separators are not as large as receiver separators.

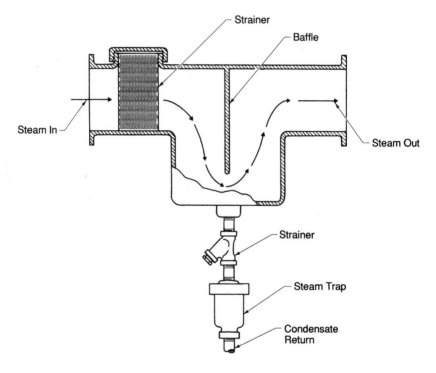

Figure 3-21. Steam strainer separators remove condensate and impurities from steam to be used in the steam turbine.

Horizontal steam lines should always be pitched in the direction of the steam flow. Any condensation flows with and not against the steam flow. When condensate flows against the steam flow, water hammer and pipe rupture could occur. Steam traps should be installed whenever the steam line dead ends or makes a 90° turn up.

Steam lines must be designed to allow for expansion. Steam lines should be set on rollers and have expansion bends and joints to allow for movement. **See Figure 3-22.** There should be as few expansion bends and joints as possible. Steam lines should also be designed to permit repairs without shutting down the whole plant. Valves should be used at locations that allow isolation of parts of the system for repairs or maintenance procedures.

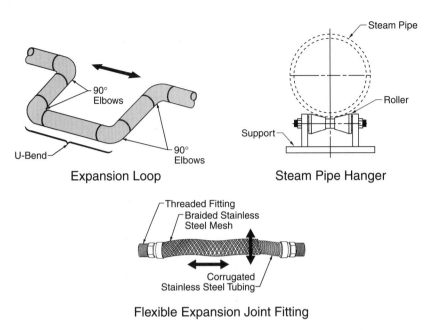

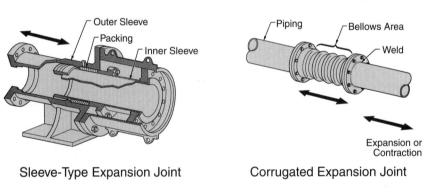

Figure 3-22. Expansion loops, steam pipe hangers, flexible expansion joint fittings, sleeve-type expansion joints, and corrugated expansion joints provide for expansion and contraction of piping.

Expansion of Piping

Pipes that carry fluids at high temperatures must allow for expansion and contraction. The amount of expansion that occurs depends on the temperature of the material in the pipe and the type of metal used to construct the pipe. Expansion of piping is compensated for by using pipe bends, expansion joints, or swing joints. Pipe bends are often used whenever possible because they require little or no maintenance. The amount of expansion of piping, in inches, is found by using the following equation:

Expansion = Length of pipe × Coefficient of linear expansion ×
(Temperature$_1$ – Temperature$_2$) × 12

$E = L \times C \times (T_1 - T_2) \times 12$

E = expansion (in inches)

L = length of pipe (in ft)

C = coefficient of linear expansion (inch per inch per °F)

T_1 = final temperature (in °F)

T_2 = initial temperature (in °F)

12 = inches per foot

The coefficients of linear expansion per inch of length per °F of some common piping materials are 23.1×10^{-6} m/m°C for aluminum and 17.8×10^{-6} m/m°C for steel. **See Figure 3-23.**

Material	Expansion Coefficients	
	10^{-6} in/in°F	10^{-6} m/m°C
Alumnium	12.8	23.1
Carbon steel	6.5	11.7
Cast iron	5.9	10.6
Copper	9.3	16.8
Stainless steel	9.9	17.8
Acrylonitrile butadiene styrene (ABS)	35.0	63.0
High-density polyethylene (HDPE)	67.0	120.0
Polyethylene (PE)	83.0	150.0
Chlorinated polyvinyl chloride (CPVC)	44.0	79.0
Polyvinyl chloride (PVC)	28.0	50.4

Figure 3-23. Expansion coefficients of piping depend on the piping material used and the amount of expansion that occurs.

EXAMPLES

1. A steel pipe 300′ long carries steam at 500 psia that is superheated 250°F. What is the amount of expansion from an initial temperature of 60°F? *Note:* From steam tables, the temperature of saturated steam at 500 psia is 467.01°F. Since steam is superheated 250°F, 250°F must be added to the saturation temperature.

$E = L \times C \times (T_1 - T_2) \times 12$

$E = 300 \times 0.0000067 \times [(467.01 + 250) - 60] \times 12$

$E = 300 \times 0.0000067 \times 657.01 \times 12$

$E = \mathbf{15.85''}$

2. A wrought iron pipe 185' long carries steam at 285 psi. The outside temperature is 90°F. What is the amount of expansion? *Note:* Psi must be converted to psia to read the correct value from the steam tables; 285 + 14.7 = 299.7 or 300 psia. The temperature is 417.33°F.

$E = L \times C \times (T_1 - T_2) \times 12$

$E = 185 \times 0.0000068 \times (417.33 - 90) \times 12$

$E = 185 \times 0.0000068 \times 327.33 \times 12$

$E = \textbf{4.94''}$

Valves must be installed for easy access. Overhead valves must be equipped with a chain wheel and chains or some system of gearing and extension stems. *Note:* One of the largest losses in a boiler room is radiant heat loss that occurs in steam lines where superheated steam temperatures reach and exceed 1050°F.

Superheaters

Superheaters are used to increase the amount of heat in the steam, superheating it. A *superheater* is a nest of tubes used to raise the temperature of steam but not its pressure. *Superheated steam* is steam at a temperature higher than its corresponding pressure. *Corresponding pressure* is the pressure of steam at saturation temperature. Superheated steam, which is heated above the saturation temperature to completely eliminate water carryover, has higher heat content than saturated steam. Therefore, superheated steam can produce more work than saturated steam can produce at the same pressure.

Initially, steam power plants using reciprocating engines were designed to operate with saturated steam. Steam separators were introduced between the boiler and the engine to remove condensate from the steam. This improved the quality of the steam but did not remove all moisture. Definite drawbacks to using this wet steam included the erosion of pipes, fittings, and engine parts caused by impingement of the steam and the loss of heat to metal, causing condensation of steam.

These drawbacks were almost eliminated by the use of superheated steam. To produce superheated steam, heat is added to the steam in a superheater. This heat is added after the steam has been removed from the steam and water drum. The degree of superheat depends on the difference in temperature between the saturated steam and the superheated steam.

For example, saturated steam at 600 psi is at a temperature of 489°F. If it were heated to 700°F it would have a superheat of 211°F (700 – 489 = 211). In an average steam turbine, there is a gain of 1% in efficiency for every 35°F of superheat. To find the increase in efficiency, divide the gain by 35. In this example there is a 6% increase in efficiency (211 ÷ 35 = 6.0).

Superheaters are nests of tubes located in the radiant zone (furnace area) or in the direct path of the gases of combustion. **See Figure 3-24.** Superheaters may be classified as radiant superheaters or convection superheaters. A *radiant superheater* is a superheater that receives its heat from the radiant heat zone in the boiler. A *convection superheater* is a superheater that receives its heat from the flow of the gases of combustion in the boiler. A consistent steam temperature during fluctuations in load can be maintained by using both radiant and convection superheaters in the same unit.

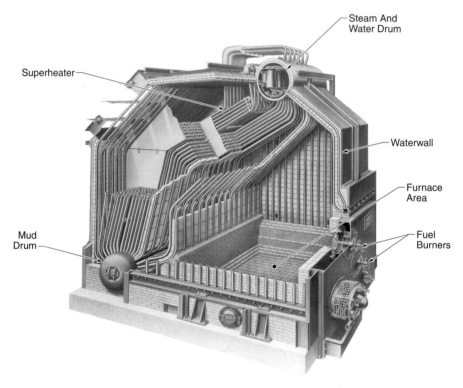

Babcock & Wilcox Co.
Figure 3-24. A superheater adds additional heat to the steam leaving the steam and water drum of a boiler.

When operating a steam boiler equipped with a superheater, a flow of steam must be maintained through the superheater at all times. Failure to maintain steam flow will result in overheating, warping, or burning of superheater tubes.

A safety valve is required at the superheater outlet header. The safety valve must be of sufficient capacity to maintain the minimum steam flow through the superheater. The superheater safety valves are set at a lower popping pressure than the safety valves on the steam and water drum of the boiler to ensure this flow of steam through the superheater.

Superheaters must also be fitted with a drain on the outlet side. This drain remains open when warming up a boiler and is left open until the boiler is cut in on the line. The drain must also be opened as soon as the boiler is taken off-line. This ensures a minimum steam flow through the superheater at all times.

This flow of steam keeps the superheater tubes from overheating during startup and shutdown. Superheaters must be kept free of soot (carbon from unburned fuel) to maintain the transfer of heat. For optimum efficiency, follow the manufacturer recommendations for startup and shutdown.

When superheater tube failure occurs, it almost always occurs from overheating. Overheating is caused by insufficient steam or by the insulating effects of deposits carried over from boiler water. These deposits originate in the boiler water and are carried over because of foaming in the steam drum. The foaming is caused by high concentrations of solids, oil, or other organic contaminants. The potential for foam and carryover can be minimized by proper water treatment and boiler blowdown as required.

Soot Blowers

A *soot blower* is a device used to remove soot deposits from around tubes and to permit better heat transfer in the boiler. **See Figure 3-25.** In order to have a transfer of heat, heat must conduct through the tube walls. Soot acts as insulation and interferes with heat conduction. Boiler efficiency increases with better heat transfer, thus reducing fuel consumption.

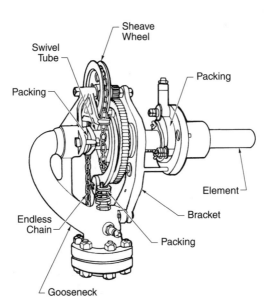

Figure 3-25. Soot blowers are used on some watertube boilers to maintain efficiency by cleaning heat-transfer surfaces.

Some watertube boilers are equipped with permanently installed soot blowers. Soot blowers vary depending on boiler type, fuel type, and soot blower application. They can be fixed or retractable and the cleaning medium can be saturated steam, superheated steam, compressed air, or water.

Fixed soot blower elements are located in the direct path of the gases of combustion. These elements are made of a steel alloy to withstand the high temperature of the gases of combustion. The nozzles in the element are spaced to allow steam or air to blow between the rows of tubes. The nozzles are held in place by bearings clamped or welded to the boiler tubes. These bearings allow the element to rotate. Element alignment is important. If the element were to shift, steam impinging on the boiler tubes would cut through the tubes in a very short time.

Steam Traps

A *steam trap* is an automatic device that increases the overall efficiency of a plant by removing air and water from steam lines without loss of steam. When steam releases heat, it starts to condense. Steam traps are located in the system to remove the maximum amount of condensate efficiently. **See Figure 3-26.**

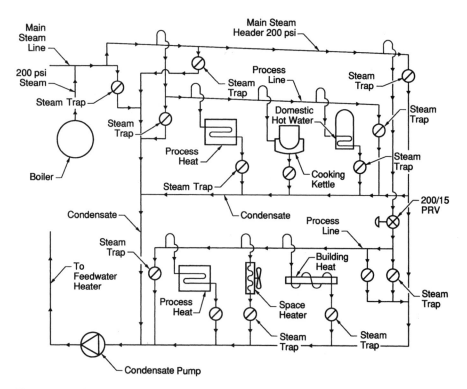

Figure 3-26. Steam traps are located in the system to remove condensate and air with maximum efficiency.

A steam trap stuck in the closed position causes condensate to build up, resulting in the line or heater filling with water. When the heater is full of condensate there is no heat transfer taking place. A steam trap stuck in the open position allows steam to blow through and enter the return lines. This results in a loss of plant efficiency and an increase in the condensate return line pressure and temperature. Common steam traps include thermostatic steam trap, float thermostatic steam trap, inverted bucket steam trap, thermodynamic steam trap, and impulse steam trap.

A *thermostatic steam trap* is a steam trap that has a bellows filled with a fluid that boils at steam temperature. The bellows expands or contracts depending on whether it is surrounded by steam or condensate. When the condensate level increases, the bellows contracts to open the discharge valve and release condensate. **See Figure 3-27.**

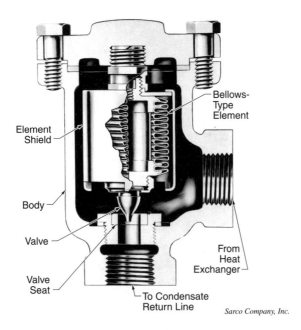

Sarco Company, Inc.

Figure 3-27. A thermostatic steam trap contains a temperature-operated device, such as a corrugated bellows, that controls a small discharge valve.

A *float thermostatic steam trap* is a steam trap that has a ball float and a thermostatic element. The ball float that opens and closes a discharge valve, depending on the amount of condensate in the float chamber. The thermostatic element opens an air vent valve when the cooler air surrounds the element and closes when steam surrounds it. **See Figure 3-28.**

An *inverted bucket steam trap* is a steam trap in which steam enters the bottom of the trap and flows into an inverted bucket. The steam holds the bucket up. When condensate fills the trap, the bucket loses buoyancy and sinks to open the discharge valve to release condensate. **See Figure 3-29.**

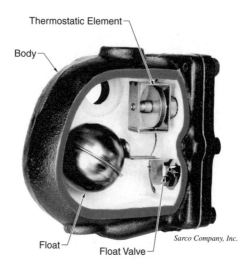

Figure 3-28. A float thermostatic steam trap contains a thermostatic bellows or other thermostatic element and also contains a steel ball float connected to a discharge valve by a linkage.

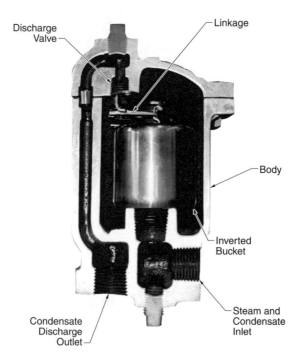

Figure 3-29. An inverted bucket steam trap contains an inverted bucket connected to a discharge valve.

A *thermodynamic steam trap* is a steam trap that has a single movable disc that raises to allow the discharge of air and cool condensate. Hot condensate creates steam at a higher pressure on top of the disc and at a lower pressure under the disc, causing it to close. Pressure in the control chamber decreases as trapped steam condenses. The disc is lifted by inlet pressure and condensate is discharged. **See Figure 3-30.**

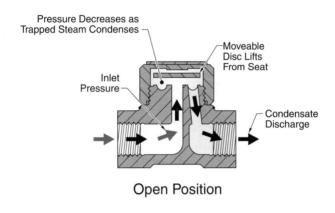

Open Position

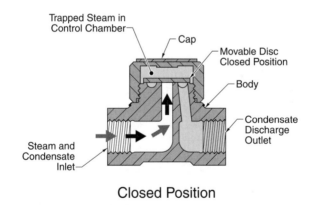

Closed Position

Figure 3-30. A thermodynamic steam trap has a single movable disc that rises to allow the discharge of air and cool condensate.

An *impulse steam trap* is a steam trap consisting of a valve with an extended, hollow stem that has a control disk attached at the top. The disk moves in a cylinder due to the pressure difference between the steam above the disk and the steam or condensate below. When condensate is present, the steam cools, decreasing the pressure in the control chamber and opening the valve to discharge condensate. As steam enters the trap and travels up the hollow stem to the control chamber, pressure increases, closing the valve. **See Figure 3-31.**

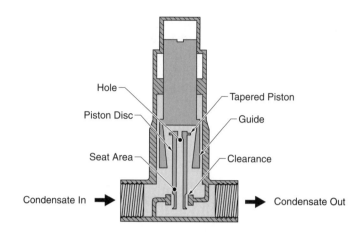

Figure 3-31. An impulse steam trap is a steam trap consisting of a valve with an extended, hollow stem that has a control disc attached at the top.

Steam traps are often neglected in both high- and low-pressure plants. A maintenance schedule for steam traps can be used to check for proper operation, and for cleaning strainers. Improperly functioning steam traps must be replaced. Steam trap manufacturers can provide information regarding steam trap function and selection.

Steam traps should be tested at the first sign of underheated heat exchangers, a temperature increase of condensate returning to the condensate return tank, and/or pressure buildup in the condensate return tank. Dirt, rust, and scale in steam lines can lead to steam trap failure. Steam trap operation can be tested by measuring the difference between the inlet temperature and outlet temperature using a contact thermometer, infrared thermometer, or a temperature-indicating crayon. The temperature drop and readings are compared with temperature readings of a properly operating steam trap.

Steam trap operation can also be tested with an ultrasonic tester that detects the sound of steam in the steam trap. A steam trap that fails in the open position emits the sound of steam blowing through it. A steam trap that fails in the closed position emits no sound of steam and/or water flowing through it. Proper steam trap function is greatly affected by steam strainer (located before the steam trap) function. Steam strainers must be routinely cleaned to prevent foreign matter from reaching the steam trap.

A relatively new method of testing steam traps is to determine the presence of steam by measuring conductivity. The conductivity of steam and the conductivity of condensate are not the same. This method uses a sensor fitted inside a sensing chamber to measure the conductivity. The sensor normally is immersed in condensate. A leaking steam trap will expose the sensor to the steam. This can be combined with a temperature measurement to prevent false alarms when the system is not operating.

Fluid Flow in Piping

The piping system is an essential part of the overall plant operation. Piping must withstand the pressures and temperatures involved, and the velocity of the fluid (gas or liquid) in the line must be carefully controlled. The velocity in saturated steam lines is limited to a flow of 6000 feet per minute (fpm). Superheated steam supplied to steam turbines may have line velocities up to 12,000 fpm. When the velocity in the lines is within the design limits, the pressure drop through the lines will be kept within the limits. It is preferable to have low velocities in steam lines. In order to reduce velocities, larger diameter steam lines are used. The need to keep cost of large diameter steam lines is balanced with the need to maintain velocities within their maximum safe limits.

The velocity of a fluid in a line is found by using the following equation:

$$Velocity\ of\ fluid\ in\ a\ line = \frac{Quantity\ of\ fluid}{Area\ of\ pipe}$$

$$V = \frac{Q}{A}$$

V = velocity of fluid (in fpm)

Q = quantity of fluid [in cubic feet per minute (cfm)]

A = cross-sectional area of pipe (in sq ft)

In steam plants, the flow or quantity of steam passing through a line is measured in lb of steam/hr rather than cfm. The lb of steam/hr must be converted into cfm. To convert lb of steam/hr to lb of steam/min, divide lb of steam/hr by 60.

$$Pounds\ of\ steam\ per\ minute = \frac{Pounds\ of\ steam\ per\ hour}{60}$$

$$W_m = \frac{W}{60}$$

$Quantity\ of\ steam\ per\ minute = Pounds\ of\ steam\ per\ minute \times Specific\ volume$

$Q = W_m \times v$

$$Velocity = \frac{Quantity\ of\ steam\ per\ minute}{Area}$$

$$Velocity = \frac{Pounds\ of\ steam\ per\ minute \times Specific\ volume}{Area}$$

$$V = \frac{W_m \times v}{A}$$

Q = quantity of steam per minute (in cfm)

W = pounds of steam per hour

W_m = pounds of steam per minute

v = specific volume of steam (in cu ft/lb; determined from steam tables relating pressure and temperature)

V = velocity of steam (in fpm)

A = cross-sectional area of pipe (in sq ft)

EXAMPLE

1. A steam line has 10,000 lb of steam/hr flowing through at 5000 fpm. The specific volume of the steam is 2.5 cu ft/lb. What is the diameter of the steam line?

$$V = \frac{W_m \times v}{A}$$

$$W_m = \frac{W}{60}$$

$$W_m = \frac{10,000}{60}$$

$$W_m = 166.66$$

$$A = \frac{W_m \times v}{V}$$

$$A = \frac{166.66 \times 2.5}{5000}$$

$$A = 0.0833 \text{ sq ft}$$

Note: To change sq ft to sq in., multiply the sq ft value by 144.

$$A = 0.0833 \times 144$$

$$A = 12 \text{ sq in.}$$

$$A = 0.7854D^2$$

$$D = \sqrt{\frac{A}{0.7854}}$$

$$D = \sqrt{\frac{12}{0.7854}}$$

$$D = \sqrt{15.28}$$

$$D = \mathbf{3.9''}$$

To find the pipe diameter required for feedwater lines, the velocity of water must be determined. The equation used to find the velocity of water in feedwater lines is the same as the equation used to find the velocity of steam in steam lines.

$$Velocity = \frac{Quantity\ of\ water}{Area\ of\ pipe}$$

$$V = \frac{Q}{A}$$

$$A = \frac{Q}{V}$$

$A = 0.7854D^2$

V = velocity of water (in fpm)

Q = quantity of water (in cu ft/min)

A = cross-sectional area of pipe (in sq ft)

Note: Water weighs 62.4 lb/cu ft at standard conditions. To convert lb of water/hr to lb of water/min, divide by 60. If water is heated, it will expand in volume and lose weight.

EXAMPLE

1. A boiler feedwater pump is feeding 25,000 lb of water to the boiler per hour. The discharge pipe velocity is 400 fpm. What is the required pipe size?

$$A = \frac{Q}{V}$$

$$A = \frac{25,000 \div (60 \times 62.4)}{400}$$

$$A = \frac{25,000}{60 \times 62.4 \times 400}$$

$A = 0.0166933$ sq ft

Note: To change sq ft to sq in., multiply the sq ft value by 144.

$A = 0.0166933 \times 144$

$A = 2.404$ sq in.

$A = 0.7854D^2$

$$D = \sqrt{\frac{A}{0.7854}}$$

$$D = \sqrt{\frac{2.404}{0.7854}}$$

$D = 1.75''$

FEEDWATER FITTINGS AND ACCESSORIES

One lb of water is required to make 1 lb of steam. A boiler producing 60,000 lb of steam/hr requires at least 60,000 lb of water/hr or 60,000/8.33, or 7203 gallons of water per hour (gph). Operating with less can result in boiler plant shutdown and possible damage to the boiler. A stationary engineer must be familiar with all feedwater fittings and accessories in the plant. A stationary engineer must know every possible way of getting water into the boiler by using the required valves, bypass lines, crossover connections, pump tie-ins, and auxiliary lines. In an emergency, there is no time to trace lines or check prints.

Vacuum Condensate Tank

A *vacuum condensate tank* is a condensate return tank equipped with a vacuum pump that creates a vacuum on the return line. This vacuum helps draw condensate from the return line back to the condensate or vacuum tank. Any air or noncondensable gases are discharged into the atmosphere, and the condensate is discharged into a feedwater heater. In low-pressure heating plants, condensate is discharged directly into the boiler. **See Figure 3-32.**

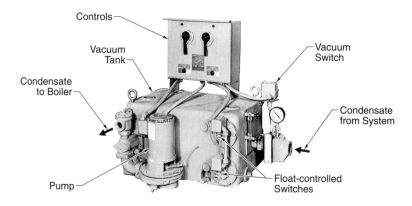

Controls

Vacuum Tank

Vacuum Switch

Condensate to Boiler

Condensate from System

Pump

Float-controlled Switches

ITT Bell & Gossett

Figure 3-32. A vacuum pump produces negative pressure to draw condensate from the system.

Feedwater Valves

A *feedwater valve* is a valve located on the feedwater line to control feedwater flow to a boiler. The feedwater line to a boiler has a check valve and stop valve located close to the boiler. The stop valve is located between the check valve and the boiler. A globe valve used on the feedwater line must be installed so that pressure from the feedwater pump is under the seat of the valve. **See Figure 3-33.**

When two or more boilers receive feedwater from a common source, a globe valve is located on each branch line between the check valve and the main feedwater line. All valves must meet specifications for the pressures and temperatures at which they operate. The stop valve located closest to the boiler is required in case the check valve fails and causes boiler water to flow back through the feedwater line. This stop valve is also used when repairing the check valve. The boiler must be taken off line if there is no stop valve. The globe valve on the branch line can be used to regulate feedwater flow if the feedwater regulator is not functioning. **See Figure 3-34.**

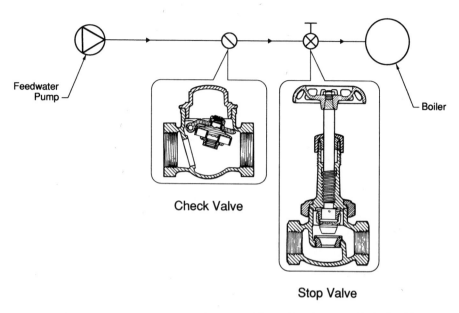

Check Valve

Stop Valve

Figure 3-33. Feedwater valves are used to control flow of feedwater to the boiler. The stop valve is located between the check valve and boiler.

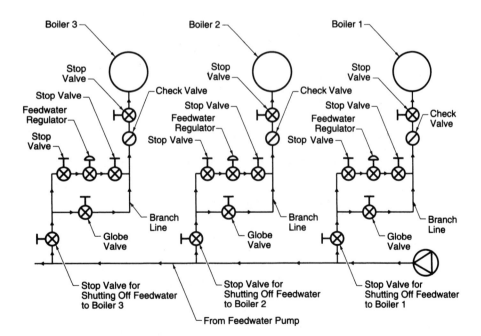

Figure 3-34. Globe valves, by design, can be used to regulate the flow of feedwater manually if the feedwater regulator malfunctions.

Feedwater System Operation

The *feedwater system* is a collection of valves, piping, control system, and auxiliary equipment that provides feedwater at the correct temperature and pressure to the boiler. **See Figure 3-35.** Condensate returns from the steam systems enter the open feedwater heater where oxygen and other noncondensable gases are vented into the atmosphere. All steam does not return to the feedwater heater as condensate because some steam is lost as a result of leaks or process. A float controls a valve to add makeup water.

To prevent the feedwater heater from overfilling, an internal overflow line discharges excess feedwater to waste. The feedwater pump may be electrically driven, steam-driven, or both, allowing the plant more flexibility. Each feedwater pump has its own set of suction and discharge valves, which allow either pump to be taken out of service for repairs. Each feedwater pump has its own discharge valve and check valve to prevent water from backing up if the check valve located close to the boiler fails. Feedwater leaves the feedwater pump through a discharge line and enters the closed feedwater heater where it is heated. The closed feedwater heater has an inlet and outlet valve and a bypass line and valve so the feedwater heater can be taken out of service without shutting down.

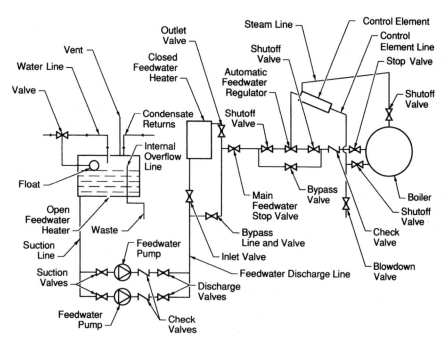

Figure 3-35. The feedwater system uses water from a city water line and condensate returns to provide the required amount of feedwater to the boiler.

The main feedwater line from the closed feedwater heater branches to each boiler in the plant. Each branch line has a main feedwater stop valve. To maintain a proper level in each boiler, an automatic feedwater regulator is used. It may be the thermoexpansion or thermohydraulic type. The control element is located at the NOWL. The top of the control element is connected by a line that goes to the highest part of the steam side of the boiler. A shutoff valve allows the regulator to be taken out of service for repairs. The bottom of the control element is connected to the boiler by a line that is well below the NOWL. The control element is connected to a regulating valve located on the feedwater line. The feedwater regulating valve has two shutoff valves that allow the feedwater regulator to be taken out of service for maintenance. The feedwater would then pass through the bypass valve. This prevents taking a boiler off line in the event of a feedwater regulator failure.

A stop valve and check valve are located on the main feedwater line. The stop valve is located closest to the boiler shell so that the check valve can be repaired without dumping the boiler. The feedwater regulator is equipped with a blowdown valve to ensure the water and steam lines are free of sludge and sediment. Feedwater piping is commonly constructed of steel and must conform to ASME code for temperature and pressure.

Feedwater Regulators

To maintain the efficiency of a steam plant, the amount of feedwater to the boilers must be controlled. A *feedwater regulator* is a control used to maintain a normal operating water level (NOWL) in a boiler by modulating the feedwater flow rate, which reduces the danger of low or high water. A consistent minimum water level can reduce fuel consumption, thermal shock, possibility of carryover, and possibility of low water conditions that could lead to overheated, burned-out tubes, or even a boiler explosion. Common types of feedwater regulators include one-element feedwater regulators, such as thermoexpansion, thermohydraulic, and float feedwater regulators; two-element feedwater regulators, and three-element feedwater regulators.

One-element feedwater regulators monitor the water level to determine the amount of feedwater to add to a boiler. Thermoexpansion and thermohydraulic feedwater regulators are used on larger field-erected and package watertube boilers where feedwater pumps run continuously. Float feedwater regulators are used on smaller package firetube or watertube boilers where feedwater pumps run intermittently.

It takes time for the boiler controls to change the rate of steam generation when there are changes in steam demand. When the steam demand increases, steam is removed faster than it is being generated. This results in a temporary drop in steam pressure and causes swell.

Swell is the process in which the water level in a boiler momentarily rises with an increase in steam demand. When the steam load suddenly increases, more steam leaves the boiler. This causes a sudden decrease in steam pressure. With the lower pressure, more water turns to steam, making the steam bubbles larger and taking up more space in the boiler. This causes the water level in the boiler to momentarily rise from the loss in steam pressure. With a one- or two-element feedwater regulator, a false signal is sent to the regulator to decrease feedwater flow to the boiler. In this instance, the feedwater regulator should be directing more feedwater to the boiler as more steam is leaving the steam and water drum. Before recovering from the false signal, the boiler water level could fall below the NOWL.

When the steam demand decreases, steam is not removed as fast as it is being generated. This results in a temporary increase in pressure and causes shrink. *Shrink* is the process where the water level in a boiler momentarily drops with a decrease in steam demand. This causes a sudden increase in steam pressure. With the higher pressure, less water turns to steam, making the steam bubbles smaller and taking up less space in the boiler. Less steam leaves the boiler and the water level drops, resulting in a false signal to a one-element or two-element feedwater regulator to increase feedwater flow to the boiler. Before recovering from the false signal, the boiler water level could rise above the NOWL, which could cause priming and/or carryover.

Thermoexpansion Feedwater Regulators. A *thermoexpansion feedwater regulator* is a feedwater regulator with a thermostat that expands and contracts when a steam space increases or decreases with changes in water level. A thermoexpansion feedwater regulator is connected to the boiler and main feedwater line.

The thermostat is located at the NOWL and is connected to the steam and water side of the boiler. If the water level drops, the steam space within the thermostat increases. This increases the temperature in the thermostat, causing it to expand.

The increase in thermostat length moves mechanical linkage, which opens the regulator valve. As the regulator valve on the main feedwater line opens, water enters the boiler, raising the water level. The opposite occurs if the water level is too high. The thermostat contracts and the linkage moves the regulator valve toward the closed position. **See Figure 3-36.**

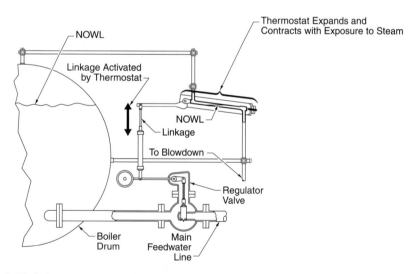

Figure 3-36. A thermoexpansion feedwater regulator has a thermostat that expands and contracts when a steam space increases or decreases with changes in water level.

Thermohydraulic Feedwater Regulators. A *thermohydraulic feedwater regulator* is a feedwater regulator consisting of a regulating valve, bellows, a generator, and stop valves. **See Figure 3-37.** The control element is the generator, which is a tube within a tube. The generator is located at the NOWL and the inner tube is connected to the steam and water side of the boiler. Stop valves are needed to secure the steam and water to the feedwater regulator generator.

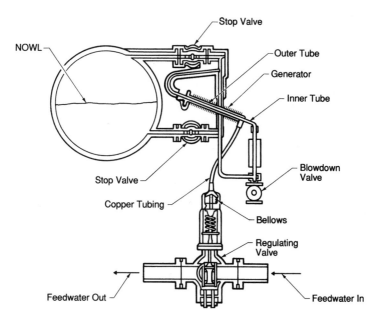

Figure 3-37. Thermohydraulic feedwater regulators use heat differences in the generator to actuate bellows, which control the regulating valve.

A blowdown valve is used to keep the inner tube of the feedwater regulator generator free of sludge and sediment. The outer tube of the feedwater regulator generator is connected to the bellows in the regulating valve with copper tubing and is filled with distilled water, making a closed system. When the boiler water level drops, the inner tube of the generator has a large steam space.

The heat from the steam in that steam space is released to the distilled water in the outer tube. The distilled water expands. Since the bellows in the regulating valve is the only moveable part in the closed system, the bellows expands, which moves the regulating valve toward the open position. When the water level in the boiler rises, the reverse action takes place. The connection to the steam and water side of the boiler should be blown down once a month to remove sludge or sediment. After the regulator is blown down, approximately one hour must be allowed for the regulator to stabilize before returning it to service.

Float Feedwater Regulators. A *float feedwater regulator* is a feedwater regulator consisting of a float chamber, float, and mercury switch or microswitch. **See Figure 3-38.** A float feedwater regulator can be incorporated into the water column or stand alone. Float feedwater regulators are the simplest type of feedwater regulator and are commonly used on small package boilers. In the float feedwater regulator, the float chamber is connected to the steam and water side of the boiler and is located at the NOWL. A blowdown line and valve are installed to remove sludge and sediment from the float chamber.

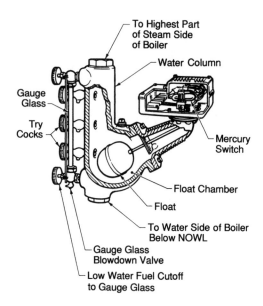

Figure 3-38. Float feedwater regulators use a float that actuates a mercury switch or microswitch to control the feedwater pump.

When the boiler water level drops, the float in the float chamber drops and mechanically moves the mercury switch or microswitch. The switch is connected to a feedwater pump starter relay that turns on the feedwater pump motor. The feedwater pump continues to run until the float rises and moves the switch in the opposite direction.

A float feedwater regulator should be blown down once every shift to prevent sludge or sediment from accumulating. Float feedwater regulators should be opened once a year for cleaning and inspection.

Two-Element Feedwater Regulators. A *two-element feedwater regulator* is a feedwater regulator that controls the water level in a boiler with a superheater by monitoring the boiler water level (the first element) and the steam flow through the superheater (the second element). A steam-flow sensing element consists of a diaphragm assembly on the top part of the regulator valve.

By connecting the top and bottom of the diaphragm across the superheater, the diaphragm moves when pressure difference changes to indicate steam flow. The movement of the diaphragm is amplified through mechanical linkage, which moves the regulator valve. For optimum efficiency, the lines to the thermostat and diaphragm should be blown down once a month. When blowing down, the regulator must be removed from service. **See Figure 3-39.**

Three-Element Feedwater Regulators. A *three-element feedwater regulator* is a feedwater regulator that controls the amount of feedwater fed to the boiler by monitoring the water level (first element), steam flow (second element) and, feedwater flow (third element). Three-element feedwater regulators are used on large watertube boilers where sudden and wide variation in steam demand requires more precise feedwater control. Watertube boilers have a small percentage of total water capacity in the steam and water drum. This requires a feedwater regulator that is capable of quick reaction during sudden changes in steam demand to prevent low water conditions. Sudden steam load fluctuation is common in food processing plants, paper mills, and textile plants. This fluctuation can cause feedwater control problems.

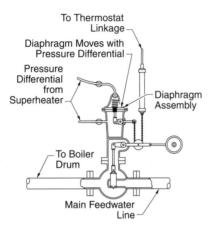

Figure 3-39. A two-element feedwater regulator controls boiler water level by monitoring the steam flow through the superheater and by monitoring the boiler water level.

Three-element feedwater regulators compensate for rapid changes in the steam load by using separate elements to sense steam flow, feedwater flow, and water level. The feedwater regulator controller reacts to signals received from the three sensing elements. On older boilers, the sensing elements and feedwater controller are typically pneumatically controlled. **See Figure 3-40.** On newer boilers, the sensing elements and feedwater controller are typically digitally controlled.

Steam flow is balanced against the feedwater flow. A change in steam flow will cause a pilot valve mechanism to respond to the steam flow change and actuate the feedwater regulator valve to restore the required balance. The response of the feedwater controller is not instantaneous when reacting to steam and water flow changes. A relay receives signals from the water level sensing element and compensates for this lag by adding a loading pressure to prevent the boiler water level from radically fluctuating up or down.

Feedwater Heaters

A *feedwater heater* is an auxiliary component on a boiler that is used to increase plant efficiency by preheating the feedwater with exhaust steam or heat from the gases of combustion leaving the boiler. The two basic types of feedwater heaters are the open feedwater heater and closed feedwater heater. Deaerating feedwater heaters are a variation of open feedwater heaters.

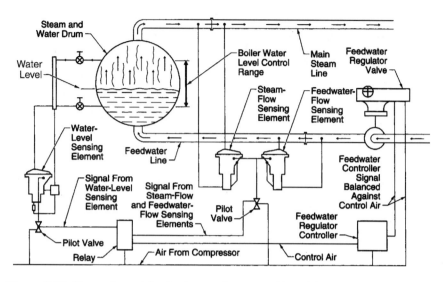

Figure 3-40. Three-element feedwater regulators provide quick reaction during sudden changes in steam loads.

Open Feedwater Heaters. An *open feedwater heater* is a feedwater heater that is open to the atmosphere, and in which steam is added to the water to raise the water temperature. In addition, when open to the atmosphere, feedwater heaters vent out oxygen and other noncondensable gases. Oxygen in the boiler leads to pitting and corrosion of the boiler drum and tubes. If exhaust steam from steam turbines, steam engines, or reciprocating pumps is used for this purpose, large fuel savings and higher plant efficiency can be obtained. **See Figure 3-41.** The percentage of fuel savings from feedwater heating is found by using the following equation:

Fuel savings =

$$\frac{Temperature\ of\ water\ leaving - Temperature\ of\ water\ entering}{Btu\ in\ exhaust\ steam - \left(Temperature\ of\ water\ entering - 32\right)} \times 100$$

$$Fuel\ savings = \frac{T_l - T_e}{H_e - \left(T_e - 32\right)} \times 100$$

Fuel savings = fuel savings (in %)

T_l = temperature of water leaving (in °F)

T_e = temperature of water entering (in °F)

H_e = Btu in exhaust steam

32 = base temperature to determine enthalpy (heat in substance)

Note: The equation is multiplied by 100 to obtain a percent value.

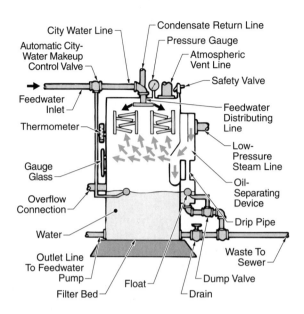

City Water Line

Automatic City-
Water Makeup
Control Valve

Condensate Return Line

Pressure Gauge

Atmospheric
Vent Line

Safety Valve

Feedwater
Inlet

Thermometer

Feedwater
Distributing
Line

Low-
Pressure
Steam Line

Gauge
Glass

Overflow
Connection

Oil-
Separating
Device

Drip Pipe

Water

Outlet Line
To Feedwater
Pump

Filter Bed

Float

Drain

Dump Valve

Waste To
Sewer

Cleaver-Brooks

Figure 3-41. An open feedwater heater is open to the atmosphere and steam is added to the water to raise the water temperature.

EXAMPLE

1. The heat in the exhaust steam is 1125 Btu/lb. The temperature of the water entering the heater is 110°F and the temperature leaving the heater is 195°F. What is the fuel savings?

$$Fuel\ savings = \frac{T_l - T_e}{H_e - (T_e - 32)} \times 100$$

$$Fuel\ savings = \frac{195 - 110}{1125 - (110 - 32)} \times 100$$

$$Fuel\ savings = \frac{85}{1047} \times 100$$

$$Fuel\ savings = \mathbf{8.12\%}$$

All condensate returns to the open feedwater heater so it acts as a reservoir for the feedwater pump to get its suction. The open feedwater heater is equipped with a float-controlled valve, and makeup water can be added to the system automatically. Some open feedwater heaters have an oil baffle and a coke filter bed to trap out oil that may have been picked up by the return condensate or exhaust steam. The open feedwater heater also picks up some of the scale-forming salts contained in the makeup water. These scale-forming salts start to settle out at about 150°F.

Although the open feedwater heater is vented to the atmosphere, it still carries 2 lb to 7 lb of steam pressure. Steam pressure in the open feedwater heater is controlled by a balanced valve vented to the atmosphere. If steam pressure gets too high, the water temperature could rise and cause the feedwater pump to become steambound. The open feedwater heater has a pressure gauge, safety valve, gauge glass, and thermometer, which are all necessary for proper operation and safety.

Condensate enters at the top and cascades down over trays. The steam coming in from the side breaks up condensate returns to drive off oxygen and other entrained gases. In addition, steam coming into the open feedwater must pass through an oil baffle where any entrained oil separates and discharges to waste. In the open feedwater heater, steam and water come in intimate contact with each other.

The open feedwater heater is located on the suction side above the feedwater pump. This is necessary so the feedwater pump can be supplied with a head pressure to permit higher feedwater temperatures. For each vertical foot of water, there is a pressure of 0.433 psi. Higher pressures result in higher boiling points. The open feedwater heater should be opened once a year, the trays removed, and it should be thoroughly cleaned. After the heater has been cleaned with a wire brush and flushed, all internal parts should be painted with a corrosion inhibitor.

Deaerators. A *deaerator* is a feedwater heater that operates under pressure and is used to separate oxygen and other gases from steam before releasing the gases to the atmosphere through a vent. **See Figure 3-42.** Because a deaerator operates under pressure, the saturated temperature corresponds to the operating pressure. For example, a deareator operating at 5 psi has a saturated temperature of 227°F. Deaerators typically use steam to heat the water.

There are two reasons to use a deaerator. First, a deaerator holds and preheats boiler feedwater before it is pumped into the boiler. This increases the sensible heat within the boiler water. Second, a deaerator removes dissolved gases from the feedwater. Dissolved oxygen and carbon dioxide are corrosive and attack the boiler metal, steam piping, and condensate return lines. Deaerators can reduce dissolved oxygen levels to as low as 5 ppm, greatly reducing the amount of chemical treatment required.

Closed Feedwater Heaters. A *closed feedwater heater* is a feedwater heater that is closed and pressurized to heat feedwater to a much higher temperature than an open feedwater heater. Closed feedwater heaters are located after the feedwater pump. The final outlet temperature is determined by the steam pressure in the feedwater heater and rate of water flow through it. In closed feedwater heaters, steam and water do not come in direct contact with each other. Water passes through tubes, and steam is in the shell (around the tubes). Closed feedwater heaters are on the discharge side of the pump; therefore, the feedwater temperature can be raised because there is no danger of the feed pump becoming steambound, which is possible with open feedwater heaters. When a closed feedwater heater is used, it reduces thermal shock on the boiler parts where the feedwater is introduced. It can also increase overall plant efficiency if the steam is bled or extracted from a steam turbine and used in the heater. **See Figure 3-43.**

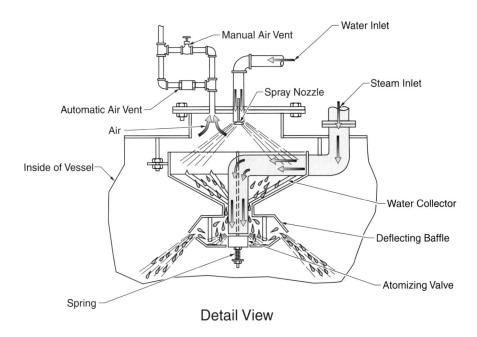

Detail View

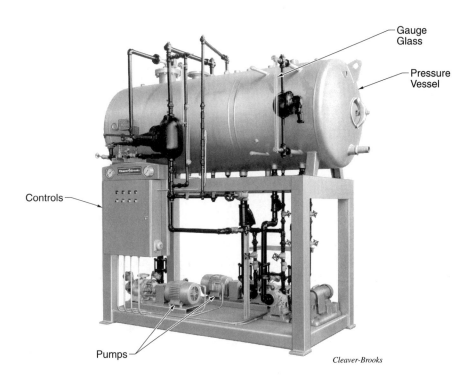

Cleaver-Brooks

Figure 3-42. Deaerators separate air and other gases from feedwater.

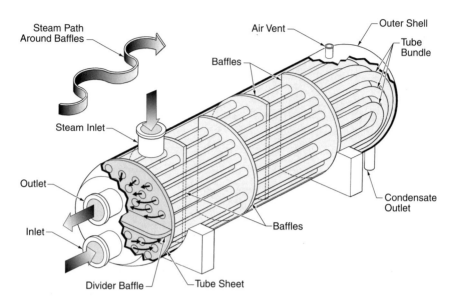

Figure 3-43. A closed feedwater heater is closed and pressurized to heat feedwater to a much higher temperature than an open feedwater heater.

Closed feedwater heaters can be vertical or horizontal, single-pass or multipass, and consist of a steel shell containing many small tubes that are secured into two tube sheets. Expansion and contraction of the shell and tubes must be carefully considered when designing closed feedwater heaters. Steam enters the upper part of the shell, and condensate is removed at the lowest part of the shell. A closed feedwater heater must be equipped with a safety relief valve, stop valves, and a bypass arrangement to facilitate taking the heater out of service for cleaning or repair.

Feedwater Economizers

An *economizer* is a boiler accessory used to reclaim heat from the gases of combustion to preheat the feedwater. When the temperature of the feedwater is increased, the overall efficiency of the plant can be increased by 7% to 20%. Two basic types of economizers are unit economizers and integral economizers. A *unit economizer* is a feedwater economizer used with the gases of combustion and feedwater for one boiler. An *integral economizer* is a feedwater economizer that is part of the boiler and is used to replace some of the boiler heating surface. **See Figure 3-44.**

Economizers require large breeching and draft fans and are commonly used in large plants with a constant steam load. A fluctuating steam load can cause rapid changes in feedwater temperatures. Economizers, because of their location, restrict the flow of the gases of combustion. Therefore, induced draft fans are needed to prevent a back pressure of gases on the boiler. The high feedwater temperatures produced reduce the thermal shock to boiler metal and decrease wear and tear on the boiler.

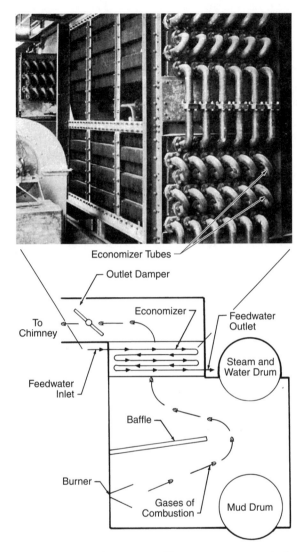

Figure 3-44. A feedwater economizer is an auxiliary component on a boiler used to reclaim heat from the gases of combustion before they enter the chimney.

Economizer heating surfaces cost less to construct than boiler heating surfaces. The trend in high-pressure boiler construction is to reduce boiler heating surface to a minimum and install large economizers. Economizers are a necessity on boilers that operate at or above 400 psi or in plants using electric auxiliaries where exhaust steam is not available for feedwater heating.

Any heat not removed from the gases of combustion by the economizer or air preheater (used to preheat air for combustion) is lost to the atmosphere. The amount of heat lost in the gases of combustion per pound of fuel burned is determined by using the following equation:

Quantity of heat lost = Weight of gases × Specific heat ×
(Temperature₁ – Temperature₂)

$$Q = W \times Sph \times (T_1 - T_2)$$

Q = quantity of heat lost to the atmosphere per lb of fuel burned (in Btu)

W = weight of gases of combustion going up chimney

Note: For complete combustion, approximately 15 lb of air is required per lb of fuel burned: 15 lb + 1 lb = 16 lb. Therefore, 16 lb of gases of combustion are entering the chimney per lb of fuel burned.

Sph = specific heat of gases of combustion (obtained from specific heat tables)

T_1 = temperature of gases entering chimney

T_2 = temperature of air entering furnace

EXAMPLE

1. A boiler chimney temperature is 325°F and the temperature of the air entering the furnace is 85°F. How much heat is lost to the atmosphere?

$Q = W \times Sph \times (T_1 - T_2)$

$Q = 16 \times 0.24 \times (325 - 85)$

$Q = 16 \times 0.24 \times 240$

$Q = 921.6$ Btu/lb of fuel burned

Feedwater Pumps

A *feedwater pump* is a pump used to deliver the correct amount of water to the boiler at the proper temperature and pressure. Feedwater pumps can be reciprocating, electric centrifugal, or steam centrifugal. The ASME code states that boilers fired with solid fuels (not in suspension) and boilers having large furnace volumes must have two ways of supplying feedwater to the boiler. One of these ways must be steam operated. This requirement is necessary because of the heat that could cause damage to the boiler if feedwater supply were interrupted.

Reciprocating Feedwater Pumps. A *reciprocating feedwater pump* is a positive-displacement pump where a piston stroke moves a specific amount of water. The amount of water discharged depends on the size of the feedwater pump. The discharge line must be open whenever the feedwater pump is operating to prevent pressure increase over design pressure. A safety relief valve is installed between the discharge outlet of the pump and the discharge stop valve to limit the pressure on the discharge line. **See Figure 3-45.**

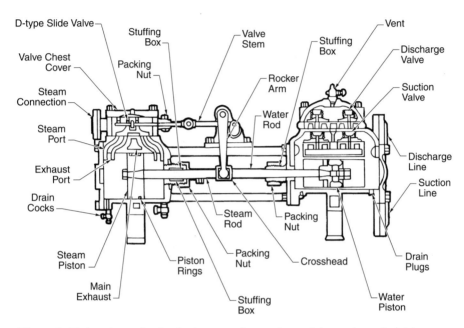

Figure 3-45. A reciprocating feedwater pump has a steam piston and a water piston.

A reciprocating feedwater pump has at least one steam piston and water piston. The steam piston must be 2 to 2½ times larger in area than the water piston. This design allows a sufficient supply of water at the proper pressure to be delivered to a boiler while using steam from the boiler to operate the feedwater pump. The location of the reciprocating feedwater pump is below the open feedwater heater or any other source of feedwater supply. This gives the pump a positive suction pressure. A reciprocating feedwater pump will fail to supply sufficient feedwater if the following occurs:

- Closed suction or discharge stop valves. This condition could result in damage to the feedwater pump unless a safety relief valve is installed.

- No water in open feedwater heater. This condition indicates that makeup water is not being supplied to the feedwater heater.

- Suction or discharge pump valves open or leaking. This condition requires reconditioning of pump valves.

- Water plunger rings worn, allowing leakage. This condition requires the replacement of worn plunger rings.

- Steam piston rings or cylinders worn or broken. This condition requires the replacement of worn or broken rings or cylinders.

- Steam valves worn or improperly set. This condition requires the replacement or resetting of steam valves.

- Water cylinder vapor bound. This condition requires reducing the temperature of feedwater in the feedwater heater to prevent flashing in the feedwater pump cylinder.

All reciprocating feedwater pumps have a data plate. This data plate includes the pump size and operating information a stationary engineer needs to determine how much steam pressure is required to obtain a certain discharge pressure. Discharge pressure may also be determined by the amount of steam pressure available to drive the pump. Gallons per minute discharged from a reciprocating feedwater pump can be determined using data from the data plate. The data plate mounted on all reciprocating feedwater pumps includes three numbers. For example, a data plate with $4 \times 2 \times 6$ indicates the following information (in inches):

4 = diameter of steam piston

2 = diameter of water piston

6 = length of stroke

This information is used to determine the total force occurring in the pump. Total force acting on the steam piston is steam pressure times the area of the steam piston. Total force acting on a water piston is water pressure times the area of the water piston.

Total force steam side = Total force water side

$TF_s = TF_w$

Area steam side $\times$ Steam pressure = Area water side $\times$ Water pressure

$A_s \times P_s = A_w \times P_w$

Note: The area of the piston is found by multiplying $0.7854 \times D^2$.

$0.7854 \times D_s^2 \times P_s = 0.7854 \times D_w^2 \times P_w$

A_s = area of steam side of reciprocating feedwater pump

P_s = pressure on steam side of reciprocating feedwater pump

A_w = area of water side of reciprocating feedwater pump

P_w = pressure on water side of reciprocating feedwater pump

D_s^2 = diameter of piston on steam side of reciprocating feedwater pump

D_w^2 = diameter of piston on water side of reciprocating feedwater pump

The total force steam side is equal to the total force water side of a reciprocating feedwater pump. For example, a $4 \times 2 \times 6$ reciprocating feedwater pump is driven with steam at 100 psi and develops 400 psi on the discharge side.

$TF_s = TF_w$

$A_s \times P_s = A_w \times P_w$

$0.7854 \times D_s^2 \times P_s = 0.7854 \times D_w^2 \times P_w$

$(0.7854 \times 4^2) \times 100 \text{ psi} = (0.7854 \times 2^2) \times 400 \text{ psi}$

$(0.7854 \times 16) \times 100 = (0.7854 \times 4) \times 400$

$12.5664 \times 100 = 3.1416 \times 400$

1256.6400 lb = 1256.6400 lb

If the water pressure developed is known, the steam pressure required can be found.

$TF_s = TF_w$

$A_s \times P_s = A_w \times P_w$

$0.7854 \times D_s^2 \times P_s = 0.7854 \times D_w^2 \times P_w$

$$\frac{0.7854 \times D_s^2 \times P_s}{0.7854 \times D_s^2} = \frac{0.7854 \times D_w^2 \times P_w}{0.7854 \times D_s^2}$$

$$P_s = \frac{D_w^2 \times P_w}{D_s^2}$$

EXAMPLES

1. A pump 6 × 3 × 6 develops 600 psi on the water side. How much steam pressure is required to operate this pump?

$TF_s = TF_w$

$A_s \times P_s = A_w \times P_w$

$$\frac{0.7854 \times D_s^2 \times P_s}{0.7854 \times D_s^2} = \frac{0.7854 \times D_w^2 \times P_w}{0.7854 \times D_s^2}$$

$$P_s = \frac{D_w^2 \times P_w}{D_s^2}$$

$$P_s = \frac{3 \times 3 \times 600}{6 \times 6}$$

$$P_s = \frac{5400}{36}$$

$P_s =$ **150 psi**

2. A reciprocating feedwater pump $4 \times 3 \times 6$ develops 250 psi on the water side. How much steam pressure is required to operate this pump?

$$TF_s = TF_w$$

$$A_s \times P_s = A_w \times P_w$$

$$\frac{0.7854 \times D_s^2 \times P_s}{0.7854 \times D_s^2} = \frac{0.7854 \times D_w^2 \times P_w}{0.7854 \times D_s^2}$$

$$P_s = \frac{D_w^2 \times P_w}{D_s^2}$$

$$P_s = \frac{3 \times 3 \times 250}{4 \times 4}$$

$$P_s = \frac{2250}{16}$$

$$P_s = \textbf{140.6 psi}$$

If the steam pressure available is known, the discharge pressure can be found.

$$TF_s = TF_w$$

$$A_s \times P_s = A_w \times P_w$$

$$0.7854 \times D_s^2 \times P_s = 0.7854 \times D_w^2 \times P_w$$

$$P_W = \frac{D_s^2 \times P_s}{D_W^2}$$

EXAMPLES

1. A pump $6 \times 3 \times 6$ is supplied with 150 psi steam. What pressure will be developed on the water side?

$$TF_s = TF_w$$

$$A_s \times P_s = A_w \times P_w$$

$$\frac{0.7854 \times D_s^2 \times P_s}{0.7854 \times D_w^2} = \frac{0.7854 \times D_w^2 \times P_w}{0.7854 \times D_w^2}$$

$$P_w = \frac{D_s^2 \times P_s}{D_w^2}$$

$$P_w = \frac{6 \times 6 \times 150}{3 \times 3}$$

$$P_w = \frac{5400}{9}$$

$$P_w = \textbf{600 psi}$$

2. A reciprocating feedwater pump $4 \times 3 \times 6$ uses a steam pressure of 140.6 psi. How much pressure will be developed on the water side?

$$TF_s = TF_w$$

$$A_s \times P_s = A_w \times P_w$$

$$\frac{0.7854 \times D_s^2 \times P_s}{0.7854 \times D_w^2} = \frac{0.7854 \times D_w^2 \times P_w}{0.7854 \times D_w^2}$$

$$P_w = \frac{D_s^2 \times P_s}{D_w^2}$$

$$P_w = \frac{4 \times 4 \times 140.6}{3 \times 3}$$

$$P_w = \frac{2249.6}{9}$$

$$P_w = \textbf{250 psi}$$

The capacity (in gpm) of any reciprocating pump is found by using the following equation:

$$gpm = \frac{L \times A \times N \times E}{231}$$

gpm = gallons per minute
L = length of stroke of piston found on data plate
A = area of water piston (in sq in.)
N = number of strokes per minute
E = pump efficiency (sometimes expressed as % of slip)

The length of the stroke times the area of piston is the number of cubic inches of displacement in the cylinder.

$$cu. in. = L \times A$$

The number of times the cylinder is cleared per minute times the cubic inches is the total cubic inches displaced per minute.

$$cu. in./min = L \times A \times N$$

No pump is 100% efficient; therefore, it discharges only a percentage (E) of its theoretical capacity.

$$actual\ cu.\ in./min = L \times A \times N \times E$$

There are 231 cubic inches in a gallon. Using this information, cubic inches per minute can be converted into gallons per minute.

$$gpm = \frac{L \times A \times N \times E}{231}$$

EXAMPLES

1. A double-acting, simplex reciprocating pump is seven-eighths full on each stroke and has a water piston 4″ in diameter with an 8″ stroke. What is the gpm discharged when the pump runs at 120 strokes/min?

$$gpm = \frac{L \times A \times N \times E}{231}$$

$A = 0.7854 \times 4 \times 4$

$A = 12.5664$ sq in.

$$gpm = \frac{8 \times 12.5664 \times 120 \times 0.875}{231}$$

$gpm = \mathbf{46}$

2. A simplex pump runs at 50 strokes/min. The data plate shows 8 × 10 × 12, and the pump has an efficiency of 80%. What is the gpm discharged?

$$gpm = \frac{L \times A \times N \times E}{231}$$

$A = 0.7854 \times 10 \times 10$

$A = 78.54$ sq in.

$$gpm = \frac{12 \times 78.54 \times 50 \times 0.80}{231}$$

$gpm = \mathbf{163}$

Centrifugal Feedwater Pumps. A *centrifugal feedwater pump* is a feedwater pump that uses centrifugal force to develop a rise in pressure for moving a liquid. *Centrifugal force* is the force that tends to move an object away from a center. A centrifugal feedwater pump imparts kinetic energy to the water, which overcomes potential energy in the boiler to cause feedwater to enter the boiler. *Kinetic energy* is energy due to a body in motion. An example of kinetic energy is a boulder rolling down a hill. *Potential energy* is energy that is due to position, not to motion. An example of potential energy is a boulder resting at the top of the hill.

Centrifugal feedwater pumps have fewer moving parts than reciprocating feedwater pumps. As the water passes into the rotating impeller, kinetic energy overcomes potential energy and gives it centrifugal force. An *impeller* is the rotating element in a centrifugal pump through which water passes and is the means by which energy is imparted to the water. Once water is thrown from the impeller, the casing guides the water to its destination. Most of the larger feedwater pumps are of the split casing centrifugal feedwater pump design. **See Figure 3-46.**

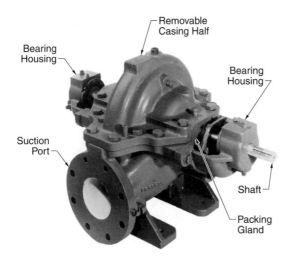

Bearing
Housing

Removable
Casing Half

Bearing
Housing

Suction
Port

Shaft

Packing
Gland

Dresser Industries, Inc.

Figure 3-46. The top half of a horizontal split-case pump is removable.

With a split casing design, the upper casing is removable for service to the pump. The lower casing is stationary (not removable) and directs water flow into and from the pump. The impeller shaft is connected at one end to a motor or steam turbine and is used to rotate the impeller. Shaft bearings support and allow the impeller shaft to rotate. A coupling attaches the pump shaft to the motor or steam turbine shaft. The packing gland or mechanical seals prevent leakage between the pump casing and shaft.

A centrifugal feedwater pump must meet certain pressure and volume conditions. A centrifugal feedwater pump should have the capacity to deliver two times the maximum steam capacity of the boiler it is supplying at the MAWP of the boiler. The advantages of using centrifugal feedwater pumps are that they

• can be used effectively with in-line feedwater regulators

• have variable drives—electric motor or steam turbine

• are reliable

• have fewer moving parts than a reciprocating feedwater pump

• are adaptable for intermittent or continuous operation

• have fairly constant efficiency over long running periods

• are capable of handling high pressure and large capacities

The disadvantages of using centrifugal feedwater pumps are that they

• require skilled workers for servicing and aligning

• must be monitored regularly because of their high speed

• lose efficiency on high or low loads

The water HP of a centrifugal feedwater pump is found by using the following equation for mechanical HP:

$$HP = \frac{ft\text{-}lb\,/\,min}{33,000\;ft\text{-}lb\,/\,min}$$

Note:

Total head of centrifugal feedwater pump =
  Discharge head + Friction head + Suction lift
or
Total head of centrifugal feedwater pump =
  Discharge head + Friction head – Suction head

Centrifugal feedwater pumps have water supplied by a head pressure or must lift water. To convert psi into foot head, divide the psi value by 0.433. (There are 0.433 psi per vertical foot.)

EXAMPLE

1. A centrifugal feedwater pump delivers 1000 gpm against a head of 120′. What is the water HP (disregard frictional loss)? *Note:* There are 8.33 lb of water in 1 gal.

$$HP = \frac{ft\text{-}lb\,/\,min}{33,000\;ft\text{-}lb\,/\,min}$$

$$HP = \frac{1000 \times 8.33 \times 120}{33,000}$$

$$HP = \mathbf{30.29}$$

This is the water HP. To find the HP needed to drive the pump, the efficiency of the unit must be considered.

$$HP_{drive} = \frac{HP_{water}}{Efficiency}$$

$$HP_d = \frac{HP_w}{E}$$

HP_d = drive horsepower
HP_w = water horsepower
E = efficiency

With an efficiency of 75%,

$$HP_d = \frac{HP_w}{E}$$

$$HP_d = \frac{30.29}{0.75}$$

$$HP_d = \mathbf{40.39}$$

EXAMPLE

1. What is the HP required to feed 600 gal. of water/hr to a boiler operating at 112 psi (disregard friction loss)? *Note:* Change gph to gpm.

$$HP = \frac{ft\text{-}lb}{33,000}$$

$$HP = \frac{600 \times 8.33 \times 112}{60 \times 0.433 \times 33,000}$$

$$HP = \mathbf{0.6529}$$

With an efficiency of 75%,

$$HP_d = \frac{HP_w}{E}$$

$$HP_d = \frac{0.65}{0.75}$$

$$HP_d = \mathbf{0.87}$$

Suction lift, or, *vertical lift,* is the vertical distance from the level of the water below a pump to the centerline of the pump inlet. *Suction head* is the vertical distance from the centerline of a pump inlet to the surface of the water above the pump. **See Figure 3-47.** The theoretical lift is approximately 34′ and changes with the atmospheric pressure. When the atmospheric pressure is 14.7 psi (equivalent to about 30″ on a barometer), the theoretical lift is

$$\frac{14.7 \text{ psi}}{0.433 \text{ psi/ft}} = 33.9'$$

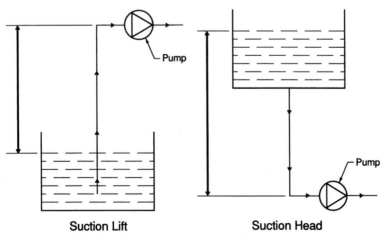

Suction Lift   **Suction Head**

Figure 3-47. Suction lift and suction head must be considered when determining the force required to pump water.

A vertical lift of 33.9′ is not possible in practice because leaks in the suction line and packing glands reduce the actual lift to approximately 24′. Theoretical lift may go as low as 10′ to 15′ because of friction loss in the suction line.

The *discharge head* is the vertical distance between the centerline of the pump outlet and the level of the water above the pump. *Friction head* is equal to a loss caused by the friction of the water on piping, valves, and fittings. The total head against which a pump operates is found by using the following equations:

Total head = Discharge head + Friction head + Suction lift

$H_t = H_d + H_f + SL$

Total head = Discharge head + Friction head – Suction head

$H_t = H_d + H_f - H_s$

H_t = total head (in ft)

H_d = discharge head (in ft)

H_f = friction head (in ft)

SL = suction lift (in ft)

H_s = suction head (in ft)

EXAMPLES

1. The suction head of a pump is 35′. It discharges into a pressure vessel 40′ above the centerline of the pump against a friction head of 17′ with a pressure in the vessel of 30 psi. What is the total head on the pump? **See Figure 3-48.**

$H_t = H_d + H_f - H_s$

$H_t = 40 + \dfrac{30}{0.433} + 17 - 35$

$H_t = 40 + 69 + 17 - 35$

$H_t = \mathbf{91'}$

2. The suction lift of a pump is 15′. It discharges into a tank 40′ above the centerline of the pump against a friction head of 17′. What is the total head of the pump? **See Figure 3-49.**

$H_t = H_d + H_f + SL$

$H_t = 40 + 17 + 15$

$H_t = \mathbf{72'}$

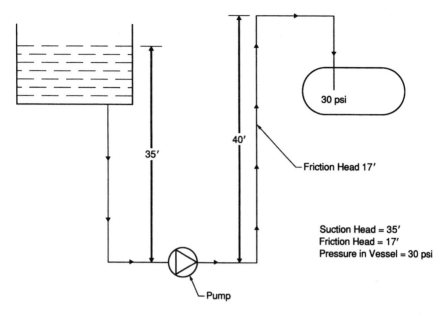

Figure 3-48. Total head is equal to discharge head plus friction head minus suction head $(H_t = H_d + H_f - H_s)$.

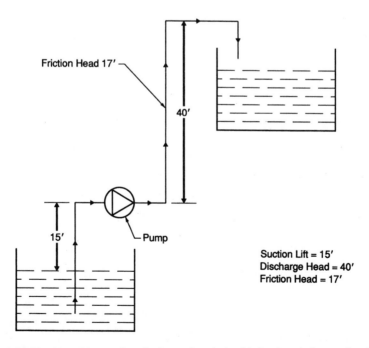

Figure 3-49. Total head is equal to discharge head plus friction head plus suction lift ($H_t = H_d + H_f + SL$).

Feedwater Injectors

A *feedwater injector* is an auxiliary or secondary means of feeding water to a boiler. Package watertube boilers usually come equipped with a feedwater injector and an electric feedwater pump. The feedwater injector is not an efficient means of feeding a boiler with water. Feedwater injectors are designed to operate at maximum feedwater temperatures of 130°F to 150°F. Temperatures above 150°F cause feedwater injectors to become steambound. They can lift water a maximum of 20′ under the best conditions. They also tend to kick out under a fluctuating steam load.

A feedwater injector should be placed close to the boiler it is feeding. The steam line to the feedwater injector should come from the highest part of the boiler as possible to ensure dry steam. The feedwater injector works on the principle that kinetic energy is greater than potential energy. Steam passing through a nozzle drops in pressure, expands, and increases in velocity. Steam picks up the water, and the velocity or kinetic energy forces open the check valve, discharging the water into the boiler.

Low-pressure feedwater injectors are used for steam pressures up to 40 psi, and high-pressure feedwater injectors are used for steam pressures over 40 psi. There are single-tube and double-tube feedwater injectors. **See Figure 3-50.** Feedwater injectors are commonly used on firetube boilers, but they can be used on smaller watertube boilers.

A stationary engineer must know where the feedwater injector is located in the plant and use it periodically to ensure its operation in the event of an emergency. If the feedwater injector is being fed by a city water line, a check valve must be placed between the feedwater injector and the city water meter. This prevents the city water meter from being damaged and the city water from being contaminated in case the feedwater injector kicks out and the steam tries to back up into the city water line.

Feedwater Lines

Feedwater lines must be designed to withstand pressures that are equal to and often higher than the boiler pressures. Feedwater lines must also be able to withstand high feedwater temperatures. As with steam lines, the velocity of feedwater in the feedwater lines must be carefully controlled. Feedwater velocity kept within the design limits will result in the proper pressure drop in the line. Standard design limits for feedwater lines are as follows:

• Water lines – 500 fpm

• Pump discharge – 500 fpm

• Pump suction – 250 fpm

In order to reduce velocity in feedwater lines, larger pipes are required, which result in higher costs. By reducing pipe sizes, the velocity of the feedwater will be closer to maximum safe limits.

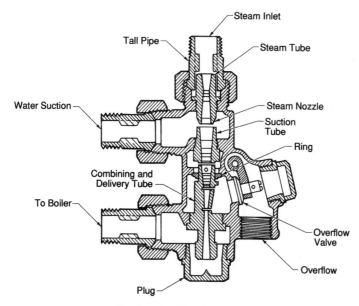

Single-tube Feedwater Injector

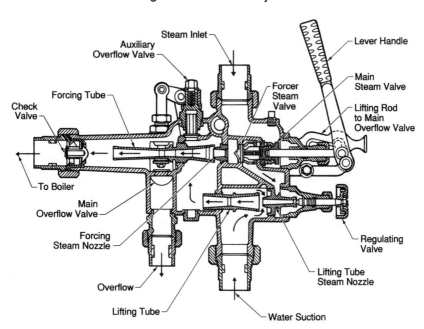

Double-tube Feedwater Injector

Figure 3-50. Feedwater injectors are an auxiliary or secondary means of supplying feedwater to the boiler.

Condensate Return System

A *condensate return system* is a collection of valves, a piping control system, and auxiliary equipment that reclaims uncontaminated condensate. After steam has left the boiler and released heat, it condenses (turns back to water). This condensate is pure and should be returned to the boiler room to be reused. If more condensate is reclaimed, less untreated makeup water has to be used, reducing the amount of chemicals needed to control scale. It is not always possible to return the condensate directly to the feedwater heater, which makes a condensate return tank necessary. **See Figure 3-51.**

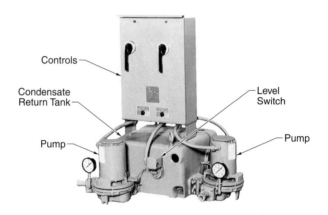

Figure 3-51. A condensate return system is a collection of valves, piping, control system, and auxiliary equipment that reclaims uncontaminated condensate.

The condensate return tank is usually located in the mechanical spaces where the steam is being used. It is fitted with a gauge glass. The condensate tank has a suction line going to a condensate pump. This pump discharges water through a line to the feedwater heater. In a large plant, there may be one or more return tanks feeding one main return tank, called a surge tank, located in the boiler room.

Surge Tanks. In larger steam boiler plants, condensate returns to the boiler room from a number of sources. In these plants, a surge tank provides the extra capacity required to handle changing loads and peak flows of condensate. Normally, the surge tank is fitted with raw makeup water fittings and controls along with transfer pumps and controls. The surge tank is vented to the atmosphere and is equipped with a manhole for cleaning and repairs. **See Figure 3-52.**

A feedwater regulator, also referred to as a pump controller, is used to operate the pump. The raw makeup water system generally has a float chamber located at the minimum water level. The float operates a makeup water valve to maintain the minimum water level in the tank. If the water level drops below the minimum water level, a low-water alarm is activated. If the water rises to a maximum acceptable water level, a high-water alarm is activated.

The surge tank discharges to transfer pumps located below the tank. Transfer pumps are used to transfer the mixture of condensate and raw makeup water to the deaerator or directly to the boiler feedwater pumps.

Gauge Glass

Level Controls

Low-Water Alarm

Figure 3-52. A surge tank provides the extra capacity required to handle changing loads and peak flows of condensate in larger steam boiler plants.

CHAPTER 3 LEARNER RESOURCES
ATPeResources.com/QuickLinks
Access Code: 735728

Fuels and Combustion

Fuels commonly used for combustion in boilers include coal, fuel oil, and natural gas. Combustion of a fuel is required to generate the heat necessary to produce steam in a boiler. Combustion accessories and fuel-burning equipment are designed to store, transport, and burn fuel safely and efficiently. The type of fuel burned determines the combustion accessories and fuel-burning equipment required. Four main types of fuel-burning equipment are fuel oil burners, gas burners, stokers (coal burners), and pulverized coal burners. A stationary engineer is responsible for achieving maximum efficiency of the fuel-burning equipment.

FUELS

Fuels commonly used to generate steam are coal, fuel oil, and natural gas. Other fuels such as sawdust, wood chips, bagasse (residue of sugarcane), pulverized coffee, industrial waste, and refuse can be used, depending on the availability and cost. Use of other fuels eliminates the need for other means of disposal. **See Figure 4-1.**

Coal, fuel oil, and natural gas are fossil fuels. Fossil fuels are derived from organic matter trapped below the earth's crust and subjected to great pressure over millions of years. Organic matter is formed by a natural chemical process in which vegetation absorbs carbon dioxide from the atmosphere. Plants use sunlight and moisture to convert these ingredients to carbon, hydrogen, and oxygen, the basic ingredients of fossil fuels. When plants are buried by natural processes, they can eventually become fossil fuels.

Figure 4-1. Wood waste, a byproduct of manufacturing wood and paper products, can be used as a fuel for boilers.

143

Coal

Coal is the most widely used fuel for electrical power generation. Bituminous coal and anthracite coal are the most common types of coal mined in the United States. Bituminous coal, or soft coal, is the most commonly used coal for industrial uses. Anthracite coal, or hard coal, has a very high carbon content and requires different burning procedures from bituminous coal. The different conditions under which coal is formed create a variety of coal characteristics. These variations require that coal be classified into different types. Coal is classified according to rank and grade. *Rank* is the degree of hardness of coal. *Grade* is the size, heating value, and ash content of coal. Proximate and ultimate analysis are used to determine characteristics of coal, such as amount of moisture, heating value, and chemical content of coal.

Coal is classified by rank based on the amount of carbon in the following types of coal:
• anthracite
• bituminous
• subbituminous
• lignite

Coal is classified by grade in the following commercial sizes:

Bituminous coal
• run of mine
• lump
• egg
• nut
• stoker
• slack

Anthracite coal
• broken
• egg
• stove
• nut
• pea
• buckwheat
• rice
• barley

Proximate analysis is a mechanical process used to determine the moisture, volatile matter, ash, fixed carbon, and sulfur content in a coal specimen. Moisture content is determined by weighing a coal specimen, then heating this specimen until the moisture is removed. The difference in weight of the coal specimen before and after heating is the moisture content of the coal specimen.

Volatile matter (gases in the coal) is determined by using the coal specimen that was heated to determine the moisture content. The coal specimen is weighed. It is then placed in a crucible with a tight cover and heated to remove the volatile gases. The coal specimen is weighed again. The difference in weight before and after heating is the amount of volatile matter in the coal specimen. *Note:* Bituminous coal has a high amount of volatile matter.

Ash content is determined by burning completely the same coal specimen that was used to determine moisture content and volatile matter and weighing the ash. The weight of the ash is the ash content.

Fixed carbon content is determined by subtracting the weight of the moisture, ash, and volatile matter from the weight of the original coal specimen. *Note:* Anthracite coal has a high fixed carbon content.

Sulfur content is determined by taking a portion of a coal specimen and mixing it with a chemical that combines with sulfur when burned. The sulfur compound is separated and weighed to determine the sulfur content of the coal specimen.

Ultimate analysis is a chemical process used to determine the nitrogen, oxygen, carbon, ash, sulfur, and hydrogen in a coal specimen. These elements can be remembered easily with the words *NO CASH.*

N Nitrogen

O Oxygen

C Carbon

A Ash

S Sulfur

H Hydrogen

Ultimate analysis expresses each element present in the coal specimen as a percentage of the total weight. Total weight of the coal specimen is equal to 100%. Ultimate analysis of coal requires the use of laboratory equipment and trained laboratory personnel.

Heating Value. *Heating value* is the amount of Btu per pound or gallon of fuel. **See Figure 4-2.** The heating value of a fuel is determined by using a calorimeter. A *calorimeter* is a laboratory instrument used to measure the amount of heat developed by combustion. However, data from the ultimate analysis of the coal specimen and Dulong's formula can be used to find the heating value of coal in Btu per pound. There are many similar formulations of Dulong's formula. A common form is as follows:

$$\text{Heating value, } Btu/lb = 14,540C + 62,000\left(H - \frac{O}{8}\right) + 4050S$$

C = % carbon

H = % hydrogen

O = % oxygen

S = % sulfur

Coal Type	Heating Values*
Anthracite	12,700—13,600
Bituminous	11,000—13,800
Medium-volatile bituminous	14,000 (standard)

* in Btu/lb

Figure 4-2. The content of coal determines its characteristics and furnace requirements.

For every pound of carbon burned completely, 14,540 Btu is released. For every pound of hydrogen minus one-eighth the oxygen, 62,000 Btu is released. The reason for this loss of oxygen is that some of the hydrogen and oxygen unite to form water vapor. For every pound of sulfur burned completely, 4050 Btu is released. *Note:* Coal with as low a sulfur content as possible must be selected because sulfuric acid formed in the combustion process attacks boiler metal. In addition, sulfur dioxide in the flue gas is a major cause of air pollution resulting from burning coal.

EXAMPLES

1. Using the ultimate analysis given, find the heating value of the fuel.

   Carbon 68%   Oxygen 8%

   Hydrogen 5%   Sulfur 4%

   Nitrogen 7%   Ash 8%

   $Heating\ value, Btu/lb = 14,540C + 62,000\left(H - \dfrac{O}{8}\right) + 4050S$

   $Heating\ value, Btu/lb = 14,540 \times 0.68 + 62,000\left(0.05 - \dfrac{0.08}{8}\right) + 4050 \times 0.04$

   $Heating\ value, Btu/lb = 9887.2 + [62,000(0.05 – 0.1)] + 162$

   $Heating\ value, Btu/lb = 9887.2 + (62,000 \times 0.04) + 162$

   $Heating\ value, Btu/lb = 9987.2 + 2480 + 162$

   $Heating\ value, Btu/lb = $ **12,529.2 Btu/lb**

2. Using the ultimate analysis given, find the heating value of the fuel.

   Carbon 85%   Oxygen 5%

   Hydrogen 5%   Sulfur 1%

   Nitrogen 2%   Ash 2%

   $Heating\ value, Btu/lb = 14,540C + 62,000\left(H - \dfrac{O}{8}\right) + 4050S$

   $Heating\ value, Btu/lb = 14,540(0.85) + 62,000\left(0.05 - \dfrac{0.05}{8}\right) + 4050(0.01)$

Heating value, Btu/lb = 12,359 + [62,000(0.05 − 0.00625)] + 40.5

Heating value, Btu/lb = 12,359 + (62,000 × 0.04375) + 40.5

Heating value, Btu/lb = 12,359 + 2712.5 + 40.5

Heating value, Btu/lb = **15,112 Btu/lb**

Coal Storage and Handling. Coal is stored in sufficient quantities to meet fuel requirements of the plant. The stockpiled coal must be stored and handled properly to prevent the possibility of spontaneous combustion. Anthracite coal has a high carbon content with less volatile matter than bituminous coal and is not affected by spontaneous combustion as much as bituminous coal. Bituminous coal must be closely monitored when stockpiled.

To reduce the possibility of spontaneous combustion, coal should be used as soon as possible. Coal stored should be rotated with new coal, and the old coal should be used first. In addition, coal in storage must be tightly packed to prevent air pockets from developing. Air in the stockpile promotes spontaneous combustion.

Coal temperature in the stockpile should be checked at various points. Pipes are inserted at various points in the stockpile to allow heat-sensing thermocouples to be lowered. If the coal temperature is above the safe storage temperature or is rising, carbon dioxide (CO_2) should be introduced into the stockpile using perforated pipes to cool the coal.

Coal stored in overhead bunkers should also be rotated and should not be stored for prolonged periods. Ambient temperatures surrounding the overhead bunkers can reach high temperatures. In addition, overhead bunkers should be protected with shielding if a possibility of heat radiating from the boilers exists.

Fuel Oil

Fuel oil used for the generation of steam is obtained from petroleum. Crude oil from an oil well is distilled into individual products, such as gasoline, diesel fuel, lubricating oils, and heating fuels. Distillate fuel oil used for the lighter grades of fuel oil has a consistency between kerosene and lubricating oil. Blended fuel oil is produced by mixing fuel oil to certain specifications. Residual fuel oil is produced by removing hydrocarbons. After hydrocarbons have been removed, the flash point is lower, allowing residual fuel oil to be safely stored and burned.

Fuel oil is also classified by grades, which are No. 1, No. 2, No. 4, No. 5, and No. 6. **See Figure 4-3.** The heating value of fuel oil is expressed in Btu/gal. or Btu/lb. The Btu content of a fuel indicates how much heat can be produced by the fuel. Heavy fuel oils have a higher Btu content than light fuel oils, allowing the specific gravity (SG) of fuel oils to be used to determine heating value. Heating value is found by using the following equation:

$$Btu/lb = 17,780 + (54 \times °API)$$

FUEL OIL CHARACTERISTICS	No. 1 Fuel Oil	No. 2 Fuel Oil	No. 4 Fuel Oil	No. 5 Fuel Oil	No. 6 Fuel Oil
Type	Distillate kerosene	Distillate	Very light residual	Light residual	Residual
Color	Light	Amber	Black	Black	Black
American Petroleum Institute (API) 60°F	40	32	21	17	12
Specific gravity	0.8250	0.8654	0.9279	0.9529	0.9861
Lb/U.S. gal.	6.87	7.206	7.727	7.935	8.212
Btu/gal.	137,000	141,000	146,000	148,000	150,000
Btu/lb	19,850	19,500	19,100	18,950	18,750

Figure 4-3. Fuel oil characteristics are considered when determining proper use and treatment of fuel oil in high-pressure boiler plants.

Specific gravity is the ratio of the weight of any volume of fuel oil at 60°F to the weight of an equal volume of water at 60°F. It is designated as SG 60°/60°F and is carried to four decimal places. A *hydrometer* is an instrument used to measure specific gravity. **See Figure 4-4.** A hydrometer is usually made of glass and consists of a cylindrical stem and a bulb weighted to make it float upright. The liquid to be tested is poured into a graduated cylinder, and the hydrometer is gently lowered into the liquid until it floats freely. The point at which the surface of the liquid touches the stem of the hydrometer is noted. It is read direct or expressed in degrees American Petroleum Institute (°API). Specific gravity is found by using the following equation:

$$SG\ 60°/60°F = \frac{141.5}{131.5 + °API}$$

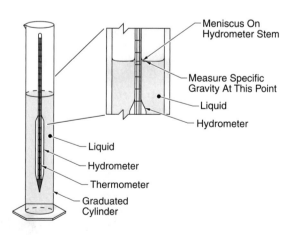

Figure 4-4. Hydrometers are used to measure the density of a liquid.

EXAMPLES

1. What is the specific gravity of fuel oil that is 40°API?

$$SG\,60°/60°F = \frac{141.5}{131.5 + °API}$$

$$SG\,60°/60°F = \frac{141.5}{131.5 + 40}$$

$$SG\,60°/60°F = \frac{141.5}{171.5}$$

$$SG\,60°/60°F = \mathbf{0.8251}$$

2. What is the specific gravity of fuel oil that is 20°API?

$$SG\,60°/60°F = \frac{141.5}{131.5 + °API}$$

$$SG\,60°/60°F = \frac{141.5}{131.5 + 20}$$

$$SG\,60°/60°F = \frac{141.5}{151.5}$$

$$SG\,60°/60°F = \mathbf{0.9340}$$

When the SG 60°/60°F is known, °API is found by using the following equation:

$$°API = \frac{141.5}{SG\,60°/60°F} - 131.5$$

EXAMPLES

1. What is the °API when the specific gravity of the fuel oil is 0.9930?

$$°API = \frac{141.5}{SG\,60°/60°F} - 131.5$$

$$°API = \frac{141.5}{0.9330} - 131.5$$

$$°API = 142.5 - 131.5$$

$$°API = \mathbf{11°API}$$

2. What is the °API when the specific gravity of the fuel oil is 0.8900?

$$°API = \frac{141.5}{SG\ 60°/60°F} - 131.5$$

$$°API = \frac{141.5}{0.8900} - 131.5$$

$$°API = 159 - 131.5$$

$$°API = \mathbf{27.5°API}$$

If the °API of a fuel oil is known, the approximate Btu content per pound of fuel oil is found by using the following equation:

Btu/lb of fuel oil = 17,780 + (54 × °API)

EXAMPLES

1. A fuel oil has a specific gravity of 10°API. What is the Btu content per pound of fuel oil?

    Btu/lb = 17,780 + (54 × °API)

    Btu/lb = 17,780 + (54 × 10)

    Btu/lb = 17,780 + 540

    Btu/lb = **18,320 Btu/lb**

2. A fuel oil has a specific gravity of 15°API. What is the Btu content per pound of fuel oil?

    Btu/lb = 17,780 + (54 × °API)

    Btu/lb = 17,780 + (54 × 15)

    Btu/lb = 17,780 + 810

    Btu/lb = **18,590 Btu/lb**

Fuel oil consists mainly of carbon and hydrogen with some moisture, sulfur, nitrogen, arsenic, phosphorous, and silt. The sulfur content permitted in fuel oil is controlled by law in some states to reduce sulfur dioxide emitted into the air.

Fuel oil characteristics include viscosity, flash point, fire point, and pour point. *Viscosity* is the internal resistance of fuel oil to flow. By raising the fuel oil temperature, viscosity is lowered, making it easier to pump. *Flash point* is the temperature at which fuel oil gives off a vapor that will flash when exposed to an open flame. Fuel oil with a low flash point can be dangerous to use. *Fire point* is the temperature at which fuel oil gives off a vapor that will burn continuously. *Pour point* is the lowest temperature at which fuel oil flows as a liquid.

Fuel oil must be handled properly to prevent hazards. Heavy fuel oils (some No. 5 and No. 6) must be heated to lower their viscosity for pumping. Fuel oil must not be overheated. Fuel oil overheated in a storage tank can cause sludge, sediment, and other impurities to settle out in the fuel oil tank. This buildup eventually requires the fuel oil tank to be opened and cleaned. In addition, if the fuel oil is overheated, it can reach its flash point, which can lead to a fire.

When using fuel oil, the proper fire extinguishers should be located in strategic locations. Fire extinguishers are not meant to take the place of plant or local fire departments. Fire extinguishers are only designed to put out small fires or help contain large fires until help arrives. The number and type of fire extinguishers are determined by the authority having jurisdiction and are based on how fast the fire may spread, potential heat intensity, and accessibility to the fire.

In addition to fire extinguishers, buckets of sand should be readily available for use in case of spills or small fires. Foam or dry-chemical fire extinguishers are used for fuel oil fires. Water or water-type fire extinguishers must never be used for fuel oil fires. Whenever a fuel oil spill occurs, it should be cleaned up immediately. All fuel oil rags should be disposed of immediately.

Natural or Manufactured Gas

Natural gas, or *manufactured gas,* is a gaseous fuel consisting of methane, ethylene, and small amounts of hydrogen, nitrogen, carbon monoxide, carbon dioxide, oxygen, and heavy hydrocarbons. Natural gas usage has gained in popularity with stricter environmental laws because it burns cleaner than coal and fuel oil and causes less pollution.

Natural gas has a Btu content that ranges from 950 Btu/cu ft to 1050 Btu/cu ft. Heating value is expressed in Btu/cu ft. Heating values are measured in therms. A *therm* is an energy equivalent of 100,000 Btu. The heating value of natural gas is determined using a gas calorimeter or by calculations based on the chemical analysis of the gas. By chemical analysis, the percentage of Btu content in each gas is determined. These percentages are then used to determine the overall heating value of the gas.

Natural gas is wet or dry, depending on where it is found. Wet natural gas is found with oil deposits, and dry natural gas is found away from oil-producing wells. Wet and dry refer to the gasoline content of natural gas. Natural gas can also be sweet or sour. Sweet natural gas does not contain hydrogen sulfide. Sour natural gas contains a high percentage of hydrogen sulfide. Hydrogen sulfide causes natural gas to smell like rotten eggs.

Manufactured gas is produced from industrial combustibles and varies in chemical composition. Manufactured gas may be blast furnace gas, coke-oven gas, or refinery gas. The Btu content of manufactured gas varies, depending on how it is manufactured.

Blast furnace gas is a by-product of furnaces used to smelt iron ore and contains carbon dioxide, nitrogen, oxygen, carbon monoxide, and hydrogen. The Btu content varies between 85 Btu/cu ft and 100 Btu/cu ft. Coke-oven gas is a by-product of making steel and contains carbon dioxide, carbon monoxide, oxygen, methane, hydrogen, and nitrogen. The Btu content varies between 400 Btu/cu ft to 500 Btu/cu ft. Refinery gas, or liquefied petroleum gas (LPG), is a by-product of the refinery process of petroleum. Two common refinery gases, propane and butane, are often used as alternate fuels if other fuel supplies are unavailable. The Btu content for refinery gas varies from approximately 2500 Btu/cu ft to 3200 Btu/cu ft.

All types of gases used to generate steam are toxic and expensive. All gas lines should be tested for leaks by using a liquid soap solution or an electronic gas detector. Vent lines from regulators, reducing valves, or governors should be piped out of the boiler room to an area where they can be discharged safely. Any gas odor must be checked immediately. If gas is allowed to build up and mix with air, contact with a spark or open flame can cause a serious explosion.

COMBUSTION PROCESS

The combustion process occurs when elements in coal, fuel oil, or gas combine with oxygen to produce heat. The goal of a stationary engineer is to achieve complete combustion-the burning of all the fuel with a minimal amount of excess air. The four requirements for complete combustion are the proper mixture, atomization, temperature, and time to complete combustion. These requirements can be easily remembered by the acronym *MATT.*

M—Proper *mixture* of air and fuel is required. The air-fuel ratio is controlled for all firing rates. High fire requires more air and fuel proportionally than when the burner is in low fire. *Note:* It is the oxygen in the air that is required for combustion. Air is made up of approximately 20% oxygen and 80% nitrogen. Nitrogen does not enter the combustion process.

A—Proper *atomization* of fuel is required. *Atomization* is the process of breaking up fuel into small particles to allow intimate contact with the air to improve combustion.

T—Proper *temperature* of air, fuel, and zone temperature must be maintained to achieve complete combustion.

T—Proper *time* must be provided to complete the combustion process before the gases of combustion come in contact with the heating surface. *Note:* The heating surface is where there is water on one side and gases of combustion on the other. If the gases of combustion come in contact with the heating surface before combustion is complete, they will cool and cause the formation of soot and smoke.

Coal is burned on a grate or in suspension. Anthracite coal is burned on grates with a fuel bed 2″ to 5″ deep. Air is introduced under the fuel bed and is forced through it. Combustion is completed just above the fuel bed. Care must be taken to prevent holes from forming in the fuel bed. These holes allow air to pass through without combining with the fuel. Furnace temperature is maintained at approximately 2750°F.

Bituminous coal when burned on grates requires a thicker bed than anthracite coal. Bituminous coal also requires air to be introduced over the fire as well as under the fire for complete combustion. Because of the large amount of gases released from the fuel bed and the combustion taking place over the bed, a larger air space is required over the fire of a bituminous coal fire than an anthracite coal fire. Proper temperature must be maintained in both cases to achieve complete combustion. When burning bituminous coal in suspension, it first has to be pulverized. *Pulverized coal* is coal that has been ground into a fine dust by a pulverizer to the consistency of talcum powder and is highly explosive. The coal dust is mixed with heated air as it passes through the pulverizer and is blown into the furnace to burn in suspension. Additional air (secondary air) is introduced to complete the combustion process. Temperatures are relatively high when burning pulverized coal. Furnace temperature when burning coal in suspension should be approximately 3000°F. If combustion is not completed before reaching the first pass, slagging of the tubes could result. The furnace volume must be large enough for the high rate of combustion.

The combustion of fuel oil in a furnace depends on the ability of the burner to deliver the correct quantity of fuel oil, which must also be properly atomized and at the correct temperature. *Atomize* is to break up a liquid into a very fine mist. Fuel oil is burned in a manner similar to burning pulverized coal. Fuel oil is introduced in a conical patterned spray with air blown in around it. The correct air-fuel ratio can be easily maintained. Furnace temperature with a fuel oil burner (approximately 2500°F) is lower than a bituminous coal burner temperature. Combustion air is introduced in a rotary fashion to mix with fuel oil in sufficient quantities for complete combustion. The furnace temperature must be high enough to allow the combustion process to be completed. In addition, the furnace volume must be large enough for the complete combustion of all gases before entering the first pass or touching the boiler heating surfaces.

Natural or manufactured gas can be readily mixed with air to obtain proper air-fuel ratio and intimate contact. Gas burners are classified into two basic types. One has air that mixes with the gas outside the furnace and one has air that mixes with the gas inside the furnace. Both types require the proper air-fuel ratio. In addition, efficient combustion is achieved with the correct furnace temperature, with the volume of the furnace large enough to allow complete combustion before the first pass. Furnace temperature of gas burners is slightly less than fuel oil burners.

Combustion Chemistry

Combustion is the rapid union of an element or a compound with oxygen that results in the production of heat. Combustion occurs when the elements in a fuel combine with oxygen in the air to produce heat. Elements in a fuel that are involved in combustion include carbon, sulfur, and hydrogen.

Oxygen supports combustion, but it is not combustible by itself. Nitrogen is neither combustible nor supports combustion at temperatures occurring in the combustion process. Carbon, hydrogen, and sulfur are the combustibles in a fuel that combine with oxygen from air to form compounds of combustion. When carbon (C) combines with oxygen (O_2), compounds of carbon monoxide (CO) or carbon dioxide (CO_2) are formed. This process is shown in the following combustion equations:

$$2C + O_2 \rightarrow 2CO + Heat$$
$$C + O_2 \rightarrow CO_2 + Heat$$

When sulfur (S) combines with oxygen, sulfur dioxide (SO_2) is formed:

$$S + O_2 \rightarrow SO_2 + Heat$$

When hydrogen (H) combines with oxygen, water vapor is formed:

$$2H_2 + O_2 \rightarrow 2H_2O + Heat$$

These chemical reactions take place in a furnace during the burning of a fuel, provided there is sufficient air (oxygen) to completely burn the fuel.

Combustion Types

The three combustion types are perfect, complete, and incomplete combustion. *Perfect combustion* is achieved when all the fuel is burned using only the theoretical amount of air. *Theoretical amount of air* is the amount of air used to achieve perfect combustion in a laboratory setting. Perfect combustion cannot be achieved in a boiler. It is only possible in a laboratory setting where the combustion process can be carefully controlled. *Complete combustion* is achieved when all the fuel is burned using the minimal amount of air above the theoretical amount of air needed to burn the fuel. Complete combustion is the goal of a stationary engineer. With complete combustion, the fuel is burned at the highest combustion efficiency with minimal pollution. *Incomplete combustion* occurs when all the fuel is not burned, which results in the formation of soot and smoke.

Air required in combustion is classified as primary, secondary, and excess air. *Primary air* is air supplied to a burner that controls the rate of combustion, which determines the amount of fuel that can be burned. *Secondary air* is air supplied to a burner that controls combustion efficiency by controlling how completely the fuel is burned. *Excess air* is air supplied to the burner that exceeds the theoretical amount needed to burn the fuel.

A stationary engineer must be familiar with the equipment and principles involved in combustion. Many states have passed laws that limit products of combustion discharged from the chimney. Failure to comply with air pollution standards can result in costly fines. By controlling the combustion process, soot and smoke discharged by the chimney can be reduced or eliminated.

Combustion Air Requirements

Combustion air requirements are based on the composition of the fuel used. Fuels commonly used contain nitrogen, ash, oxygen, sulfur, carbon, and hydrogen. An ultimate analysis of the fuel used determines the percentage of each element in the fuel. The pounds of air needed per pound of fuel is found by using the percentage of each element in the fuel and the following equation:

$$Pounds\ of\ air\ per\ pound\ of\ fuel = 11.53C + 34.56\left(H - \frac{O}{8}\right) + 4.32S$$

$$Lb\ air/Lb\ fuel = 11.53C + 34.56\left(H - \frac{O}{8}\right) + 4.32S$$

C = % carbon

H = % hydrogen

O = % oxygen

S = % sulfur

Note: For every pound of carbon burned completely, 11.53 pounds of air are required. For every pound of hydrogen minus one-eighth the oxygen, 34.56 pounds of air are required. This allows for the amount of hydrogen uniting with oxygen to form water vapor. For every pound of sulfur burned completely, 4.32 pounds of air are required.

The pounds of air per pound of fuel found from the previous equation is the theoretical amount of air required. Excess air above the theoretical amount is needed to achieve complete combustion but must be kept to a minimum. Once the pounds of air per pound of fuel is known, pounds of oxygen in that air is determined by dividing by 4.32.

EXAMPLES

1. Using the ultimate analysis given, find the pounds of air per pound of fuel and the pounds of oxygen per pound of fuel.

   Carbon 70%   Oxygen 8%

   Hydrogen 7% Sulfur 2%

   Nitrogen 5%   Ash 8%

   $$Lb\,air/Lb\,fuel = 11.53C + 34.56\left(H - \frac{O}{8}\right) + 4.32S$$

   $$Lb\,air/Lb\,fuel = (11.53 \times 0.7) + 34.56\left(0.07 - \frac{0.08}{8}\right) + (4.32 \times 0.02)$$

   $$Lb\,air/Lb\,fuel = 8.071 + (34.56 \times 0.06) + 0.0864$$

   $$Lb\,air/Lb\,fuel = 8.071 + 2.0736 + 0.0864$$

   $$Lb\,air/Lb\,fuel = \textbf{10.23 lb}$$

   $$Lb\,oxygen/Lb\,fuel = \frac{10.23}{4.32}$$

   $$Lb\,oxygen/Lb\,fuel = \textbf{2.37 lb}$$

2. Using the ultimate analysis given, find the pounds of air per pound of fuel and the pounds of oxygen per pound of fuel.

   Carbon 65%   Oxygen 11%

   Hydrogen 6% Sulfur 4%

   Nitrogen 2%   Ash 12%

   $$Lb\,air/Lb\,fuel = 11.53C + 34.56\left(H - \frac{O}{8}\right) + 4.32S$$

   $$Lb\,air/Lb\,fuel = (11.53 \times 0.65) + 34.56\left(0.06 - \frac{0.11}{8}\right) + (4.32 \times 0.04)$$

   $$Lb\,air/Lb\,fuel = 7.4945 + (34.56 \times 0.04625) + 0.1728$$

   $$Lb\,air/Lb\,fuel = 7.4945 + 1.5984 + 0.1728$$

   $$Lb\,air/Lb\,fuel = \textbf{9.27 lb}$$

   $$Lb\,oxygen/Lb\,fuel = \frac{9.27}{4.32}$$

   $$Lb\,oxygen/Lb\,fuel = \textbf{2.15 lb}$$

In day-to-day operation of a plant, it is not always possible to obtain an analysis of gases of combustion at any given time. This is true even though some plants have carbon dioxide and oxygen recorders. If an analysis of gases of combustion is possible, it only provides the condition of the combustion process. A simpler analysis that considers steam generated per amount of fuel is a standard widely used in the industry. The number of pounds of steam generated by a boiler over a given period of time is divided by the amount of fuel used during that period. It can be pounds of steam per gallon of fuel oil, pounds of steam per pound of coal, or pounds of steam per 100 cubic feet of gas.

Using this information in a day-to-day comparison will indicate boiler performance. If values start to drop, it must be determined whether it is a result of combustion inefficiency or reduced heat transfer within the passes. Using pounds of steam generated per unit of fuel is only accurate for plants with relatively constant feedwater temperatures.

When using coal, the reading from the coal scale is read every 8 hours, and the steam generated is read on the steam flow integrator at the same time.

$$\frac{Pounds\ of\ steam}{Pounds\ of\ coal} = \frac{Total\ steam\ per\ 8\ hours}{Total\ coal\ per\ 8\ hours}$$

$$\frac{Lb\ steam}{Lb\ coal} = \frac{TS}{TF}$$

When using fuel oil, gallons are used instead of pounds of fuel. Gallons of fuel oil used are read directly from the fuel oil meter.

$$\frac{Pounds\ of\ steam}{Gallons\ of\ fuel\ oil} = \frac{Total\ steam\ per\ 8\ hours}{Total\ fuel\ oil\ per\ 8\ hours}$$

$$\frac{Lb\ steam}{Gal.\ fuel\ oil} = \frac{TS}{TF}$$

When using gas, the calculations are based per 100 cu ft.

$$\frac{Pounds\ of\ steam}{100\ cubic\ feet\ of\ gas} = \frac{Total\ steam\ per\ 8\ hours}{Total\ cubic\ feet\ per\ 100\ cubic\ feet\ per\ 8\ hours}$$

$$\frac{Lb\ steam}{100\ cu\ ft\ gas} = \frac{TS}{TG}$$

It is also useful to find the pounds of steam generated per unit of fuel over 1-hour periods on minimum and maximum loads. By maintaining a running record of the daily pounds of steam generated per unit of fuel, daily boiler performance can be determined, which has a direct bearing on thermal efficiency. Thermal efficiency is found by using the following equation:

$$Thermal\ efficiency =$$

$$\frac{Pounds\ of\ steam/hr\left[Btu\ content\ of\ steam - \left(Feedwater\ temperature - 32\right)\right]}{Units\ of\ fuel/hr \times Btu\ content\ per\ unit\ of\ fuel}$$

$$TE = \frac{W_s \left[H_s - \left(T_{fw} - 32 \right) \right]}{W_f \times C}$$

W_s = pounds of steam per hour

H_s = enthalpy of steam (in Btu/lb)

T_{fw} = feedwater temperature

32 = base temperature from which total heat is calculated

W_f = units of fuel per hour

C = Btu content per unit of fuel

Note: Units of fuel per hour are expressed in pounds, gallons, or cubic feet. Btu is also expressed per pound, per gallon, or per cubic feet. Comparisons between different plants cannot be made unless feedwater temperature, steam pressure, and steam temperature are the same.

Analysis of Gases of Combustion

An analysis of gases of combustion is used to determine combustion efficiency. Using this data, a stationary engineer can make the necessary adjustment to the burner for optimum efficiency. Carbon dioxide in gases of combustion is an indication of complete combustion. Carbon monoxide in gases of combustion is an indication of incomplete combustion. Oxygen in gases of combustion is an indication of the presence of excess air.

Fyrite® Analyzers. Fyrite analyzers are used to measure the percentage of oxygen or carbon dioxide in gases of combustion. **See Figure 4-5.** When using the Fyrite analyzer to measure the percentage of carbon dioxide in gases of combustion, the following procedure is used:

1. Vent the Fyrite analyzer to the atmosphere by depressing the top of the Fyrite analyzer, and set the indicator on 0%.

2. Draw a sample of gases of combustion into the Fyrite analyzer. *Note:* The sample of gases of combustion should be taken as close to the boiler outlet as possible so the sample is not diluted with air.

3. The Fyrite analyzer is inverted a couple of times to thoroughly mix with the carbon dioxide absorbing agent.

4. The absorbing agent increases in volume, and a direct reading in percentage of carbon dioxide can be taken.

5. Take the temperature of the boiler room and the temperature of the gases of combustion. Follow the directions on the slide rule calculator supplied with the Fyrite analyzer to determine combustion efficiency and percentage of chimney loss.

Orsat Analyzers. Orsat analyzers are used for more complete analysis of gases of combustion. The Orsat analyzer can be used to measure the percentage of carbon dioxide, oxygen, and carbon monoxide in the gases of combustion. The absorbing agent in the carbon dioxide burette is a solution of caustic potash. The absorbing agent in the oxygen burette is pyrogallic acid. The absorbing agent for carbon monoxide is cuprous (copper) chloride. Using the data from the Orsat analyzer, the amount of excess air needed in the combustion process can be determined. The percentage of excess air over the theoretical requirements is found by using the following equation:

$$\% \text{ of excess air} = \frac{O_2 - \frac{1}{2}CO}{0.263N_2 + \frac{1}{2}CO - O_2} \times 100$$

O_2 = oxygen
CO = carbon monoxide
N_2 = nitrogen

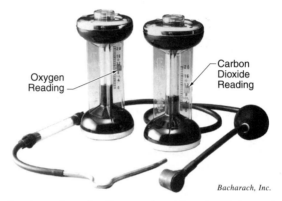

Oxygen Reading

Carbon Dioxide Reading

Bacharach, Inc.

Figure 4-5. Sample gases of combustion are drawn into the Fyrite analyzer close to the boiler outlet for accurate readings.

EXAMPLES

1. What is the percentage of excess air required with the following analysis of gases of combustion?

   Carbon dioxide 12%   Oxygen 5%
   Carbon monoxide 2%   Nitrogen 81%

$$\% \text{ of excess air} = \frac{O_2 - \frac{1}{2}CO}{0.263N_2 + \frac{1}{2}CO - O_2} \times 100$$

$$\% \text{ of excess air} = \frac{0.05 - \left(\frac{1}{2} \times 0.02\right)}{(0.263 \times 0.81) + \left(\frac{1}{2} \times 0.02\right) - 0.05} \times 100$$

$$\% \text{ of excess air} = \frac{0.05 - 0.01}{0.21303 + 0.01 - 0.05} \times 100$$

$$\% \text{ of excess air} = \frac{0.04}{0.17303} \times 100$$

$$\% \text{ of excess air} = \frac{4}{0.17303}$$

$$\% \text{ of excess air} = \mathbf{23.12\%}$$

2. What is the percentage of excess air required with the following analysis of gases of combustion?

Carbon dioxide 12%   Oxygen 6%

Carbon monoxide 0.6%   Nitrogen 81.4%

$$\% \text{ of excess air} = \frac{O_2 - \frac{1}{2}CO}{0.263N_2 + \frac{1}{2}CO - O_2} \times 100$$

$$\% \text{ of excess air} = \frac{0.06 - \left(\frac{1}{2} \times 0.006\right)}{\left(0.263 \times 0.814\right) + \left(\frac{1}{2} \times 0.006\right) - 0.06} \times 100$$

$$\% \text{ of excess air} = \frac{0.06 - 0.003}{0.214082 + 0.003 - 0.06} \times 100$$

$$\% \text{ of excess air} = \frac{0.057}{0.157082} \times 100$$

$$\% \text{ of excess air} = \frac{5.7}{0.157082}$$

$$\% \text{ of excess air} = \textbf{36.29\%}$$

Note: The amount of excess air required depends on the type of fuel used and how it is burned. Coal burned on a grate requires 50% to 75% excess air. Coal burned in suspension as in a pulverized coal burner requires 20% to 40% excess air. Fuel oil requires 10% to 30% excess air, and natural gas requires only 5% to 10% excess air, or even lower in well-designed boilers. These are average values and efficient combustion is still determined by MATT.

Electronic Combustion Testers. Combustion efficiency can be checked quickly and accurately using electronic testing equipment. Electronic combustion testers maximize boiler performance but can also be used as a management tool to keep records and control plant operating costs. Electronic combustion testers measure two temperatures, five gases, draft, and smoke. The electronic combustion tester also computes combustion efficiency, excess air, and carbon dioxide. This information can be stored, printed, or plotted. In addition, this information can be sent to remote locations by direct connection, via modem over telephone lines, or through an Internet connection. **See Figure 4-6.**

The electronic combustion tester uses a probe that is inserted in the breeching as close to the boiler as possible for each test, or a probe that is mounted permanently in the breeching. A pump inside the instrument draws a small sample of gases of combustion. Sensors in the instrument analyze the flue gas for contents and temperature. The results are displayed and may be printed or plotted as part of the logging shift activities. The date, time, and test results recorded document exact plant functions at a specific time. After reviewing the test results, a stationary engineer can make adjustments to optimize boiler performance.

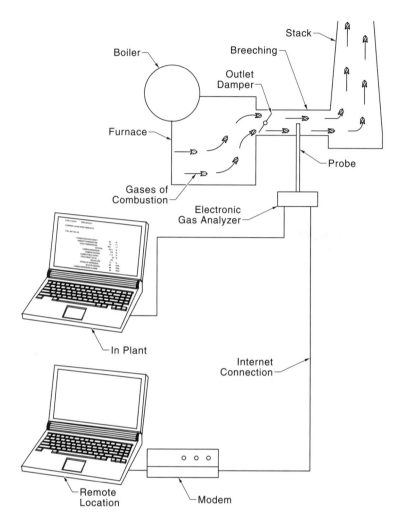

Figure 4-6. Data output from electronic combustion testers is used to monitor plant performance.

Boiler Heat Balance

Fuel fed to a burner is converted to heat. The heat is absorbed by the water in the boiler. Not all heat released from the fuel is used to heat the water. Some heat is wasted in the process. The *heat balance* of a boiler consists of accounting for all the heat units in the fuel used or wasted. It is a balance because it is the sum of all the heat consumed. Heat balance of a boiler is found by using the following equation:

Heat available in fuel = Heat absorbed by water in boiler + Heat losses

$A = B + C$

A = heat available in the fuel

B = heat absorbed by water in the boiler

C = heat losses

$$Efficiency\ of\ unit = \frac{B}{A}$$

For example, Btu losses that occur in a boiler could result from the following:

• gases of combustion to atmosphere (about 11%)

• radiation (about 4%)

• water vapor produced from burning the hydrogen (about 3%)

• unburned combustibles (about 3%)

• incomplete combustion (about 1.5%)

• moisture in the fuel (less than 1%)

• moisture in the air used for combustion (less than 1 %)

The total heat losses (Btu losses) in this example equal 24.5%.

Preventable losses are losses that a stationary engineer has control over when operating a boiler, including the following:

1. heat carried away in the dry chimney gases

2. incomplete combustion of fuel

3. unburned combustibles in the ash

For example, a boiler with heat losses totaling approximately 18% has an efficiency of 82%. Conditions in a boiler that could affect the heat losses in gases of combustion to the atmosphere include the following:

• soot on the heating surfaces of the boiler, reducing the heat transfer rate from gases to water

• scale or deposits on the heat transfer surfaces on the water side of the boiler, affecting heat transfer

• broken baffles in the boiler, which short-circuit gases of combustion without contacting the heating surfaces, reducing heat transfer

• too much air for combustion, which carries heat past the transfer surfaces faster, causing less heat to be transmitted

Soot, scale, and broken baffles are indicated by excessive outlet temperatures of gases of combustion and a lower evaporation rate. In addition, with broken baffles hot spots can develop in gases of combustion passages.

Draft

Draft is the difference in pressure between two points that causes air or gases to flow. The primary purpose of a draft system on a boiler is to supply sufficient air to complete the combustion process and remove products of combustion. Approximately 15 lb of air is required to burn 1 lb of fuel. To burn fuel completely, sufficient quantities of air with pressures high enough to overcome the resistance caused by breeching, dampers, baffles, tube passes, economizers, and superheaters are required. Draft is measured in inches or tenths of an inch of a vertical water column and is classified as natural draft or mechanical draft.

Natural Draft. Natural draft is produced by a chimney alone, without mechanical means. It is caused by the difference in weight between the column of hot gas inside the chimney and a column of cold outside air of the same height and cross section. The intensity of draft produced by the chimney is determined by the height of the chimney and the temperature difference between the inside and outside of the chimney. Gases of combustion, because they are lighter than the outside air, tend to rise in the chimney. Heavier outside air replaces this air in the furnace area of the boiler.

The units used to measure draft are inches of water column (WC). When the pressure is below atmospheric pressure, it is designated with a minus sign (–). Typical draft pressures are atmospheric pressure before the inlet damper; –0.1″ WC (below atmospheric pressure) inside the furnace area; –1″ WC before the outlet damper; and –1.5″ WC on the chimney side of the outlet damper. **See Figure 4-7.** *Note:* Atmospheric pressure, or barometric pressure, is measured in inches of mercury (in. Hg). To convert inches of mercury to inches of water column, inches of mercury is multiplied by 13.6.

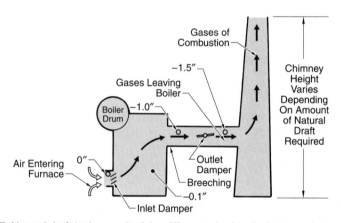

Figure 4-7. Natural draft is the result of the difference in density between hot and cold air.

The amount of natural draft is determined by measuring the difference between two points, such as the atmosphere and the inside of the boiler setting. Air leaks in the boiler, breeching, and/or boiler casing dilute the gases of combustion, increase the volume, lower the temperature, and lower the effective draft. Natural draft is controlled by using dampers, which control the air flow to the burners and gases of combustion leaving the boiler setting.

The applications of natural draft chimneys are limited by the chimney size and draft requirements. The trend in design of modern boiler units is toward high rates of heat transfer, which increase draft loss. This requires a higher chimney, which becomes physically and economically impractical. Theoretical natural draft produced by a chimney is found by using the following equation:

$$D = 0.52HP\left(\frac{1}{T_1} - \frac{1}{T_2}\right)$$

D = draft at base of chimney (in inches of water)

0.52 = pressure constant

H = height of chimney (in ft)

P = atmospheric pressure (14.7 psia)

T_1 = outside temperature (in absolute °R)

T_2 = gas temperature (in absolute °R)

Note: To change °F to °R, add 460 to °F; that is, $°R = °F + 460$.

EXAMPLES

1. A plant has a chimney 200′ high. The chimney temperature is 450°F, and the outside temperature is 100°F. What is the theoretical draft at the base of the chimney?

$°R = °F + 460$

$°R = 450°F + 460$

$°R = 910$

$°R = 100°F + 460$

$°R = 560$

$$D = 0.52HP\left(\frac{1}{T_1} - \frac{1}{T_2}\right)$$

$$D = 0.52 \times 200 \times 14.7 \times \left(\frac{1}{560} - \frac{1}{910}\right)$$

$D = 1528.8(0.0017 - 0.0010)$

$D = 1528.8 \times 0.0007$

$D = \mathbf{1.0702″}$ **of water**

2. A steam plant has a chimney 175′ high. The temperature of the gases of combustion is 540°F as it enters the chimney. The outside temperature is 60°F. What is the theoretical draft at the base of the chimney?

$°R = °F + 460$

$°R = 540°F + 460$

$°R = 1000$

$°R = 60°F + 460$

$°R = 520$

$$D = 0.52HP\left(\frac{1}{T_1} - \frac{1}{T_2}\right)$$

$$D = 0.52 \times 175 \times 14.7 \times \left(\frac{1}{520} - \frac{1}{1000}\right)$$

$D = 1337.7(0.001923 - 0.001)$

$D = 1337.7 \times 0.000923$

$D = \mathbf{1.2347″}$ **of water**

Natural draft is controlled by an outlet damper located after the last pass of the boiler and an inlet damper located before the furnace. The outlet damper opens or closes to maintain a constant furnace pressure of 0.05″ to 0.1″ of water below the atmospheric pressure. The inlet damper opens or closes to allow the proper amount of air for the amount of fuel being used (air-fuel ratio).

Mechanical Draft. Mechanical draft is produced by power-driven fans. The two types of mechanical draft are forced and induced. *Forced draft* is produced when the fan or blower located in front of the boiler setting forces (pushes) air into the furnace. **See Figure 4-8.** *Induced draft* is produced when the fan or blower located between the boiler and chimney removes (pulls) gases of combustion from the boiler and discharges them into the chimney. **See Figure 4-9.** *Combination forced and induced draft* uses both forced and induced draft fans on large boiler installations. **See Figure 4-10.**

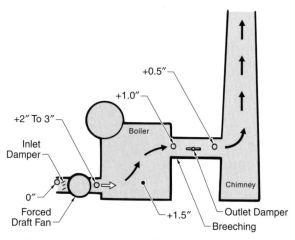

Figure 4-8. A forced draft system uses a draft fan at the inlet damper of the boiler.

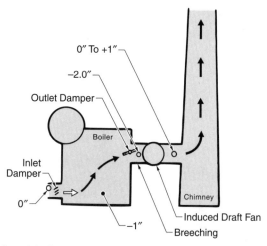

Figure 4-9. An induced draft system uses a draft fan at the outlet damper of the boiler.

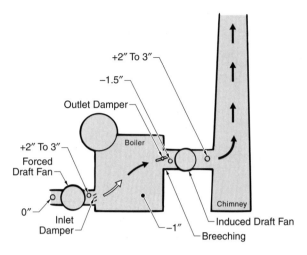

+2″ To 3″

−1.5″

Outlet Damper

+2″ To 3″
Forced
Draft Fan

Boiler

Chimney

0″

Inlet
Damper

−1″

Induced Draft Fan

Breeching

Figure 4-10. A combination draft system uses draft fans at both the inlet damper and the outlet damper of the boiler.

To maintain correct air-fuel ratios for optimum combustion efficiency, draft must be controlled at all times, especially during changing load conditions. Draft is controlled by the following:
• speed of the induced or forced draft fans
• adjustment of inlet and outlet dampers
• use of inlet vanes on the fan

To determine the overall plant efficiency, the HP required to operate the draft fans (air HP) must be determined. The HP required to operate the draft fans is found by using the following equation:

$$Air\ horsepower = \frac{Quantity \times Pressure}{6356}$$

$$Air\ HP = \frac{QP}{6356}$$

Q = quantity of air (in cu ft/min)

P = inches of water pressure

Note: For simplicity, 6356 is given as the constant derived from the reduction of the horsepower formula.

To determine the power needed for the drive on the fan, not only must the HP of the air output be considered, but also the efficiency of the fan. A value of 55% efficiency is assumed for the fan.

$$Drive\ HP = \frac{Air\ HP}{Efficiency}$$

EXAMPLES

1. A boiler uses 100 lb of fuel/min, and each lb of fuel requires 200 cu ft of air. What is the HP needed to drive the unit if the forced draft fan has a discharge pressure of 3" of water?

$$Air\ HP = \frac{QP}{6356}$$

$$Air\ HP = \frac{100 \times 200 \times 3}{6356}$$

$Air\ HP =$ **9.44**

$$Drive\ HP = \frac{Air\ HP}{Efficiency}$$

$$Drive\ HP = \frac{9.44}{0.55}$$

$Drive\ HP =$ **17.16**

2. A boiler generates 36,000 lb of steam/hr and each lb of fuel requires 200 cu ft of air/lb of fuel. The boiler has an evaporation rate of 12 lb of steam/lb of fuel. What is the HP needed to drive the unit if the forced draft fan has a discharge pressure of 3" of water? *Note:* First find Q.

$$Q = \frac{Lb/hr}{Min/hr} \times \frac{Lb\ of\ fuel}{Lb\ of\ steam} \times \frac{Cu\ ft\ of\ air}{lb\ of\ fuel}$$

$$Q = \frac{36,000}{60} \times \frac{1}{12} \times \frac{200}{1}$$

$Q = 10,000$ cu ft/min

$$Air\ HP = \frac{QP}{6356}$$

$$Air\ HP = \frac{10,000 \times 3}{6356}$$

$Air\ HP =$ **4.72**

$$Drive\ HP = \frac{Air\ HP}{Efficiency}$$

$$Drive\ HP = \frac{4.72}{0.55}$$

$Drive\ HP =$ **8.58**

Inlet vanes on constant speed forced draft fans control inlet air to the furnace. A *balanced draft system* is a type of draft that occurs when draft pressure within the furnace is kept constant. The gases of combustion in a balanced draft system leave the boiler at a rate that maintains pressure in the furnace slightly below that of the atmosphere.

Draft Systems. Draft systems used vary depending on the fuel burned. Air entering the furnace of a boiler is either primary or secondary air and is controlled by the air-fuel ratio controller. Coal stokers that burn anthracite coal need underfire air but very little overfire air because of the small amount of volatile matter present in anthracite coal. Coal stokers that burn bituminous coal need underfire and over fire air because of the larger amount of volatile gases produced by the coal. Both forced and induced draft are used with stokers.

Pulverized coal systems use a combination of forced and induced draft. Air is heated as it passes through an air preheater that is placed at the end of the last pass of the boiler where the heat in the gases of combustion can warm it. The heated air increases combustion efficiency and is also used in the mill to dry the coal. **See Figure 4-11.**

The draft system used when burning fuel oil commonly includes the following:

• Natural draft using a relatively high chimney. The burner is commonly an air atomizing or steam atomizing unit.

• Forced draft with a high chimney. The burner is a rotary cup, pressure atomizing, or steam atomizing unit.

• Forced and induced draft with a low or high chimney (depending on the draft losses through accessories such as air preheaters and economizers). The burner is commonly a pressure atomizing unit.

The draft system used when burning gas is very similar to the draft system used for fuel oil furnaces. Gas burners can replace fuel oil burners and use the same draft systems with little modification.

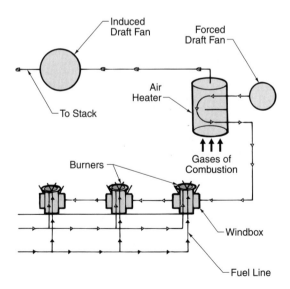

Figure 4-11. A natural gas and fuel oil draft system uses a forced draft fan to supply the fuel to the boiler and an induced draft fan to remove the gases of combustion.

Pressurized Furnace

Most plants use balanced draft in which gases of combustion leave the boiler at a rate that maintains pressure in the furnace slightly lower than atmospheric pressure. However, boilers with pressurized furnaces operate with pressure in the furnace slightly above atmospheric pressure. The primary purpose of using a pressurized furnace is to prevent infiltration of air into the boiler setting.

Note: A pressurized furnace will cause gases of combustion to enter the boiler room if there are leaks in the boiler setting. A pressurized furnace will also cause a fire to blowout any doors or ports that are opened to inspect the condition of the fires.

COMBUSTION ACCESSORIES

Combustion accessories are pieces of equipment required to store and deliver fuels to the burner safely and efficiently. In addition, fuels must be prepared in a manner to provide complete combustion to minimize environmental pollution. The type of combustion accessories required is determined by the fuel used. In addition, different fuels require different storage and handling procedures. All fuels can be dangerous if safety procedures are not followed.

Air Heaters

As fuel costs have risen, there has been a demand for higher plant efficiency. One large loss in steam boilers is the heat in the gases of combustion going up the chimney. To reduce heat loss, economizers and air preheaters (air heaters) are used. Air heaters are located between the boiler and the chimney. The two basic types of air heaters are convection (tubular or plate) and regenerative. **See Figure 4-12.** All air heaters are designed to retrieve some of the heat from the gases of combustion before the gases of combustion go up the chimney. Air heaters must be constructed and insulated to allow the temperature of the air to be controlled.

Plants using pulverizers require heated air in the mill to dry out the coal. The temperature of the heated air must be closely monitored to prevent problems from occurring in the mill. Usually a source of outside air can be blended automatically to maintain an air temperature of about 200°F to 210°F at the mill. Air heaters are most successfully used in plants having a fairly constant load.

The temperature of the gases of combustion entering the chimney must also be closely monitored. If the temperature gets too low, it will cause sweating in the outlet side of the air heater, which can lead to corrosion of the metal.

Air heaters are located in the direct path of the gases of combustion where they are continually bombarded with soot and flyash (fine ash particles). The surfaces of the air heater must be kept clean. Soot blowers are commonly used for cleaning air heaters. Air heaters must also be examined annually for signs of corrosion because tubes or plates that are worn allow incoming air and gases of combustion to mix. Bypass dampers are used in some convection air heaters to accurately control the temperature of the air for combustion and the temperature of the gases of combustion. The load on the boiler determines when the bypass damper must be opened or closed.

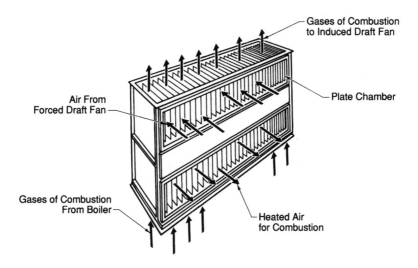

Convection Air Heater

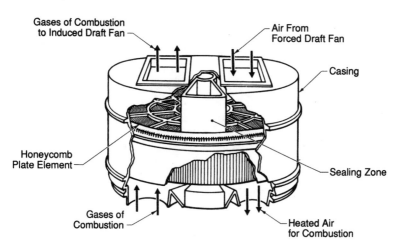

Regenerative Air Heater

Figure 4-12. Air heaters collect heat from gases of combustion and release heat to air for combustion to increase plant efficiency.

Air heaters are located between the boiler and chimney where they will not cause a restriction to the flow of the gases of combustion when they are kept clean. Air heaters that become plugged with soot cause a backpressure buildup in the boiler. This could result in gases of combustion leaking into the boiler room and/or dangerous pulsation of the boiler. As a preventive measure, design engineers specify using large induced draft fans in specific locations to maintain the flow of gases of combustion.

Coal Accessories

The location and size of the plant are factors that determine the amount of coal to be stockpiled. In large plants, a conveyor system transports coal from the stockpile to the storage bin, then to an overhead coal bunker. The coal is weighed from the overhead bunker and dumped into coal chutes directed to each boiler.

Stokers

Stokers are mechanical feeding devices for feeding coal to the burner. Stokers were developed because of the limitation of boiler size and inefficiency of hand firing. Stokers are classified by the method of coal feed. Stokers are designed to do the following:
• feed coal continually or intermittently
• provide a means of igniting the coal
• provide adequate air for combustion of fuel and a means of distributing gases of combustion over the heating surface and out to the chimney
• provide for ash removal

The three main types of stokers are the underfeed (ram or screw), spreader (sprinkler), and chain-grate (traveling grate) stoker.

Underfeed Stokers. Underfeed stokers feed coal in from the bottom. **See Figure 4-13.** Underfeed stokers can be screw fed or ram fed. Underfeed stokers are used in small, medium, and large boilers by using single retort, double retort, or multiretort stokers.

Many types of underfeed stokers follow the same basic principle. Coal is introduced into a hopper and delivered into the front of a retort where pusher blocks distribute the coal back and up. The coal is delivered by either a piston-type ram or a screw. The lateral movement of the moving grate bars slowly moves the ash toward the dump grates. Underfeed stokers can burn anthracite and bituminous coal. Coal with a high volatile content requires overfire air to control smoke.

Spreader Stokers. Spreader stokers (sprinkler stokers) introduce coal to the furnace above the grates. Some of the coal is burned in suspension. The remainder of the coal falls to the grates where the combustion process is completed. **See Figure 4-14.**

Spreader stokers are designed for firing small-size coal. Coal is sprinkled or showered on the grates. Fine coal particles burn in suspension, while large coal particles fall to the grates. This prevents the grates from being covered with green coal to keep the firebed hot. Spreader stokers are used in plants with less than 10,000 sq ft of boiler heating surface.

The main parts of a spreader stoker are the coal hopper, which is a conveyor type of feeder, distributor, and drive unit. The coal hopper supplies coal to the conveyor feeder, then to the distributor, which sprinkles it out to be burned in suspension or on the grates. Air for combustion is supplied both from under the grates and over the fire to complete combustion. Spreader stokers use bituminous coal and operate with a very thin fuel bed. The rate of combustion is controlled by the speed of the drive unit on the feeder and the amount of underfire and overfire air.

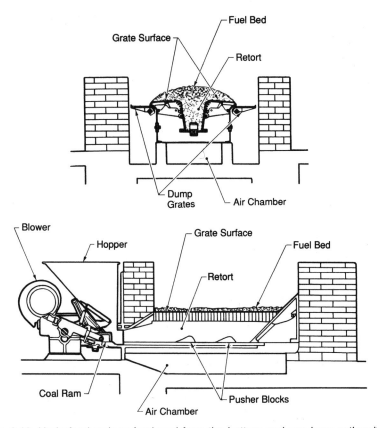

Figure 4-13. Underfeed stokers feed coal from the bottom and can burn anthracite and bituminous coal.

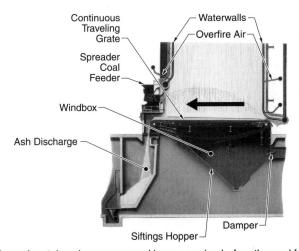

Figure 4-14. Spreader stokers burn some coal in suspension before the coal falls to the grate.

Chain-grate Stokers. Chain-grate stokers use a moving grate to transport coal to the furnace. **See Figure 4-15.** Chain-grate stokers can burn high volatile, high ash, noncaking coal with a grate surface area of 25 sq ft to 600 sq ft. Chain-grate stoker burning rates vary from 20 lb to 35 lb of anthracite coal per sq ft/hr and 30 lb to 50 lb of bituminous coal per sq ft/hr when using forced draft.

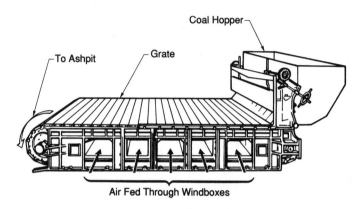

Figure 4-15. Coal is fed on a moving grate in a chain-grate stoker. Coal consumption is expressed in cu ft/hr.

Chain-grate stokers use a chain that runs around two sprockets, one of which is attached to a variable-speed drive. Coal is deposited on the front of the grate and travels past an ignition arch that starts the combustion process. Air is fed through windboxes under the grates in sufficient quantities to complete combustion of the coal before the end of the grate is reached. Ash with a minimal amount of unburned fuel falls off the grate into the ashpit. The amount of coal burned depends on the thickness of the fuel bed, rate of grate travel, and amount of air supplied. Coals best suited for chain-grate stokers are noncaking coals with a relatively high ash content.

Coal consumption on chain-grate stokers can be determined if the following is known:
• width of stoker in feet
• depth of coal bed in feet
• speed of stoker in feet per minute (fpm)

The number of cubic feet of coal used per minute is found by using the following equation:

Cubic feet burned per minute = Width × Depth × Speed

Cu ft/min = W × D × S

Cubic feet per minute is converted to pounds per hour by using the following conversions:
• 48 lb/cu ft for bituminous coal
• 60 lb/cu ft for anthracite coal

The number of pounds of bituminous coal burned per hour is found by using the following equation:

Pounds per hour = Cubic feet per minute × 60 minutes per hour ×
48 pounds per cubic foot

Lb/hr = Cu ft/min × 60 min/hr × 48 lb/cu ft

Note: The number of pounds of anthracite coal burned per hour is found by using the following equation:

Lb/hr = Cu ft/min × 60 min/hr × 60 lb/cu ft

Cubic feet per hour is converted to tons per hour by using the following equation:

$$Tons\ per\ hour = \frac{Pounds\ per\ hour}{2000}$$

$$Tons/hr = \frac{Lb/hr}{2000}$$

EXAMPLES

1. A chain-grate stoker 6′ wide is burning bituminous coal. The grate speed is 4″ per minute and the coal is 4″ deep. At what rate is coal being used, in tons per hour?

Cu ft/min = W × *D* × *S*

$$Cu\ ft/min = 6 \times \frac{4}{12} \times \frac{4}{12}$$

Cu ft/min = 0.667

Lb/hr = Cu ft/min × 60 min/hr × 48 lb/cu ft

Lb/hr = 0.667 × 60 × 48

Lb/hr = 1921

$$Tons/hr = \frac{Lb/hr}{2000}$$

$$Tons/hr = \frac{1920}{2000}$$

Tons/hr = **0.96**

2. A chain-grate stoker 10′ wide is burning anthracite coal. The grate speed is 6″ per minute and the coal is 6″ deep. At what rate is coal being used, in tons per hour?

$Cu\ ft/min = W \times D \times S$

$Cu\ ft/min = 10 \times \dfrac{6}{12} \times \dfrac{6}{12}$

$Cu\ ft/min = 2.5$

$Lb/hr = Cu\ ft/min \times 60\ min/hr \times 60\ lb/cu\ ft$

$Lb/hr = 2.5 \times 60 \times 60$

$Lb/hr = 9000$

$Tons/hr = \dfrac{Lb/hr}{2000}$

$Tons/hr = \dfrac{9000}{2000}$

$Tons/hr = \mathbf{4.5}$

Pulverizers

Pulverizers are machines used to grind and pulverize coal so that it is suitable for combustion. Pulverized coal was used as early as 1824. In 1890, Rudolf Diesel tried to use pulverized coal in his internal combustion engine. During this time, the cement industry had achieved some success with pulverized coal in cement kilns and had experience in grinding and pulverizing the materials used in the manufacture of cement. Thomas A. Edison made additional improvements that increased the output and efficiency of cement kilns using pulverized coal. However, it was not until after World War I that great strides were made when pulverized coal was used for power-generating stations.

Pulverizers are classified by the speed at which they operate and the method used to pulverize the coal. Methods used in the pulverizing process include crushing, impact, and attrition (grinding). Pulverizers may use one, two, or all three of these methods according to their design. Classification of pulverizers include the following:

- High-speed (over 300 RPM) impact pulverizer: Coal is pulverized by the impact of hammers on large pieces of coal, attrition of smaller pieces on each other, and the grinding surface of the mill.

- Medium-speed (between 70 RPM and 300 RPM) ball and race: Coal is pulverized by crushing and attrition with some impact. The pulverizing action takes place between two surfaces, one rolling over the other. **See Figure 4-16.**

- Low-speed (under 70 RPM) ball tube mill: Coal is pulverized by the impact of falling balls on the coal. Crushing takes place as balls roll over each other and the liner, as the coal slides over other coal, and as coal comes in contact with the liner.

Attrition

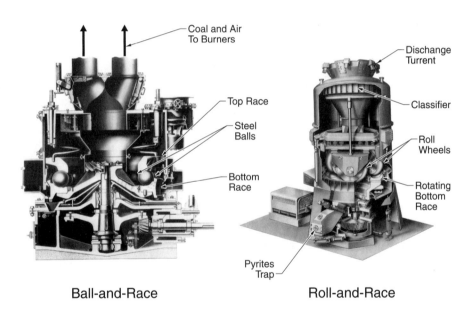

Ball-and-Race   Roll-and-Race

Figure 4-16. Pulverizers grind coal into a very fine powder.

When selecting coal for pulverizing, the following characteristics must be considered:
• grindability
• moisture content
• rank
• volatile matter
• ash content

Pulverized coal systems are the bin storage system and direct-fired unit system. In the *bin storage system,* the coal is prepared in a separate space and stored in a coal bunker. The coal is first crushed and stored in a raw coal bunker. It passes from the bunker, through a dryer, to the pulverizer, then to the pulverized coal bunker. The coal feeder then supplies the burner with the necessary amount of pulverized coal to maintain the plant load. This system requires more equipment and personnel to operate and has mostly been replaced by direct-fired unit systems.

Direct-fired unit systems use an overhead coal bunker where coal is stored. A belt conveyor at the base of the bunker feeds coal to the coal scale. **See Figure 4-17.** The belt conveyor is controlled by the position of the scale. When the scale is in an up position, the conveyor runs until the scale is filled with the proper amount of coal. When the scale moves down to dump, the conveyor stops. The coal moves from the coal scale to the feeder, which controls the amount of coal that is fed into the pulverizer. The feeder commonly uses a rotating blade, which by RPM, controls the amount of coal delivered to the pulverizer.

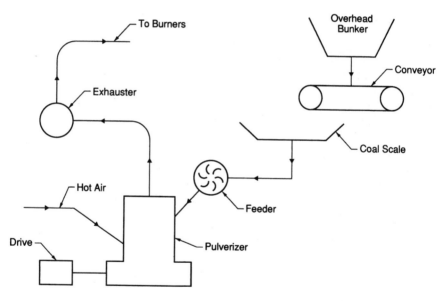

Figure 4-17. In a direct-fired unit system, bituminous coal is fed from an overhead bunker.

Once the coal is in the pulverizer, the process of grinding it to the consistency of talcum powder begins. This is accomplished by steel balls or rollers rotating around a bowl. The coal is crushed as it passes between the surfaces. Hot air is introduced into the pulverizer mill to dry the coal and prevent caking. The air mixes with the powdered coal and passes through the exhauster on its way to the burner throat. The air that is introduced into the mill is primary air. Air passing through the windbox is secondary air.

The coal is burned in suspension. It requires a relatively high furnace temperature to maintain good combustion, which is approximately 3000°F. Pulverized coal is highly explosive and care must be taken when burning it. A flame in the furnace and also a negative pressure must be maintained at all times. If the pressure in the furnace becomes positive, the flame may travel back to the mill, which would cause an explosion in the mill.

Burning pulverized coal requires special considerations:

- Operators must be trained in the handling of pulverized coal because pulverized coal is highly explosive.

- Large furnace volumes are necessary so that complete combustion takes place before any gases of combustion come in contact with lower boiler tubes. This could result in slagging of tubes and smoke.

- Since flame temperatures are higher in furnaces burning pulverized coal, waterwalls are used. Waterwalls help cut down on refractory maintenance and allow for higher heat release rates. Using waterwalls will, to some extent, allow reduction of the furnace volume.

- Ash, flyash, and slag present problems. In small pulverizers, ash may be removed in a dry state. A stationary engineer must monitor furnace temperatures to prevent formation of slag. If the furnace temperature is too high, the ash will reach the point of fusion and slagging will result. In large pulverizers with high furnace temperatures, the floor of the furnace is designed so that slag can be directed (tapped) to flow from the furnace in a liquid state into a water bath where it solidifies. The furnace can be designed for intermittent or continuous slag tapping.

- Because of the danger involved in furnace explosions when burning pulverized coal, explosion doors are placed in strategic locations in the furnace and in the path that the gases of combustion must pass through on the way to the chimney. If sudden buildup of pressure results from a furnace explosion, explosion doors will blowout first to relieve the pressure and protect the boiler casing and brickwork from damage.

Advantages and disadvantages of using pulverized coal as a fuel include the following:

Advantages

- Because air is used to deliver pulverized coal to the burner, excess air requirements are kept to a minimum.

- Coal is burned in suspension, and variation in plant loads can be compensated for easily.

- Pulverized coal equipment is suitable for automatic combustion controls.

- Pulverized coal is burned in suspension and there are no losses resulting from banking.

Disadvantages

- Because of the explosive nature of pulverized coal, qualified operators trained in the handling of pulverized coal are required.

- The initial cost is high.

- There is a higher maintenance cost than for other types of stokers.

- Sophisticated equipment is needed to remove flyash from flue gases.
- A gas or fuel oil burner is needed to warm up the furnace before introducing pulverized coal.
- When there are large fluctuating loads, a secondary means of fuel is required to prevent loss of fire.

Flyash Precipitators. Flyash travels with flue gases and must be removed before being discharged into the atmosphere through the chimney. Small amounts of flyash are trapped in soot hoppers located before the air heaters. This helps keep the passage in the air heaters cleaner and helps prevent the clogging of air passages. The bulk of flyash is removed by mechanical flyash precipitators, electrostatic flyash precipitators, or combination mechanical-electrostatic flyash precipitators.

Mechanical flyash precipitators sharply change the direction of flow of the gases of combustion to cause flyash to drop out. The change of direction is accomplished by using baffles or cyclone separators, which cause gases of combustion to spin, imparting a centrifugal force to the flyash. Mechanical flyash precipitators are considered 70% to 85% efficient. However, because of their construction, mechanical flyash precipitators can cause a restriction to the flow of the gases of combustion. This draft loss must be compensated for with the proper induced draft fans.

Electrostatic flyash precipitators are initially more expensive but do not restrict the flow of the gases of combustion. There is little or no draft loss, and they are considered 90% to 95% efficient. Electrostatic flyash precipitators send high-voltage charges across the area that the gases of combustion must pass through. Flyash particles are collected on plates with the flyash removed to suitable hoppers for proper disposal. In some instances, combination mechanical-electrostatic precipitators are used. **See Figure 4-18.**

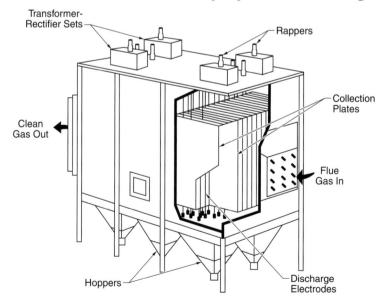

Figure 4-18. Flyash precipitators are air cleaning systems that are used to remove flyash from flue gases.

Fuel Oil Burners

Fuel oil burners deliver fuel oil to the furnace in a fine spray that is mixed with air for efficient combustion. The types of fuel oil burners commonly used in industrial plants include pressure atomizing, steam atomizing, air atomizing, and rotary cup burners. Atomization is accomplished by using air under pressure, forcing the fuel oil through a nozzle assembly. Breaking the fuel oil up into a fine spray allows the air to come in closer contact with the fuel oil, which results in efficient combustion.

A *pressure atomizing burner* is a fuel oil burner used to atomize fuel oil without using steam or air. Fuel oil is pumped to the burner at a pressure above 100 psi. A nozzle consisting of tangential slots, a whirl chamber, and an orifice sprays a fine mist of fuel oil into the combustion chamber. The two types of pressure atomizing burner tips are the plug-and-tip and the sprayer plate types. **See Figure 4-19.**

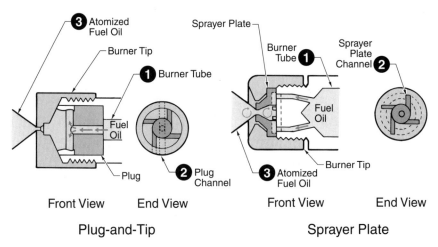

Front View   End View   Front View   End View

Plug-and-Tip   Sprayer Plate

Figure 4-19. A pressure atomizing burner requires high temperatures and pressures to atomizes the fuel oil.

Pressure atomizing burners used in small installations consist of a motor-driven fuel oil pump, strainer, pump, pressure regulating valve, nozzle, fan and housing, and an air diffuser. The duplex strainer is located in the fuel oil line before the fuel oil pump to protect it from foreign particles and sediment in the fuel oil. The fuel oil pump is a rotary positive displacement pump that supplies fuel oil at a minimum pressure of 100 psi to the burner nozzle. The pressure regulating valve consists of a spring-loaded diaphragm, bellows, or a piston that controls the flow of fuel oil to the nozzle and diverts the excess fuel oil back to the fuel oil tank. The nozzle atomizes the fuel oil. The fan is a squirrel-cage type coupled to a motor shaft that supplies air needed for combustion. The air diffuser controls the air pattern of a pressure atomizing burner after the fuel oil has been atomized.

In large installations, the fan supplying air for combustion may supply air to more than one burner and/or boiler through a system of wind boxes. One fuel oil pump may also supply fuel oil to more than one burner and/or boiler. In addition, ignition is commonly automatic using gas pilots for burners in large installations.

The amount of fuel oil delivered by the pressure atomizing burner is controlled by fuel oil pressure, tip orifice diameter, and plug channel dimensions. Pressure atomizing burners include the following:

• an atomizer that delivers fuel oil to the furnace in a fine mist

• an air register that admits air to the furnace

• valves and fittings that connect the atomizer to the fuel oil lines

A *steam atomizing burner* is a fuel oil burner that uses steam at a pressure approximately 20 psi higher than the fuel oil pressure to atomize fuel oil. The two basic types of steam atomizing burners are the outside-mixing and inside-mixing burners. **See Figure 4-20.** In the outside-mixing burner, fuel oil and steam come in contact outside the burner. In the inside-mixing burner, fuel oil and steam come in contact inside the burner.

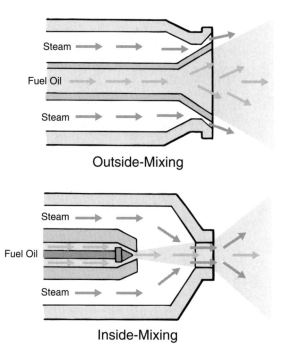

Outside-Mixing

Inside-Mixing

Figure 4-20. Steam is used to atomize fuel oil and can mix with the fuel oil outside or inside the burner.

The inside-mixing burner is the most common steam atomizing burner used. No. 6 fuel oil must still be heated, but it is heated to a much lower temperature (approximately 120°F to 150°F). The fuel oil is supplied at a pressure of approximately 50 psi, and the steam pressure is approximately 20 lb higher. Live steam is used to atomize the fuel oil, and the steam consumption varies with operating conditions. Under ideal conditions, 1% to 3% of the total steam is used for atomization.

In plants using steam atomizing burners, pressure regulators are used on both the fuel oil and steam lines to the burner. The pressure differential between the fuel oil pressure and steam pressure is approximately 20 lb, with steam having the higher pressure. This spread will be maintained automatically over a fair range as the load increases or decreases.

The steam atomizing burner is commonly used in small boilers. It can burn a poorer grade of fuel with a shorter fire. Initially, it costs less than a rotary cup burner, but it costs more to operate because of the need for live steam to atomize the oil. It is easy to change the flame pattern to accommodate the furnace. In addition, the steam atomizing burner allows easy adjustment for efficient air-fuel ratios.

Air atomizing burners function in a similar manner to steam atomizing burners. *Air atomizing burners* are inside-mixing burners and use compressed air from an air compressor for atomization. Air atomizing burners must control the following:

• atomizing of fuel oil

• flow of fuel oil

• ignition of air-fuel mixture

• correct air-fuel mixture

• correct amount of air for atomizing fuel oil

The burner controls the flow of fuel oil to the nozzle assembly by the use of a fuel oil solenoid valve. Ignition of the air-fuel mixture is accomplished by a high-voltage electric spark igniting a gas pilot. The gas pilot is programmed in the burner cycle to ignite the main burner flame at the proper time. Secondary air is allowed to mix with the burning fuel by using an air shutter or damper. The amount of secondary air depends on the firing rate of the burner. Atomizing air at the correct pressure is supplied by an air compressor assembly.

A *rotary cup burner* is a fuel oil burner consisting of an atomizing or a spinning cup, a blower to supply air for combustion (primary air), an air nozzle to mix the air and fuel oil, a means of driving the spinning cup and blower, and a means for delivering fuel oil to the spinning cup. **See Figure 4-21.** The spinning cup rotates at 3450 RPM or higher. Fuel oil spreads over the inner surface of the spinning cup and is eventually thrown off by centrifugal force. Air supplied by the blower is directed into the fuel oil thrown from the spinning cup to atomize the fuel oil. Primary air supplied by the burner fan provides only 15% of the air required for combustion. Secondary air is required to complete combustion.

An *air flow switch* is a safety device used with rotary cup burners. It is located at the burner fan housing and proves (verifies) primary air pressure. If the primary pressure is not proven, the air flow switch prevents the firing sequence from starting. With a failure of primary air pressure, fuel oil to the burner is secured while the burner motor continues to run, purging the furnace. Rotary cup burners, once used only in low-pressure plants, are now used in high-pressure package boilers for large buildings and process plants. In addition, rotary cup burners are ideal for automatic operation.

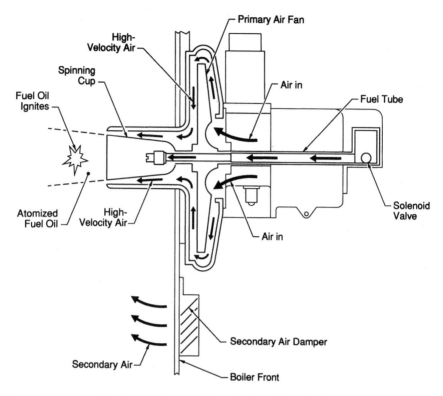

Figure 4-21. Fuel oil is atomized by centrifugal force of the spinning cup and primary air in a rotary cup burner.

Fuel Oil Tanks and Lines

Fuel oil tanks and lines store and transport fuel oil safely and efficiently for use in the plant. A stationary engineer must know the location of all fuel oil lines, valves, pumps, regulators, and crossover connections. No two plants are alike. However, each plant has the same basic equipment with varying locations. A stationary engineer must have a flow chart including lines, valves, regulators, strainers, heaters, and pumps. If there is a problem in supplying fuel oil to the burner, a flow chart will help a stationary engineer take decisive action. If the boiler goes down, all production goes down. If the plant is generating electricity and the steam turbine goes down, the entire plant is dead. This will not happen if a stationary engineer is prepared.

Some plants require more than one fuel oil tank for plant flexibility. Fuel oil from one fuel oil tank can be used while the other fuel oil tank is down for cleaning or repairs. Process plants cannot afford plant shutdowns for tank maintenance or repair.

The size and number of fuel oil tanks are determined by plant needs and local regulations. In some locations, fuel oil tanks must be installed underground with a limitation on the number of gallons allowed. **See Figure 4-22.** In other locations, fuel oil tanks must be installed aboveground with a limitation on the number of gallons allowed.

Figure 4-22. Fuel oil tanks installed underground provide fuel oil storage without occupying land space.

EPA Regulations

The U.S. Environmental Protection Agency (EPA) estimates that there are about 704,717 federally regulated underground storage tanks (UST) buried at over 269,000 sites nationwide (as of September 30, 2001). A UST system is a tank and any underground piping connected to it in which at least 10% of their combined volume is underground. Federal UST regulations apply only to underground tanks and piping that store petroleum or certain hazardous substances. The following USTs do not need to meet federal requirements:

• Farm and residential tanks of 1100 gallons or less capacity holding motor fuel used for noncommercial purposes

• Tanks storing heating oil used on the premises where it is stored

• Tanks on or above the floor of underground areas, such as basements or tunnels

• Septic tanks and systems for collecting storm water and wastewater

• Flow-through process tanks

• Tanks of 110 gallons or less capacity

• Emergency spill and overfill tanks

In 1988, the EPA issued UST regulations divided into three sections: technical requirements, financial responsibility requirements, and state program approval objectives. The technical regulations are designed to reduce the probability of releases from USTs, improve detection of leaks and spills, and secure prompt cleanup. The financial responsibility regulations are designed to ensure that UST owners and operators, in the event of a leak or spill, will have the monetary resources to pay for cleanup. Subtitle I of the Resource Conservation and Recovery Act (RCRA) allows state UST programs approved by the EPA to operate in lieu of the federal program, and the EPA's state program approval regulations set standards for state programs to meet.

These regulations specify that all newly installed USTs require monthly leak detection monitoring. The leak detection system must be capable of sensing a leak as small as 0.2 gallons per hour, with a 95% detection rate and a 5% false alarm rate.

Leak detection equipment for newly installed USTs is divided into three categories: internal monitoring, interstitial monitoring, and external monitoring. Internal monitoring uses a sensor located within the tank to detect changes in fluid volume. Interstitial monitoring uses a sensor located between the walls of a double-wall tank to identify the presence of water or hydrocarbons. External monitoring uses sensors to identify the presence of vapor or hydrocarbons in the excavation surrounding the tank.

Interstitial monitoring is required on all new USTs for fuel oil. Sensors are located in a well connected to the bottom of the double wall of the fuel oil tank. The sensors detect water or hydrocarbons if a leak occurs. A control panel with visual indicators is installed in the boiler room to alert a stationary engineer of any fuel oil tank malfunction. **See Figure 4-23.**

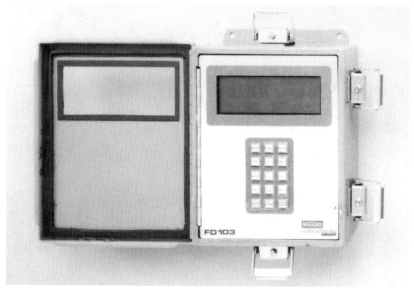

Pollulert Systems/Mallory Components Group

Figure 4-23. A leak detection system monitors fuel oil leakage and indicates the status on a control panel in the boiler room.

All new USTs have spill, overfill, and corrosion protection. Spill protection requires a spill bucket to catch spills at the fill pipe when the delivery truck hose is disconnected. Overfill protection requires equipment to restrict or stop the flow of fuel during delivery before the tank reaches full capacity such as an automatic shutoff device, overfill alarm, or ball float valve. Corrosion protection requires the addition of equipment on steel USTs subject to corrosion to the tank and piping. This equipment includes cathodic protection and tank interior lining. Cathodic protection equipment uses the flow of electrical current to protect against corrosion.

Fuel Oil Consumption

Fuel oil consumption by the plant is recorded as part of standard boiler operation procedures. A stationary engineer is responsible for maintaining boiler room records and must know how to calculate fuel oil consumption per hour, day, week, or year based on readings taken.

Fuel oil tanks may be equipped with a fuel gauge, which indicates the gallons of fuel oil present in the tank. If a fuel gauge is not installed or is faulty, a stationary engineer must be able to sound (measure the fuel oil) the tank and calculate the quantity of fuel oil from a chart. When the tank is sounded, the reading is in feet and inches. **See Figure 4-24.** One inch of fuel oil does not contain the same number of gallons when measured at different levels in the tank.

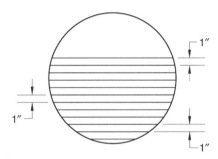

Figure 4-24. The quantity of fuel oil measured in the fuel oil tank is dependent on the level of fuel oil present.

EXAMPLES

1. A fuel oil tank is stick sounded at 5'-6" on Monday. The same fuel oil tank is stick sounded at 5'-2½" on Tuesday. How much fuel oil is in the fuel oil tank on Monday? How much fuel oil is in the fuel oil tank on Tuesday? How much fuel oil is consumed? A tank chart is normally provided by the tank manufacturer.

Monday:

Change feet to inches:

5 × 12 = 60"

60 + 6 = 66"

Using the tank chart, 66" = 11,322 gal.

Fuel oil in tank Monday = **11,322 gal.**

Tuesday:

Change feet to inches:

5 × 12 = 60"

60 + 2½ = 62½"

Note: There is no reading on the tank chart for ½" of fuel oil; therefore, interpolation is needed.

Gal. at 63" = 10,681

Gal. at 62" = 10,468

10,681 – 10,468 = 213 gal.

½ of 213 gal. = 106.5 gal.

Gal. at 62" = 10,468

Gal. at ½" = 106.5

10,468 gal. at 62" + 106.5 gal. at 62½" = 10,574.5

Fuel oil in tank Tuesday = **10,574.5 gal.**

Fuel oil consumed:

Fuel oil consumed = 11,322 – 10,574.5

Fuel oil consumed = **747.5 gal.**

2. A fuel oil tank is stick sounded at 7'-11" on Wednesday. The same fuel oil tank is stick sounded at 7'-4½" on Thursday. How much fuel oil is in the fuel oil tank on Wednesday? How much fuel oil is in the fuel oil tank on Thursday? How much fuel oil is consumed?

Wednesday:

Change feet to inches:

7 × 12 = 84"

84 + 11 = 95"

Using the tank chart, 95" = 17,077 gal.

Fuel oil in tank Wednesday = **17,077 gal.**

Thursday:

Change feet to inches:

7 × 12 = 84"

84 + 4½ = 88½"

Gal. at 89" = 15,997

Gal. at 88" = 15,810

15,997 – 15,810 = 187 gal.

½ of 187 gal. = 93.5 gal.

Gal. at 88" = 15,810

Gal. at 88½" = 15,810 + 93.5

Fuel oil in tank Thursday = **15,903.5 gal.**

Fuel all consumed:

Fuel oil consumed = 17,077 – 15,903.5

Fuel oil consumed = **1173.5 gal.**

Fuel Oil Correction

Fuel oil correction is the allowance for expansion of fuel oil when it is heated, which affects the volume of fuel oil. No. 6 fuel oil must be heated for pumping and burning efficiently. The volume of fuel oil is based on the temperature of the fuel oil. This presented a problem to a stationary engineer because the amount of fuel oil delivered could vary with temperature. A standard was accepted, establishing that regardless of the fuel oil temperature at delivery, it would be converted to 60°F to determine the number of gallons billed.

The American Petroleum Institute has published a standard, API MPMS, *Manual of Petroleum Measurement Standards*, that gives methods to correct for volume changes caused by temperature. This standard takes the place of a previous publication from the National Bureau of Standards. The methods included in this standard can be used for all hydrocarbons. Specifically for No. 6 fuel oil, a correction factor of 0.000345 is used in the following equation:

Corrected gallons =
Gallons delivered[1.0000 – 0.000345(*Temperature* – 60°F)]

The corrected number of gallons is determined using the following information:

Step 1: *Temperature of fuel oil delivered* – 60°F
Step 2: Answer from Step 1 × 0.000345
Step 3: 1.0000 – Answer from Step 2
Step 4: Answer from Step 3 × *Gallons of fuel oil delivered* = *Corrected gallons*

EXAMPLES

1. A steam plant received 5415 gal. of fuel oil at 130°F. How many gallons are there when corrected to 60°F?

   Step 1: *Temp. of fuel oil delivered* – 60°F

   130 – 60 = 70

   Step 2: Answer from Step 1 × 0.000345

   70 × 0.000345 = 0.024150

   Step 3: 1.0000 – Answer from Step 2

   1.0000 – 0.02415 = 0.9758

   Step 4: Answer from Step 3 × *Gal. of fuel oil delivered*

   Corrected gal. = 0.9758 × 5415

   Corrected gal. = **5284 gal. at 60°F**

2. A steam plant received 5247 gal. of fuel oil at 138°F. How many gallons are there when corrected to 60°F?

Step 1: *Temp. of fuel oil delivered* – 60°F

138 – 60 = 78

Step 2: Answer from Step 1 × 0.000345

78 × 0.000345 = 0.02691

Step 3: 1.0000 – Answer from Step 2

1.0000 – 0.02691 = 0.9731

Step 4: Answer from Step 3 × *Gal. of fuel oil delivered*

Corrected gal. = 0.9731 × 5247

Corrected gal. = **5106 gal. at 60°F**

A table can be prepared to make it easier to calculate the fuel oil correction. **See Figure 4-25.** This is a partial listing of a fuel oil correction table.

NO. 6 FUEL OIL CORRECTION	
Temperature	Correction Factor
115	0.9809
116	0.9806
117	0.9802
118	0.9799
119	0.9795
120	0.9792
121	0.9789
122	0.9785
123	0.9782
124	0.9779
125	0.9775
126	0.9772
127	0.9768
128	0.9765
129	0.9762
130	0.9758
131	0.9755
132	0.9751
133	0.9748
134	0.9745
135	0.9741
136	0.9738
137	0.9735
138	0.9731
139	0.9728

Figure 4-25. A correction factor is required when converting fuel oil readings to the standard of 60°F.

Fuel Oil Strainers and Pressure Gauges

A *fuel oil strainer* is a strainer used to remove foreign matter in a fuel oil system before the fuel oil reaches the burner assembly. Duplex fuel oil strainers (strainers that contain two screens) should be located on the suction line before the fuel oil pump and sometimes on the discharge line before the burner. The strainer on the suction line has a coarse mesh screen. Most often, the strainer on the discharge line is a single strainer. The strainer on the discharge line before the burner has a fine mesh screen. The burner can operate while a strainer is being cleaned because only one strainer is cleaned at a time. Fuel oil strainers should be cleaned at least once every 24 hours. When cleaning the strainers, care must be taken to prevent air from entering the system on the suction side or fuel oil leakage on the discharge side. **See Figure 4-26.**

A *fuel oil pressure gauge* is a pressure gauge used to indicate when a strainer is dirty. The pressure gauge on the suction line shows an increase in vacuum when the suction strainer is dirty. The pressure gauge at the burner shows a decrease in pressure when the discharge strainer is dirty.

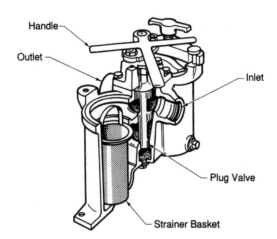

Figure 4-26. Duplex fuel oil strainers allow cleaning of one strainer without removing the burner from service.

Fuel Oil Pumps

A *fuel oil pump* delivers fuel oil from storage tanks aboveground or underground to the burners at a controlled pressure. Fuel oil pumps are positive displacement pumps. Fuel oil pumps always have a positive pressure on the discharge side of the fuel oil pump. Fuel oil pumps can be reciprocating, gear, screw, or lobe types. Gear fuel oil pumps are most commonly used for pumping fuel oil.

Gear Fuel Oil Pumps. Gear fuel oil pumps are a type of rotary pump and provide a constant discharge pressure like a centrifugal pump. However, gear fuel oil pumps are positive displacement pumps and must have a relief valve between the pump and the pump discharge valve. Every rotation of the pump gear produces a definite quantity of fuel oil to be discharged. **See Figure 4-27.**

The following are advantages of gear fuel oil pumps:
• They have few moving parts.
• There is a low initial cost.
• They are small in size.
• There is constant pressure.
• They are self-priming.
• They have a high suction lift.
• They can handle high-viscosity liquids.

The following are disadvantages of gear fuel oil pumps:
• They are slow.
• They are noisy.
• Their efficiency drops quickly when the pump gears are worn.
• Low-viscosity liquids are difficult to pump at high pressures.

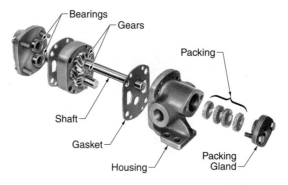

Worthington Pump

Figure 4-27. A gear-type fuel oil pump is used in small- to moderate-sized boiler plants.

Fuel Oil Heaters

Fuel oil heaters are required for pumping and burning No. 6 fuel oil and No. 4 and No. 5 fuel oil in some cases. Fuel oil in the tank is kept warm by heating coils, a heating bell, or electric pipeline heaters. When fuel oil leaves the fuel oil pump, it passes through steam or hot water fuel oil heaters, then to electric fuel oil heaters at the burners that bring the fuel oil to its firing point. The temperature of fuel oil in storage tanks must be carefully monitored. If the temperature drops below the pour point of the fuel oil, it will solidify and become impossible to pump. If the fuel oil gets too hot, impurities will start to settle out in the fuel oil tank. As the fuel oil reaches its flash point, a fire could result.

The heating coil or heating bell can use either steam or hot water as a source of heat. Regardless of which is used, the condensate returns coming from the fuel oil tank heater must be either dumped to waste or carefully monitored to prevent contamination of boiler water with fuel oil.

Electric pipeline heaters are connected to the fuel oil suction line and fuel oil discharge line to warm the fuel oil in the fuel oil tank. Electric pipeline fuel oil heaters use low voltages and high amperages to heat fuel oil. **See Figure 4-28.** Using electric pipeline fuel oil heaters eliminates the danger of contaminating feedwater.

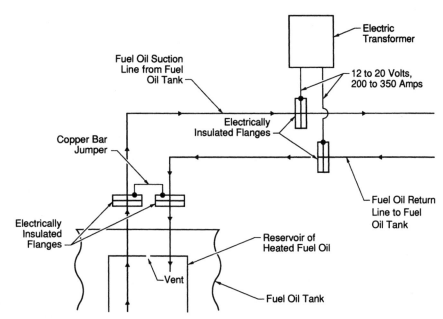

Figure 4-28. Electric pipeline heaters are commonly used to heat No. 6 fuel oil and eliminate the possibility of contaminating the feedwater.

Steam fuel oil heaters located on the discharge side of the fuel oil pump are usually the shell-and-coil or shell-and-tube type heater. In either fuel oil heater, the fuel oil passes through the tubes that are surrounded by either steam or hot water. Both fuel oil heaters are protected from excess fuel oil pressure by a relief valve. The relief valve is located on the fuel oil line leaving the heater, as close to the heater as practical. The discharge from the relief valve is piped to the fuel oil return line to the tank. When fuel oil heaters are removed from service for cleaning or repairs, steam or hot water to the fuel oil heaters must be secured, the fuel oil inlet and outlet valves must be closed and tagged out, and any fuel oil in the heater must be drained to prevent pressure buildup in the fuel oil heater.

Note: Many large plants do not have electric fuel oil heaters at the burners. The fuel oil temperature is raised to its proper firing temperature leaving the steam or hot water fuel oil heaters. The fuel oil lines going to the burners are carefully insulated to prevent the fuel oil temperature from decreasing.

Natural Gas Burners

Natural gas burners have cleaner stack emissions, higher combustion efficiencies, and fewer accessories than coal or fuel oil, and have no storage requirements. Natural gas is highly explosive and toxic and requires careful handling. All gas lines should be color coded for quick identification in leak detection procedures.

The gas pressure available is an important factor in choosing the type of gas burner to be used in a plant. The two gas burners used are low-pressure and high-pressure gas burners. In low-pressure gas burners, gas is supplied to the burner reduced to 0 lb pressure. The gas and primary air mix together outside the combustion chamber and are forced along to the nozzle by a blower. The gas line is fitted with a gas cock, which allows a stationary engineer to shut off the gas supply to the system when making repairs. The solenoid valve controls gas to the pilot. A manual reset valve is an electric valve that cannot be opened until the gas pilot is lighted. The pressure-reducing governor reduces the pressure of the gas to 0 lb pressure. The small line just before the pressure-reducing governor goes to the vaporstat. The vaporstat is a switch that is turned on by gas pressure in the line or turned off when there is no pressure. The main gas solenoid valve opens at the proper time, allowing gas to be drawn down to the mixjecter. The forced draft blower sends air through the butterfly valve. Air passing through the venturi draws gas with it to the mixing chamber.

In a high-pressure gas system, gas is supplied through a high-pressure gas main at a higher pressure than a residential gas supply line. **See Figure 4-29.** Gas is routed from the supplier through a pressure regulating valve and utility meter. The plant pressure-regulating valve reduces gas pressure to line pressure used in the system. The main gas shutoff cock isolates the gas supply from the gas burner system during service. Gas is directed to a separate pilot gas line.

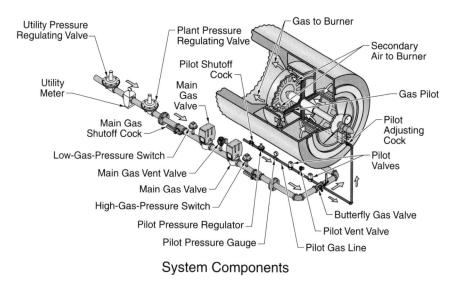

System Components

Figure 4-29. High pressure gas systems use high gas pressures in the system from the main supply to the burner.

The pilot gas line has a pilot shutoff cock that shuts off flow in the pilot gas line. The pilot pressure regulator reduces gas pressure for use in the pilot flame. A pilot pressure gauge indicates the pressure in the line. The pilot valve opens during the ignition period to admit gas to the pilot. The pilot valve closes after the main flame is established. The pilot vent valve vents gas to the atmosphere if gas is present in the pilot line when the pilot valves are closed. The pilot adjusting cock regulates the size of the pilot flame.

Main gas valves open simultaneously to admit gas to the burner. The low-gas-pressure switch is a manual reset pressure-actuated switch that is closed whenever main gas line pressure is above a preselected pressure. The low-gas-pressure switch opens when the gas pressure drops below a preselected minimum requirement and de-energizes the main gas valves, which close the normally closed main gas valves, and the main gas vent valve, which opens the normally open main gas vent valve. The low-gas-pressure switch is located before the main gas valves.

A main gas vent valve vents gas to the atmosphere if gas is present when the main gas valves are de-energized. The high-gas-pressure switch is normally closed whenever gas line pressure is below a preselected pressure. It opens when gas pressure rises above a preselected maximum requirement and de-energizes the main gas valves and the vent valve. The high-gas-pressure switch is a manually reset type of switch. The high-gas-pressure switch is located after the main gas valves but before the butterfly gas valve. A second main gas shutoff cock shuts off gas supplied during an emergency or service. The butterfly gas valve is actuated by the connecting linkage from the gas modulating cam. The butterfly gas valve and air damper in the burner are controlled by the modulating motor that receives varying signals from the modulating pressure control.

Combination Gas-Fuel Oil Burners

Combination gas-fuel oil burners have an advantage over other burners because they can burn gas or fuel oil. The advantage of burning the least expensive fuel is an important consideration. However, it may be necessary to burn only gas to comply with air pollution regulations.

A combination gas-fuel oil burner is basically the same as having a gas and fuel oil burner system connected together at the front of the furnace. The parts of the system are the same as if they were separate. **See Figure 4-30.**

In some package boiler plants, changing from fuel oil to gas only requires opening the manual gas cock and turning the selector switch to gas. The program clock then puts the boiler through a firing cycle—purge, pilot, ignition—then the burner is controlled by the pressure control and the modulating pressure control.

COMBUSTION CONTROLS

Combustion controls regulate fuel supply, air supply, air-to-fuel ratio, and removal of gases of combustion to achieve optimum boiler efficiency. The three basic types of combustion controls are ON/OFF/modulating, positioning, and metering combustion controls. Modern boiler management systems use the latest communication methods, such as DeviceNet, Ethernet IP, and Internet technologies.

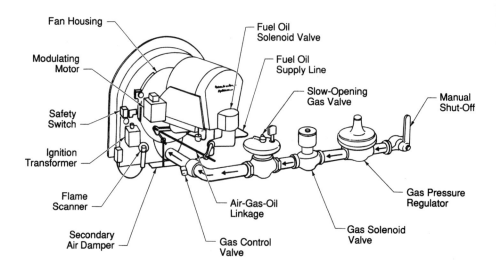

Combination Gas-Fuel Oil Burner

Figure 4-30. Combination gas-fuel oil burners allow switching of fuel for maximum economy or compliance with air pollution regulations.

ON/OFF/Modulating Combustion Controls

ON/OFF/modulating combustion controls are found on small package boilers that have pressure controls and program clocks. Larger package boilers also have modulating pressure controls.

- Pressure control controls the operating range of the boiler by starting and stopping the burner on steam pressure demand.
- Program clock controls the starting sequence of a burner.
- Modulating pressure control controls high and low fire. A burner should always start in low fire and shut down in low fire.

ON/OFF control is a combustion control strategy that is used to start and stop a burner without any modulation of the flame. ON/OFF control can be used in plants where the steam load is steady and allows the burner to be started and stopped. Simple ON/OFF control consists of an operating pressure control that signals a programmer to start and stop the burner and a high-pressure limit control with a manual reset.

High/low/OFF control is a combustion control strategy that senses the steam pressure and sends a control signal to start and stop the burner to maintain the steam pressure. High/low/OFF control is used in plants that have a variable steam demand large enough to require the burners to switch between high and low fire.

Modulating pressure control provides local control of the firing rate proportional to the steam demand.

When the load increases, more fuel must be supplied to the burner. This requires more air for complete combustion, and more gases of combustion must be released to the chimney. For example, a typical plant has a boiler with an operating range of 85 psi to 100 psi. When the pressure in the boiler drops to 85 psi, the program clock will operate, putting the burner through a firing cycle. The burner lights off in low fire. The modulating pressure control will start to bring the burner up to high fire by energizing the modulating motor through linkage.

The modulating motor is connected to the fuel oil valve, primary air damper, and secondary air damper. As the burner passes from low to high fire, the air-fuel ratio must change. When the boiler starts to pick up plant load and steam pressure increases, the modulating pressure control puts the burner back toward the low fire position. The burner modulates between high and low fire until the boiler reaches its cut-out pressure of 100 psi. The pressure control shuts the burner off, and the cycle repeats when the boiler calls for steam. The burner should always be on for longer periods than it is off. This keeps the furnace from uneven cooling, which could cause spalling of the brickwork. This is done by proper setting of the pressure control and modulating pressure control.

Solid-State ON/OFF/Modulating Combustion Controls. Solid-state ON/OFF/ Modulating combustion controls use solid-state circuitry and function in a similar fashion to a conventional ON/OFF/modulating combustion control. Solid-state ON/ OFF/modulating combustion controls provide for ignition and main flame failure protection on automatically fired gas or fuel oil-fired burners by controlling the burner blower motor, ignition, main fuel valves, and modulating motor. The programmer cycles automatically when the limit operating control (pressure control) closes, but must be reset following a safety shutdown.

The solid-state programmer initiates the same firing sequence as a conventional ON/OFF/modulating combustion control. With the power on and the limit operating control circuit and the main fuel valve interlock closed, the burner blower motor is energized, purging the furnace. The running interlock circuit (air flow switch) closes.

The blower and purge indicator lights are illuminated on the programmer. The modulating motor opens the damper to high fire position. The open damper interlock closes, initiating a 30-second prepurge of the furnace. When prepurge is complete, the purge indicator light goes out and the modulating motor returns to the low fire position.

The spark ignition and pilot valve are then energized and the pilot indicator lights. Following a 10-second period after the pilot flame is proven, the main fuel valve opens and the pilot indicator light goes out. After a 10- to 15-second main flame trial for ignition, the pilot ignition is de-energized to shut off the pilot. The modulating motor is switched to automatic control and the auto indicator lights. The burner is now controlled by the pressure control and the modulating pressure control.

Microprocessor-based ON/OFF/Modulating Combustion Controls. Microprocessor-based ON/OFF/modulating combustion controls use microprocessor technology to control the burner operation. An example of a microprocessor-based ON/OFF/modulating combustion control is the microprocessor-based burner management control system (MBBMCS) developed by the Fireye Products Division of the Allen-Bradley Company. The MBBMCS is a self-diagnostic, microprocessor-based burner control system with a nonvolatile memory. *Nonvolatile memory* is a memory that cannot be lost if the power is interrupted. Similar microprocessor-based ON/OFF/modulating combustion control systems are available from other manufacturers.

The MBBMCS has a vocabulary of 42 different messages that scroll out on the message center, providing the operator with operating data and failure-mode information in a simple, easy-to-understand manner. The messages provide clear, concise information concerning system timing, present burner sequence position, lockout causes, and burner history.

The MBBMCS provides for direct connection of limit and operating controls, fuel valve interlock, damper position interlocks, running interlocks (such as air flow, fuel pressure, and temperature), burner motor, ignition, pilot valves, main fuel valves, modulating motor, and alarms.

The three separate components in the MBBMCS are the programmer, amplifier, and display modules. A self-diagnostic circuit within the control identifies module failure. For example, if the amplifier module is malfunctioning, the message center will display the message, "Lockout-Replace Amplifier."

The MBBMCS does more than control the burner flame. It also provides a stationary engineer or service personnel with important burner operation history. The MBBMCS takes the guesswork out of troubleshooting a burner problem by identifying the malfunction and the corrective action to be taken.

The MBBMCS performs a safe-start component check during each firing sequence. In the firing sequence, a purge period of not less than 30 seconds is performed with a damper actuated to the open position. An interlock circuit is provided to prove air flow rate during the purge period. A starting interlock circuit is required to ensure that the burner equipment is in the low fire position at the time of ignition.

In addition, an interlock is provided to prove air flow during the purge and firing cycle. Proof of the ignition is restricted to 10 seconds for the pilot flame and 10 seconds to 15 seconds for the main flame for fuel oil or gas. Safety shutdown following flame failure, with fuel and ignition circuits de-energized, is achieved in 4 seconds or less. A postpurge period of 15 seconds follows a burner shutdown.

This system was designed to replace older models of programming controls with minimal rewiring. In some instances, the MBBMCS can be plugged into the old chassis. In older models, an adapter must be used.

Positioning Combustion Controls

Positioning combustion controls are used on medium-size boilers. Positioning combustion controls operate from the steam header pressure, which actuates a master control unit, and then relays pressure signals to slave units that control the fuel and air supplied to the furnace. The entire system must be adjusted so that any change in the steam pressure produces a proportional change in the air and fuel supply. The positioning method of control assumes that one setting of a control always produces the same results. However, this is not necessarily true because of wear on the linkage and pins of the controls. Although it is possible to maintain a fairly constant steam pressure, combustion efficiency at times can drop off.

The positioning control requires clean, dry air for trouble-free operation. An air compressor equipped with special filters and driers supplies control air to the system. **See Figure 4-31.** A stationary engineer can operate the master control or relay units by hand or automatically. To shift from hand to automatic operation or from automatic to hand operation, the pressure signals on the hand and automatic controls of all units must be synchronized before changing over. Failure to synchronize leads to improper air-fuel ratio.

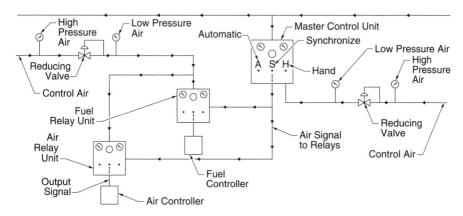

Figure 4-31. Positioning combustion controls use air pressure signals to control air and fuel supplied to the furnace.

Metering Combustion Controls

Metering combustion controls are used on large-size, high-capacity boilers. Metering combustion controls are more sensitive to load demands and load changes. Metering combustion controls are a further development of the positioning control and do not follow the principle that air and fuel flow is directly proportional to steam pressure. Metering combustion controls measure the flow of steam (difference of pressure across an orifice) and balance it against signals for more or less fuel and air. This provides combustion efficiency at a high level. The outlet damper on both of these systems is controlled by the furnace pressure during automatic operation or manually from the control panel when on hand control. The advantages of using combustion controls include the following:

• improved air-fuel ratios, resulting in improved efficiency and less smoke and soot;

• improved water circulation, increasing life of boiler tubes;

• improved feedwater regulation, resulting from more even firing rates (less fluctuation of water level);

• consistent superheat temperature; and

• elimination of excessively high combustion rates.

Boiler Management and Control Systems

Some control manufacturers have designed boiler management and control systems that integrate the functions of programmable logic controllers (PLC) and burner management control systems. Boiler management and control systems include complete boiler firing rate controls for steam and hot water boilers with the latest communication methods, such as DeviceNet, Ethernet, and the Internet. These communication systems enable remote monitoring of a boiler control system from anywhere in the world.

Boiler management and control systems can be used on most types of steam and hot water boilers, including firetube boilers, industrial watertube boilers, and commercial watertube boilers. Most are designed to operate with a gas, oil, or combination gas/oil burner using a single-point modulating control. In addition to installation on new boilers, most can be added as a retrofit to existing boilers.

Boiler management and control systems typically are modular designs, providing flexibility for expansion with easily serviceable components. They often incorporate a user-friendly, graphical touch screen human-machine interface (HMI) that displays boiler parameters, alarm history, and provide fault annunciation and access to boiler configuration and control functions. **See Figure 4-32.**

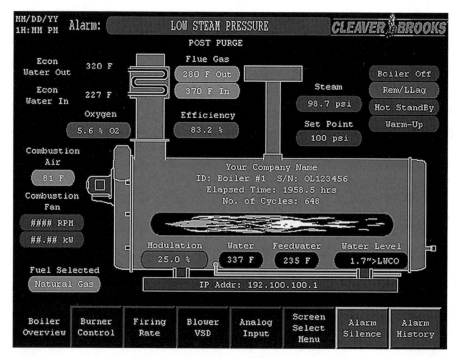

Cleaver-Brooks

Figure 4-32. A graphical touch screen human-machine interface (HMI) provides easy access to boiler configuration and control functions.

Standard features of boiler management and control systems include integrated control function of burner sequencing with firing rate and on-off control, screens showing overview and detail views of boiler operation, and interlocks with auxiliary devices. The systems can be programmed for many warm-up and start-up routines and extensive logging of alarms and faults.

Stack Emissions

Local, state, and federal governments tightly regulate the emissions from the stacks of larger industrial, commercial, and utility boilers. These regulations are continually changing and vary regionally. The most significant stack emissions are sulfur dioxide (SO_2), oxides of nitrogen (NO_X), carbon monoxide (CO), and airborne particulate matter. They are primarily by-products of the combustion process and are exhausted from the stack.

Most sulfur dioxide (SO_2) in stack emissions comes from high sulfur content coal used as boiler fuel. A smaller amount of sulfur is in fuel oils that, when burned, add pollutants to the air. When in the presence of particulate matter, SO_2 reduces visibility, increases materials corrosion, irritates the human respiratory system, and has varying effects on vegetation. Once SO_2 has been released into the atmosphere, it can react with the moisture in the air and contribute to the acid rain effect.

Oxides of nitrogen (NO_X) are emissions that contain nitrogen and oxygen. Nitric oxide (NO) and nitrogen dioxide (NO_2) are released into the atmosphere in stack emissions in greater quantities than other oxides of nitrogen. When nitrogen in the fuel and combustion air combine with oxygen, NO is formed. When NO reacts with oxygen, NO_2 is formed. The amount of NO_X formed in the combustion process depends on the quantities of nitrogen and oxygen present, the temperature, how well the reactants are mixed, and the amount of time in which the reaction can occur. NO_X is associated with corrosion and degradation of materials, damage to vegetation, and respiratory disorders. It has also been identified as one of the possible causes of ozone depletion and smog formation.

Carbon monoxide (CO) is formed by the incomplete combustion of fuels. Generally, stack emissions contain small amounts of CO; most CO is formed by internal combustion engines. Exposure to CO adversely affects the health of humans and animals. When breathed in, CO is absorbed into the bloodstream through the lungs and reduces the oxygen carrying capacity of the blood. Depending on the concentration and the length of exposure time, CO can cause motor skill impairment, physiological stress, or death.

Airborne particulate matter is solid or liquid matter that is suspended in flue gas and discharged to the atmosphere. The particle size of airborne particulate matter is generally small (1 to 100 μm). Airborne particulate matter can impair visibility, spread to surrounding areas as fallout, and adversely affect the human respiratory system. When combined with SO_2, airborne particulate matter has additional adverse effects.

Air pollution regulations are enacted at the federal level and/or at the state and local levels. Federal regulations, which primarily establish outdoor (ambient) air quality standards, are the primary impetus behind state and local air pollution regulations.

However, with a few exceptions, federal regulations only set the ambient air quality standards. They do not detail how to comply with them. The actions required to comply with federal standards must be developed and implemented by state and local air quality agencies.

A *pollutant* is matter that contaminates air, soil, or water. Air pollutants are airborne contaminants that produce unwanted effects on humans and the environment. They occur as solids, liquid droplets, gases, or combinations of these forms. Air pollutants are classified as primary pollutants and secondary pollutants. A *primary pollutant* is a pollutant emitted directly from identifiable sources. A *secondary pollutant* is a pollutant formed by interaction between two or more primary pollutants.

National Ambient Air Quality Standards. To protect humans and the environment from the adverse effects of air pollutants, the U.S. Environmental Protection Agency (EPA) has established the National Ambient Air Quality Standards. The *National Ambient Air Quality Standards (NAAQS)* are a set of pollution standards for six priority pollutants set by the EPA through the Clean Air Act. The six pollutants addressed in the NAAQS are the following:

- ozone (O_3)
- carbon monoxide (CO)
- nitrogen dioxide (NO_2)
- sulfur dioxide (SO_2)
- particulate matter with a diameter of less than 10 μ (PM_{10})
- lead

The amendments to the Clean Air Act require that regulations be implemented to reduce the ambient pollution levels of any process that increases one or more of the pollutants addressed in the NAAQS. All six pollutants addressed in the NAAQS are directly or indirectly related to the combustion process. The 1990 Clean Air Act Amendment comprises 11 titles that have the potential to affect nearly every source of air pollution. Although several titles affect boilers, the title having the most impact is Title I, *Attainment and Maintenance of the National Ambient Air Quality Standards*.

Clean Air Act. Nearly all air pollution regulations originate from the Clean Air Act, which was enacted in 1963. The act improved and strengthened pollution prevention programs and was the first major step toward increased federal control of air pollution. The first major amendments to the Clean Air Act occurred in 1970. The 1970 amendments set national air quality standards and established performance standards for new sources of pollution. As a result of the 1970 amendments, standards were set for sulfur oxides and nitrogen oxides for several sources, including boilers.

The next significant amendment to the Clean Air Act occurred in 1977. The 1977 amendment enhanced many aspects of the Clean Air Act by implementing a more comprehensive permit program, establishing emission limitations on existing sources, and imposing stricter emission standards on new sources. Most important, the 1977 amendment extended compliance deadlines because many geographic areas had not achieved compliance with the ambient air quality standards. After almost 15 years of air pollution regulation, nationwide compliance still had not been achieved.

The most recent amendment to the Clean Air Act occurred in 1990. The 1990 amendment has been labeled the most complex, comprehensive, and far-reaching environmental law Congress has ever enacted. Under the 1990 Clean Air Act amendments, comprehensive permit programs, enforcement provisions, and miscellaneous programs have been established. Also under the amendments, air pollution from stationary and mobile sources, air toxins, and acid rain pollutants (sulfur oxides and nitrogen oxides) are regulated.

Because of new scientific evidence, the EPA made revisions to two areas of the Clean Air Act in 1997. The first revision placed restrictions on fine particulate matter (PM_{25}), which contributes to health problems. At the same time, the EPA is developing a new program to control regional haze, which is largely caused by particulate matter. The second revision placed further restrictions on ground-level ozone, which contributes to smog.

State Implementation Plan Program. Title V of the 1990 amendment to the Clean Air Act established the State Implementation Plan program. The *State Implementation Plan (SIP)* is a plan that gives state governments the responsibility for developing their own programs to reduce air pollution. The federal government has restrained from imposing uniform regulations for the states because of the differences in weather, demographics, and geography. It is important that owners and operators of boilers producing pollutants become familiar with state regulations. Violations of state regulations can result in fines and criminal penalties.

New Source Performance Standards. The federal government has established national minimum standards called New Source Performance Standards. *New Source Performance Standards (NSFS)* are a set of environmental standards for ambient pollutant emissions from new sources.

The standards differ from one installation to another, depending on the size of the boiler and fuel type being used. Boilers are classified by their burner input rating: between 10 million Btu/hr and 100 million Btu/hr, between 100 million Btu/hr and 250 million Btu/hr, and more than 250 million Btu/hr. All small boilers built since June 9, 1989, must meet NSPS requirements. The emissions of new source facilities must be tested to verify compliance.

Classification of Emissions Sources. The EPA has different regulations for different emissions sources. A *major source facility* is a facility that emits 10 tons or more per year of any single air toxic or 25 tons or more per year of any combination of air toxics. Most boilers in major source facilities are located in industrial facilities such as refineries, chemical plants, manufacturing plants, and paper mills. Major source facilities have more stringent regulations than area source facilities.

An *area source facility* is a facility that emits less than 10 tons per year of any single air toxic and less than 25 tons per year of any combination of air toxics. Boilers in area source facilities are located in commercial facilities, institutional facilities, and a small number of industrial facilities. Commercial facilities include stores, malls, laundries, apartments, restaurants, hotels, and motels. Institutional facilities include hospitals, clinics, nursing homes, schools, churches, amusement parks, courthouses, and prisons. Area source boilers emit small amounts of pollutants, primarily because most burn natural gas.

Control of stack emissions is not easy and the equipment is very expensive to install and operate. Three generalized methods used to control stack emissions are pre-combustion, combustion modification, and post-combustion.

Pre-combustion techniques utilize the selection of cleaner-burning fuels. For example, low sulfur content coal or natural gas could be used to replace high sulfur content coal.

Combustion modification is the alteration of a combustion process in an attempt to lower stack gas emissions. For example, fluidized-bed combustion could be used when limestone is used as the bed material. This is done in large utility and industrial boilers. The limestone absorbs the NO_X and SO_2 that are released from the fuel during combustion.

Post-combustion control is common in medium-to-large boilers. In this technique, the pollutants are removed after the combustion process. The most common method of post-combustion control is the implementation of wet or dry scrubbers. Injection of sorbents or catalysts can be used for gas removal. Dry scrubbing for removal of SO_2 and flyash, which result from the combustion of coal, involves spraying aqueous sorbent slurry into the hot flue gasses. The slurry droplets dry into a particulate to be carried in the combustion gas stream. During the drying process, SO_2 is adsorbed into the droplets. The dry particulate and flyash are then collected downstream by particulate control equipment such as cyclone separators, fabric filters, or electrostatic precipitators.

CHAPTER 4 LEARNER RESOURCES

ATPeResources.com/QuickLinks
Access Code: 735728

BOILER OPERATION AND MAINTENANCE

A stationary engineer routinely makes decisions that can affect a company and its employees, or whether people in a community receive electricity, gas, water, or sewage disposal. Steam produced in the boiler room supplies many necessities and must provide steam consistently. In addition, the plant must be operated efficiently to minimize operation costs. Plant conditions must be accurately recorded and relayed to the appropriate personnel. Safety hazards must be addressed promptly to avoid injury to personnel or damage to equipment. Established plant procedures for emergency situations and routine maintenance must be followed.

STATIONARY ENGINEER'S RESPONSIBILITIES

The primary responsibility of a stationary engineer is to operate the plant in a safe and efficient manner. Equipment may vary from plant to plant. Operation of the equipment may require specialized duties and procedures. **See Figure 5-1.** Standard operating procedures common to all plants include the following:

1. *Maintain proper water level in the boiler at all times.* A 1000 HP boiler generates 34,500 pounds of steam per hour and needs a minimum of 34,500 pounds of water per hour, or approximately 4142 gallons of water per hour. One gallon of water weighs 8.33 pounds. Failure to maintain the proper water level could cause a low or high water condition in the boiler. A low water condition can result in overheating of the heating surface. A high water condition can result in priming or carryover. Both conditions are dangerous and must be prevented. A stationary engineer must be able to respond quickly and effectively to either of these conditions.

2. *Maintain control of the boiler at all times.* The primary responsibility of any boiler operator is to safely operate and maintain the boilers in their care. But, attendance requirements for local jurisdictions have allowed employers to assign other duties to the boiler operator, such as other maintenance and operation duties throughout the facility. The attendance requirements typically require boiler checks at set intervals depending on the size of the burners and the controls and interlocks that are installed on the boiler. The intervals can vary from 20 minutes to 24 hours. Constant attendance is normally required for nonautomatic boilers. An operating

nonautomatic boiler should never be left unattended for any extended period of time. The length of time a boiler can be left unattended is determined by the time it would take for the boiler water level in the gauge glass to drop from one half to empty. The length of time may vary from 2 to 8 minutes as determined by the pounds of steam the boiler is producing per hour.

3. *Follow instructions from the chief engineer.* The shift engineer reports to the chief engineer. All written or verbal instructions from the chief engineer must be carried out. For example, specific instructions regarding boiler blowdown and boiler water treatment may be included in the instructions.

4. *Maintain proper steam pressure.* Variations in steam pressure can have an adverse effect on plant process operation.

5. *Test low water fuel cutoffs.* All low water fuel cutoffs must be tested for proper operation at least once a shift.

6. *Blow down feedwater regulators.* Boilers equipped with float-type feedwater regulators should be blown down once a shift to remove impurities from the float chamber and lines going to the feedwater regulator. Boilers equipped with thermoexpansion or thermohydraulic feedwater regulators should have the sensing element connections blown down on the feedwater regulator once a month to remove impurities. In this procedure, the feedwater regulator is removed from service. Water is fed to the boiler by hand. After the feedwater regulator has stabilized (approximately 1 hour), it can be returned to service.

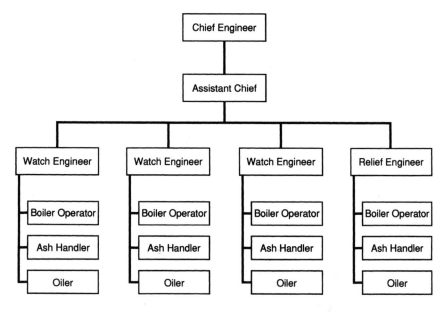

Figure 5-1. Boiler operation requires coordination of personnel for safety and efficiency.

7. *Maintain the burner.* Burner maintenance procedures include cleaning burner tips and ignition electrodes and strainers, checking ignition gap, and keeping pilot lines and flame scanner ports clear. In rotary cup burners, proper lubricating oil levels in sumps must be maintained. In addition, lubricating oil in sumps should be periodically dumped, flushed, and replaced. The forced draft fan of rotary cup burners must be kept clean. The line going to the air switch must be removed and blown out at regular intervals.

8. *Maintain proper fuel oil temperature.* Fuel oil that is too hot causes white smoke from the chimney. Fuel oil that is too cold causes black smoke and soot deposits on the boiler heating surface.

9. *Record fuel burned.* Fuel on hand at the start of the shift, fuel on hand at the end of the shift, and the amount of fuel consumed during the shift must be recorded. If the pounds of steam per hour is constant and fuel consumption varies, there is a problem that must be corrected.

10. *Record makeup water used.* Record the amount of makeup water used on the shift. Any extraordinary increase in makeup water indicates a loss of condensate returns, which affects fuel consumption and feedwater treatment requirements.

11. *Monitor feedwater temperatures.* Feedwater temperatures that are too high can cause the feedwater pump to become steam bound. Feedwater temperatures that are too low can increase the possibility of oxygen pitting in the boiler and thermal shock to the boiler heating surface.

12. *Monitor boiler room auxiliaries.* Early detection of a problem reduces downtime and maintenance costs.

13. *Inspect the fire in the furnace.* Periodically inspect the condition of the furnace fires for combustion quality. *Note:* Use proper eye protection.

14. *Monitor chimney temperature and use soot blowers.* Buildup of soot on boiler heating surfaces acts as an insulator and reduces efficiency.

15. *Log unusual conditions occurring during a shift.* Documenting unusual conditions can help in identifying and correcting a problem.

BOILER CONTROLS MAINTENANCE SCHEDULE

Several national, state, and local codes recommend setting up a formal system of periodic preventive maintenance and testing of boiler controls. Maintenance, repairs, and testing should be conducted on a regular basis and recorded in a boiler logbook for future reference. ASME's *Controls and Safety Devices for Automatically Fired Boilers* (ASME CSD-1) provides a checklist of minimum daily, weekly, monthly, semiannual, annual, and as-required requirements, the recommended tests, and who should perform those tests. Some tests are done by the boiler operator and some by a service technician. In many larger installations, the boiler operator does most of the service technician's tests. According to ASME CSD-1, checks should be performed as follows:

Daily
1. Inspect all gauges, monitors, and indicators.
2. Inspect all instruments and equipment settings.
3. Verify the operation of the low water fuel cutoff device and alarm on high pressure boilers.
4. Visually inspect the burner flame.

Weekly
1. Verify the operation of the low water fuel cutoff device and alarm on low pressure boilers.
2. Inspect the igniter.
3. Verify flame signal strength.
4. Test the flame failure detection system.
5. Verify the operation of the fire rate control.
6. Inspect pilot and main flame valves visually and aurally.

Monthly
1. Inspect flue, vent, stack, or outlet dampers.
2. Test high and low oil pressure and temperature interlocks, high and low gas pressure interlocks, and low draft fan air pressure and temperature interlocks.

Semiannually
1. Calibrate all indicating and recording gauges.
2. Perform a slow drain test of the low water fuel cutoff device on steam boilers.
3. Inspect all flame failure detection system components.
4. Verify the operation of the fire rate control.
5. Inspect the piping and wiring of all interlocks and shutoff valves.
6. Inspect all burner components.

Annually
1. Use a pilot turndown test to test the flame failure detection system.
2. Test the flame failure detection system with a test for hot refractory hold-in.
3. Verify the operation of the dual-fuel changeover control.
4. Test the high limit, operating temperature, and steam pressure controls.
5. Replace the vacuum tubes, scanners, or flame rods in accordance with instructions from the manufacturer.
6. Perform a combustion test.
7. Test operating parts, such as coils and diaphragms, of all safety shutoff and control valves.
8. Test the fuel valve interlock switch in accordance with instructions from the manufacturer.

9. Test for leakage on pilot and main gas and/or oil valves.

10. Test purge air switch in accordance with instructions from the manufacturer.

11. Test all interlocks, such as air/steam, burner position, rotary cup, and low fire start interlocks in accordance with instructions from the manufacturer.

<u>As Required</u>

1. Repair, replace, or recondition the low water fuel cutoff device.

2. Clean atomizers and oil strainers on fuel oil burners.

3. Check drip-leg and gas strainers on gas-fired burners.

4. Use a pilot turndown test to test the flame failure detection system.

5. For the flame failure detection system, test for false hold-in from hot refractory.

6. Test safety and/or safety-relief valves in accordance with ASME Boiler and Pressure Vessel Code Sections VI and VII.

TAKING OVER A SHIFT

When taking over a shift, an established routine is followed to prevent overlooking any procedure. The primary objective during a shift change is to maintain safe plant operation and consistent flow of the required steam. A stationary engineer assesses the plant operation, unusual conditions, and the needs of the next shift. Established routines include the following:

1. A stationary engineer taking over a shift must report early enough to check all the operating equipment. *Note:* A stationary engineer is responsible for all existing conditions in the plant after taking over a shift. Any equipment malfunction is the responsibility of a stationary engineer in charge during that shift.

2. Check water level on all boilers on-line by blowing down the gauge glass and the water column. *Note:* This procedure may vary depending on the type of gauge glass on the boiler. Boilers with flat gauge glass and mica should be blown down according to the manufacturer's recommendations.

3. Check steam pressure gauge for proper operating pressure.

4. Check condition of fires.

5. Check all running auxiliaries for proper temperature, pressure, and lubrication.

6. Check feedwater heater for proper level, temperature, and pressure.

7. Check log for any instructions from the chief engineer. Check with the stationary engineer being relieved to identify any unusual events that happened on the previous shift.

Routine Procedures

Routine procedures are those procedures performed after the shift has been taken over and the plant is operating at the required capacity. Routine procedures include handling low and high water conditions, testing low water fuel cutoffs, and testing flame failure controls.

Handling Low and High Water Conditions. Low and high water conditions occur from boiler failure, control failure, or operator error. Low water fuel cutoffs secure the fuel to the burner. Alarms or low and high water whistles provide a stationary engineer with a warning against low or high water conditions. The extent of overheating or damage caused by low and high water conditions is determined by how quickly a stationary engineer reacts to correct the condition that exists. Any time water is below the normal operating water level (NOWL) in the gauge glass, a low water condition exists. By taking immediate corrective action, excessive boiler damage can be prevented.

The correct method for handling a low water condition in a boiler varies with the boiler type and fuel used. When in doubt about the water level in a boiler, always secure the fires immediately. The authority having jurisdiction requires that the boiler be removed from service and opened for inspection for possible overheating when a low water condition has occurred in which the water level is not visible in the gauge glass. **See Figure 5-2.**

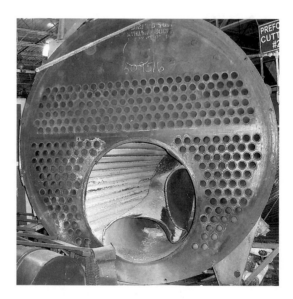

Figure 5-2. Tube and furnace distortion or an explosion can result from a low-water condition in a boiler.

The danger caused by a high water condition in a boiler is entirely different from a low water condition. With a low water condition, the boiler drum, tubes, and furnace could be damaged. With a high water condition, superheater tubes, headers, and equipment such as pumps and steam turbines could be damaged. Any time the water level in a boiler is three-fourths of a gauge glass or higher, there is a possibility that water is being carried over with the steam. The level should be corrected immediately, even if it requires using the bottom blowdown valve. An NOWL in a boiler varies from one-third to one-half of a gauge glass. High water alarms and whistles are used in most plants to alert a stationary engineer to a high water condition.

Testing Low Water Fuel Cutoffs. Low water fuel cutoffs are tested by blowing them down or by an evaporation test. For maximum safety, the low water fuel cutoff should be tested once a shift by blowing it down. **See Figure 5-3.** The burner should be firing when blowing down. An evaporation test is recommended once every 30 days. To perform an evaporation test, the feedwater pump must be secured in addition to any makeup water going to the boiler. A stationary engineer must exercise caution while performing the evaporation test because a low water condition is deliberately created. As the water level slowly drops in the gauge glass, it also drops in the low water fuel cutoff and should shut off the burner while there is still water showing in the gauge glass.

Figure 5-3. Blowing down the water column is one of the duties of a boiler operator.

Note: Many stationary engineers prefer the evaporation test because it is more reliable than blowing down. Blowing down the low water fuel cutoff drops the float more suddenly than an evaporation test. In an evaporation test, the float drops more gradually, which more accurately simulates a low water condition developing.

Testing Flame Failure Controls. When entering a boiler room, if the boiler is down with little or no pressure, it would be down because of a low water condition or flame failure. Low water fuel cutoffs shut off the burner. Flame failure is a condition in which the fire has been lost. Flame failure can lead to a furnace explosion caused by ignition of explosive or highly flammable gases, vapors, or dust accumulated in the furnace. Minor furnace explosions (puffbacks, flarebacks, or blowbacks) can also occur. Furnace explosions can cause serious injury and costly damage.

Not all boilers are equipped with automatic flame failure controls. Small package boilers are equipped with a flame scanner or similar device that shuts the main fuel valve in case of pilot failure or loss of the main flame. Many large plants rely on a stationary engineer to handle flame failures safely. Common reasons for flame failures in a plant vary, depending on the fuel used.

Fuel oil-fired plants:
• Cold fuel oil
• Water in fuel oil
• Air in fuel lines
• Clogged strainers
• Clogged burner tip
• Loss of fuel oil pressure

Gas-fired plants:
• Low pressure in gas lines
• Water in gas

Pulverized coal plants:
• Wet coal
• Loss of primary air
• Loss of coal to pulverizing mill resulting from feeder failure or blockage

In plants equipped with automatic controls, a stationary engineer checks the furnace for signs of excess fuel and resets the controls to put the burner through a purge and firing cycle. In a plant without automatic controls, a stationary engineer must manually secure the fuel. The burner must be manually purged to prevent a furnace explosion. A stationary engineer must be thoroughly familiar with the equipment and follow the manufacturer's recommendations regarding length of purges and number of air changes required before attempting to fire the burner.

To test flame failure control, the burner must be firing. The scanner is removed and the scanner eye is covered. **See Figure 5-4.** This simulates a flame failure, which should shut off the fuel valve. The programmer puts the burner through a postpurge cycle then shuts off. The scanner is reinstalled and the programmer reset. A stationary engineer should verify the operation of a normal firing cycle. For maximum safety, the flame failure controls should be tested once a week.

Pump Maintenance

A stationary engineer is responsible for feedwater pumps, vacuum pumps, fuel pumps, sump pumps, and similar equipment. **See Figure 5-5.** In addition, a stationary engineer is expected to perform minor repairs and service on that equipment. Packing glands and mechanical seals require proper lubrication or flushing to assure good operation. When shaft pump seals start leaking, the seals must be replaced. Shaft packing can be adjusted with the packing gland nuts to ensure minimal leakage from the pump. If there is excessive leakage from the packing or shaft wear, the packing must be replaced or the shaft must be repaired. Bearings, unless factory-sealed, require proper lubrication. Manufacturer's specifications should be followed for the proper type of lubricant to use.

Worn bearings must be replaced. If not replaced, damage to the shaft or bearing housing could result. A stationary engineer should note the operating condition of each pump. If a pump starts to drop in efficiency because of mechanical wear, it is taken out of service for a general overhaul. Many plants have a preventive maintenance schedule for all pump overhauls.

Figure 5-4. A flame scanner is tested to verify proper operation.

Figure 5-5. Feedwater pumps are inspected for leaks, unusual noises, and overheating.

Valve Maintenance

Valve maintenance must be performed on a regular basis. Valves are opened, closed, and throttled to control the flow of various liquids and gases. **See Figure 5-6.** All valves must be in good working order for safety and efficiency. Recommended valve usage and maintenance procedures include the following:

- Never use excessive force when first opening or closing a valve. Valves that require excessive force need servicing.
- Never use a gate valve for throttling service because it usually leaks around the gate.
- Use the manufacturer's recommended lubricants for packings.
- Never overtighten packing glands. Service and maintenance should be performed as required to replace packing or gland seals, gaskets, or stem, or to regrind seats, gates, or discs.

On automatic valves, diaphragms and pilot valves also require routine service. In addition, gaskets should be replaced before assembling reconditioned valves, line flanges, or flange connected valves.

Figure 5-6. The boiler main steam stop valves must be closed on boiler shutdown.

Setting Reciprocating Pump Valves. Reciprocating pump valves are set to maximize efficiency and prevent damage to the pump. Before the valves can be set, the piston must be placed in mid-position. **See Figure 5-7.** The procedure for setting reciprocating pump valves is as follows:

1. Secure steam and exhaust suction and discharge valves (four valves).
2. Open drains on steam and water sides.

3. Bar pump toward steam side as far as it will go.

4. Scribe a mark on steam piston rod flush with packing gland.

5. Bar pump in opposite direction as far as it will go. Mark steam piston rod again, flush with packing gland.

6. With dividers, find center between the two marks and mark the steam piston rod.

7. Bar the pump toward steam side, lining the last mark flush with packing gland.

8. The rocker arm should be perpendicular to steam piston rod. One side of pump is now in mid-position.

9. Follow same procedure on other side of pump. The pump is now ready to have valves set. *Note:* When barring pump over, bar it against the crosshead or spool piece, never against the rocker arm.

10. With the piston in mid-position, mark and remove the valve chest cover.

11. Move the slide valve so it covers all ports.

12. Move the valve in one direction as far as it will go. Measure port opening.

13. Move the valve in the opposite direction as far as it will go. Measure port opening. *Note:* Steam port openings should be equal, with approximately one-half of a port open on each side when the valve is at its end position. This distance is known as lost motion. The lost motion allows the pump to make a full stroke, and it also prevents the pump from striking the head. After setting the lost-motion nut, move one valve to uncover a port, otherwise the pump cannot start. Once in motion, the reciprocating pump can never stop with all the ports covered.

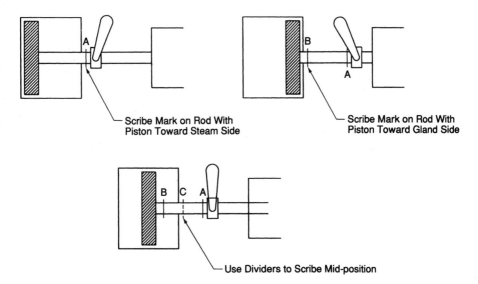

Figure 5-7. The reciprocating pump must be in mid-position before the valves are set.

BOILER STARTUP

Boiler startup procedures vary from plant to plant. Startup procedures are determined by the number of boilers, size of plant, type of fuel burned, and if the plant is operated automatically or manually. Standard procedures for boiler startup include the following:

1. Prepare the fuel before starting boilers. This procedure is determined by the type of fuel used. When gas is used, little or no preparation is necessary except to ensure that the correct pressure is at the burner. When coal is used, an adequate supply of coal must be on hand at the stoker or at the pulverizing mill. In stoker-fired boilers, a wood fire must be started in the furnace to ignite the bed of coal.

   When fuel oil is used, the fuel oil must be circulating and at the proper temperature before it can be burned. This requires the use of electric fuel oil heaters on No. 5 or No. 6 fuel oil when starting up the burner. All fuel oil strainers and the fuel oil burner assembly should be cleaned. The furnace should be purged of all combustible gases before the burner is fired.

2. Boilers should be checked for missing or open inspection openings. All chimney coverings should be removed and hand-operated dampers should be opened. The correct water level should be showing in the gauge glass. The automatic nonreturn valve and the main steam stop valves in the main steam line should be checked for ease of operation. All drains in the main steam lines should be opened.

3. The feedwater system should be checked to see that the system is ready with the proper valves open. Feedwater pumps and feedwater regulators should be inspected. There should be an adequate supply of water in the feedwater heater.

4. Check all valves on boilers to ensure their proper position. Bottom blowdown valves should be closed. The water column, gauge glass, steam pressure gauge, air cocks, and superheater drain valves should be open.

5. Warm up the boiler slowly to prevent uneven expansion and contraction. Maintain the NOWL in the boiler. Close the air cocks when the steam pressure has reached approximately 10 psi or 15 psi. Open the steam line drain between the nonreturn valve and main steam stop valve. Open the equalizing line around the main steam stop valve and open the main steam stop valve. When the boiler is up to about 80% of line pressure, open the automatic nonreturn valve. Slowly bring up steam pressure and let the automatic nonreturn valve cut the boiler in on-line. All superheater drain valves should be closed when approximately 20% of boiler capacity steam flow has been established. The automatic combustion controls and feedwater regulator should be checked for proper operation, along with all boiler room auxiliaries.

   This point in startup is very critical. A stationary engineer must be alert for sudden and unusual occurrences. All boiler room auxiliaries must be monitored closely. In addition, automatic controls should be put into operation and also carefully monitored.

BOILER SHUTDOWN

Boiler shutdown procedures vary from plant to plant, depending upon the size of the plant, length of shutdown, and time of year. Proper boiler shutdown procedures prevent equipment from being damaged when the plant is started up again.

Fuel system shutdown procedures are determined by the type of fuel used. A gas system requires little or no attention. A coal stoker requires cleaning and lubrication. The fuel oil strainers and the burner of a fuel oil system are required to be cleaned. **See Figure 5-8.**

Figure 5-8. Fuel oil burners must be cleaned to prevent a carbon deposit buildup.

The main steam stop valves should be closed and all steam line drains opened. The superheater drain should be opened as soon as the automatic nonreturn valve cuts the boiler off-line. When the steam pressure is down to 25 psi, the air cock should be opened.

While the boiler is cooling down, the water level should be maintained at the NOWL. When the boiler has cooled sufficiently, the feedwater pump should be shut down and all valves in the system closed.

When the boiler has cooled off enough to enter, the fire side should be opened up and cleaned. The breeching and base of the chimney should also be cleaned. After cleaning the fire side, all access doors and panels should be reinstalled and the fire side closed up. *Note:* Make sure all tools are removed before closing up the fire side of the boiler.

BOILER INSPECTION

A stationary engineer is responsible for preparing the boiler for annual inspection by the boiler inspector. Prior to inspection, the boiler must be opened up on the fire and water sides and be thoroughly cleaned. Cleaning the boiler also increases efficiency and prevents possible damage from neglect.

To prevent corrosion caused by sulfur in the ash, coal, or soot, the fire side must be cleaned. Cleaning the fire side reduces fuel consumption because soot acts as an insulator and retards the transfer of heat. Whenever the boiler is off-line, the fire side should be inspected and cleaned. The water side of a boiler, after a period of time, builds up deposits of sludge and sediment. If feedwater treatment is neglected, the boiler may have scale deposits built up on the heating surfaces.

All plugs on cross connections to the water column, feedwater regulator, gauge glass, and low water fuel cutoff must be removed. The feedwater regulator and low water fuel cutoff float chambers must also be opened for internal examination. A stationary engineer should be present to assist the boiler inspector and point out possible problems. General guidelines used in preparation for a boiler inspection include the following:

- Cool the boiler slowly.
- Open vent to prevent a vacuum from forming on the boiler coming off-line.
- Secure, lock, and tag out the main steam stop valves.
- Secure, lock, and tag out the feedwater valves to the boiler.
- Never dump a hot boiler.
- Never dump a boiler unless it is ready to be opened and flushed.
- Always check to make sure the air cock is open before opening the manhole cover.
- After dumping, secure, lock, and tag out bottom blowdown valves, or remove the valves and blank-flange the blowdown lines to the blowdown tank.
- Never enter the steam and water side of the boiler until steam stop valves, feedwater lines, and bottom blowdown lines are properly secured.
- Never enter the steam and water side of a boiler with a high-voltage droplight. Use a battery-operated spotlight or a low-voltage droplight.
- Always use proper ladders or scaffolding for cleaning.
- The gauge glass should be cleaned or replaced if it is dirty or leaking. The procedure used is determined by the type of gauge glass.

Flat gauge glasses are usually kept as a spare unit. The holder and glass are installed by breaking a union on the steam connection and water connection. The new holder and glass are then installed. The old holder is fitted with new gaskets, mica, and glass and stored for future use. **See Figure 5-9.**

Tubular gauge glasses are changed by removing the gauge glass nuts, old gauge glass, and old gauge glass washers. **See Figure 5-10.** New gauge glass washers and a new gauge glass are then installed. The gauge glass nuts are hand-tightened, then a quarter turn is made with a wrench.

Note: If a gauge glass must be cut to size, it is cut ¼" shorter than the inside measurements to allow for expansion.

Figure 5-9. The boiler water level must be maintained at the NOWL during boiler startup.

Figure 5-10. A gauge glass is removed by loosening the nuts that hold the glass in place.

If a gauge glass must be changed on a boiler under pressure, the water and steam valve to the gauge glass must first be shut, and the gauge glass blowdown valve opened. After the gauge glass is installed, it must be warmed up as follows:

1. Crack the steam valve to the gauge glass.
2. After the gauge glass has had a chance to warm up, open water and steam valve to gauge glass.
3. Close gauge glass blowdown valve.
4. Check for leaks.

Note: Always wear eye protection when working on a gauge glass. Serious injury can result if a gauge glass shatters under pressure.

Accumulation Test

An *accumulation test* is used to establish the relieving capacity of boiler safety valves. All boilers having over 500 sq ft of boiler heating surface must have two or more safety valves. The safety valves must be large enough or there must be enough safety valves to prevent the boiler pressure from rising over 6% above the MAWP or highest safety valve setting.

When performing an accumulation test, a boiler inspector should be present. Secure all steam outlets leaving the boiler except those that might be necessary to run the boiler. Turn up the burner to maximum firing rate. With all the safety valves popping, the pressure on the boiler should not exceed 6% of the MAWP or the highest safety valve setting. If the pressure exceeds 6% of the MAWP, the safety valves are not large enough or there are not enough safety valves to protect the boiler.

Hydrostatic Test

A *hydrostatic test* is a water test made on a new boiler or after it has had a low water condition or repairs made to their steam or water side. A hydrostatic test is used to test the strength of materials and construction of the boiler. When performing a hydrostatic test, the main steam stop valves must be secured. The drains between the main steam stop valves must be left open. This prevents water buildup in case the valve closest to the boiler leaks.

Treatment of gauge glasses must follow manufacturer's recommended procedures during a hydrostatic test. Some pressures will cause flat or tubular gauge glasses to leak. The safety valves are blanked or gagged, and the air cock is closed after the boiler is full of water at a temperature of 100°F. Water pressure is increased by using a hand pump to 1½ times the safety valve setting. Water pressure is then lowered to the safety valve setting, and the boiler is inspected for leaks or bulging. When the hydrostatic test and inspection are completed, remove gags or replace safety valves.

BOILER LAY-UP

Boiler lay-up procedures are required for boilers that are to be out of service. Improper lay-up procedures can result in damage caused by oxygen pitting. The two methods of laying up a boiler are the wet and dry methods. Regardless of the method used, the boiler must be thoroughly cleaned on both the fire and water sides. All repairs must be made.

The lay-up method used is determined by the length of lay-up time and plant conditions. If the boiler is needed on short notice, the wet method is recommended. If the boiler is not needed for a long time or if the boiler may freeze up, the dry method is recommended.

Wet Method

The wet method of boiler lay-up is as follows:

1. Clean fire side and steam and water side.
2. Close up steam and water side (use new gaskets).
3. Fill the boiler with warm, treated water until filled to the boiler vent.
4. Close the boiler vent.
5. Maintain a pressure in the boiler slightly above atmospheric pressure.

Dry Method

The dry method of boiler lay-up is as follows:

1. Clean fire side and steam and water side.
2. Dry out boiler completely.
3. Close all valves on the boiler to prevent moisture from entering drums.
4. Lay trays of moisture-absorbing chemicals in the boiler drums.
5. Close steam and water side.
6. Check moisture-absorbing chemicals periodically. Replace when needed.

CHAPTER 5 LEARNER RESOURCES

ATPeResources.com/QuickLinks
Access Code: 735728

STEAM TURBINES

6

Steam turbines convert the heat energy in steam to kinetic energy in order to perform work. The heat energy and steam velocity determine the amount of work performed. Steam turbines are commonly used to drive generators for producing electricity. Fluctuation in the speed of the steam turbine affects the frequency of the electricity being generated. The development of hydraulic governors allowed the accurate speed control of steam turbines required for generating electricity. Other uses of the steam turbine in industry include driving auxiliaries such as feedwater pumps, condensate pumps, and draft fans. Steam turbines have virtually replaced reciprocating steam engines because they have fewer moving parts and are more efficient.

STEAM TURBINE DEVELOPMENT

Steam turbines were first used in approximately 120 B.C. Hero of Alexandria produced the first reaction steam turbine by heating water in a closed vessel. Steam was directed through vents, which caused the vessel to rotate. In 1629, Giavanni Branca, an Italian physicist, designed the first working impulse steam turbine. A closed vessel containing water was heated. Steam created was directed through a single vent. Steam leaving the vessel was directed against vanes of a rotating disc. The rotating disc actuated gears to drive plungers up and down. This motion of the plungers was used to pulverize chemicals. The process was functional, but not very efficient. **See Figure 6-1.**

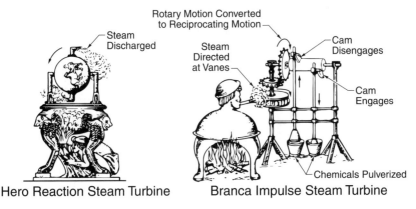

Hero Reaction Steam Turbine   Branca Impulse Steam Turbine

Figure 6-1. Steam turbines used today operate on the same principles as those of early steam turbines.

221

Further development of steam turbines was slowed by the lack of materials that could withstand stresses caused by high speeds and temperature changes. Metals developed in the late 1800s spurred the development of the steam turbine as it is known today. This allowed higher steam pressures and temperatures for greater efficiency.

Steam turbines in the 1920s commonly used steam at about 325 psi. By 1940, the steam pressure used increased to 1300 psi and to 2300 psi in 1941. Steam turbines now use steam pressures as high as 2800 psi. A typical steam turbine used today may be 20′ long, operating at 1200 psi at 1050°F. Steam turbines are commonly used for generating electricity at power plants. **See Figure 6-2.** At that temperature, the metal glows dull red at the steam inlet side. In approximately ⅓₀ of a second, the steam exhausts into a vacuum of 28″ mercury (Hg) at a temperature of about 102°F at the exhaust end. The difference in temperature causes high levels of stress and thermal shock. Continuing experimentation with new materials has resulted in further developments to increase durability and efficiency of steam turbines.

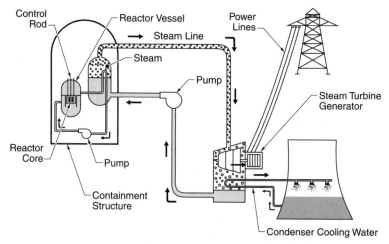

Figure 6-2. Pressurized water reactors contain a steam generator that produces steam to drive the steam turbine generator.

Efficiency of a steam turbine can be determined by its thermal efficiency. *Thermal efficiency* is the ratio of the heat energy used in the turbine to the heat energy available in the steam. The enthalpy of steam supplied to the steam turbine minus the enthalpy of exhaust steam indicates how much heat was used in the steam turbine. Thermal efficiency is expressed as a percentage. Thermal efficiency is found by using the following equation:

Thermal efficiency =

$$\frac{Enthalpy\ of\ steam\ at\ throttle - Enthalpy\ of\ steam\ at\ exhaust}{Enthalpy\ of\ steam\ at\ throttle - Enthalpy\ of\ condensate}$$

$$TE = \frac{H_a - H_e}{H_a - H_c}$$

TE = thermal efficiency

H_a = enthalpy of steam at throttle

H_e = enthalpy of steam at exhaust

H_c = enthalpy of condensate

EXAMPLES

1. A 2500 kW steam turbine operates with a steam pressure of 200 psia at 600°F. It exhausts into a condenser with a vacuum of 28½" Hg. What is the thermal efficiency of the steam turbine? *Note:* All values are obtained from Dry Saturated and Superheated Steam tables. To find the enthalpy of the condensate, the vacuum in inches of Hg must be converted to psia.

   $psia = 14.7 - (vacuum \times 0.491)$

   $psia = 14.7 - (28.5 \times 0.491)$

   $psia = 14.7 - 13.99$

   $psia = 0.71$

   $H_a = 1322$ Btu/lb

   $H_e = 1100.9$ Btu/lb

   $H_c = 90 - 32$ Btu/lb

   Note: The enthalpy of the condensate is based on a 5% to 10% condensation of steam at the exhaust.

   $$TE = \frac{H_a - H_e}{H_a - H_c}$$

   $$TE = \frac{1322 - 1100.9}{1322 - (90 - 32)}$$

   $$TE = \frac{221.1}{1264}$$

   $TE = \textbf{17.49\%}$

2. A 10,000 kW steam turbine operates with a steam pressure of 450 psia at 900°F. It exhausts into a condenser with a vacuum of 28" Hg. What is the thermal efficiency of the steam turbine?

   $H_a = 1467.7$ Btu/lb

   $H_e = 1105.2$ Btu/lb

   $H_c = 100 - 32$ Btu/lb

   $$TE = \frac{H_a - H_e}{H_a - H_c}$$

   $$TE = \frac{1467.7 - 1105.2}{1467.7 - (100 - 32)}$$

   $$TE = \frac{362.5}{1399.7}$$

   $TE = \textbf{25.9\%}$

STEAM TURBINES

Steam turbines convert enthalpy of steam to kinetic energy. Steam, when vented through a small opening, attains a high velocity. The amount of heat energy in the steam determines the amount of work performed. Steam turbines have fewer moving parts, require less maintenance, and are more efficient than reciprocating steam engines. Common uses of steam turbines include generating electricity and driving auxiliaries such as pumps, fans, and compressors. Steam turbines can be designed for use with high-pressure, superheated and/or saturated steam and can operate condensing or noncondensing. A steam turbine operating condensing operates at a lower exhaust pressure, allowing for greater steam turbine efficiency. A steam turbine operating noncondensing uses exhaust steam for heating and process work. The greater the heat energy in the steam, the greater the amount of work that can be performed by the steam turbine. A steam turbine classified as a prime mover is a large steam turbine driving an electric generator. Smaller steam turbines are used to drive auxiliary equipment. Three types of steam turbines commonly used are the impulse, reaction, and impulse-reaction steam turbines. Special-purpose steam turbines are designed for specific plant requirements.

Impulse Steam Turbines

Impulse steam turbines use steam velocity as a force acting in a forward direction on a blade or bucket mounted on a wheel. **See Figure 6-3.** To produce the force, steam is routed through a first-stage nozzle (1) and gains velocity before striking the blades (2) on the first-stage wheel. This causes the shaft (3) to rotate. Steam is then routed through fixed blades (4), which redirect the flow of steam to the blades (5) on the same wheel of the first stage. The steam then enters the second-stage nozzle (6), gaining velocity before striking the blades (7) of the second-stage wheel. Steam exits through the exhaust opening (8). *Note:* Most steam turbines are designed to have a steam velocity that is twice as great as the blade velocity.

Steam turbines may contain many sets of wheels that can be divided into stages. A nozzle or nozzles located before each stage cause the steam to expand and increase in velocity. With the increase in velocity, there is a corresponding decrease in pressure. The steam velocity returns to its initial value after passing through the blades before the next stage. A set of fixed blades is used to reverse the flow of steam between each set of revolving blades. **See Figure 6-4.**

Steam enters the first-stage nozzle (1) and drops slightly in pressure as it passes through, but the steam velocity increases. The steam strikes the revolving blades (2), imparting energy to the blades but losing velocity in the process. There is no pressure drop going through the revolving blades. The steam then enters the fixed blades (3) and changes direction without losing pressure or velocity. Once again, the steam strikes a second set of revolving blades (4), with the corresponding drop in velocity but not in pressure. This is *velocity compounding.*

The velocity drops to the initial value. To increase the velocity, the steam is passed through the second-stage nozzle (5). The velocity-pressure relationship is then repeated as the steam passes through the second-stage revolving blades (6). The pressure diagram shows a pressure drop through the first- and second-stage nozzles before exhausting after the second stage. The diameter of the wheels increases in size from the first stage to subsequent stages to allow for the drop in pressure and increase in volume.

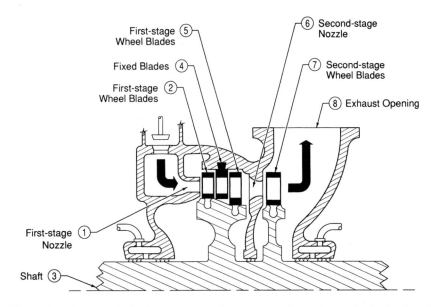

Figure 6-3. Steam velocity increases as steam passes through nozzles in the impulse steam turbine.

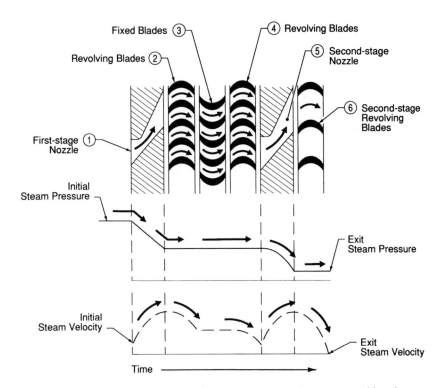

Figure 6-4. Steam pressure decreases from stage to stage in a steam turbine. Increases in wheel diameter sizes compensate for the increases in volume of steam.

Impulse Steam Turbine Parts. Impulse steam turbine parts are the same regardless of the number of stages in the steam turbine. Parts found on impulse steam turbines include the shaft, nozzles, throttle valve, governor, shaft glands, rotor, bearings, labyrinth packing seals, and bearing lubrication accessories. **See Figure 6-5.**

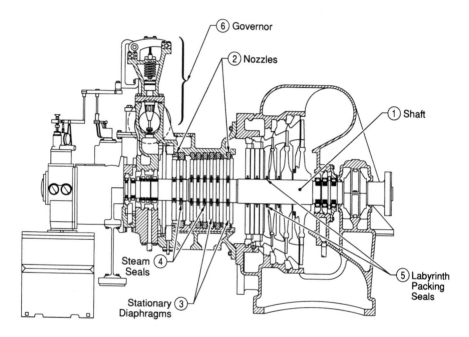

Figure 6-5. The parts of an impulse steam turbine are designed to control steam, steam turbine speed, and lubrication.

The shaft (1) supports mounted wheels with blades attached on the outer edges. Nozzles (2) are fitted before each pressure stage with stationary diaphragms (3) separating the wheels. The stationary diaphragms seal each stage against steam leakage. At the points where the stationary diaphragms join the rotating shaft, a steam seal (4) must be fitted to prevent the higher pressure of one stage from leaking through to the next stage. Labyrinth packing seals (5) are commonly used to seal stages. The throttle valve is used to govern the amount of steam admitted to the steam turbine during startup. The throttle valve is controlled by hand. In addition to the throttle valve, a governor (6) is installed. The governor is an automatic speed-regulating device that prevents a steam turbine from exceeding safe operating speed if the load suddenly changes.

Small auxiliary impulse steam turbines are equipped with mechanical governors that control the speed of the steam turbine and also prevent it from overspeeding. Large impulse steam turbines are equipped with hydraulic governors that maintain a constant speed. A low-oil pressure cutoff and an overspeed trip are also installed on large steam turbines.

Shaft glands are located where the shaft emerges from the casing or where steam leakage could occur unless sealed by glands. The shaft glands at the high-pressure end prevent steam leakage to the atmosphere. The shaft glands at the low-pressure end prevent air from entering the steam turbine when running condensing or steam leakage when running noncondensing. Carbon rings, carbon ring and labyrinth packing seals, or water and labyrinth packing seals can be used. **See Figure 6-6.**

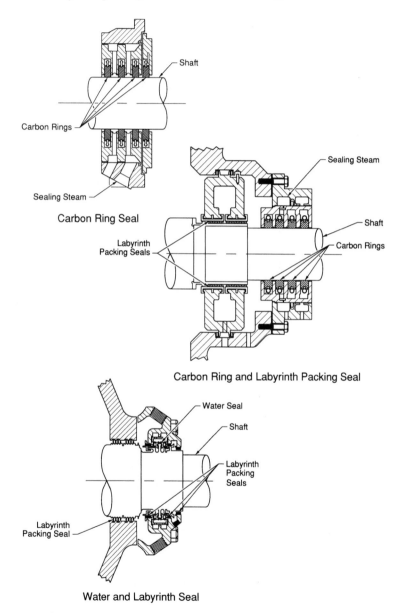

Figure 6-6. Shaft seals and packing are required to prevent shaft steam leaks.

The rotor consists of the shaft and its rotating wheels, with the blades mounted on the outer rims of the wheels. The blades are attached to a shroud ring, which strengthens the uniform positioning of the blades. Bearings are used to maintain the rotating rotor in a fixed position within the casing. The two types of bearings used are the radial (main) and thrust bearings. Radial bearings maintain the vertical position of the rotor. Thrust bearings maintain the horizontal position of the rotor. *Note:* If radial or thrust bearings exceed the specified tolerance, contact between the rotor and casing will occur. Contact between the rotor and casing results in overheating, distortion, and subsequent damage.

Radial bearings are commonly constructed of two cast iron shells lined with babbitt metal with spherical self-aligning pads. All radial bearings are bored and scraped to the shaft before use. The journal is the part of the bearing in contact with the shaft. The clearance allowance between the journal and the shaft is 0.001″ per inch of the shaft diameter. For example, a 10″ shaft rotates in a bearing with an inside diameter of 10.010″. The minimum bearing clearance is 0.005″ for shafts less than 5″ in diameter. Other clearances in an impulse steam turbine include radial, axial, and diaphragm gland clearances.

Radial clearance is the distance between the top of the revolving blades and casing. Radial clearances commonly vary from 0.180″ to 0.250″. *Axial clearance* is the distance between the nozzle exits and leading edge of the rotor blades. Axial clearances commonly vary from 0.100″ to 0.200″. *Diaphragm gland clearance* is the distance between the shaft and labyrinth packing on the diaphragm. Diaphragm gland clearances are usually approximately 0.002″. Radial, axial, and diaphragm gland clearances are measured cold. When the steam turbine is warmed up and running, these clearances are reduced.

The most common thrust bearing used is the movable bearing or Kingsbury thrust bearing. Thrust bearings compensate for variable axial thrust generated by the steam turbine and maintain the axial (horizontal) position of the rotor.

Labyrinth packing is a seal used to prevent leakage of steam from the steam turbine. As steam leaks through the narrow spaces between packing strips and the shaft, pressure decreases and velocity increases. This leak forms a nozzle. Upon entering the pocket or space between the packing strips, the velocity acquired by the expansion of steam is reduced as a result of the swirling (eddy) motion. Therefore, the velocity with which the steam begins the next throttling is very low. The more packing strips there are, the smaller the leakage is. The smaller the pressure drop is, the smaller the velocity increase at each throttling is.

Bearing lubrication accessories are required for impulse steam turbines. Lubrication is commonly force-fed on larger steam turbines. Deflectors must be fitted to prevent leaking gland steam from contaminating lubricating oil. Wear of the radial bearing causes the shaft and rotor to drop. Radial bearings must be checked frequently for wear. During steam turbine shutdown and overhaul, the radial bearings are examined for wear using a bridge gauge or a bearing micrometer gauge.

Reaction Steam Turbines

Reaction steam turbines function similarly to a lawn sprinkler in which water pressure is converted to kinetic energy. Water under pressure is introduced at the center of the sprinkler. From the center, the water branches to the arms and leaves the openings in the arms. As the water leaves, the arms move backward in reaction to the forward force of the water.

Newton's law states that for every action there is an equal and opposite reaction. Reaction steam turbines operate on this principle for part of their action. Instead of nozzles, a reaction steam turbine uses fixed blades for its first stage. Fixed blades are designed so each blade pair acts as a nozzle. Steam expands between each blade pair. The expanding steam gains velocity in the same manner as the impulse nozzle. This steam, with its high velocity, enters the blades of the moving element, imparting a direct impulse to it in the same manner as the impulse steam turbine. Steam is then directed into the nozzle-shaped blade passages instead of being exhausted through the remainder of the blade passage without further expansion as in the impulse steam turbine. The steam expands in the nozzle-shaped passages and gains additional velocity. When leaving the steam turbine blades at a high velocity, a backward kick, or reaction, is applied to the blades. This is where the term reaction steam turbine comes from. **See Figure 6-7.**

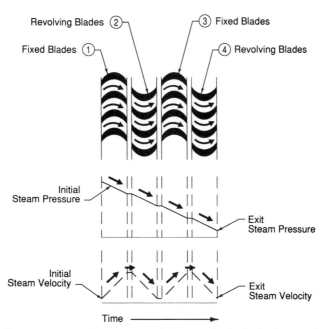

Figure 6-7. Steam pressure decreases and steam velocity fluctuates as steam passes through fixed and revolving blades in a reaction steam turbine.

The steam at its initial pressure enters the fixed blades (1) and increases in velocity while losing pressure. The steam strikes the revolving blades (2), giving up energy and losing velocity. As the steam leaves, it gives a reactive force to the revolving blades, and a loss of pressure occurs. The second stage begins as the steam enters the next row of fixed blades (3). Again, velocity increases and some pressure is lost in the fixed blades. Then velocity and pressure decrease through the revolving blades (4). Because of the difference in pressure between the entrance and exit sides of both fixed and revolving blades, the reaction steam turbine is a full-admission steam turbine. Admission of steam takes place completely around the wheel. In impulse steam turbines, only certain nozzles are opened to produce the flow required.

Because certain characteristics of reaction steam turbine operation differ from impulse steam turbine operation, physical characteristics of the reaction steam turbine differ from the impulse steam turbine. **See Figure 6-8.** The following are physical characteristics of reaction steam turbines:

- Close mechanical clearance between tips of fixed blades and shaft; close mechanical clearance between moving blades and casing
- Relatively large number of elements, permitting small pressure drop per row of blades
- Moving blades mounted on drums
- Full admission of steam all around the blades
- Large axial thrust, resulting from difference in pressure between stages, must be balanced by (a) balancing pistons of various sizes; (b) dummy piston, which is a single balance piston; and (c) double flow of steam.

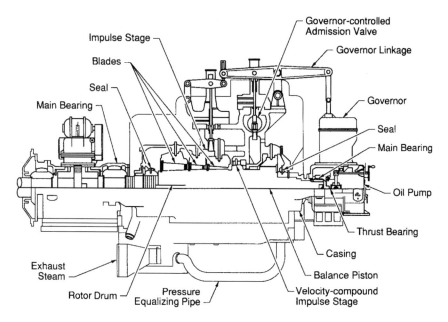

Figure 6-8. The physical characteristics of an impulse steam turbine differ from those of a reaction steam turbine.

Reaction Steam Turbine Parts. Reaction steam turbines commonly include the following parts: fixed blades, moving blades, throttle and governor system, shaft glands, radial and thrust bearings, lubrication accessories, labyrinth packing seals, and sealing strips. Fixed blades are fastened to the housing and act as nozzles when steam enters. Moving blades mounted on drums form the rotor. Steam releases energy to the blading as it passes through. The throttle and governor system functions the same as in an impulse steam turbine. Shaft glands, radial and thrust bearings, lubrication accessories, and labyrinth packing seals are used for the same purposes in reaction steam turbines as in impulse steam turbines. Sealing strips are located in the casing to reduce leakage between stages and warn of excessive radial tolerance. If the radial tolerance is exceeded, clearance between the sealing strips and rotor blades is eliminated. Rubbing of the rotor blades on the sealing strips results in a squealing noise. This alerts the operator to a problem before significant damage occurs. **See Figure 6-9.**

The impulse steam turbine wheel revolves in chambers with equal pressure on each side. There is no axial thrust resulting from unbalanced pressures if the shaft is approximately the same diameter at the high-pressure and the exhaust ends.

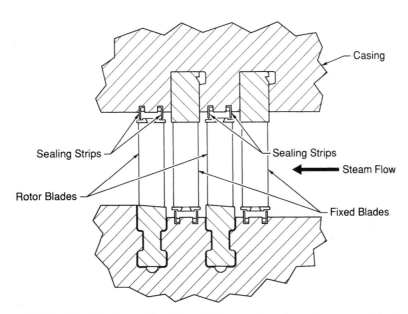

Figure 6-9. Sealing strips in reaction steam turbines reduce steam loss around blades and warn of possible damage from improper radial clearance.

Axial thrust is an important consideration in reaction steam turbines. First, there is a cumulative pressure resulting from the drop in pressure across each row of blades. Second, there is also an axial thrust resulting from pressure on the end of each drum when the diameter changes. The Kingsbury thrust bearing controls axial thrust by using one collar on the shaft, which bears against pivoted shoes. When the shaft is rotated, the shoes pivot to form a wedge-shaped film of lubricating oil on the bearing surfaces. **See Figure 6-10.**

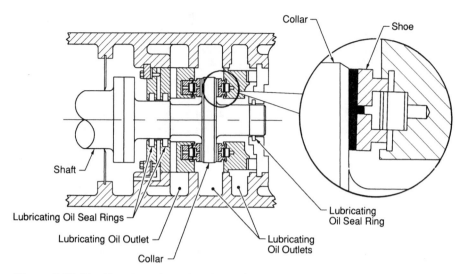

Figure 6-10. The Kingsbury thrust bearing maintains axial position of the steam turbine rotor within specified tolerances.

Reaction steam turbines commonly use one balance piston or dummy piston to balance all thrust at full load on the machine. Any inequalities of thrust at other loads are compensated for by the thrust bearing itself. Because the dummy piston is exposed to high pressure, some antileakage packing must be provided. Labyrinth packing is placed between the dummy piston and dummy cylinder. **See Figure 6-11.**

Pressure on each side of every row of blades is different. The pressure difference is necessary for steam to gain velocity in each set of blades. This pressure difference can cause steam leakage around the tops of the revolving blades and under the bottoms of the fixed blades. To minimize steam leakage, clearances should be reduced, and a small steam pressure difference should be present across each row of blades.

Impulse-Reaction Steam Turbines

Impulse-reaction steam turbines combine impulse and reaction blading. **See Figure 6-12.** Usually there are several stages of impulse blading in the high-pressure end of the steam turbine to utilize large pressure drops through the nozzles. The reaction blading in the low-pressure end has small pressure drops across each row of blades.

Impulse steam turbines are primarily high-speed machines that utilize large pressure drops. They are smaller and weigh and cost less than reaction steam turbines of the same capacity. Reaction steam turbines are primarily a low-pressure steam turbine because of losses around the blades at high pressures. Impulse-reaction steam turbines combine the advantages of each steam turbine. In addition, impulse-reaction steam turbines are commonly run on a high vacuum for even greater efficiency.

Reaction steam turbines have small pressure drops per stage, resulting in small amounts of steam expansion with low output of work per stage. Many stages must be used to obtain the full range of pressure drops throughout the steam turbine.

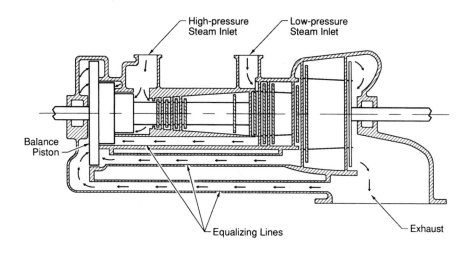

Balance Piston

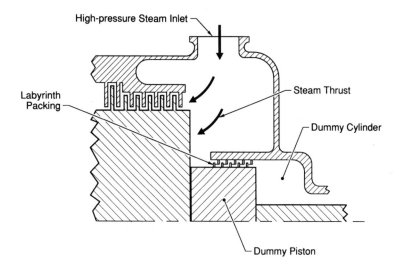

Dummy Piston

Figure 6-11. Balance pistons or dummy pistons are used to minimize the effects of axial thrust in reaction steam turbines.

In some impulse-reaction steam turbines, nozzles are used. In others, the fixed reaction blades act as a series of nozzles directing steam against the impulse blades. The same parts are required in an impulse-reaction steam turbine as those in the separate steam turbines. The governor controls steam admission to the first stage. The overspeed and low-oil pressure trip shuts off steam to the steam turbines if necessary.

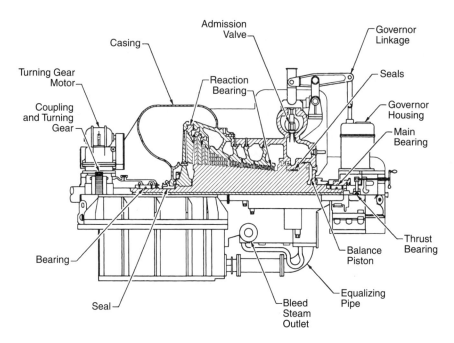

Figure 6-12. Impulse-reaction steam turbines combine the operating features of the impulse steam turbine and the reaction steam turbine.

Special-Purpose Steam Turbines

Special-purpose steam turbines may be required in plants with special generating or process needs. Steam turbines designed for special needs are used where standard steam turbines would not work. For example, a mixed-pressure steam turbine is used in plants that have an abundance of low-pressure steam. **See Figure 6-13.** The steam turbine is designed to operate using low-pressure steam. High-pressure steam is used only when plant demands require it.

For example, a plant that has three boilers, two boilers at 200 psi MAWP and one boiler at 400 psi MAWP, is a candidate for a mixed-pressure steam turbine. A double-flow steam turbine is used to balance the axial thrust on the rotor. **See Figure 6-14.** Steam is admitted in the center of the casing and flows in two directions to equalize the axial thrust on the rotor.

Extraction and Bleed Steam Turbines. Extraction and bleed steam turbines are special-purpose steam turbines used to capture the heat in steam for use in process or feedwater heating rather than being sent to the condenser. *Extraction steam* is steam withdrawn from a steam turbine that is under pressure control. *Bleed steam* is steam withdrawn from a stage of the steam turbine without pressure control. **See Figure 6-15.**

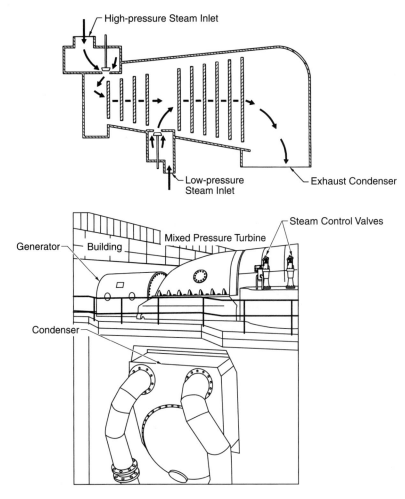

Figure 6-13. Mixed-pressure steam turbines are designed to operate using two or more different steam pressures.

Extraction steam can be taken from one or more pressure stages of the steam turbine. The flow is controlled by a positioning control that actuates a servo-motor connected to a grid extraction valve and the governor system controlling the steam turbine speed. **See Figure 6-16.** Extraction steam at various pressures is used for all types of steam process loads. Bleed steam is used for heating feedwater or constant-load use, such as process work in a paper mill. Bleed steam can also be taken from one or more stages of the steam turbine for heating feedwater.

The primary advantage of using extraction or bleed steam turbines is the ability to use latent heat in the steam rather than lose it to condensing water or the atmosphere. Efficiency of extraction or bleed steam turbines can be improved by maintaining the proper balance between the output load and steam demand.

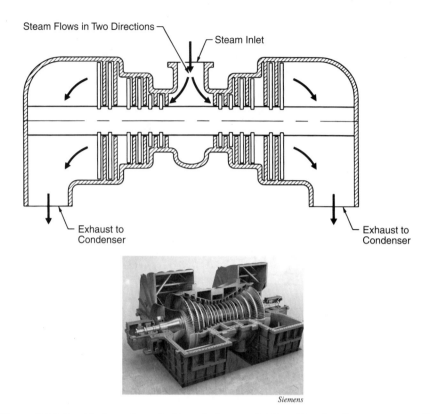

Figure 6-14. Double-flow steam turbines balance axial thrust by directing steam equally in two directions.

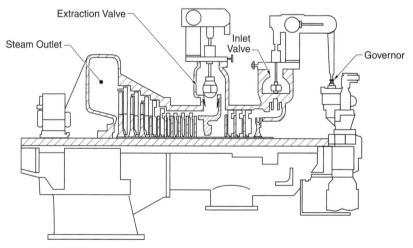

Figure 6-15. Extraction and bleed steam turbines are selected for industrial use when process operations demand steam at fixed pressure.

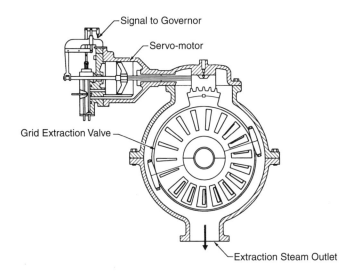

Figure 6-16. Grid extraction valves control steam flow from the extraction steam turbine.

Lubrication Systems

The lubrication system is a very important part of steam turbine operation. The steam turbine rotor is supported on a film of lubricating oil to prevent wear from friction. Because of the high operating speeds of the steam turbine, the proper lubricating oil free of foreign matter must be used. Lubricating oil must have the proper viscosity to flow properly and handle high temperatures without breaking down. Steam turbine manufacturers recommend specific lubricating oils that have been tested and meet operating requirements.

Ring Oilers. Ring oilers are used on small auxiliary steam turbines. Ring oilers consist of a loose-fitting ring that revolves in a small oil reservoir. This supplies lubricating oil to bearing grooves for even distribution over bearing surfaces. **See Figure 6-17.**

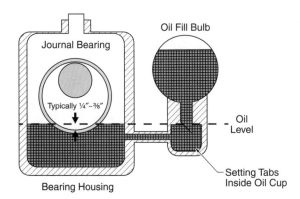

Figure 6-17. A ring oiler is an oil-lubrication system that is used on bearings.

Gravity-Fed Lubrication Systems. Gravity-fed lubrication systems use oil cups or reservoirs to feed lubricating oil to the bearings on steam turbines. Oil cups can also be used on small auxiliary steam turbines. Oil cups are mounted on the top of bearings and can be adjusted for rate of feed of lubricating oil to the bearings. Oil cups can also be mounted on the side of bearings with a cap that lifts up to show the amount of lubricating oil in the cup. **See Figure 6-18.**

Trico Corporation

Figure 6-18. Oil cups are used in gravity-fed lubrication systems to lubricate bearings, chains, and other machinery parts.

Force-Fed Lubrication Systems. Force-fed lubrication systems are used on large steam turbines. **See Figure 6-19.** Lubricating oil is supplied to the steam turbine bearings and hydraulic governor system that controls the steam turbine speed. The lubricating oil must be cooled to prevent it from breaking down and losing its lubricating ability. In addition, it must be clean and free of impurities that could plug up an oil line or cause a bearing to become scratched. It must also be free of water, which causes emulsion or sludge to develop.

The auxiliary lubricating oil pump must be started so that lubricating oil is supplied to the steam turbine bearing and governor before the steam turbine is started. The main lubricating oil pump is driven off the shaft, and as the steam turbine reaches operating speed, the auxiliary lubricating oil pump cuts out. The main lubricating oil pump delivers oil through the duplex oil strainer to the lubricating oil cooler. Before it reaches the lubricating oil cooler, a small amount of oil is taken from the line and passes to a centrifuge. The centrifuge removes dirt and water from the lubricating oil, discharging the clean oil to the oil reservoir and water and dirt to waste. A portion of lubricating oil in the lubrication system is constantly being cleaned. The duplex oil strainer should be cleaned at least once each shift with the steam turbine in operation. When the steam turbine is taken off-line, an oil pressure signal switch automatically starts up the auxiliary lubricating oil pump to circulate lubricating oil to the steam turbine bearings until properly cooled.

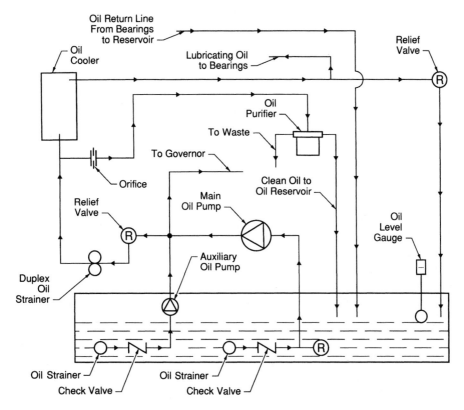

Figure 6-19. Force-fed lubrication systems supply clean lubricating oil at the proper temperature to the steam turbine.

Lubricating Oil Coolers. Lubricating oil coolers are commonly shell-and-tube coolers. Lubricating oil passes through the tubes while the cooling water discharges out the top and enters the bottom of the shell. Lubricating oil coolers should be removed from service, dismantled, and thoroughly cleaned at regular intervals. If necessary, the tube bundle should be removed and boiled in cleaning solution to remove scale. *Note:* The lubrication system should have two lubricating oil coolers and two centrifuges to allow maintenance without having to take the steam turbine off-line.

Lubricating Oil Reconditioning. Lubricating oil is reconditioned as needed to maintain the required amount for the lubrication system. The batch system requires two tanks large enough to handle all the lubricating oil in the lubrication system. One batch of clean lubricating oil is ready for service at all times. The dirty lubricating oil is removed from the system and pumped to a used lubricating oil storage tank. The lubricating oil is then pumped to a centrifuge. The centrifuge separates the impurities from the lubricating oil for removal. From the centrifuge, the lubricating oil is pumped to the clean lubricating oil storage tank until it is ready to be put back in service.

Governors

Governors control the maximum speed reached by the steam turbine. In addition, the steam turbine requires an automatic overspeed tripping device. As the steam turbine rotor (shaft) rotates, it builds up centrifugal force, which pulls the blades or buckets away from the shaft. Centripetal force holds the buckets or blades on the shaft. Centripetal force is exerted inward toward the shaft opposite centrifugal force. **See Figure 6-20.** If the centrifugal force exceeds the centripetal force, the steam turbine will destroy itself. *Note:* Centrifugal and centripetal forces also occur on the flywheel on a reciprocating engine. Overspeed tripping devices on reciprocating steam engines and steam turbines must be tested as required to ensure proper operation.

The two types of governors are mechanical and hydraulic. Mechanical governors are used on small auxiliary steam turbines. Hydraulic governors are used on large steam turbines used with generators.

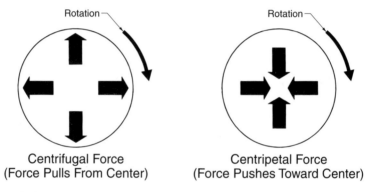

Centrifugal Force
(Force Pulls From Center)

Centripetal Force
(Force Pushes Toward Center)

Figure 6-20. Rotating elements in a steam turbine account for centrifugal and centripetal forces present.

Mechanical Governors. Mechanical governors commonly consist of a speed-sensing element and linkage that control the governor valve by the force acting on the speed-sensing element. **See Figure 6-21.** The speed-sensing element consists of weights (1) mounted on the shaft (2). The centrifugal force of the weights is opposed by a spring (3), which moves the linkage of the governor valve (4) toward the closed position.

The speed of the steam turbine is controlled by the speed-sensing element and the governor valve. Most mechanical governors can be adjusted for speed while the steam turbine is operating. The set speed is changed by tightening or loosening an adjusting spring on the linkage.

The overspeed trip functions as a limit switch and is separate from the governor. **See Figure 6-22.** A bolt is mounted in the shaft and opposed by an adjustable spring. It remains in one position until the shaft speed reaches the point where the centrifugal force on the bolt overcomes the tension of the spring. The bolt, as it moves outward, trips a latch device that releases a butterfly valve. The butterfly valve shuts off the steam supply. *Note:* Overspeed trips are designed to trip at 10% over rated RPM. For example, a steam turbine rated for 3600 RPM trips at 3600 + 10%, or 3960 RPM.

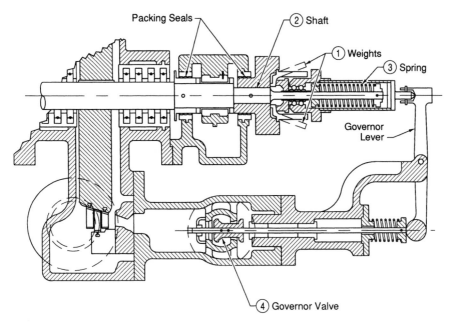

Packing Seals
② Shaft
① Weights
③ Spring
Governor Lever
④ Governor Valve

Figure 6-21. Auxiliary steam turbines commonly use mechanical governors to maintain recommended speed (in RPM).

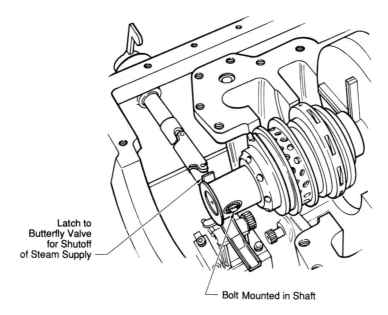

Latch to Butterfly Valve for Shutoff of Steam Supply
Bolt Mounted in Shaft

Figure 6-22. Overspeed tripping devices are required to prevent damage from speed above the maximum rated speed (in RPM).

Mechanical governors are primarily used on small auxiliary steam turbine drives for pumps, fans, and compressors. Mechanical governors control the speed of the steam turbine within 2% to 3% of its rated speed. Mechanical governors are not used on large steam turbines for driving generators where close control of speed is required.

Hydraulic Governors. Hydraulic governors are very similar to mechanical governors and consist of a speed-sensing element and speed changer. However, instead of linkage, hydraulic governors have a servo-motor, which amplifies the force to actuate admission valves.

The basic Westinghouse oil-relay governor has a closed hydraulic system, using a single source of oil for governing and steam turbine lubrication. It has four essential elements: oil pump, governor weights, governing oil valve, and servo-motor. The oil pump supplies lubricating oil for governing and lubrication. A pair of governor weights is secured to a spring steel strap mounted across the governor hub and driven from the steam turbine rotor shaft. The governor weights are not in contact with other parts and operate frictionless. The governing oil valve transforms force received from the governor weights into oil pressure changes. The governing oil valve functions as an oil relief valve, with the pressure setting dependent on the steam turbine speed. The servo-motor actuates the steam admission valve in response to governing oil pressure changes. **See Figure 6-23.**

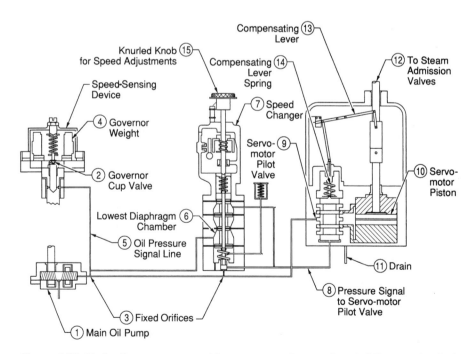

Figure 6-23. Hydraulic governors provide more accurate speed control than mechanical governors.

The main oil pump (1) supplies oil at a constant high pressure (approximately 135 psi) for operation of the governor. The small portion of oil used by the speed element enters below the governor cup valve (2) through a fixed orifice (3). An increase in speed causes an increase of centrifugal force on the governor weights (4), which causes a decrease in the governing oil pressure signal. The decreased governing oil pressure signal (5), which acts on the lowest diaphragm chamber (6) in the speed changer (7), indicates a decreased oil pressure signal (8) in the servo-motor pilot valve (9). The decreased oil pressure signal causes a decrease in the upward force on the servo-motor pilot valve. This causes a downward movement of the pilot valve, which positions the ports, admitting high-pressure oil to the chamber beneath the servo-motor piston (10) and simultaneously connecting the chamber above the piston to the drain (11). As the piston moves upward, it repositions the steam admission valves (12). The compensating lever (13) and spring (14) return the pilot valve to mid-position. The steam turbine speed can be adjusted by hand using the knurled knob (15) on top of the speed changer or automatically from a remote location.

The Westinghouse overspeed trip is separate from the governor. It operates by centrifugal force and is triggered by a spring-held plunger in the governor end of the steam turbine shaft. At the proper setting, approximately 10% overspeed, it trips and closes the throttle valve.

The basic GE oil-relay governor, like the Westinghouse oil-relay governor, is also a closed hydraulic system, but uses separate oil pumps for lubrication and governor control. Speed is controlled by a centrifugal governor, using an oil-operated valve gear that cuts out steam admission valves of the first stage of the steam turbine. The governor is a centrifugal weight governor. The oil for operating the governor is supplied by an oil pump driven from the same secondary shaft that drives the governor weights. Movement of the governor weights positions the pilot valve, controlling the flow of oil to and from the primary relay. **See Figure 6-24.**

As the speed changes, the governor weights (1) move the pilot valve (2) that controls the oil to the primary relay cylinder (3). This moves the governor lever (4) and secondary relay (5), which controls the oil to the operating piston (6) to reposition the steam admission valves.

An emergency governor is separate from the main governor valve system. It consists of a chamber in which a bolt and controlling spring are mounted in an extension of the main shaft. A speed increase of 10% above normal increases centrifugal force enough to cause the bolt to strike a trip. This releases the latch that holds the throttle valve open.

Steam turbines that have extraction steam for process must have a governor that controls the amount of extraction steam removed from the steam turbine. Extraction steam can be removed from the steam turbine at one or more stages at a controlled pressure for plant process. If the steam turbine requires more steam to maintain its speed, the governor will throttle back on the extraction valves.

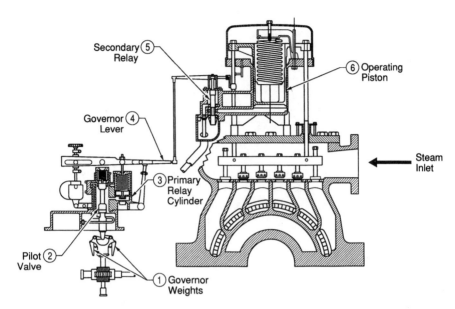

Figure 6-24. The GE oil-relay governor uses a closed hydraulic system for maintaining constant steam turbine speed (in RPM).

Noncondensing and Condensing Steam Turbines

A *noncondensing steam turbine* is a steam turbine that exhausts at atmospheric pressure or above. **See Figure 6-25.** Noncondensing steam turbines are normally used in steam plants that require large amounts of low-pressure process or heating steam. For example, a paper mill that heats a considerable amount of water for process uses a noncondensing steam turbine. If the exhaust steam is not used and allowed to go to waste into the atmosphere, the steam turbine would have a very low thermal efficiency and would waste a considerable amount of energy. The exhaust steam from small, noncondensing steam turbines is sometimes used to heat feedwater in the open feedwater heater. Noncondensing steam turbines must use a large amount of the available exhaust steam to maintain maximum efficiency. In addition, noncondensing steam turbines have no condensate returns.

A *condensing steam turbine* is a steam turbine that allows condensate to be reclaimed for use in the system. **See Figure 6-26.** The condensate is returned with some heat still in it. The heat is then reclaimed at the open feedwater heater. Condensing steam turbines are commonly used on medium to large units and operate with an exhaust pressure of 26″ Hg to 29″ Hg vacuum. The higher the vacuum is, the greater the efficiency is. This exhaust pressure allows the steam to expand to a greater volume and release more heat to perform more work. An increase in the vacuum on the exhaust of a steam turbine increases the efficiency of the steam turbine. Plant conditions determine the steam turbine best suited. Plants are designed for maximum plant efficiency and may use both condensing and noncondensing steam turbines.

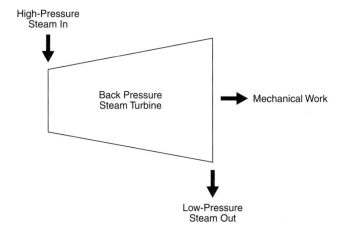

Figure 6-25. Noncondensing steam turbines exhaust at or above atmospheric pressure.

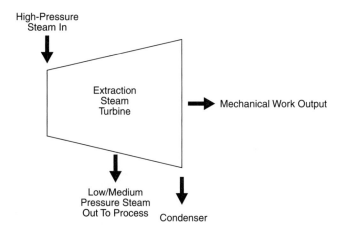

Figure 6-26. Condensing steam turbines extract steam inside turbines at intermediate pressure and the remaining steam is expanded and condensed.

Condensers

A *condenser* is a boiler auxiliary used to reduce the back pressure on the steam turbine, thereby reducing its water rate. *Water rate* (steam rate) is the amount of steam needed per unit of output. The reduction of back pressure is accomplished by creating a vacuum at the point of exhaust. A higher vacuum produced results in less back pressure. Steam turbines can operate at higher vacuums than reciprocating steam engines. To achieve the same vacuum in reciprocating steam engines, the low-pressure cylinder and valve gear must be very large. This results in large friction losses and high construction costs. Condensers commonly used include surface condensers and barometric and jet condensers.

Surface Condensers. Surface condensers are similar to a closed feedwater heater in that the steam and cooling water do not mix. The cooling water passes through the tubes that are surrounded by the exhaust steam. **See Figure 6-27.** Surface condensers are classified as singlepass or multipass, depending on how many passes the cooling water makes before leaving the condenser.

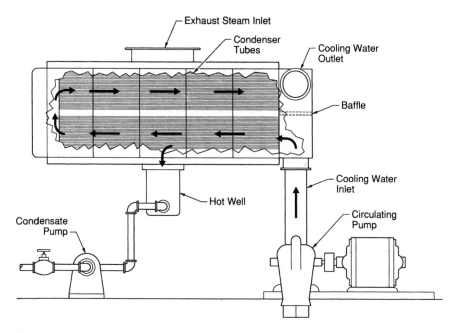

Figure 6-27. Cooling water passes through condenser tubes and does not come in contact with steam in a surface condenser.

Barometric and Jet Condensers. Barometric and jet condensers have cooling water and steam in direct contact with each other. **See Figure 6-28.** Barometric and jet condensers condense exhaust steam by routing steam into direct contact with cooling water in a process similar to an open feedwater heater. The primary advantage of barometric and jet condensers over surface condensers is cost. The price and operating and maintenance costs are less for barometric and jet condensers than for surface condensers. Barometric and jet condensers use approximately 50% less cooling water than do surface condensers. In addition, barometric and jet condensers can operate with condensing water that contains more impurities than in surface condensers.

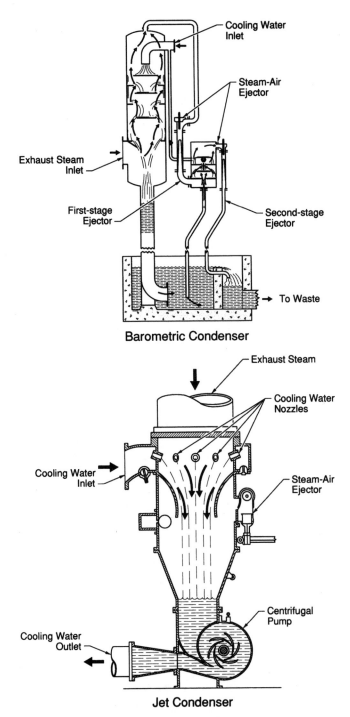

Figure 6-28. Barometric and jet condensers can operate using water with more impurities than in a surface condenser.

CONDENSER COOLING WATER

An adequate supply of cooling water is required for condensing steam in condensers. Cooling water can come from lakes, rivers, wells, ponds, or municipal water supplies. As much as 50 lb of water are used to condense 1 lb of steam. This amount varies with air temperature and cooling water temperature. When using lake or river water, the point of intake (suction) and outlet (discharge) must be considered. The cooling water suction must be upstream and the cooling water discharge must be downstream to prevent using water that already has been discharged from the plant. **See Figure 6-29.**

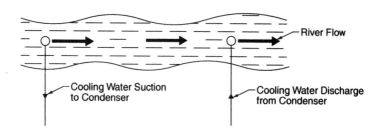

Figure 6-29. Steam plants with condensers require large quantities of cooling water and commonly use rivers as a water source.

For years it was necessary to build generating stations next to large bodies of water to supply cooling water to condensers. In time, with increasing industrial demands and population growth, it became necessary to build plants where there was no free, available water. If there is an insufficient water supply or the water is unsuitable for use in condensers, a water recirculating system is required. In the cooling process, approximately 5% to 10% of the cooling water evaporates. This loss is made up by adding water to the system. Cooling water must then be supplied, cooled, and recirculated using cooling ponds, spray ponds, or cooling towers.

Cooling Ponds

A *cooling pond* is a body of water used to provide cooling. Cooling ponds are simpler and cost less initially than cooling towers, but they are less effective and require large amounts of land area. Cooling ponds must be large enough to cool water, through surface contact of the water to the air, to the temperature desired. This may require a pond covering too large an area to be practical, in which case a spray pond can be used.

Spray Ponds

A *spray pond* is a cooling pond in which water is sprayed in order to increase the surface area of water in contact with the air by breaking up the water into a fine spray. A spray pond can be smaller than a cooling pond and have similar cooling efficiency. **See Figure 6-30.** Water from the plant is forced through nozzles under pressure ranging from 3 psi to 15 psi and broken up into a fine spray. The spray should not exceed the boundaries of the spray pond. The nozzles provide a rotating motion, which maximizes the amount of water atomized and exposed to the evaporation and cooling effect of the atmosphere. The water can then be recirculated through the condenser.

Spray pond cooling systems reduce the water temperature by 20°F to 40°F, depending on the temperature and humidity of the surrounding air. Under ideal conditions, the water temperature drops below the air temperature, but not below the wet bulb temperature. Spray pond depths are commonly 3' to 4'. Surface area, rather than depth, has the greatest effect on the cooling ability of the spray pond. Spray ponds require less land area than cooling ponds. One sq ft of surface area accommodates approximately 150 lb of water per hour.

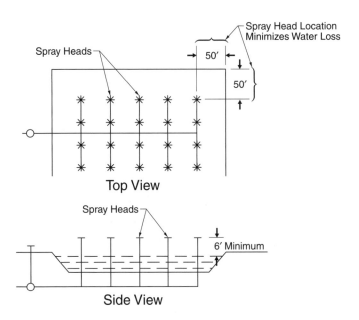

Figure 6-30. Spray ponds cool water by contact between air and a fine spray of water.

Cooling Towers

A *cooling tower* is a large auxiliary device where air moves upward to mix with falling water, resulting in cooling of the water. Cooling towers cool large amounts of water more efficiently than do cooling ponds and spray ponds. Water releases heat to the air by evaporation, convection, and radiation. Evaporation absorbs 75% to 85% of the heat. The heat that must be extracted in the cooling tower is equal to the latent heat absorbed by the cooling water while the steam is being condensed in the condenser. The water flows by gravity onto horizontal rows of cypress or redwood slats, breaking into a fine spray. The water is cooled as it falls, via gravity, through the entire cooling surface to a tank at the base of the tower. The cooled water may be lower in temperature by 30°F to 40°F. Cooling towers are classified as natural- or mechanical-draft cooling towers, based on how the air is delivered.

A *natural-draft cooling tower* is a cooling tower that cools water with air provided by natural draft. Natural-draft cooling towers are classified as atmospheric (open) or hyperbolic cooling towers. Atmospheric cooling towers are constructed with louvers or slats around the outside, with openings between the slats for maximum air flow. Hyperbolic cooling towers are constructed with closed sides and an additional height above the portion of the towers containing the cooling surfaces. **See Figure 6-31.** Hyperbolic cooling towers are built of concrete and are commonly several hundred feet tall. For optimum stability, the base can be as wide as the height. The hyperbolic shape creates a venture effect, which provides air flow without the use of draft fans. Air is drawn through grating at the base. Water is cooled inside the tower as water flows downward with gravity, and air flows upward. Hyperbolic cooling towers are expensive to build but save operating costs by eliminating draft fans.

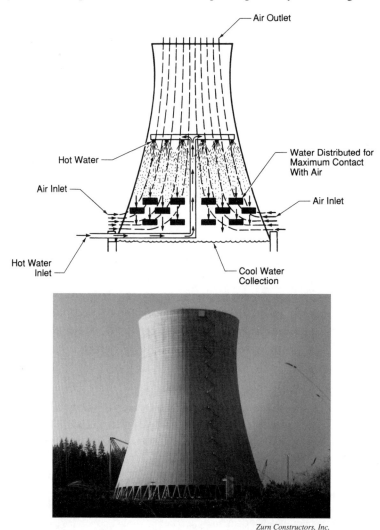

Zurn Constructors, Inc.

Figure 6-31. Hyperbolic cooling towers use air flow created by the venturi effect.

A *mechanical-draft cooling tower* is a cooling tower in which the air is forced or induced through the cooling tower. Mechanical-draft cooling towers require less land area and less head pressure and are more efficient than natural-draft cooling towers during warm weather. In warm weather, the difference between the temperature of the air inside a cooling tower and the air outside a cooling tower is not as great. This results in less draft creation, making mechanical-draft cooling towers more efficient than natural-draft cooling towers during warm weather. Mechanical-draft cooling towers are constructed as units, or cells, with the sides closed to form an airtight, watertight structure. The two types of mechanical-draft cooling towers are forced-draft and induced-draft cooling towers. Forced-draft cooling towers have a fan at the base of the tower delivering air under pressure to the interior, forcing air though the outlet at the top of the tower. **See Figure 6-32.**

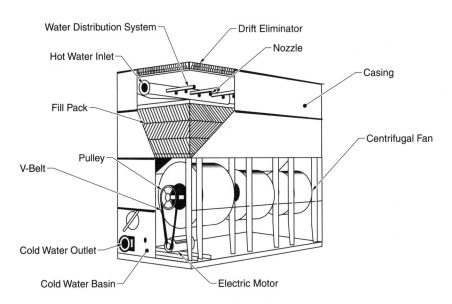

Figure 6-32. Forced-draft cooling towers force air used to cool water from the base to the top of the cooling tower.

Induced-draft cooling towers have a fan located at the top of the tower. The air is drawn in through louvers at the base of the tower. Water is cooled as it flows downward, via gravity, in a fine spray through the cooling tower. As air passes upward through the cooling tower, the water releases heat to the rising column of air by evaporation, convection, and radiation. **See Figure 6-33.** In combination systems, air is both forced and induced through the cooling tower.

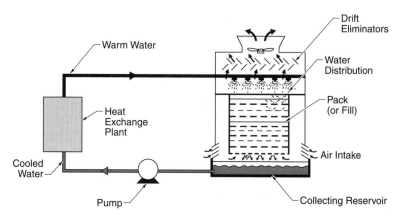

Figure 6-33. Induced-draft cooling towers pull air used to cool water with fans located at the top of the tower.

Water Flow Rates

The flow rate of water used for cooling is measured in gallons per minute (gpm). The amount of water required for condensing steam in a condensing steam turbine varies, depending on the amount of steam to be condensed, temperature of the steam at the condenser, temperature of the cooling water available, and type of condenser used.

On a surface condenser, to find the pounds of cooling water needed per pound of steam, the following equation is used:

Pounds of water per Pound of steam =

$$\frac{Enthalpy\ of\ exhaust\ steam - Enthalpy\ of\ condensate}{Temperature\ of\ cooling\ water\ leaving - Temperature\ of\ cooling\ water\ entering}$$

$$lb\ water/lb\ steam = \frac{H_e - H_c}{T_1 - T_2}$$

lb water/lb steam = pounds of cooling water needed per pound of steam

H_e = enthalpy of exhaust steam

H_c = enthalpy of condensate

T_1 = temperature of cooling water leaving

T_2 = temperature of cooling water entering

To find the total amount of water needed to condense steam, the following equation is used:

Total pounds of water per hour =
  Pounds of steam per hour × Pounds of water per Pound of steam

total lb water/hr = lb steam/hr × lb water/lb steam

EXAMPLES

1. A condenser must condense 2000 lb of steam/hr. The condensate temperature is 110°F, and the steam has a heat content of 1023 Btu/lb. The cooling water enters at a temperature of 70°F and leaves at 100°F. How much water is needed to condense 1 lb of steam? What is the total amount of water needed per hour to condense 2000 lb of steam?

$$lb\,water/lb\,steam = \frac{H_e - H_c}{T_1 - T_2}$$

$$lb\,water/lb\,steam = \frac{1023 - (110 - 32)}{100 - 70}$$

$$lb\,water/lb\,steam = \frac{945}{30}$$

$lb\,water/lb\,steam = \mathbf{31.5}$

$total\,lb\,water/hr = lb\,steam/hr \times lb\,water/lb\,steam$

$total\,lb\,water/hr = 2000 \times 31.5$

$total\,lb\,water/hr = \mathbf{63,000}$

2. A condenser must condense 3600 lb of steam/hr. The steam has a heat content of 1022 Btu/lb and the condensate temperature is 126°F. The cooling water enters at a temperature of 60°F and leaves at 75°F. How much water is needed to condense 1 lb of steam? What is the total amount of water needed per hour to condense 3600 lb of steam?

$$lb\,water/lb\,steam = \frac{H_e - H_c}{T_1 - T_2}$$

$$lb\,water/lb\,steam = \frac{1022 - (126 - 32)}{75 - 60}$$

$$lb\,water/lb\,steam = \frac{928}{15w}$$

$lb\,water/lb\,steam = \mathbf{61.9}$

$total\,lb\,water/hr = lb\,steam/hr \times lb\,water/lb\,steam$

$total\,lb\,water/hr = 3600 \times 61.9$

$total\,lb\,water/hr = \mathbf{222,840}$

To find the approximate pounds of cooling water needed per pound of steam when using barometric or jet condensers, the following equation is used:

Pounds of water per Pound of steam =

$$\frac{Enthalpy\ of\ exhaust\ steam - \left(Temperature\ of\ hot\ well - 32\right)}{Temperature\ of\ hot\ well - Temperature\ of\ cooling\ water}$$

$$lb\ water/lb\ steam = \frac{H_e - \left(T - 32\right)}{T - t}$$

lb water/lb steam = pounds of cooling water needed per pound of steam

H_e = enthalpy of exhaust steam

T = temperature of hot well

t = temperature of cooling water

Note: The hot well is located at the bottom of the barometric or jet condenser. The condensate collects in the hot well and is removed by a centrifugal pump.

EXAMPLES

1. The temperature of a hot well is 95°F and the enthalpy of the exhaust steam is 1050 Btu/lb. The cooling water temperature is 70°F. How much cooling water is needed to condense 5000 lb of steam/hr exhausting into a barometric condenser?

$$lb\ water/lb\ steam = \frac{H_e - \left(T - 32\right)}{T - t}$$

$$lb\ water/lb\ steam = \frac{1050 - \left(95 - 32\right)}{95 - 70}$$

$$lb\ water/lb\ steam = \frac{987}{25}$$

$$lb\ water/lb\ steam = \mathbf{39.48}$$

Total lb water/hr = 39.48 × 5000

Total lb water/hr = **197,400**

2. The temperature of a hot well is 104°F and the enthalpy of the exhaust steam is 1075 Btu/lb. The cooling water temperature is 75°F. How much cooling water is needed per pound of steam?

$$lb\ water/lb\ steam = \frac{H_e - \left(T - 32\right)}{T - t}$$

$$lb\ water/lb\ steam = \frac{1075 - \left(104 - 32\right)}{104 - 75}$$

$$lb\ water/lb\ steam = \frac{1003}{29}$$

$$lb\ water/lb\ steam = \mathbf{34.59}$$

This amount shows the large quantities of water required when operating a steam turbine condensing. Failure to provide adequate cooling water will cause a reduction in vacuum, possibly resulting in plant shutdown.

Heat Loss in Condensers

A large amount of the total heat supplied to a steam turbine is lost in the condenser. This loss is a result of the high latent-heat value of water vapor. A method commonly used to reduce this loss is to extract or bleed steam. Extracting or bleeding steam reduces the amount of steam to be condensed in the condenser, which reduces heat loss. To find the amount of heat lost in the condenser, the following equation is used:

$$Heat\ loss = \frac{Enthalpy\ of\ exhaust\ steam - Enthalpy\ of\ condensate}{Enthalpy\ of\ steam\ at\ throttle}$$

$$Heat\ loss = \frac{H_e - H_c}{H_a}$$

Heat loss = heat loss (expressed as a percentage)

H_e = enthalpy of exhaust steam

H_c = enthalpy of condensate

H_a = enthalpy of steam at throttle

Note: Steam at the exhaust end of the steam turbine is not 100% evaporated. In practice, there is 5% to 10% moisture content, which decreases the heat lost to the cooling water.

EXAMPLES

1. A 2500 kW steam turbine is operated with a steam pressure of 200 psia at 600°F. It exhausts into a condenser with a vacuum of 28½″ Hg. How much heat is lost to the cooling water in the condenser?

From Superheated Steam tables:

H_e = 1100.9 Btu/lb

H_c = 90 − 32 Btu/lb

H_a = 1322 Btu/lb

$$Heat\ loss = \frac{H_e - H_c}{H_a}$$

$$Heat\ loss = \frac{1100.9 - 58}{1322}$$

Heat loss = **78.89%**

2. A 10,000 kW steam turbine is operated with a steam pressure of 450 psia at 900°F. It exhausts into a condenser with a vacuum of 28" Hg. How much heat is lost to the cooling water in the condenser?

From Superheated Steam tables:

H_e = 1105.2 Btu/lb

H_c = 100 – 32 Btu/lb

H_a = 1468 Btu/lb

$$Heat\ loss = \frac{H_e - H_c}{H_a}$$

$$Heat\ loss = \frac{1105.2 - 68}{1468}$$

Heat loss = **70.65%**

Condenser Auxiliaries

A steam turbine that operates condensing can reduce the back pressure on the steam turbine. This reduces the water rate, which increases the steam turbine efficiency. However, to operate the condenser, certain auxiliaries are required, including circulating pumps, condensate pumps, air ejectors, and atmospheric relief valves.

The circulating pump for a surface condenser is commonly a centrifugal pump that can handle large volumes of water. Circulating pumps can be driven by an electric motor or a steam turbine. Suction for the pump can come from a river or lake, cooling pond, spray pond, or a cooling tower. The circulating pump discharges through the condenser before returning the water to the cooling water source.

The condensate pump for a surface condenser is a centrifugal unit. Suction comes from the hot well of the condenser and discharges the condensate into the feedwater heater. In some installations, this condensate is used to condense the vapor in air ejector condensers before going to the feedwater heater. Condensate can also be used for sealing purposes on the low-pressure end of the steam turbine.

Air ejectors remove air and other noncondensable gases from the condenser, which results in a higher vacuum on the condenser. In most cases, two or more air ejectors are necessary. **See Figure 6-34.**

Atmospheric relief valves function in the same way as a back pressure valve. In noncondensing plants, an atmospheric relief valve is a back pressure valve. **See Figure 6-35.** In condensing plants, it is an atmospheric relief valve. The purpose of an atmospheric relief valve is to prevent pressure buildup on the exhaust side of reciprocating engines and steam turbines. Atmospheric relief valves are located on the main exhaust line or directly at the top of the condenser as close to the exhaust steam inlet as possible. The atmospheric relief valve stays in the closed position when there is a vacuum in the condenser. If the vacuum is lost, the atmospheric relief valve opens, allowing the reciprocating engine or steam turbine to exhaust into the atmosphere. This protects the condenser from excessive steam pressure. Most atmospheric relief valves are designed to close automatically when the vacuum in the condenser has been restored. This action is similar to boiler safety valves after pressure has been released. Atmospheric relief valves have a water seal to prevent air leakage into the condenser. A stationary engineer must adjust the water seal so there is a constant drip coming from the overflow to ensure the valve is sealed. The atmospheric relief valve should be inspected whenever the plant is down to check for proper operation.

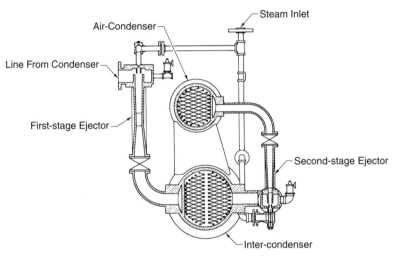

Figure 6-34. Air ejectors remove air and noncondensable gases and require minimal maintenance.

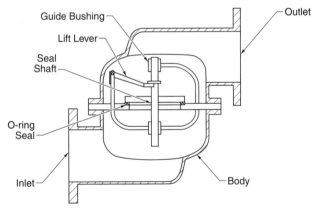

Figure 6-35. Atmospheric relief valves are used to prevent excessive back pressure in the exhaust piping of reciprocating steam turbines.

CHAPTER 6 LEARNER RESOURCES

ATPeResources.com/QuickLinks
Access Code: 735728

STEAM TURBINE OPERATION AND MAINTENANCE

7

Steam turbines operate at very high speeds. For safe and efficient operation, close tolerances must be maintained. Axial and radial clearances prevent contact between moving blades, stationary blades, and the steam turbine casing. Proper lubrication and startup and shutdown procedures must be followed to prevent damage to equipment and/or injury to plant personnel. Operating principles of steam turbines include those applied to torque, steam volume changes, water rate, and thermal efficiency.

STEAM TURBINE OPERATION

Steam turbines must be operated using manufacturer's recommendations regarding the proper lubricating oil, warm-up time, and cooling time when starting and securing the steam turbine. Specific starting procedures must be followed for startup and cutting in steam turbines and auxiliary steam turbines. The stationary engineer uses information regarding rim speed, steam volume changes, water rate, and thermal efficiency to assess steam turbine operating efficiency.

Steam Turbine Startup

The first step in starting up a steam turbine that generates electricity and supplies process steam is preparing the boiler and bringing it on-line. When the boiler room is ready, the main steam line to the steam turbine or turbo-generator should be warmed up and all condensate removed through a free-blowing drain. A *turbo-generator* is a steam turbine connected to a generator. The main steam line is warmed up by cracking the bypass valve to allow steam flow around the main steam stop valve going to the steam turbine. This step is crucial because moisture must be prevented from striking the steam turbine blades even on startup.

The auxiliary lubricating oil pump is used to supply lubricating oil to the bearings of the steam turbine during startup. **See Figure 7-1.** Steam from the same steam line to the steam turbine is used to drive the auxiliary lubricating oil pump. A steam control valve regulates the pressure in the lubricating oil line to the steam turbine and cooler. The auxiliary lubricating oil pump is kept running after the steam turbine is off-line. This allows for the lubrication of bearings and sufficient cooling after shutdown.

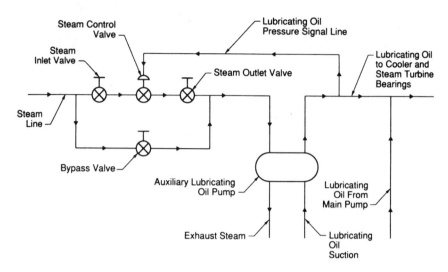

Figure 7-1. An auxiliary lubricating pump supplies lubricating oil to the steam turbine during startup and shutdown procedures.

After the main steam lines are warmed up, the steam turbine manufacturer's recommendations for startup should be followed. A standard startup procedure includes the following:

1. Start the auxiliary lubricating oil pump and check the oil pressure.
2. Check the level in the lubricating oil reservoir.
3. Check the cylinder drain valves. The drain valves should be in the open position.
4. Open the gland bleed valves.
5. Drain condensate from the steam header. Admit steam to the main steam header up to the throttle valve.
6. Establish cooling water flow through the condenser.
7. Establish seals on the high-pressure and low-pressure ends for starting.
8. Admit enough steam to start the rotor turning, then shut off the steam.
9. Listen for rubbing noises at the blades and clearance strips (squealer rings) and at the casing and seal locations.
10. If no rubbing noise is present, admit enough steam to maintain 200 RPM for ½ hour to warm the casing and rotor evenly.
11. Turn on cooling water to the lubricating oil cooler to maintain a 100°F outlet temperature.
12. Close the cylinder drains.
13. Adjust the high-pressure and low-pressure seals for operating condition.
14. Start the condensate pump.

15. Seal the atmospheric valve.
16. Start the air ejectors.
17. Bring the steam turbine slowly up to rated speed and allow the governor to take over.
18. Activate the overspeed trip to check operation.
19. Close the steam throttle.
20. Reset the overspeed trip. Bring the steam turbine up to rated speed and verify that the governor is working.
21. Check the lubricating oil, air ejector cooling water, and generator coolant temperatures.
22. Place the steam turbine on-line as quickly as possible with at least 20% of the rated load.
23. Cut in bleed or extraction steam lines.
24. Check turbo-generator operation frequently during the first few hours of operation. Follow the established schedule for checking proper operation.

Auxiliary Steam Turbine Operation

Auxiliary steam turbines are small steam turbines used to drive boiler and engine room auxiliaries such as feedwater pumps, draft fans, blowers, circulating pumps, compressors, and condenser cooling water pumps. Auxiliary steam turbines are smaller than steam turbines used for generating electricity and require different startup procedures. Like large steam turbines, startup procedures must be followed to protect auxiliary steam turbines from damage. Damage to auxiliary steam turbines is commonly caused by improper drainage of condensate, insufficient lubricating oil, improper warm-up time, or governor malfunction. A standard startup procedure includes the following steps:

1. Open exhaust valve.
2. Check that all drains are open.
3. Check all lubrication.
4. Admit enough steam to start the rotor. Listen for rubbing or extraordinary noises.
5. Open the steam valve slowly, bring the steam turbine up to speed, and let the governor take over.
6. Activate the overspeed trip. **See Figure 7-2.**
7. Close the steam throttle.
8. Reset the overspeed trip.
9. Open the throttle valve. Bring the steam turbine up to rated speed. **See Figure 7-3.**
10. Close all steam drains.
11. Verify that the steam turbine governor has taken over. Bring the steam turbine up to rated speed.

Figure 7-2. The overspeed trip is activated to check for proper operation when starting an auxiliary steam turbine.

Figure 7-3. The throttle valve is opened to bring the steam turbine up to rated speed after the overspeed trip has been checked and reset.

Rim Speed

Rim speed is the distance covered by the surface of a rotating element in a given unit of time. Steam turbines can operate at much higher speeds than reciprocating steam engines because they do not have to change direction. The motion in a steam turbine is continuous rotation rather than reciprocating motion. The diameter of a steam turbine rotor has an effect on rim speed. The larger the steam turbine rotor is, the less RPM is required to obtain the necessary rim speed. Steam turbine rotors are tested at 120% of their rated RPM. The emergency governor (overspeed trip) shuts off steam to the steam turbine when it reaches 10% over its rated RPM. For any rotating element, the weight and radius are constant, whereas the rim speed can vary. To find the rim speed (measured in feet per minute) of any rotating element, the following equation is used:

$$Rim\ speed = Circumference \times RPM$$
$$Circumference = \pi \times Diameter$$
$$Rim\ speed = \pi \times D \times RPM$$
$$Rim\ speed = 3.1416D \times RPM$$

EXAMPLES

1. A steam turbine with a 3′ diameter wheel is running at 300 RPM. What is the rim speed?

   Rim speed = 3.1416*D* × *RPM*

   Rim speed = 3.1416 × 3 × 300

   Rim speed = **2827 fpm**

2. A steam turbine with an 18″ wheel diameter is running at 3600 RPM. What is the rim speed?

   Rim speed = 3.1416*D* × *RPM*

   Rim speed = 3.1416 × 1.5 × 3600

   Rim speed = **16,965 fpm**

Torque

Torque is the force on a body that causes it to turn. To find the amount of torque on a rotating element, the force acting on the element and the radius of the element must be known. The following equation is used to find torque:

Torque = *Force* × *Radius*

$$T_q = F \times R$$

With turbo-generators, torque on the shaft is found by using the following equation:

$$Horsepower = \frac{2\pi \times RPM \times Torque}{33,000}$$

$$HP = \frac{2\pi \times RPM \times T_q}{33,000}$$

$\pi = 3.1416$

RPM = number of revolutions per minute

T_q = torque (in ft-lb)

By solving for T_q

$$HP = \frac{2\pi \times RPM \times T_q}{33,000}$$

$$T_q = \frac{HP \times 33,000}{2\pi \times RPM}$$

$$T_q = \frac{HP \times 33,000}{2 \times 3.1416 \times RPM}$$

$$T_q = \frac{5252 \times HP}{RPM}$$

This equation can now be used to solve problems involving torque.

EXAMPLES

1. What is the torque on a 2500 HP steam turbine running at 3600 RPM at full load?

$$T_q = \frac{5252 \times HP}{RPM}$$

$$T_q = \frac{5252 \times 2500}{3600}$$

$T_q = \textbf{3647 ft-lb}$

2. What is the torque on a 10,000 kW steam turbine running at 1800 RPM at full load? *Note: There are 1.34 HP per kW.*

$$T_q = \frac{5252 \times HP}{RPM}$$

$$T_q = \frac{5252 \times 1.34 \times 10,000}{1800}$$

$T_q = \textbf{39,098 ft-lb}$

Steam Volume Changes

Steam volume changes as it is used in a steam turbine. As steam passes through a steam turbine, steam pressure drops but steam volume increases. The steam volume increase is caused primarily by the difference in pressure between exhaust and admission steam. The ratio between exhaust steam volume to admission steam volume is the *number of expansions*. The greater the number of expansions is, the more energy there is released from the steam. To find the number of expansions that occur in a steam turbine, the following equation is used:

$$Number\ of\ expansions = \frac{Specific\ volume\ of\ exhaust\ steam}{Specific\ volume\ of\ admission\ steam}$$

$$Exp = \frac{SV_E}{SV_A}$$

Exp = number of expansions

SV_E = specific volume of exhaust steam (in cu ft/lb)

SV_A = specific volume of admission steam (in cu ft/lb)

Note: Assume admission steam and exhaust steam are saturated, with a quality of 100%.

EXAMPLES

1. How many times is steam expanded going through a steam turbine from 120 psi to 28″ vacuum?

From Dry Saturated Steam tables:

SV_A at 120 psi = 3.34 cu ft/lb

SV_E at 28″ vacuum = 349.67 cu ft/lb

$$Exp = \frac{SV_E}{SV_A}$$

$$Exp = \frac{349.67}{3.34}$$

$Exp = \mathbf{104.69}$

2. How many times is steam expanded going through a steam turbine from 400 psia to 29″ vacuum?

From Dry Saturated Steam tables:

SV_A at 400 psi = 1.16 cu ft/lb

SV_E at 28″ vacuum = 708 cu ft/lb

$$Exp = \frac{SV_E}{SV_A}$$

$$Exp = \frac{708}{1.16}$$

$Exp = \mathbf{610.34}$

Water Rate

The *water rate* of a steam turbine is the amount of steam or water required per unit of power output. Water rate is expressed as pounds of steam per horsepower-hour (HP hr) or kilowatt-hour (kWh). A given amount of steam flow is required to produce each horsepower-hour or kilowatt-hour. Admission steam and exhaust steam conditions can affect the water rate. Admission steam can be saturated or superheated, and exhaust steam can be of high or low quality. Both superheated admission steam and low-quality, high-condensate exhaust steam lower the water rate.

Operating a steam turbine condensing has the effect of lowering the water rate compared to operating a steam turbine of the same size noncondensing. The thermal efficiency of a steam turbine is the ratio of the heat energy used in the turbine to the heat energy available in the steam. Calculating thermal efficiency is not required if a water rate reading is taken regularly. Any extraordinary change in the water rate reading indicates a change in thermal efficiency. The average water rate for a noncondensing steam turbine is 24 lb/kWh and for a condensing unit, 14 lb/kWh. The water rate of any steam turbine can be found by using the following equations:

$$Water\ rate = \frac{Pounds\ of\ steam\ per\ hour}{Kilowatts}$$

$$WR = \frac{W}{kW}$$

$$Water\ rate = \frac{Pounds\ of\ steam\ per\ hour}{Horsepower}$$

$$WR = \frac{W}{HP}$$

WR = water rate, in lb steam/kWh or lb steam/HP hr

W = pounds of steam per hour

EXAMPLES

1. A steam turbine uses 40,000 lb of steam/hr while generating 2000 kW. What is the water rate?

$$WR = \frac{W}{kW}$$

$$WR = \frac{40,000}{2000}$$

WR = **20 lb steam/kWh**

2. A steam turbine uses 25,000 lb of steam/hr to produce 2000 HP. What is the water rate?

$$WR = \frac{W}{HP}$$

$$WR = \frac{25,000}{2000}$$

WR = **12.5 lb steam/HP hr**

Water rate may also be found by using the following equation (for simplification, all heat losses have been omitted):

$$WR = \frac{Btu/kW}{H_a - H_e}$$

WR = water rate, in lb steam/kW

Btu/kW = 3413 Btu/kW, heat in 1 kW

H_a = enthalpy of steam at throttle

H_e = enthalpy of steam at exhaust

Note: Values of enthalpy of steam at throttle and at exhaust are found in Superheated Steam tables and Dry Saturated Steam tables.

EXAMPLES

1. A condensing turbo-generator is operating with a steam pressure of 400 psia at a temperature of 600°F. It exhausts into a condenser at 28″ Hg vacuum. These amounts allow for a 10% condensation of exhaust steam. What is the water rate?

$$WR = \frac{Btu/kW}{H_a - H_e}$$

Btu/kW = 3413 Btu/kW, heat in 1 kW

H_a = 1306.9 Btu/lb

H_e = 1105.2 Btu/lb

$$WR = \frac{3413}{1306.9 - 1105.2}$$

$$WR = \frac{3413}{201.7}$$

WR = **16.92 lb steam/kW**

2. A noncondensing turbo-generator is operating with a steam pressure of 400 psia at a temperature of 600°F. It exhausts at 30 psia. What is the water rate?

$$WR = \frac{Btu/kW}{H_a - H_e}$$

Btu/kW = 3413 Btu/kW, heat in 1 kW

H_a = 1306.9 Btu/lb

H_e = 1164 Btu/lb

$$WR = \frac{3413}{1309.9 - 1164}$$

$$WR = \frac{3413}{142.9}$$

WR = **23.88 lb steam/kW**

Thermal Efficiency

Thermal efficiency is the ratio of the heat energy used in the turbine to the heat energy available in the steam. Thermal efficiency is expressed as a percentage. Thermal efficiency of a steam turbine varies with the type of steam turbine and whether it is running condensing or noncondensing. In addition, steam extracted is a factor affecting thermal efficiency. Thermal efficiency of a steam turbine can be found by using the following equation, which incorporates water rate:

Thermal efficiency =

$$\frac{Btu\ per\ kilowatt}{Water\ rate\left(Enthalpy\ of\ steam\ at\ throttle - Enthalpy\ of\ condensate\right)}$$

$$TE = \frac{Btu/kW}{WR\left(H_a - H_c\right)}$$

TE = thermal efficiency

Btu/kW = 3413 Btu/kW, heat in 1 kW

WR = water rate, in lb of steam/kW

H_a = enthalpy of steam at throttle

H_c = enthalpy of condensate

Note: Values of enthalpy of steam at throttle and at exhaust are found in Dry Saturated and Superheated Steam tables.

EXAMPLES

1. A 5000 kW turbo-generator has a water rate of 10 lb of steam/kW and operates at a steam pressure of 500 psia at 750°F. It exhausts into a condenser with a vacuum of 28½″ Hg. What is the thermal efficiency of the turbo-generator?

$$TE = \frac{Btu/kW}{WR(H_a - H_c)}$$

Btu/kW = 3413 Btu/kW, heat in 1 kW

WR = 10 lb steam/kW

H_a = 1384 Btu/lb

H_c = 90

$$TE = \frac{3413}{10(1384 - 90)}$$

$$TE = \frac{3413}{10(1290)}$$

$$TE = \frac{3413}{12,900}$$

TE = **26.5%**

2. A 10,000 kW turbo-generator has a water rate of 14 lb of steam/kW and operates at a steam pressure of 400 psia at 600°F. It exhausts into a condenser with a vacuum of 28″ Hg. What is the thermal efficiency of the turbo-generator?

$$TE = \frac{Btu/kW}{WR(H_a - H_c)}$$

Btu/kW = 3413 Btu/kW, heat in 1 kW

WR = 14 lb steam/kW

H_a = 1306.9 Btu/lb

H_c = 100 Btu/lb

$$TE = \frac{3413}{14(1306.9 - 100)}$$

$$TE = \frac{3413}{14(1206.9)}$$

$$TE = \frac{3413}{16,896.6}$$

TE = **20.2%**

The enthalpy of the condensate is subtracted from the enthalpy of the admission steam because it is not used, but reclaimed. Therefore, it cannot be charged to the steam turbine efficiency.

Heat Recovered From Condensate

When a steam turbine operates condensing, not all the heat in the exhaust steam is lost to the cooling water. Some of the heat is recovered in the condensate. The amount of heat recovered in the condensate, which is a percentage of the total heat admitted, is found by using the following equation:

Percentage of heat recovered from condensate =

$$\frac{Enthalpy\ of\ condensate}{Enthalpy\ of\ admission\ steam}$$

$$HR = \frac{H_c}{H_a}$$

HR = percent heat recovered from condensate

H_c = enthalpy in condensate

H_a = enthalpy in admission steam

EXAMPLES

1. A steam turbine operates with a pressure of 450 psia at 600°F and exhausts into a 28″ vacuum. What percentage of heat is recovered in the condensate? *Note:* Convert 28″ vacuum to psia using values from Dry Saturated or Superheated Steam tables.

$$HR = \frac{H_c}{H_a}$$

$$HR = \frac{100}{1302.8}$$

HR = **0.0768**, or **7.68%**

2. A steam turbine operates with a pressure of 300 psia at 500°F and exhausts into a 28½″ vacuum. What percentage of heat is recovered in the condensate?

$$HR = \frac{H_c}{H_a}$$

$$HR = \frac{90}{1257.6}$$

HR = **0.0716**, or **7.16%**

Steam Turbine Shutdown

When shutting down a steam turbine, follow the manufacturer's recommended procedure. A standard shutdown procedure includes the following:

1. Remove extraction valves from service.

2. Reduce load to zero and quickly remove steam turbine from service.

3. Activate the overspeed trip.

4. Open the vacuum breaker.

5. Shut off the air ejector.

6. Check to make sure the auxiliary oil pump starts at the correct pressure.

7. Shut off the water to the seals.

8. Shut down the condensate pump.

9. Open all drains.

10. After the unit is cool, secure condenser cooling water.

11. Keep the auxiliary oil pump in operation until the unit is cool.

12. Secure cooling water to the oil cooler.

The main steam stop valve to the turbo-generator can be closed. Record all shutdown operations performed in the boiler room and engine room in the daily log. *Note:* Some steam turbines are equipped with a turning gear to keep the rotor moving slowly until it cools. This prevents the rotor from warping.

STEAM TURBINE MAINTENANCE

Establishing and following a preventive maintenance program reduces operation costs, prolongs equipment life, and prevents costly downtime. **See Figure 7-4.** Every plant, regardless of how large or small, should have established procedures for routine maintenance for plant personnel to follow.

Figure 7-4. Preventive maintenance includes equipment checks and established procedures for plant personnel.

Lubrication System

The steam turbine rotor is supported on a thin film of lubricating oil. The lubricating oil must be high-quality and capable of sustaining high operating temperatures without breaking down and losing the required lubricating qualities. Lubricating oil must also have the ability of having water easily separated from it. Lubricating oil that mixes with water can form a thick emulsion that affects its ability to keep the steam turbine bearings cool. The steam turbine manufacturer supplies a lubrication schedule that covers daily, weekly, monthly, and yearly schedules. In addition, the type and grade of lubricating oil are also specified by the manufacturer.

Normally a steam turbine is shut down once a year for routine servicing. At that time the auxiliary lubricating oil pump is given careful attention. The following servicing procedures are performed:

1. The steam control valve is cleaned and the diaphragm checked and replaced if bad. The valve stem packing is replaced, and the valve is checked for correct operation.

2. The stop valves located before and after the steam control valve are checked for wear and replaced if worn. The stop valve stems are repacked.

3. The bypass valve around the steam control valve is cleaned and repacked.

4. Any oil leaks on the piping or flange connections are repaired. *Note:* Oil used in steam turbine governors is flammable.

5. The steam and oil sections of the auxiliary lubricating oil pump are cleaned and checked for leaks.

6. The auxiliary lubricating oil pump unit is ready to be tested for correct operation.

Condensers

A *condenser* is a type of heat exchanger that turns exhaust steam into condensate. Surface condensers turn exhaust steam into condensate that can be reused. **See Figure 7-5.** The transfer of heat from exhaust steam to circulating water requires the condenser tubes be kept free of foreign matter. Condenser tubes must be kept from leaking and contaminating the condensate.

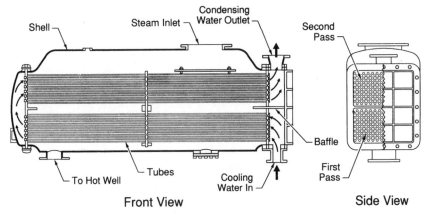

Figure 7-5. A condenser is a heat exchanger that removes heat from high-pressure refrigerant vapor.

When downtime is unacceptable, leaking tubes may be plugged until the condenser can be taken out of service for retubing. Plant size and the condition of the cooling water determine the frequency of condenser service. **See Figure 7-6.**

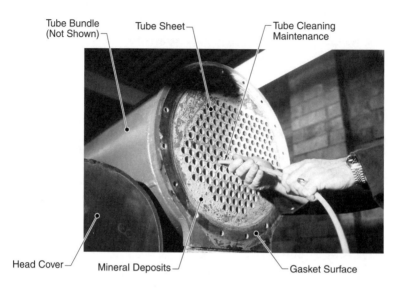

Tube Bundle (Not Shown)   Tube Sheet   Tube Cleaning Maintenance

Head Cover   Mineral Deposits   Gasket Surface

Figure 7-6. Surface condensers are serviced as required to remove sediment and scale from tubes and baffles.

The water side of a steam turbine surface condenser should be cleaned once a year. The cooling water is secured and the condenser covers are removed. The covers and tubes are flushed with water to remove all loose sediment and scale. The tubes are then brushed or turbined with a scraper to remove any scale adhering to the metal surface. When the tubes have been thoroughly cleaned, the tube sheets and condenser covers are wire-brushed. New gaskets are used when replacing the condenser covers. The nuts and bolts are wire-brushed and lubricated before they are reassembled.

After the condenser is reassembled, it should be pressure-tested for leaks. The inlet and outlet water valves should be repacked and checked for ease of opening and closing. In addition, the atmospheric relief valve is inspected and repaired if necessary.

When the condenser is opened for cleaning, the baffles should be inspected and repaired as necessary to prevent cooling water from short-circuiting the flow path. Servicing procedures for steam turbine surface condensers vary from plant to plant. In all cases, follow the manufacturer's recommended procedures and/or established plant procedures. Jet or barometric condensers are not affected by leaking tubes because the exhaust steam and cooling water mix. They require little maintenance, but the nozzles must be inspected and cleaned periodically and all joints inspected for leaks.

Air Ejectors. Air ejectors used with a surface condenser are steam-driven. They have no moving parts and require little or no maintenance. **See Figure 7-7.** Steam enters a nozzle where there is a drop in pressure and an increase in velocity. High-velocity steam passes through the suction chamber and draws air and noncondensable gases from the surface condenser. The air and noncondensable gases are discharged to the first-stage air ejector (inter-condenser). When higher vacuums are needed, a second-stage air ejector (after-condenser) is used. The second-stage air ejector takes its suction from the inter-condenser, discharging to the after-condenser. Air and noncondensable gases discharge to the atmosphere, and condensate is returned to the condensate return line. The cooling water for the inter- and after-condensers comes from the main condenser condensate pump. This cooling water prevents the tubes from becoming clogged with foreign matter. The air ejector nozzles must be clean, and the tubes in the inter- and after-condensers must be checked for leaks. During plant shutdown, the gaskets on the air ejectors should be checked and replaced if necessary. Steam ejectors are also used on disc flow barometric condensers.

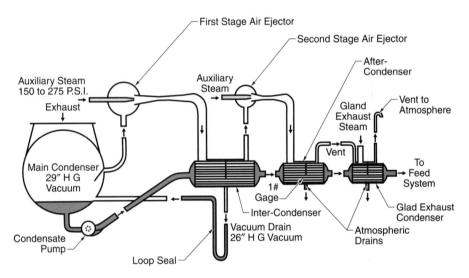

Figure 7-7. Air ejectors are steam-driven devices used on condensers to remove air and other noncondensable gases to maintain a high vacuum.

Cooling Water Pumps. The condenser cooling water pump service and maintenance are determined by the source of the cooling water. Water taken from a river can contain mud, sand, or other corrosive or abrasive materials. Some tidal river water contains salt that can cause corrosion or leave deposits that can affect pump operation. Screens in the suction inlet help prevent most solid material from entering the cooling water pump, thereby preventing wear.

A routine preventive maintenance schedule should be set up for servicing the cooling water pump. A suggested service includes examination and replacement if necessary of the following:

1. Check suction and discharge stop valves for free operation and check if new stem packing is needed.

2. Clean suction screens as required. Cleaning may have to be done more frequently than once a year, depending on the source of cooling water.

3. Remove casing to flush cooling water pump and examine rings and shaft sleeves for wear.
4. Check shaft packing or mechanical seals.
5. Clean and examine bearings.
6. Flush and replace lubricating oil.
7. Check coupling between the drive and cooling water pump for wear and alignment. When assembling the pump, new gaskets should be used. Also, the nuts and bolts for fastening the cover should be cleaned and lubricated.

Condensate Pump. Condensate pumps should be serviced once a year, or more frequently as required. The methods used for servicing condensate pumps may vary from plant to plant, based on service and performance records. Centrifugal pumps are the most common condensate pump used. A standard service procedure for condensate pumps includes examination and replacement if necessary of the following:
1. All stop valves (suction, discharge, and drain) are checked for proper operation and the stem packings renewed.
2. The pump bearings are cleaned and lubricated. Replace if worn.
3. Lift condensate pump cover to check wear on rings and shaft sleeves. Replace if worn.
4. Shaft gland packing is renewed. If the condensate pump is equipped with shaft seals, they must be checked and replaced if they show excessive wear.
5. The coupling between the drive and condensate pump must be checked for alignment and wear.

While the condensate pump is being serviced, the recirculating line and valves should also be checked. The level control on the condensate pump system maintains a constant level of condensate in the hot well. Normally, only stem packing is required for the valves. The gauge glass on the hot well must be cleaned and the gauge glass valves checked for leaks. Repack valves if necessary.

Steam Turbine Internal Inspection

Steam turbines should be opened for internal inspection every five years, or more frequently if requested by the insurance company inspector. An internal inspection normally requires a one- to two-week shutdown. All work is usually done under the supervision of the steam turbine manufacturer's representative. An internal inspection of a steam turbine includes the following:
1. Note any shaft vibration at load and no load.
2. Check bearing clearances.
3. Drain and thoroughly clean the lubrication system.
4. Check nozzle and blade clearances and make sure they are within the manufacturer's tolerances.
5. Check thrust bearings for clearance, wear, and any other faults.
6. Check rotors for trueness and cracks.
7. Check blades for cracks, and clean. They should also be checked for looseness in their attachment to the rotor or casing and for wear.

8. Clean nozzles and diaphragms of all deposits and inspect for wear, erosion, and water cutting.
9. Remove packing rings from their cells and clean and inspect.
10. Inspect the inlet steam strainer and remove all accumulated dirt and sediment.
11. Clean throttle and governing valves and regrind discs and seats if necessary.
12. Clean valve operating and governing mechanisms and remove any lost motion in the linkage by inserting new pins.
13. Inspect lubricating oil pumps for wear. Lubricating oil relief valves should be cleaned and reset.
14. Clean steam joints and repair steam and water cuts.
15. Clean oil lines and repair all leaks.

Governors

During the service shutdown of a steam turbine the governor must be checked and maintained as follows:

1. Check for lost motion in pins and bushings. This can cause hunting, which is a continual opening and closing of the governor steam valves. To correct this, new pins and bushings should be installed.
2. Check for oil leaks and repair immediately. Leaks often are at a flange or joint.
3. Inspect and lap in the governor valve or admission valves wherever necessary.
4. Remove and clean the steam inlet strainer.
5. On small turbines, check the butterfly valve for proper tripping operation. This can be set and tripped manually.
6. Check the throttle valve for oil leaks.
7. Check springs for wear and breaks.
8. When the steam turbine is test-run, check the overspeed trip and low-oil pressure trip for proper operation.
9. When painting is done around or on the steam turbine make sure the linkage and pins are not painted. This can cause poor or sluggish operation.

Always follow the manufacturer's recommended maintenance procedures when servicing governors.

Modern governors are either mechanical-hydraulic or mechanical-hydraulic with integrated digital control. The mechanical-hydraulic governors are self-contained, speed-droop governors for use on small steam-turbine driving pumps, compressors, or generators in which constant speed operation is not required. **See Figure 7-8.** These governors are typically directly coupled to the steam turbine's rotor or auxiliary shaft to sense and control turbine speed.

The mechanical-hydraulic governor that has an integrated digital controller provides accurate, constant speed control for steam turbines and engines. **See Figure 7-9.** It combines all the advantages of traditional mechanical governors with state-of-the-art digital control algorithms for optimal steam turbine operation.

Woodward, Inc.

Figure 7-8. Mechanical-hydraulic governors are self-contained, speed-droop governors that do not require constant speed operation when used in small steam-turbine driving pumps, compressors, or generators.

Woodward, Inc.

Figure 7-9. Mechanical-hydraulic governors with integrated digital controllers provide constant speed control for steam turbines using digital control algorithms.

CHAPTER 7 LEARNER RESOURCES

ATPeResources.com/QuickLinks
Access Code: 735728

WATER TREATMENT

8

Untreated water, or raw water, is never pure. Water contains varying amounts of gases, solids, and pollutants from industrial waste. Water in the form of rain absorbs gases from the air as it falls to the ground. Certain gases in water, if left untreated, can lead to corrosion and pitting of metal. Water passing through soil absorbs solids that are dissolved and carried along with the water. These solids, if left untreated, can settle on the heating surfaces of heat exchangers in the form of scale. The buildup of scale reduces efficiency by retarding heat transfer. The heat built up in the heating surfaces can cause metal failure and result in the formation of blisters, bags, and/or burned out heating surfaces. The purpose of water treatment is to provide the plant with properly treated water in sufficient quantities to meet plant needs.

BOILER WATER

Boiler water must be treated to achieve maximum plant efficiency and prevent damage to the boiler and auxiliaries. To condition water before it enters the boiler, oxygen and other noncondensable gases must be removed to prevent corrosion and pitting of the boiler metal. Calcium carbonate and magnesium carbonate found in raw water must also be treated. Both carbonates cause scale. **See Figure 8-1.** Scale can be prevented by using an alkali that changes the carbonate compounds to a precipitate. Surface tension in the steam and water drum must be reduced to prevent foaming and carryover.

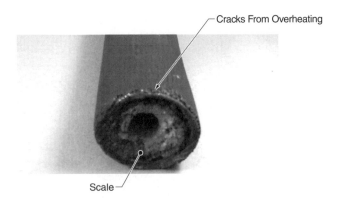

Cracks From Overheating

Scale

Figure 8-1. Scale acts like an insulator that can lead to overheating and damaged heating surfaces.

Oxygen and Noncondensable Gases

Oxygen and noncondensable gases are removed from feedwater and condensate by heating in the open feedwater heater or a deaerating feedwater heater. The oxygen and noncondensable gases are then vented to the atmosphere. In addition, heating the feedwater reduces thermal shock to the boiler and improves the thermal efficiency of the plant when using exhaust steam or steam turbine extraction steam. The temperature of the feedwater should be as high as possible. However, the temperature of the water in the feedwater heater must be carefully monitored to prevent the feedwater pump from becoming steambound.

Alkalinity

Water with a pH value of 7 is neutral, below 7 is acidic, and above 7 is alkaline. **See Figure 8-2.** The pH value of boiler water typically should be maintained between 10 and 11.5 to prevent scale formation. Acidic water will remove metal, leading to the weakening of the tubes and tube sheets, causing them to leak.

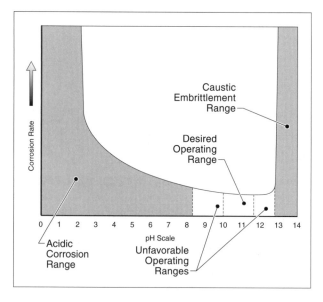

Figure 8-2. Recommended boiler water pH varies from 10 to 11.5 on the pH scale.

To test pH, a sample of the boiler water is cooled and drawn into a clean container. The pH meter is calibrated, inserted into the sample, and read directly. **See Figure 8-3.** High alkalinity can lead to caustic embrittlement. *Caustic embrittlement* is the collection of material with high alkalinity, which leads to the breakdown and weakening of boiler metal. Caustic embrittlement can cause metal to crack along the seams and at the ends of tubes in a boiler. Either case, if undetected, could lead to a boiler explosion. Alkalinity of boiler water is controlled by adding sodium hydroxide (caustic soda) to the boiler water.

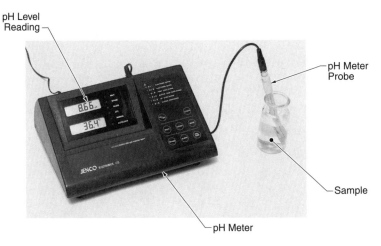

Figure 8-3. The pH value of boiler water should be kept within the range recommended by the boiler manufacturer and the water treatment company.

Surface Tension

Surface tension is a condition caused by impurities floating on top of water in the steam and water drum. Excessive surface tension prevents steam bubbles from breaking through the surface into the steam space. This causes the water level in the gauge glass to fluctuate rapidly. The gauge glass fills and the water level then drops out of sight.

Surface tension can be controlled by using a surface blowdown to remove the impurities on the water surface. In addition, a continuous blowdown (if equipped) also is used to control the proper total dissolved solids in the boiler water.

Foaming

Foaming is the formation of layers of steam bubbles trapped below the water surface in the steam and water drum. Impurities floating on the water surface, similar to a membrane stretched across the water surface, trap steam bubbles, which causes foaming. By removing floating impurities, surface tension is reduced. This allows the steam to be released, eliminating foaming and carryover resulting from foaming.

Carryover

Carryover is particles of water that leave the boiler with steam and enter the main steam line. Carryover is caused by foaming, sudden changes in steam flow from the boiler, and/or a water level that is too high. Carryover is a very dangerous condition because it causes water hammer in the steam lines, which can result in rupture of the steam header or steam line. **See Figure 8-4.** Carryover can also damage reciprocating pumps, steam engines, and steam turbines.

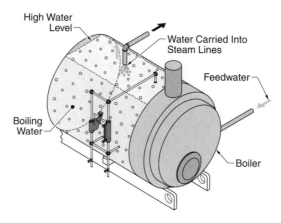

Figure 8-4. Carryover can occur when a high water level causes water particles to be carried into steam lines.

BOILER WATER ANALYSIS

Boiler water analysis is required to determine the condition of boiler water and the necessary treatment. **See Figure 8-5.** To determine the frequency and amount of blowdown and chemicals required, boiler water and condensate returns are tested. The tests are performed daily and the results are recorded in a water treatment log similar to the boiler room log. All boilers are equipped with blowdown connections to remove sludge and sediment from the boiler. The bottom blowdown connection is designed for intermittent use, which is usually once in every 8- to 24-hour period.

Figure 8-5. A boiler water test kit is part of a water analysis program.

Large boilers, instead of using a batch method of chemical treatment and a bottom blowdown one to three times a day, use a continuous chemical treatment with continuous blowdown. Continuous chemical treatment uses a chemical tank and a proportioning pump that takes its suction from the chemical tank and discharges the chemical treatment into the steam and water drum. The advantage of continuous chemical treatment with continuous blowdown is that as solids begin to form in the boiler water with the chemical treatment, they are removed through the continuous blowdown. **See Figure 8-6.** The continuous blowdown connection is located just below the low water level in the steam and water drum. The discharge from the continuous blowdown line goes to the blowdown tank or flash tank where heat recovery may be possible. Continuous blowdown allows better control of total solids in the boiler water. Bottom blowdown is still required, but is best performed when the boiler is off-line or on a light load in order to remove a greater concentration of solids.

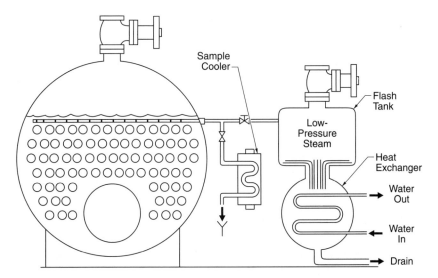

Figure 8-6. Large boilers may use a proportioning type of chemical treatment with a continuous blowdown.

Raw water contains some gases and scale-forming salts. The gases are removed by heating the water before it enters the boiler by using a deaerator and an oxygen scavenger in the boiler. The scale-forming salts, when heated, accumulate on heating surfaces as scale. When chemicals are added to the boiler water, the scale-forming salts change into a nonadhering sludge. The nonadhering sludge settles to the bottom of the boiler and is removed using the bottom blowdown valves.

Boiler water must be treated to prevent scale formation on tubes and heating surfaces, corrosion of the shell and tubes, caustic embrittlement at seams, and carryover of boiler water into superheater and steam lines. Boiler water conditions that cause scale, corrosion, caustic embrittlement, foaming, and carryover must be addressed by the stationary engineer.

1. *Scale:* Caused when calcium carbonate or magnesium carbonate is deposited on hot metal surfaces, forming a hard, brittle substance.

   Treatment: Add caustic soda and phosphate. This changes calcium carbonate and magnesium carbonate into calcium phosphate and magnesium phosphate, which are nonadhering sludges.

Tests: Hardness tests — soap hardness and phosphate tests.

Boiler water limits: Hardness — 0 parts per million (ppm); phosphate — 20 ppm to 40 ppm.

Note: The only reason for scale buildup in a boiler is improper feedwater treatment.

2. *Corrosion:* Caused by oxygen and carbon dioxide.

   Treatment: Heat water before it enters the boiler. Sodium sulfite is also added to the boiler water as an oxygen scavenger to remove oxygen from the boiler water. Oxygen present in the boiler water changes sodium sulfite to sodium sulfate.

   Test: Measure residual sulfite.

   Boiler water limits: 20 ppm to 40 ppm.

3. *Caustic embrittlement:* Caused by a high alkaline solution that works into cracks in seams of improperly caulked joints or rivets.

   Treatment: Maintain correct alkalinity levels in boiler water and control blowdowns.

   Tests: P alkalinity and M alkalinity tests.

   Boiler water limits: P alkalinity — 250 ppm to 300 ppm; M alkalinity — less than twice P alkalinity (less than 500 ppm to 600 ppm).

4. *Foaming and carryover:* Caused by high amounts of total dissolved solids, high alkalinity, or high surface tension.

   Treatment: Bottom blowdown to control solids and alkalinity. Surface blowdown to control or reduce high surface tension.

   Tests: Chlorides and conductivity meter tests.

   Boiler water limits: Chlorides — 150 ppm to 200 ppm; TDS — 2000 micromhos/cm to 2500 micromhos/cm.

Boiler water analysis is achieved by performing tests daily and recording the results. When performing boiler water analysis tests, always use the chemicals specified on test equipment labels and the recommended test procedures. Improper use can result in inaccurate test data or serious chemical reactions. Recommended chemical limits differ based on the pressures carried. Chemical companies set up test procedures best suited for boiler water treatment for the plant involved. In addition, charts listing recommended chemical levels and limits can also be obtained from chemical companies.

Boiler water treatment chemicals are added to the system based on the daily test performed. Any drastic changes in the test results should be verified and closely analyzed before any drastic change is made to the water treatment procedures. The three methods commonly used to test boiler water include the titration test, colorimetric analysis, and the instrumental test.

Titration is a common laboratory method of chemical analysis that is used to determine the concentration of an element. A reagent is prepared as a standard solution. A known concentration and volume of the water sample reacts with the solution to determine the concentration.

Colorimetric analysis is a method of determining the concentration of an element in a solution with the aid of a color reagent. As the water sample is added to the solution, a color change occurs at a given concentration.

With the instrumental test, electric/electronic instruments are used to determine the concentration of the element being tested for. With the continual improvement of electronic instruments and their capability of continuous and remote monitoring, instrumental testing is becoming more common.

The following titration tests and procedures are examples. Specific plant procedures must be followed.

Soap Hardness Test

The soap hardness test is used to determine the amount of hardness found in makeup water, condensate returns, water softener discharge, and boiler water. The hardness in condensate returns, water softener discharge, and boiler water should be maintained as close to 0 ppm as possible. The hardness in makeup water cannot be controlled. However, monitoring makeup water hardness allows for better control of the water softener and its regeneration. **See Figure 8-7.**

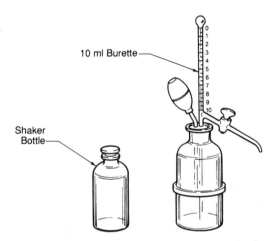

10 ml Burette

Shaker Bottle

Figure 8-7. Hardness in makeup water can be estimated using a soap hardness test.

Apparatus:
  1. 10 ml burette
  2. 58.3 ml shaker bottle
Reagent:
  1. Standard soap solution
Procedure:
  1. Fill burette to 0 mark with soap solution.
  2. Wash shaker bottle and rinse with water sample.
  3. Add 50 ml of cooled water sample to shaker bottle.

4. Add soap solution, 0.2 ml at a time. Shake after each addition of soap solution.

5. Continue adding soap solution until a permanent lather forms. (Turn shaker bottle on side. Permanent lather occurs when the lather covers entire surface of water for 5 minutes.)

6. Deduct 0.3 ml from burette reading. This is the lather factor of soap solution.

7. The burette reading minus the lather factor of soap solution is the total hardness of the sample. Multiply this reading by 20 to get ppm value and record in water treatment log.

Hardness Test

Hardness of boiler water is controlled by changing the calcium carbonate and magnesium carbonate salts to a nonadhering calcium phosphate and magnesium phosphate sludge. The sludge can then be removed from the boiler by using the continuous blowdown valves and the bottom blowdown valves. The hardness test is more accurate than the soap hardness test. **See Figure 8-8.**

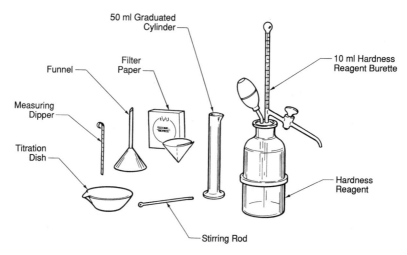

Figure 8-8. The hardness indicator used in the hardness test provides a more accurate measurement of boiler water hardness than the soap hardness test.

Apparatus:
1. 10 ml hardness reagent burette
2. 50 ml graduated cylinder
3. Titration dish
4. Stirring rod
5. Funnel and filter paper
6. Measuring dipper

Reagents:
1. Hardness indicator
2. Hardness buffer
3. Hardness reagent

Procedure:
1. Filter 50 ml of water sample into dish.
2. Add 0.2 gm of hardness indicator and stir.
3. Add 0.5 ml (13 drops) of hardness buffer and stir.
4. No color change indicates no hardness.
5. If pink appears, add hardness reagent until color changes to gray.
6. Multiply burette reading by 20 to obtain hardness in ppm. Reading should be at or near 0 ppm. Record reading in water treatment log.

Phosphate Test (PO$_4$ - Phosphate)

The amount of phosphate in the boiler water should be 20 ppm to 40 ppm to prevent scale formation. **See Figure 8-9.**

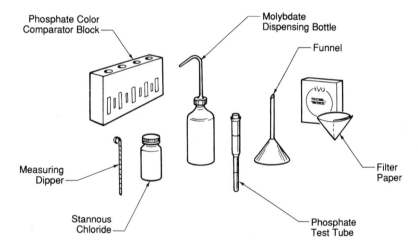

Figure 8-9. Measurements obtained with the phosphate test are used to control scale formation in boiler water.

Apparatus:
1. Phosphate test tube graduated at 5 ml and 17.5 ml
2. Measuring dipper
3. Funnel and filter paper
4. Phosphate color comparator block
5. Molybdate dispensing bottle

Reagents:
  1. Catalyzed molybdate
  2. Stannous chloride
  3. Distilled water
Procedure:
  1. Filter 2.5 ml of water sample (must be clear).
  2. Add 2.5 ml catalyzed molybdate. Mix and wait 5 minutes.
  3. Add distilled water to 17.5 ml mark.
  4. Add 0.2 gm stannous chloride.
  5. Mix and fill phosphate test tube to correct level.
  6. Match color on phosphate color comparator block for reading. Reading should be 20 ppm to 40 ppm. Record reading in water treatment log.

Alkalinity Test (P - Phenolphthalein)

The alkalinity test (P - phenolphthalein), or P alkalinity test, is used to determine the alkalinity of boiler water. The proper alkalinity prevents scale formation and caustic embrittlement. The P alkalinity range of the boiler should be 250 ppm to 350 ppm. **See Figure 8-10.**

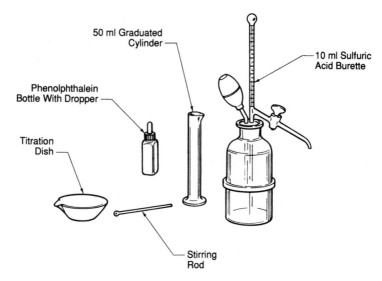

Figure 8-10. The alkalinity test uses sulfuric acid as a reagent.

Apparatus:
  1. 10 ml sulfuric acid burette
  2. 50 ml graduated cylinder
  3. Titration dish
  4. Stirring rod
  5. Phenolphthalein bottle with dropper

Reagents:
1. N/50 sulfuric acid
2. Phenolphthalein

Procedure:
1. Filter 50 ml of water sample into titration dish.
2. Add three to four drops of phenolphthalein.
3. If no pink or red appears, the P alkalinity is 0.
4. If pink or red appears, add N/50 sulfuric acid from burette. Stir until color disappears.
5. Multiply burette reading by 20 to obtain P alkalinity in ppm. Record reading in water treatment log.

Alkalinity Test (M - Methyl Purple)

The alkalinity test (M - methyl purple), or M alkalinity test, is used to determine the total alkalinity in boiler water. The proper alkalinity must be maintained to prevent corrosion from boiler water that is too acidic. **See Figure 8-11.**

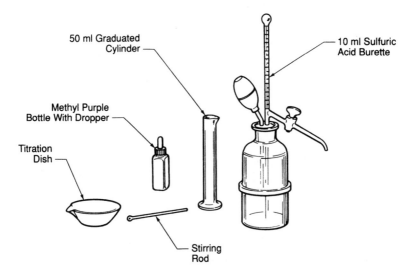

Figure 8-11. The alkalinity test using methyl purple measures alkalinity for corrosion prevention.

Apparatus:
1. 10 ml sulfuric acid burette
2. 50 ml graduated cylinder
3. Titration dish
4. Stirring rod
5. Methyl purple bottle with dropper

Reagents:
1. N/50 sulfuric acid
2. Methyl purple indicator

Procedure:

1. Using the water sample from P alkalinity test, add two or three drops of methyl purple indicator.
2. If purple appears, M alkalinity is 0.
3. If gray or green appears, add N/50 sulfuric acid from burette until color changes to purple.
4. Multiply burette reading by 20 to obtain M alkalinity in ppm. Maximum M alkalinity reading should be less than twice P alkalinity. Record reading in water treatment log.

Chloride Test (Cl - Chloride)

The chloride test determines the amount of chloride in boiler water. Some plants use the chloride measurement in boiler water, rather than the total dissolved solids present, to determine the frequency of boiler blowdowns required. **See Figure 8-12.** When using the chloride measurement test, the test is performed on a sample of boiler water and a sample of the makeup water. The *cycles of concentration* is the concentration of dissolved solids in the boiler water relative to that of the makeup water. In the case of the chloride test, the cycles of concentration is the result of the boiler water test is divided by the result of the makeup water test. The cycles of concentration should be kept as high as possible without the dissolved solids coming out of solution and collecting on the heating surface.

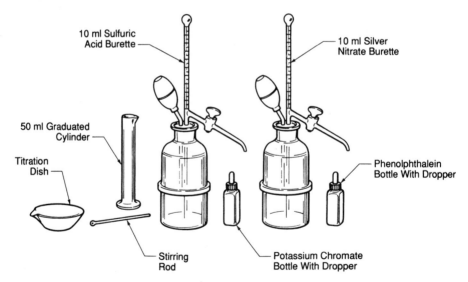

Figure 8-12. The chloride test measures chloride content to determine the frequency of blowdowns required.

Apparatus:

1. 10 ml sulfuric acid burette
2. 10 ml silver nitrate burette

3. 50 ml graduated cylinder
4. Titration dish
5. Stirring rod
6. Potassium chromate bottle with dropper
7. Phenolphthalein bottle with dropper

Reagents:
1. N/50 sulfuric acid
2. N/50 silver nitrate
3. Potassium chromate
4. Phenolphthalein

Procedure:
1. Filter 50 ml of water sample into titration dish.
2. Add three drops of phenolphthalein.
3. If red appears, add N/50 sulfuric acid until color disappears.
4. Add six drops of potassium chromate to obtain yellow.
5. Add N/50 silver nitrate until red appears (not a deep red).
6. Subtract 0.3 from burette reading, then multiply by 20 to obtain chloride in ppm. Reading should be 150 ppm to 200 ppm. Record reading in water treatment log.

Sulfite Test (SO_3 - Sulfite)

The sulfite test is used to determine the amount of sodium sulfite present in boiler water. Sodium sulfite mixes with oxygen to remove free oxygen in boiler water that may cause corrosion of the boiler metal. Sodium sulfite when combined with oxygen turns to sodium sulfate. **See Figure 8-13.**

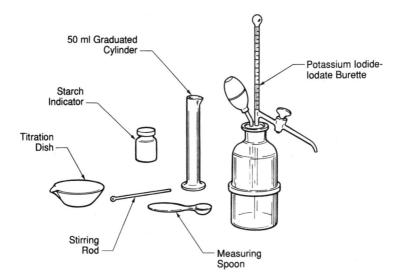

Figure 8-13. The sulfite test measures levels of sodium sulfite in boiler water used to prevent corrosion from oxygen.

Apparatus:
1. 10 ml potassium iodide-iodate burette
2. 50 ml graduated cylinder
3. Stirring rod
4. Titration dish
5. Measuring spoon

Reagents:
1. N/126 potassium iodide-iodate
2. Starch indicator

Procedure:
1. Collect water sample and protect from air.
2. Measure 50 ml of water sample into titration dish.
3. Add 0.2 or 0.3 gm of starch indicator and stir.
4. Add N/126 potassium iodide-iodate from burette until faint blue (not deep blue) appears.
5. Multiply burette reading by 10 to obtain SO_3 reading in ppm. Reading should be 20 ppm to 40 ppm. Record reading in water treatment log.

   Note: Keep the sulfite test water sample covered until ready to use to prevent the absorption of oxygen from the air.

Catalyzed Hydrazine Test

Catalyzed hydrazine is an oxygen scavenger that is primarily used on high-pressure boilers. Catalyzed hydrazine is introduced directly into the boiler and main condensate return lines to reduce or prevent return line corrosion. Catalyzed hydrazine when combined with oxygen ($N_2H_4 + O_2 \rightarrow 2H_2O + N_2$) does not leave sludge that must be removed. Boilers equipped with economizers should have an oxygen scavenging agent added in the feedwater heater to protect the economizer from corrosion. **See Figure 8-14.**

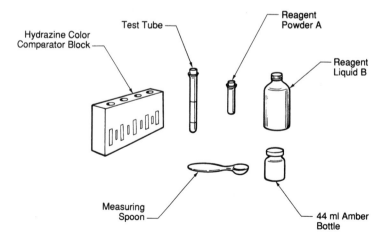

Figure 8-14. Catalyzed hydrazine is used as an oxygen scavenger in boilers equipped with economizers.

Apparatus:

1. 0 ppm to 0.5 ppm hydrazine color comparator block
2. Test tube marked 5 ml and 10 ml
3. 44 ml amber bottle for mixing hydrazine reagent
4. Measuring spoon

Reagents:

1. Reagent powder A
2. Reagent liquid B
3. Hydrazine powder A
4. Hydrazine powder B

Procedure:

1. Draw water sample. Allow bottle to overflow. Cap at once to prevent contamination from oxygen.
2. Wash, then rinse test tube with water sample.
3. Fill test tube to 5 ml mark.
4. Mix hydrazine powder A and powder B to specifications in the 44 ml amber bottle.
5. Add mixed hydrazine reagent to 10 ml mark.
6. Mix and allow to stand about 5 minutes.
7. Match color on hydrazine color comparator block for reading. Reading should be 0.05 ppm to 0.10 ppm. Record reading in water treatment log.

   Note: Hydrazine reagent must be fresh. Discard any over two weeks old. Store in a container that prevents chemical breakdown from light.

Total Dissolved Solids Test

The *total dissolved solids (TDS)* is the amount of sludge that is in suspension in boiler water. The range of TDS is between 2000 and 2500 micromhos per centimeter (micromhos/cm). Excessive TDS can cause foaming, priming, and carryover. If readings are over 2500 micromhos/cm, the boiler must be given a bottom blowdown.

The TDS meter is a portable instrument used to measure the conductance (the ability to conduct electricity) in the water being tested. **See Figure 8-15.** Conductance is the opposite of resistance. The more dissolved solids there are in the water, the higher the conductance is in micromhos/cm. This reading includes the measurement of total alkalinity, chlorides, phosphates, and other dissolved solids. TDS meter readings indicate when a boiler should be blown down. If the TDS meter readings are kept lower than the established limits, the danger of carryover and deposits of scale or sludge are reduced.

Apparatus:

1. TDS meter
2. Small brass dipper
3. 50 ml graduated cylinder
4. Celsius thermometer

Reagents:
1. Neutralizing solution (gallic acid)
2. Phenolphthalein
Procedure:
1. Measure 100 ml of water sample.
2. Add three drops of phenolphthalein.
3. Water sample will turn pink. Add 1 dipper of gallic acid until color disappears.
4. Immerse the dip cell of the TDS meter into water sample and move up and down to remove air bubbles.
5. Adjust solubridge on TDS meter for temperature of water sample and readout.
6. The readout is the specific conductance in micromhos/cm. Range should be 2000 micromhos/cm to 2500 micromhos/cm. Record reading in water treatment log.

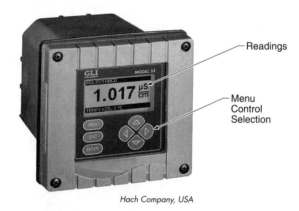

Hach Company, USA

Figure 8-15. A total dissolved solids (TDS) meter measures the amount of sludge in suspension in the boiler water.

Conductivity Test

The conductivity test measures the amount of impurities carried over with steam, and the amount of impurities returned with condensate. This test is different from the TDS test, which measures TDS in boiler water. Conductivity readings are taken using a permanently installed conductivity meter. The readings are then recorded on a chart. The chart can be used to determine when carryover is occurring and to troubleshoot the problem.

The conductivity meter functions the same as the TDS meter. It measures the amount of impurities in boiler water. Impurities in boiler water increase the conductivity of the water. The more impurities there are, the greater the amount of conductivity there is in the water. For example, a conductivity reading of raw water is a high reading. Low readings indicate low levels of impurities present. For example, a conductivity reading of distilled water is a low reading. Conductivity readings are expressed in micromhos.

BOILER WATER TREATMENT

Boiler water can be treated before it is used in the boiler (external treatment), or it can be treated in the boiler (internal treatment). Boiler water is best treated by a combination of external and internal treatment to provide the necessary protection against damage from untreated boiler water. Reducing the amount of chemicals required for boiler water treatment minimizes the cost of operating the plant. Chemicals used in boiler water treatment are often premixed by chemical companies to obtain the necessary alkalinity and boiler water characteristics.

External Treatment

External treatment is the treatment of water before it enters the boiler to remove most of the scale-forming salts and gases. The higher the boiler pressures are, the more critical the external treatment is. Boilers operating over 750 psi require feedwater to be demineralized. Large plants require more elaborate water treatment systems. External treatment of boiler water commonly includes heating feedwater, the ion-exchange (zeolite) process, the lime-soda process, and demineralizing. Although boiler water should be externally treated, it is still necessary to carry a residual amount of chemicals in the boiler for protection.

Heating Feedwater. When heating feedwater in an open or deaerating feedwater heater, some of the scale-forming salts turn into sludge and accumulate in the feedwater heater. **See Figure 8-16.** The sludge is then removed as necessary. Some feedwater heaters have a filter bed at the bottom to remove sludge. Open feedwater heaters also remove oxygen that can cause pitting of the boiler metal. For maximum efficiency, the temperature in the feedwater heater should be maintained as high as possible.

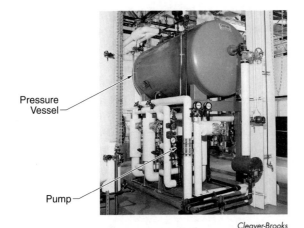

Cleaver-Brooks

Figure 8-16. A deaerator removes dissolved gases and reduces the amount of water treatment chemicals needed.

Ion-Exchange (Zeolite) Process. The ion-exchange (zeolite) process is one of the simplest methods of removing hardness from water (softening the water). **See Figure 8-17.** In the ion-exchange process, zeolite, a mined or manufactured material, is used to attract calcium carbonate, magnesium carbonate, and ferrous oxide iron. In the process, the zeolite gives up its sodium ions. After a quantity of hard water has passed over the zeolite, it no longer has any sodium ions left to exchange for the calcium, magnesium, and iron ions. The zeolite is similar to a sponge that has absorbed all the water it can. In order for it to absorb more water, the sponge must be wrung out. In the ion-exchange process, this is regeneration. Regeneration requires a solution of common salt passed over the zeolite. This releases the calcium, magnesium, and iron ions into the solution to be flushed away and replaced with sodium ions. After regeneration, the water softener is ready to be put back in service. A standard soap hardness test should be performed before and after regeneration to determine if the water softener is functioning properly.

Cleaver-Brooks

Figure 8-17. Twin unit sodium zeolite water softener removes hardness from boiler makeup water by an ion-exchange process.

Lime-Soda Process. The lime-soda process is used in large plants such as paper mills and large plants that have unusually hard water. **See Figure 8-18.** The cold lime-soda process is performed at room temperature and feedwater or makeup water passes through a lime and soda bed. It softens water by forming sludge out of calcium and magnesium compounds. The sludge is removed from water as it passes through a filter. This process requires about 6 hr to complete.

The hot lime-soda process is carried out at a temperature above 212°F. Exhaust or live steam is used. The hot lime-soda process performs the same function as the cold lime-soda process. It softens water and removes sludge and silica. The main difference between the two processes is that the hot lime-soda process can be performed continually and in a shorter period of time.

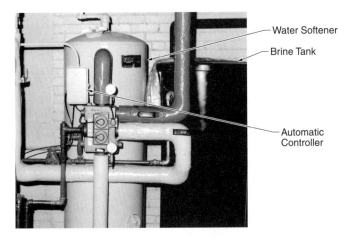

Figure 8-18. A lime-soda softener is a water softener that uses lime and soda ash to remove hardness from water.

Demineralizing. Demineralization of makeup water is necessary for boilers operating above 1000 psi. **See Figure 8-19.** Boilers operating at these high pressures require total dissolved solids be kept to a minimum. The possibility of carryover is reduced and the possibility of scale or sludge buildup is eliminated by removing the minerals before they enter the boiler. Demineralization systems are complex and require careful operation and accurate recordkeeping. The basic demineralizers are mixed-bed and two-vessel demineralizers.

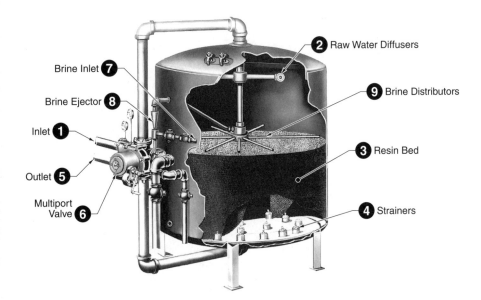

Figure 8-19. An ion-exchange water softener is commonly used to soften water.

Mixed-bed demineralizers are demineralizers that purify makeup water as it passes through a single vessel equipped with a cation and an anion solution. Mixed-bed demineralizers have a single vessel equipped with a cation (solution with positively charged ions) from a strong acid solution, and an anion (solution with negatively charged ions) from a strong alkaline solution. Makeup water is purified as it is passes through the vessel. **See Figure 8-20.**

Two-vessel demineralizers are demineralizers that purify make water as it passes through one vessel wtih a cation solution and one vessel with an anion solution. Two-vessel demineralizers have the cation in one vessel and the anion in a separate vessel. The cation is from an acid solution and the anion is from an alkaline solution. Makeup water is passed first through the cation vessel and then through the anion vessel. In the mixed-bed and two-vessel demineralizers, acid and alkaline solutions must be regenerated as they are exhausted.

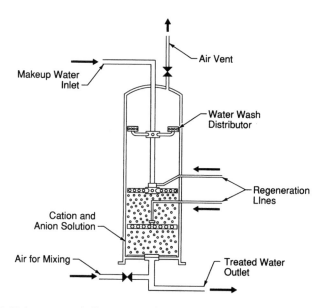

Figure 8-20. Makeup water is first passed through the mixed-bed demineralizer for boilers operating above 1000 psi.

Internal Treatment

Internal treatment is treating the water after it has entered the steam and water drum. **See Figure 8-21.** Chemicals are added directly into the boiler. Internal treatment methods required vary depending on the condition of the boiler water. Oxygen scavengers, such as sodium sulfite, mix with oxygen to form sodium sulfate, which accumulates at the bottom of the boiler. The sodium sulfate is then removed through the bottom blowdown line or the continuous blowdown line. Sodium hydroxide is added to the boiler to change calcium carbonate to sodium carbonate (nonadhering sludge). The nonadhering sludge stays in suspension and is removed by the bottom blowdown line when the boiler is at a light load. The magnesium carbonate changes to sodium carbonate and magnesium hydroxide when sodium hydroxide is added. Both of these are nonadhering sludges that are also removed using bottom blowdown or continuous blowdown.

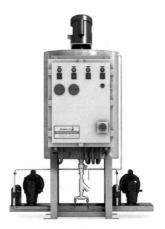

Cleaver-Brooks

Figure 8-21. Internal treatment is treating the water in the steam and water drum using a chemical feed system.

COOLING WATER TREATMENT

Like boiler water, cooling water must be properly treated to prevent scale formation, corrosion, and pitting. Cooling water must be chemically treated to remove harmful impurities that cause scale deposits on internal surfaces of piping and heat exchangers. Scale deposits, which can form at less than 100°F, coat the internal surfaces and interfere with heat transfer. These deposits, as in a boiler, reduce the capacity and efficiency of the equipment. With large accumulations of scale, the heat exchanger may have to be taken out of service for manual and/or acid cleaning. This results in costly downtime and loss of production. In addition, acid cleaning can remove some metal in the heat exchanger along with the scale, resulting in thinning of the cooling surface, potential leaks, or a costly retubing job. The treatment method used for cooling water varies according to the purpose and composition of the cooling water and whether the cooling water is to be reused.

Calcium carbonate and magnesium carbonate in cooling water are removed by external water treatment using water softening equipment before water is added to the cooling water system. Cooling water can also be treated using internal treatment by adding chemicals and compounds directly to the cooling water. This forms a sludge that is removed by bleed-off directly from the bottom of the cooling system. The *bleed-off* is a drain line connected to the cooling water system in which water is continually removed. The amount of blowdown varies and depends on the size of the unit and hardness of the water. Cooling water towers should have a continuous blowdown to control the removal of solids that continually build up. The water that is lost as a result of continuous blowdown and evaporation must be replaced with soft water.

Corrosion can occur from improperly treated cooling water. Corrosion of the cooling water tower, pipes, controls, and pumps can be controlled by monitoring the pH value and amount of oxygen and carbon dioxide in the water. Determining pH value of water is simple and is an easy method of checking the acidity of cooling water. Cooling water can be maintained at a pH of 7 to prevent corrosion and scale formation by introducing caustic soda. A slow-releasing type of phosphate is also used for scale and corrosion control. The phosphate level should be maintained at 20 ppm to 40 ppm. *Note:* These limits are suggested ranges. For maximum accuracy and efficiency, a reliable water testing firm should be consulted. A chemical control program can be designed to meet the specific needs of the plant.

Cooling water under certain conditions may also promote organic growths, such as algae. Uncontrolled algae will grow, coating condenser tubes, restricting heat transfer and obstructing water flow. **See Figure 8-22.** To control algae growth, a slow-releasing chlorine compound or copper sulfate is used. In addition, microorganisms aggravated by high alkalinity cause wood rot in cooling tower cypress or redwood slats. This is controlled by maintaining the proper pH level and chlorine content in the water.

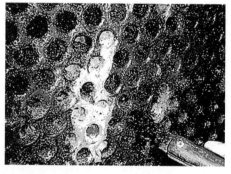

H&C Heat Transfer Solutions Inc.

Figure 8-22. Untreated or improperly treated boiler water can cause scale, corrosion, caustic embrittlement, priming, carryover, foaming, and unwanted organic growths.

If a river or lake is used as a source of cooling water, the cooling water temperature discharged must be within established limits. Environmental Protection Agency (EPA) regulations require that plants using river or lake water monitor the water's temperature. The difference between the cooling water suction temperature and cooling water discharge temperature must not exceed 10°F to prevent an adverse effect on the environment. This temperature difference may vary according to state and local requirements. Additional water cooling provisions may be required to meet the temperature limits specified. **See Figure 8-23.**

When using river or lake water, suction liner must extend well offshore. Filters are required on suction lines to provide clean cooling water. Any loss of cooling water could reduce vacuum and affect plant efficiency. A loss of cooling water can also cause the steam turbine to trip out from increased back pressure on the exhaust side.

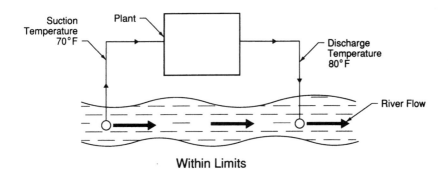

Within Limits

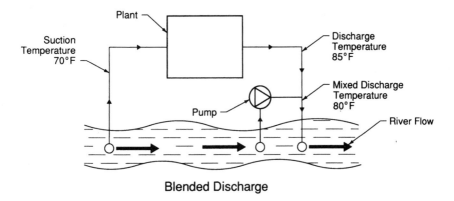

Blended Discharge

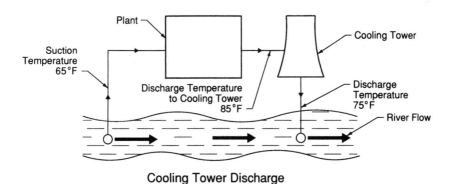

Cooling Tower Discharge

Figure 8-23. Cooling water discharged after use may have to be cooled further to comply with EPA discharge temperature regulations.

CHAPTER 8 LEARNER RESOURCES

ATPeResources.com/QuickLinks
Access Code: 735728

ELECTRICAL PRINCIPLES

9

The development of the electric generator and electric motor was a major step forward for industry. Generators used to produce electricity may be driven using steam turbines, hydroelectric power, gas turbines, diesel engines, and reciprocating steam engines. Electricity generated is then transmitted over long distances using high voltage. The voltage is then reduced at the point of use. Fundamental electrical principles are present in any electric device. A basic understanding of electrical principles is necessary for safe operation, maintenance, and troubleshooting in the plant.

ELECTRICITY THEORY

Electricity is the flow of electrons (negative charges) from an atom, which leaves the atom positively charged, to another atom, which then becomes negatively charged. **See Figure 9-1.** The most common form of electricity is static electricity. An example of static electricity is lightning. Terms used to describe various properties of electricity include the following:

- A *conductor* is a material that permits free movement of electrons, such as copper or aluminum.

- An *ampere,* or *amp* (A), is a unit of measure used to express the number of electrons flowing through a conductor per unit of time. The electron flow measured in amps is similar to the flow of water through a pipe measured in gallons per minute.

- A *volt* (V) is a unit of measure used to express the electrical pressure differential between two points in a conductor. This pressure causes electrons to move. Voltage required to make electrons move in a conductor is similar to the water pressure required to make water flow in a pipe.

- An *ohm* (Ω) is a unit of measure used to express resistance to the flow of electrons in a conductor. Resistance to flow of electrons in a conductor is similar to the friction that occurs in pipes as water is passed through.

- A *watt* (W) is the basic unit of electrical power. Power is the rate of doing work. To determine watts, multiply voltage by amperage. The term watt is also expressed as volt amps in many electrical applications.

- A *circuit* is a conductor or series of conductors through which electrons flow.
- *Current* is the flow of electrons through a circuit (measured in amps).
- *Resistance* is the opposition of a material to the flow of electrons in a circuit (measured in ohms).

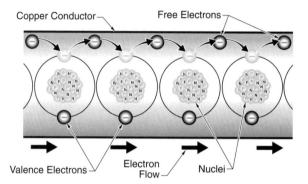

Figure 9-1. In a conductor atom, an outside force can be applied to force the atom to lose or gain valence electrons.

Electricity in use today is direct current (DC) or alternating current (AC). *Direct current* is electron flow in one direction only in a circuit. All DC circuits have a positive and a negative terminal. The positive and negative terminals establish polarity in a circuit. *Polarity* is the positive (+) or negative (−) state of an object. All points in a DC circuit have polarity. The most common power sources that directly produce DC voltage are batteries and photocells. *Alternating current* is electron flow that reverses its direction at regular intervals of time. **See Figure 9-2.**

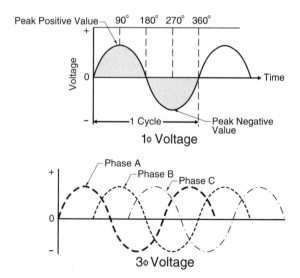

Figure 9-2. AC voltage produces a symmetrical waveform that contains 360 electrical degrees. Three-phase AC voltage produces a waveform for each phase.

AC voltage is the most common type of voltage used to produce work. AC voltage is produced by generators that produce AC sine waves as they rotate. An *AC sine wave* is a symmetrical waveform that contains 360 electrical degrees. The wave reaches its peak positive value at 90°, returns to 0 V at 180°, increases to its peak negative value at 270°, and returns to 0 V at 360°. A *cycle* is one complete positive and negative alternation of a wave form. *Frequency* is the number of cycles per second of alternating current, in hertz (Hz).

AC voltage is either single-phase (1φ) or three-phase (3φ). Single-phase AC voltage contains only one alternating voltage waveform. Three-phase AC voltage is a combination of three alternating voltage waveforms, each displaced 120 electrical degrees (one-third of a cycle) apart. Three-phase AC voltage is produced when three pairs of coils are simultaneously rotated in a generator. Three-phase AC voltage is used in industrial applications.

High AC voltages (208 V to 480 V) are used in commercial applications such as lighting, heating, and cooling. High AC voltages are also used in industrial applications to convert raw materials into usable products.

Ohm's Law

Ohm's law states that current in a DC circuit is proportional to the voltage and inversely proportional to the resistance. The formulas for Ohm's law also work with pure resistive AC circuits. They do not work with AC inductive or capacitive circuits. Ohm's law was named after George Simon Ohm, who was first to prove that this relationship existed. An easy way to remember the formulas derived from Ohm's law is to cover the unknown variable on the Ohm's law chart. The remaining variables show how to solve for the unknown variable. **See Figure 9-3.**

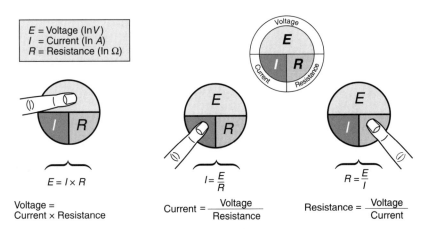

Figure 9-3. Using Ohm's law, any two known variables can be used to determine a third unknown variable.

EXAMPLES

1. Find the current when the battery voltage is 12 V and the resistance is 6 Ω.

$$I = \frac{E}{R}$$

$$I = \frac{12}{6}$$

$$I = 2 \text{ A}$$

2. Find the resistance when the voltage is 12 V and the current is 2 A.

$$R = \frac{E}{I}$$

$$R = \frac{12}{2}$$

$$R = 6 \text{ Ω}$$

3. Find the voltage when the current is 2 A and the resistance is 6 Ω.

$$E = I \times R$$

$$E = 2 \times 6$$

$$E = 12 \text{ V}$$

Power

Power is the rate at which work is performed. Power is measured in watts, based on voltage and amperage (current) in the circuit. The power of a circuit in watts is equal to the voltage times the amperage. **See Figure 9-4.**

$Watts = Voltage \times Amperage$

$W = E \times I$

The relationship between voltage and amperage is related to resistance in a circuit as stated in Ohm's law.

$W = E \times I$ and $E = IR$

Substituting IR for E,

$W = IR \times I = I^2 R$

Substituting $\dfrac{E}{R}$ for I,

$$W = \frac{E \times E}{R}$$

$$W = \frac{E^2}{R}$$

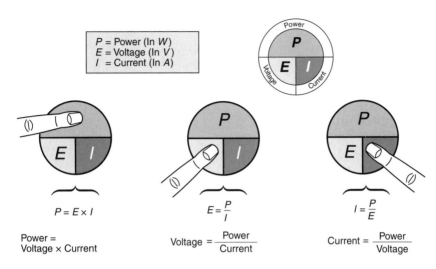

Figure 9-4. A watt is a unit of measurement for power in a circuit.

Power in a circuit is expressed by the following equations:

$$W = E \times I$$
$$W = I \times R$$
$$W = \frac{E^2}{R}$$

Power in these equations is expressed in watts. To express larger amounts of power, watts are converted to kilowatts (kW) by dividing watts by 1000.

$$kW = \frac{W}{1000}$$

1 kW = 1000 W

By multiplying the number of kilowatts used in an electrical system by the number of hours used (kWh), the total power used can be determined. If 100 kW are used continuously over 12 hours, the total power used would be

$kW \times hr = kWh$, or kilowatt-hours

$100 \times 12 = 1200$ kWh

EXAMPLES

1. How many watts would be used in each of the following circuits?

   A. 120 V at 5 A

   $W = E \times I$

   $W = 120 \times 5$

   $W = \textbf{600}$

B. 1000 Ω resistor at 1 A

$W = E \times I$

$W = (IR) \times I$

$W = I^2R$

$W = 1 \times 1 \times 1000$

$W = \mathbf{1000}$

C. 6 Ω resistor at 12 V

$W = E \times I$

$W = \dfrac{E \times E}{R}$

$W = \dfrac{E^2}{R}$

$W = \dfrac{12 \times 12}{6}$

$W = \mathbf{24}$

D. 9 A at 120 V

$W = E \times I$

$W = 120 \times 9$

$W = \mathbf{1080}$

2. What is the resistance of a 60 W, 120 V lamp?

$W = E \times I$

$W = \dfrac{E \times E}{R}$

$W = \dfrac{E^2}{R}$

$R = \dfrac{E^2}{W}$

$R = \dfrac{120 \times 120}{60}$

$R = \mathbf{240\ \Omega}$

3. Find the current for an 800 W, 120 V toaster.

$W = E \times I$

$I = \dfrac{W}{E}$

$I = \dfrac{800}{120}$

$I = \mathbf{6.66\ A}$

4. How many kilowatts are required to operate an electric motor that uses 300 A at 240 V?

$$kW = \frac{E \times I}{1000}$$

$$kW = \frac{240 \times 300}{1000}$$

$$kW = \textbf{72}$$

5. An electric motor uses 15 A at 4100 V. How many kilowatts are required to operate the electric motor?

$$kW = \frac{E \times I}{1000}$$

$$kW = \frac{4100 \times 15}{1000}$$

$$kW = \textbf{61.5}$$

Combined Ohm's Law and Power Formula

Ohm's law and the power formula may be combined mathematically and written as any combination of voltage (E), current (I), resistance (R), or power (P). *Note:* In many cases, the letter P, for power, is used instead of W, for watts, in power formula calculations. There are six basic formulas, three commonly used rearranged versions, and three other rearranged versions. **See Figure 9-5.**

Basic Formulas		
$E = R \times I$	$I = \dfrac{E}{R}$	$R = \dfrac{E}{I}$
$E = \dfrac{P}{I}$	$I = \dfrac{P}{E}$	$P = E \times I$

Rearranged Versions Commonly Used		
$R = \dfrac{P}{I^2}$	$I = \sqrt{\dfrac{P}{R}}$	$P = R \times I^2$

Rearranged Versions Typically Not Used		
$E = \sqrt{P \times R}$	$R = \dfrac{E^2}{P}$	$P = \dfrac{E^2}{R}$

Values in Inner Circle Are Equal to Values in Corresponding Outer Circle

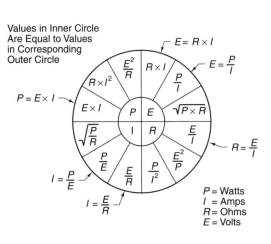

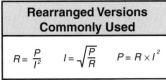

P = Watts
I = Amps
R = Ohms
E = Volts

Figure 9-5. The Ohm's law formulas can be combined with the power formulas to provide a total of 12 formulas.

Circuits

A *circuit* is a conductor or series of conductors through which electrons flow. Circuits in industrial plants commonly consist of a path through which electrons move from their source to electrically operated equipment. Conductors are commonly copper wires that connect to all parts of the circuit. Common electrical circuits are the series circuit, parallel circuit, and series-parallel circuit. **See Figure 9-6.**

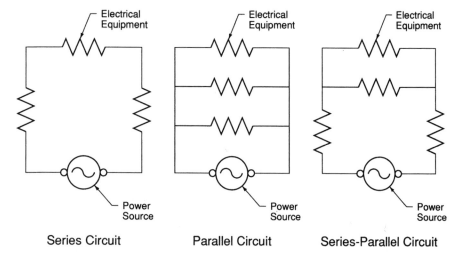

Figure 9-6. The path in which electrons flow is determined by the type of circuit.

Series Circuits. A *series circuit* is a circuit that has one path for current to flow through all devices in the circuit and that has electrical devices and/or controls in-line with each other. In a series circuit, there is only one path for the current. Each electrical device is connected together in a continuous path. Because there is only one path, current in the circuit must be the same throughout. **See Figure 9-7.**

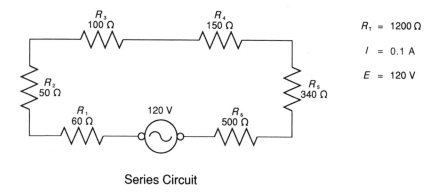

Figure 9-7. Electrons have only one path in a series circuit.

Total resistance in a series circuit is the sum of the resistances in the circuit.

$R_t = R_1 + R_2 + R_3 + ...$, etc.

R_t = total resistance, in ohms (Ω)

Voltage in a series circuit is proportionally divided between the loads in the circuit. The voltage drop for each load must be calculated by using the following formula:

$E_1 = IR_1$

E = voltage, in volts (V)

I = current, in amps (A)

R = resistance, in ohms (Ω)

The total voltage is then equal to the sum of the voltage drops.

$E_t = E_1 + E_2 + E_3 + ...$, etc.

Current in a series circuit is dependent on the voltage applied to the circuit and the total resistance within the circuit.

$$I = \frac{E}{R_t}$$

I = current, in amps (A)

E = voltage, in volts (V)

R_t = total resistance, in ohms (Ω)

EXAMPLES

1. If six resistors are connected in series, what will be the total resistance in the circuit? **See Figure 9-8.**

$R_t = R_1 + R_2 + R_3 + R_4 + R_5 + R_6$

$R_t = 60 + 50 + 100 + 150 + 340 + 500$

$R_t = \mathbf{1200\ \Omega}$

To find the current in the circuit,

$$I = \frac{E}{R_t}$$

$$I = \frac{120}{1200}$$

$I = \mathbf{0.1\ A}$

Ohm's Law can be applied to the entire circuit or to each part of the circuit.

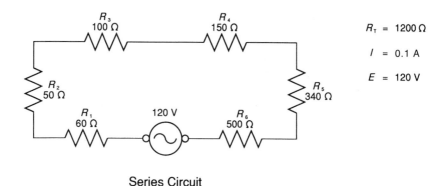

Series Circuit

Figure 9-8. Total resistance in a series circuit is determined by the sum of all resistances in the circuit.

2. What is the voltage across each resistor in the circuit?

$E = IR_1$ $E = IR_2$ $E = IR_3$
$E = 0.1 \times 60$ $E = 0.1 \times 50$ $E = 0.1 \times 100$
$E = \textbf{6 V}$ $E = \textbf{5 V}$ $E = \textbf{10 V}$

$E = IR_4$ $E = IR_5$ $E = IR_6$
$E = 0.1 \times 150$ $E = 0.1 \times 340$ $E = 0.1 \times 500$
$E = \textbf{15 V}$ $E = \textbf{34 V}$ $E = \textbf{50 V}$

Note: The sum of the voltages across all resistors is equal to the applied voltage.

$E = E_1 + E_2 + E_3 + E_4 + E_5 + E_6$

$E = 6 + 5 + 10 + 15 + 34 + 50$

$E = \textbf{120 V}$

Parallel Circuits. A *parallel circuit* is a circuit that has two or more paths for current to flow. In a parallel circuit, each path or branch is connected directly across the voltage source. The voltage is the same across each branch. The current through each branch is independent of the others and depends only on the resistance of that particular branch and voltage. The current through each branch is calculated separately. **See Figure 9-9.**

$$I_1 = \frac{E}{R_1}$$

$$I_1 = \frac{30}{5}$$

$$I_1 = \textbf{6 A}$$

$$I_2 = \frac{E}{R_2}$$

$$I_2 = \frac{30}{10}$$

$$I_2 = 3\,\text{A}$$

$$I_3 = \frac{E}{R_3}$$

$$I_3 = \frac{30}{30}$$

$$I_3 = 1\,\text{A}$$

The total current of the parallel circuit is equal to the sum of the currents through the individual branches.

$$I_t = I_1 + I_2 + I_3$$

$$I_t = 6 + 3 + 1$$

$$I_t = 10\,\text{A}$$

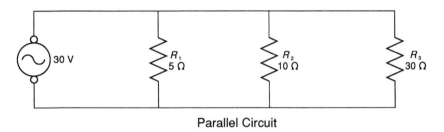

Parallel Circuit

Figure 9-9. Electrons have more than one path in a parallel circuit.

To find the total resistance of a parallel circuit, use the following equation:

$$\frac{1}{R_t} = \frac{1}{R_1} + \frac{1}{R_2} + \frac{1}{R_3} + \ldots, \text{etc.}$$

Using the values from Figure 9-9, the total resistance is equal to 3 Ω.

$$\frac{1}{R_t} = \frac{1}{R_1} + \frac{1}{R_2} + \frac{1}{R_3}$$

$$\frac{1}{R_t} = \frac{1}{5} + \frac{1}{10} + \frac{1}{30}$$

$$\frac{1}{R_t} = \frac{6}{30} + \frac{3}{30} + \frac{1}{30}$$

$$\frac{1}{R_t} = \frac{10}{30} = \frac{1}{3}$$

$$R_t = 3\,\Omega$$

As in the series circuit, the total power used is equal to the sum of the power consumed in each electrical device.

$W_1 = EI_1 = 30 \times 6 = 180$ W

$W_2 = EI_2 = 30 \times 3 = 90$ W

$W_3 = EI_3 = 30 \times 1 = 30$ W

Total power $= 180 + 90 + 30 = $ **300 W,** or

$W_t = EI_t = 30 \times 10 = $ **300 W**

Series-Parallel Circuits. A *series-parallel circuit* is a circuit that consists of a combination of series and parallel circuits. It can be complicated but the same basic rules used in solving problems in series and parallel circuits are applied. The series-parallel circuit is simplified by separating and solving the series and parallel problems within it. **See Figure 9-10.**

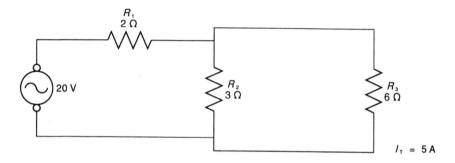

Series-Parallel Circuit

Figure 9-10. In a series-parallel circuit, current flows through all the series loads and divides through the parallel loads according to the resistance in each branch.

R_2 and R_3 are in parallel, resulting in a total resistance of

$$\frac{1}{R_{2,3}} = \frac{1}{R_2} + \frac{1}{R_3}$$

$$\frac{1}{R_{2,3}} = \frac{1}{3} + \frac{1}{6}$$

$$\frac{1}{R_{2,3}} = \frac{1}{2}$$

$R_{2,3} = $ **2 Ω**

$R_{2,3}$ and R_1 are in series, resulting in a total resistance of

$R_t = R_{2,3} + R_1$

$R_t = 2 + 2$

$R_t = $ **4 Ω**

The total current in the circuit is determined using Ohm's law:

$$I_t = \frac{E}{R_t}$$

$$I_t = \frac{20}{4}$$

$$I_t = \textbf{5 A}$$

The voltage across each resistor is

$$E_{R1} = I_t R_1$$

$$E_{R1} = 5 \times 2$$

$$E_{R1} = \textbf{10 V}$$

$$E_{2,3} = I_t R_{2,3}$$

$$E_{2,3} = 5 \times 2$$

$$E_{2,3} = \textbf{10 V}$$

The voltage across R_2 and R_3 is 10 V. The current is

$$I_2 = \frac{E}{R_2}$$

$$I_2 = \frac{10}{3}$$

$$I_2 = \textbf{3.33 A}$$

$$I_3 = \frac{E}{R_3}$$

$$I_3 = \frac{10}{6}$$

$$I_3 = \textbf{1.66 A}$$

Magnetism

Magnetism is invisible lines of force that exist between the north and south poles of all permanent magnets or electromagnets. These lines of force, or magnetic fields, connect the north and south poles of permanent magnets. The earth is a permanent magnet that causes a compass needle to point toward the north pole. Magnetic fields make possible the operation of electric motors, generators, measuring instruments, and other electrical apparatuses.

Magnetism is based on the principle that a material such as a piece of iron or steel consists of millions of tiny magnets invisible without the aid of a microscope. The material may consist of atoms or molecules so aligned as to form iron or steel crystals. Before a bar of iron or steel is magnetized, these millions of tiny magnets have no organization or order. If the north pole of an inducing magnet is drawn over the bar, it attracts the south poles of the tiny magnets and turns them so they align themselves in a given direction. This alignment of tiny magnets within the material gives the bar a definite north pole at one end and a south pole at the other end. **See Figure 9-11.**

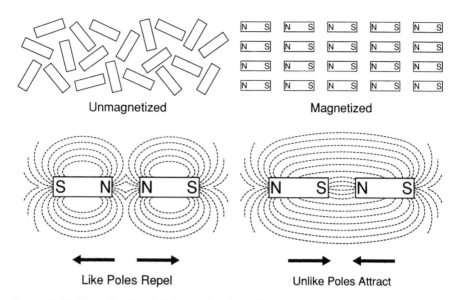

Unmagnetized   Magnetized

Like Poles Repel   Unlike Poles Attract

Figure 9-11. Magnetized materials consist of a definite organization of magnetic material.

The fundamental law of magnetic fields is that like magnetic poles repel each other. Unlike magnetic poles attract each other. The amount of magnetic force existing in the space surrounding a magnet can be estimated by measuring its lifting power. Lifting power varies depending on the material lifted, and the strength of a magnet is best measured by the force of attraction or repulsion that the magnet has on another magnet of known strength.

MOTORS

Motors convert electrical energy into mechanical energy used to drive various types of machinery. The development of the electric motor has had a profound effect on modern industry. Some believe that the electric motor is to modern industry what the steam engine was to the industrial revolution. The electric motor has been improved over the years. The development of extremely efficient insulating varnish and engineering technology has made smaller, more powerful electric motors possible. Low operating costs and ease of operation of electric motors has resulted in motors replacing steam engines and steam turbines to drive auxiliary equipment.

The following are terms associated with classifying motors:
- Horsepower—fractional horsepower (such as ⅙, ¼, ¾ horsepower) or 1 horsepower and greater (such as 1, 3, 10, 300 horsepower).
- Open casing motors—motor casing open for maximum cooling.
- Closed casing motors—motor casing closed for protection from dirt, dust, and other foreign matter from the working environment.

- Explosionproof motors—motor casing sealed to prevent arcing from ignitable materials such as dust, chemicals, and vapors in the working environment.
- Waterproof motors—motor casing sealed for protection from moisture and/or water in the working environment.
- Voltage/current—AC or DC.
- Working environment—explosive conditions, dust, temperature, moisture, water, and chemicals.
- Starting current available in circuit.
- Torque requirements (starting load on motor).
- Single speed—constant RPM.
- Variable speed—variable RPM.
- Constant operation—continuous service.
- ON/OFF operation—intermittent service.
- Efficiency—ratio of power supplied versus power output.
- Power factor—portion of energy in an AC circuit that can be used to perform work. Power factor is also defined as the phase relationship between the current and voltage.
- Full-load amperes (FLA)—The amount of amperes drawn when the motor is connected to the maximum load the motor is designed to drive.
- Locked-rotor amperes (LRA)—The amount of amperes a motor draws on startup.
- Service factor—A multiplier that represents the amount of load, beyond the rated load, that can be placed on a motor without causing damage.

Motors are designed to operate in certain ambient (surrounding air) temperatures under full-load conditions without overheating. If a motor is operated continually under overload conditions, it will lead to high winding temperatures, which shortens insulation life.

Motor Circuits

Motor circuits commonly used in industrial applications are single-phase and polyphase circuits. Single-phase circuits consist of two conductors plus a ground. Polyphase circuits consist of either three or four conductors plus a ground. Three-phase circuits are the most common polyphase motor circuits used.

In a single-phase circuit there is one phase of current for each cycle. In a polyphase circuit there are two or three separate supplies of single-phase current arranged to allow peaks of voltage to follow each other in a regular repeating cycle. Each phase is $\frac{1}{180}$ of a second or $\frac{1}{3}$ of a cycle apart in a three-phase, 60-cycle current.

Single-phase induction motors are similar in construction to the squirrel-cage, polyphase induction motor. The one set of windings used in a single-phase induction motor will not cover the internal periphery of the stator compared to a polyphase motor in which two or three sets of windings cover the entire internal periphery of the stator. Current is simultaneous in all the poles in the stator when the stator windings of the single-phase induction motor is energized. However, there is no rotating stator magnetic field produced.

Current induced in the windings of the squirrel-cage motor is such that the magnetic field established in the rotor is in-line with the magnetic field of the stator. There is no tendency for the rotor to turn. But if the rotor starts to turn, the current induced in the windings of the rotor will lag a little behind the stator winding current. The rotor field lags behind the stator field and the torque produced will keep the rotor turning. A rotating field is produced in a single-phase motor once its rotor starts. The motor then operates the same as a three-phase, squirrel-cage motor. **See Figure 9-12.**

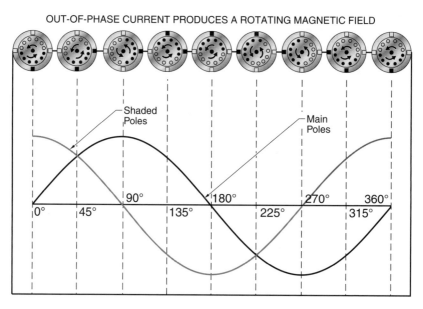

Figure 9-12. A shaded pole creates a magnetic field that is out of phase with the line current, causing rotation.

Single-Phase Motors

Single-phase motors are induction motors with squirrel-cage rotors. Single-phase motors are primarily used for fractional horsepower applications and where three-phase power is not available. Single-phase motors differ from one another in the methods used to produce the starting torque. Single-phase motors commonly used include split-phase, capacitor-start, capacitor-run, capacitor start-and-run, and shaded-pole.

A *split-phase motor* is a single-phase motor that has starting windings with a higher resistance than the running windings. **See Figure 9-13.** The current flow in the starting windings lags 30° behind and out of phase with the running windings. The starting torque produced causes the motor to rotate. At 80% of the rated speed, the centrifugal switch located in the starting winding will open. Split-phase motors have a low starting torque and are used to drive small blowers, fans, and pumps with voltages of 120 V to 240 V.

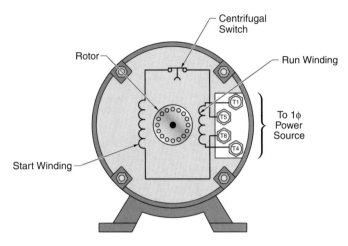

Figure 9-13. Split-phase motors do not require a high starting torque.

A capacitor-start motor is a single-phase motor that has a capacitor placed in series with the starting windings. **See Figure 9-14.** The starting windings on capacitor-start motors are wound with larger wires than in the split-phase motor. The capacitor causes the current to lead the voltage while the current lags the voltage in the running windings. The phase displacement between the two windings can be up to 90°. This high starting torque is commonly used for driving small compressors that must start under full-load. When the rotor reaches about 80% of its rated speed, the starting windings are removed from the circuit.

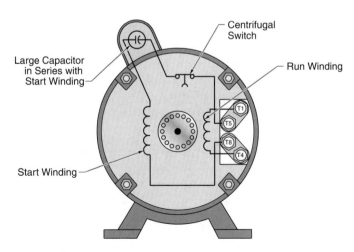

Figure 9-14. Capacitor-start motors are used in applications where a high starting torque is required.

A capacitor-run motor is a single-phase motor that has a start winding and capacitor in series at all times. **See Figure 9-15.** A capacitor-run motor is also known as a permanent split-capacitor motor. A smaller capacitor is used in a capacitor-run motor than in a capacitor-start motor because the capacitor remains in the circuit at full-load speed.

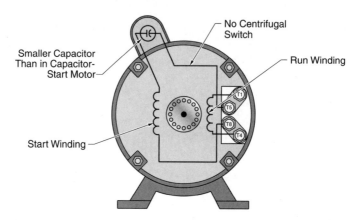

Figure 9-15. In a capacitor-run motor, the capacitor and start winding remain in the circuit at all times.

The advantage of leaving the capacitor in the circuit is that the motor has more running torque than a capacitor-start motor or split-phase motor. This allows a capacitor-run motor to be used for loads that require a higher running torque, such as to drive shaft-mounted fans and blowers ranging in size from ⅙ HP to ⅓ HP.

A capacitor start-and-run motor is a single-phase motor that has separate start-and-run capacitors for high starting torque. Once started and up to speed, this allows high running torque. This high starting torque motor is commonly used for driving small compressors that must start under full-load. When the rotor reaches about 80% of its rated speed, the starting windings are removed from the circuit. **See Figure 9-16.**

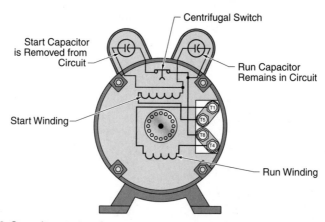

Figure 9-16. Capacitor start-and-run motors use two windings for starting and one winding for running.

A *shaded-pole motor* is a single-phase motor that operates with a two-phase magnetic field to develop the torque necessary for starting. This is accomplished by the trailing edge of each pole wound with a shaded coil. The shaded coil delays the magnetic field by about 90° and causes rotation. Shaded-pole motors have a very low starting torque and cannot be used with equipment that requires high starting torque. **See Figure 9-17.**

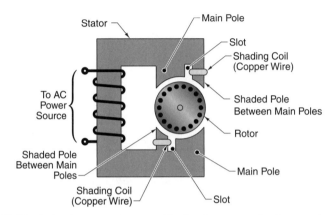

Figure 9-17. Shaded-pole motors are used for low torque applications such as fans and small loads.

Three-Phase Induction Motors

Three-phase induction motors operate on three-phase systems. All three-phase induction motors have three separate stator winding sets distributed 120° around the stator that produce a rotating magnetic field. **See Figure 9-18.** The rotating magnetic field in the stator induces magnetic poles in the rotor. The rotor magnetic poles are attracted to and follow the poles of the rotating stator magnetic field, which causes the rotor to turn.

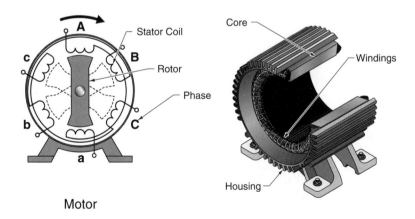

Motor

Figure 9-18. The stator consists of an iron core and windings and is enclosed within a housing.

The most common three-phase induction motor is the squirrel-cage motor. Other commonly used three-phase induction motors are wound-rotor and synchronous motors. The stators of all three types are very similar. Each type of induction motor has a different style of rotor. A *squirrel-cage motor* is a three-phase induction motor that has a rotor with iron bars embedded below the surface of the rotor. A *wound-rotor motor* is a three-phase induction motor that uses coils of wire in place of the iron bars of a squirrel-cage motor, with added slip rings, brushes, and a resistor circuit.

A *synchronous motor* is a three-phase induction motor that has iron bars like a squirrel-cage motor and a DC field winding in the rotor. A synchronous motor starts similarly to a three-phase induction motor and is excited by DC at the point of synchronization to bring it up to synchronous speed. DC power is supplied to the slip rings of the rotor and creates a stationary magnetic field. The rotating magnetic field causes the rotor to turn. **See Figure 9-19.**

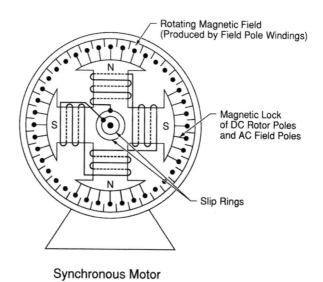

Synchronous Motor

Figure 9-19. Synchronous motors operate at high efficiency and can be used to improve the plant power factor.

Synchronous motors are used where constant speed with high degrees of efficiency are required and are primarily used in large industrial installations. Synchronous motors normally run with a power factor of 1. The power factor can be increased by increasing the voltage to the rotor windings or decreased by lowering the voltage to the rotor windings. Plants that have a poor power factor because of a large number of inductive motors can improve the power factor by having a leading power factor on the synchronous motor.

Motor Efficiency

Motor efficiency is determined by dividing output by the input. The output can never be greater than the input.

$$Motor\ efficiency = \frac{Output}{Input}$$

$$Motor\ efficiency = \frac{Rated\ HP}{HP\ utilized}$$

Note: 1 kW = 1.34 HP, 1 kW = 3413 Btu, and 1 HP = 746 W.

EXAMPLE

1. What is the efficiency of a 10 HP motor that operates with 220 V and 40 A at full load?

$$Motor\ efficiency = \frac{Output}{Input}$$

$$Motor\ efficiency = \frac{Total\ watts\ output}{Total\ watts\ input}$$

$$Motor\ efficiency = \frac{10 \times 746}{220 \times 40}$$

$$Motor\ efficiency = \frac{7460}{8800}$$

$$Motor\ efficiency = \textbf{0.85, or 85\%}$$

Motor Controls

Motor controls commonly consist of relays, contactors, motor starters, and switches. **See Figure 9-20.** A *relay* is an electromechanical device that controls one electrical circuit by opening and closing contacts in another circuit. A relay consists of contacts and a magnetic coil that closes or opens the contacts. A relay may have several pairs or only one pair of contacts, depending on its use. The coil is normally energized by a control circuit. Common types of relays are time, current, voltage, and power relays, as well as others that are designed for special functions. Typically, a relay is not rebuildable.

A *contactor* is a control device that uses a small control current to energize or de-energize a load connected to it. A contactor is usually much larger than a relay and can be rebuilt. The contacts and magnetic coil can normally be replaced for a fraction of the cost of replacing the contactor. Contactors are usually used to switch nonmotor loads larger than a relay.

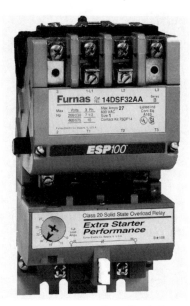

Furnas Electric Company

Rockwell Automation,
Allen-Bradley Company

Figure 9-20. Relays, contactors, and motor starters are all used for motor control.

A *motor starter* is a contactor with overload protection included. It is used to start motors sized from about 1 HP and up, through the use of a control circuit and the magnetic coil in the starter.

The switches used for opening or closing electrical circuits may be manual or automatic. Manual switches function similarly to toggle switches used to turn on lights in a home. A pressure control is an example of an automatic switch. Electric switches can be single-pole single-throw or double-throw. Three-pole, four-pole single-throw, double-throw switches are for special applications. **See Figure 9-21.**

Overload Protection

Overload protection for motors is required to protect motors from improper voltage and/or current. Overload protection for motors must comply with the provisions of the National Electrical Code® and any applicable local or state codes. In addition, motors must be protected against overheating to prevent insulation failure, moisture formation, and lubricating oil breakdown, which could result in damage to equipment or fire. The most common overload protection devices are fuses, circuit breakers, and heaters.

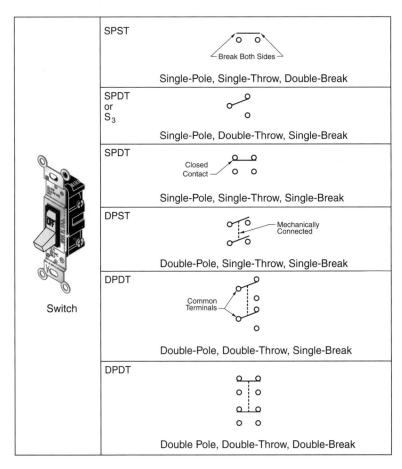

SPST	Single-Pole, Single-Throw, Double-Break
SPDT or S₃	Single-Pole, Double-Throw, Single-Break
SPDT	Single-Pole, Single-Throw, Single-Break
DPST	Double-Pole, Single-Throw, Single-Break
DPDT	Double-Pole, Double-Throw, Single-Break
DPDT	Double Pole, Double-Throw, Double-Break

Switch

Figure 9-21. Contacts are the conducting parts of a switch that operate with other conducting parts to make or break a circuit.

A *fuse* is a device that contains a fusible strip of metal inserted in series with a circuit in which the metal will melt or break if current is increased beyond a specific value for a specific amount of time. This opens the circuit to interrupt current to prevent damage to the motor. Some fuses used for motor protection have a time delay built into them. The time delay prevents the fuse from blowing on startup when there is a high current. **See Figure 9-22.**

A *circuit breaker* is an electromagnetic or thermal device that trips open to interrupt current when the current in the circuit exceeds a set point. Within normal current ranges, the circuit breaker remains closed. Circuit breakers can be equipped with a time-delay device for high current starts required for some motors. **See Figure 9-23.**

A *heater* is a thermal overload device, primarily used in motor starters, that is connected in series in the motor circuit and opens the holding coil circuit in the relay when the motor is overloaded. When the holding circuit is opened, the contacts open and stop the motor. This prevents damage or burnout of the motor. **See Figure 9-24.**

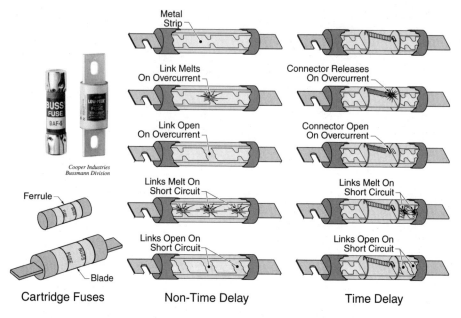

Figure 9-22. A fuse is an overcurrent protection device with a fusible link that melts and opens the circuit when an overload condition or short circuit occurs.

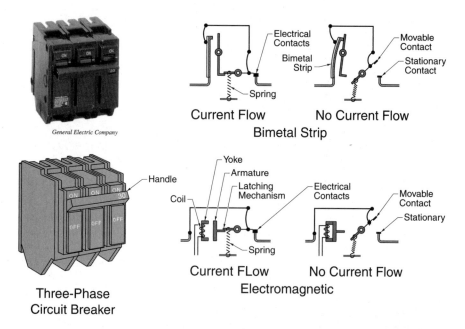

Figure 9-23. A circuit breaker is an overcurrent protection device with a resettable mechanism that automatically opens the circuit when an overload condition or short circuit occurs.

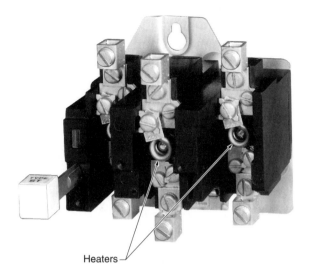

Heaters

Cutler-Hammer

Figure 9-24. A heater (thermal overload device) is an external motor protection device that uses the high temperatures caused by overcurrent conditions to open a motor circuit.

GENERATORS

Generators are either direct current or alternating current. DC generators were developed before AC generators and were very large and cumbersome. At that time, DC generators were driven by reciprocating steam engines and produced low voltages with high amperage, which required large wires and resulted in line losses. This limited the distance electricity could be distributed. To compensate for line losses, many separate generating stations were required within the city being supplied.

The introduction of the AC generator completely changed the way electricity was distributed. AC generators produce electricity at very high voltages at low amperages. This made it possible to generate electricity efficiently and transmit power over transmission lines extending for hundreds of miles. Substations are located as required to step down voltage at the point of use.

DC Generators

DC generators consist of a frame, field poles, armature, commutator, and brushes. **See Figure 9-25.** The frame is circular and is used to hold the field poles in place. The frame is usually constructed of cast iron or steel. There may be two, four, six, eight, or more field poles. They consist of a laminated steel core wound with insulated copper wire and are fitted into the generator frame. The armature consists of a soft iron laminated core with coils inserted around the core. The core is laminated to reduce the eddy currents that cause heat buildup. The coil ends are connected to the segments of the commutator.

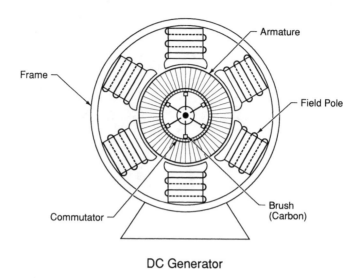

DC Generator

Figure 9-25. DC generators are operated at 240 V and have largely been replaced by AC generators.

The commutator is made of copper segments and is fastened to the armature by clamping rings. **See Figure 9-26.** The brushes are usually made of carbon. They conduct electricity from the commutator to the external circuit. The armature of the DC generator is supported by bearings and rotates within the field poles, which are stationary. The brushes are mounted on brush holders equipped with springs to maintain proper tension and make contact with the segments of the commutator. The air gap between the armature and field poles should be approximately ⅛″ to ¼″. This air gap is the same for DC and AC generators.

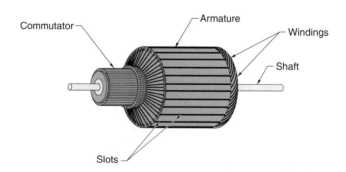

Figure 9-26. A commutator is the part of an armature that connects the armature to the insulated copper bars on which the brushes ride.

When a DC generator is energized, the field poles establish an electromagnetic field. The armature moving through the magnetic field induces an alternating current in the windings that are connected to the commutator. The brushes and the commutator rectify (change) the alternating current to direct current.

DC generators are commonly driven by reciprocating steam engines equipped with flywheel governors. The flywheel governors are needed to keep a fairly constant RPM. They normally produce 240 V and could be either self-excited or receive their excitation from an outside source. To prevent sparking at the brushes, DC generators operate at low RPM. It is also essential that the brushes and commutator be kept clean to prevent sparking. Brushes must be monitored for signs of excess wear and replaced when worn.

Paralleling DC Generators. Paralleling DC generators is the process of adding a second generator when the electrical load in the plant requires a second generator to be put in service. Care must be taken to prevent the circuit breaker from being tripped when the second generator is put in service. The reciprocating steam engine should be started using the following basic startup procedure:

1. Bring the reciprocating engine up to normal operating speed, making sure the governor takes over.

2. Close the negative and equalizer switches on the electric panelboard.

3. Using the field rheostat, adjust the voltage of the incoming generator so it is equal to or slightly below the generator on-line, then close positive switch.

4. Balance the load between both generators with the field rheostat. *Note:* A field rheostat is a variable resistor that is located in the field circuit.

Plants that generate 240 V DC often use the full voltage to drive motors in the plant, but the lighting circuits will be split into two 120 V circuits.

AC Generators

AC generators consist of a frame, stator, rotor, and slip rings. The frame is circular and holds the stator in place. The stator is made up of a core with coils spaced evenly around the core. **See Figure 9-27.**

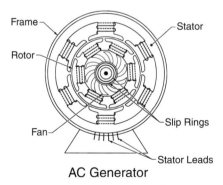

AC Generator

Figure 9-27. AC generators provide the high voltage necessary to transmit electricity over long distances.

The rotor is a revolving field consisting of a laminated core with coils of wire wound around the edges. A fan is mounted on one end of the rotor to remove heat from the coils. *Note:* The core is laminated to reduce the heat that builds up from the eddy currents. Slip rings are fastened to one end of the rotor. They are connected to the ends of the field coils and are used to excite or energize the field coils. **See Figure 9-28.**

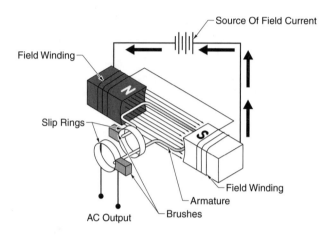

Figure 9-28. The slip rings of an AC generator transfer power to and from the generator's rotor.

In the AC generator, the rotor with its field windings rotates within the stator and is supported by bearings. The rotating field poles are energized through the slip rings by an outside DC source. This produces a rotating electromagnetic field. The electromagnetic field passes over the stationary stator windings, inducing an alternating current. The current produced is removed by way of the stator leads. The exciter used to excite the field windings may be mounted directly on the shaft of the generator or be separately driven by a steam turbine or an AC motor. *Note:* An exciter functions like a DC generator. Power used for field excitation is 1% to 2% of the total power generated. The frequency generated by an AC generator depends on its speed and the number of field poles it contains. The standard frequency in use in the United States is 60 Hz.

Paralleling AC Generators. It is more difficult to parallel AC generators than DC generators because both AC generators must have their frequencies synchronized. Plants that require large quantities of steam for process use steam turbines to drive the generators. To parallel two AC generators driven by steam turbines, the following procedure is followed:

1. Warm up the steam turbine following manufacturer's startup procedures, making all necessary safety checks.

2. Bring the steam turbine up to normal speed.

3. Using the field rheostat, adjust the voltage of the incoming generator to line voltage.

4. Use a synchroscope to synchronize the incoming generator by increasing or decreasing its RPM.

5. Close the circuit breaker.

6. Apply 20% of the generator capacity by using a governor speed changer.

7. Load the machines according to plant conditions.

8. Adjust air cooling on generator windings.

Both AC and DC generators are excited by DC. Some DC generators are self-exciting and therefore do not need a separate DC source for excitation. Also, both AC and DC generators produce alternating current. However, the commutator and brushes on the DC generator act as rectifiers to produce direct current. AC generators are usually smaller than DC generators for comparable output and have higher output voltages.

Frequency Determination. The frequency of an AC generator can be determined using the following equation:

$$Frequency = \frac{No.\ of\ poles \times RPM}{120}$$

$$F = \frac{P \times N}{120}$$

F = frequency, in hertz (Hz)

P = number of field poles in the generator

N = RPM of the generator

EXAMPLES

1. Find the frequency of an AC generator that is operating at 3600 RPM and has two field poles.

$$F = \frac{P \times N}{120}$$

$$F = \frac{2 \times 3600}{120}$$

$F =$ **60 Hz**

2. An AC generator runs at 60 Hz at a speed of 3600 RPM. How many field poles does this generator have?

$$F = \frac{P \times N}{120}$$

$$P = \frac{120 \times F}{N}$$

$$P = \frac{120 \times 60}{3600}$$

$P =$ **2**

Voltage Regulators

Voltage regulators are used to maintain a constant terminal voltage on an AC generator. Constant terminal voltage is maintained by increasing or decreasing the field current to the generator. An increase in field current increases the terminal voltage. A decrease in field current lowers the terminal voltage.

POWER DISTRIBUTION

Power is distributed using transformers for efficiency and reliability. AC electrical power is usually generated at 2300 V to 2400 V in a generating process plant. Heavy electrical equipment such as motors for driving compressors, coal mills, forced and induced draft fans, feedwater pumps, and motor-generator sets are run on high voltage and confined to the power plant. Transformers are used to reduce voltage for other plant requirements. **See Figure 9-29.**

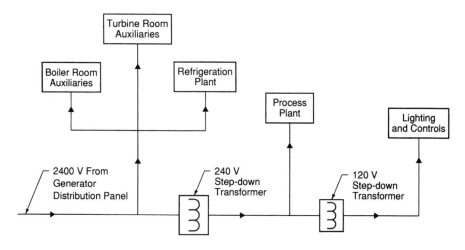

Figure 9-29. Power is distributed for various uses throughout the plant using transformers to increase or decrease voltage.

Transformers

Transformers increase or decrease voltage in an AC circuit. Transformers are simple and rugged in construction with no moving parts. Step-down transformers are used to step down, or reduce, voltage. Step-up transformers are used to step up, or increase, voltage. **See Figure 9-30.**

Step-down voltages commonly used are 440 V and 220 V for process motors and 120 V for lighting and control circuits. **See Figure 9-31.** Step-up transformers are used to transmit electricity at a high voltage and low amperage. This reduces transmission losses. Voltage must also be stepped up in ignition transformers on boilers to light gas pilots.

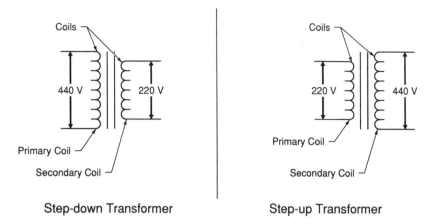

Step-down Transformer   Step-up Transformer

Figure 9-30. Windings in transformers permit AC power to be stepped up or stepped down for different voltage requirements.

Copper Development Association

Figure 9-31. A transformer is a device that steps up or steps down alternating voltages.

Transformers consist of two coils on an iron core. When alternating current is supplied to the primary coil, a magnetic field is produced. The magnetic field makes and breaks across the secondary coil, inducing a voltage in it. The ratio of the two voltages is directly proportional to the number of turns on the two coils.

Motor-Generator Sets

Motor-generator sets convert alternating current from the generating plant to direct current. The process of converting alternating current to direct current is *rectification*. Motor-generator sets consist of an AC motor that runs off line voltage and turns a DC generator. The direct current generated by the motor-generator set is then used for plant requirements such as elevator operation, to excite synchronous motors, charge batteries, and supply DC control circuits. **See Figure 9-32.** Most plants that require DC power have two motor-generator sets that are changed over once a month. When changing over, the incoming generator is paralleled with the other generator, the load is shifted, then the other generator is removed from service.

Horlick Company, Inc.

Figure 9-32. Motor-generator sets provide a means to produce DC power from AC power.

Motor-generator sets require the same maintenance as other boiler or engine room auxiliaries. Lubricating oil in the bearings must be kept at the proper level and must be changed at the required intervals. The commutator must be kept clean and dressed when there is wear or grooving. The windings must be kept clean to allow for air circulation to prevent overheating, and plants usually have two sets so they may be alternated at regular intervals. Check for arcing between the commutator and brushes. Adjust for magnetic field excitation to maintain the correct terminal voltage for either 120 V or 240 V.

INSTRUMENTS AND METERS

Instruments and meters are used to monitor and control electrical operations in the plant. Instruments and meters commonly used by the stationary engineer include voltmeters, ammeters, ohmmeters, megohmmeters, synchroscopes, and frequency meters. **See Figure 9-33.** A *voltmeter* is used to measure the difference of voltage between two points in a circuit. Voltmeters are commonly used to check the voltage, fuses, and for breaks in the circuit.

Fluke Corporation
Figure 9-33. Test instruments are used when troubleshooting electrical circuits.

An *ammeter* is used to measure the current in any part of a circuit. Ammeters are commonly used to check for excessively high current in a circuit. Ammeter readings must be carefully monitored as plant load is increased. When the plant load increases, the generator load increases. An increase in generator load will cause a corresponding increase of heat in the generator windings. If this heat is not dissipated, the windings will overheat, possibly causing damage. The generator is equipped with an air cooler to prevent the windings from building up heat. In large utility plants, generators are a completely enclosed system filled with hydrogen gas as a cooling medium.

An *ohmmeter* is a meter that is used to measure the amount of resistance in a part of a circuit or in a complete circuit. Continuity (proper connections) can also be checked using an ohmmeter. A *megohmmeter* (megger®) is a meter that is used to measure resistance in a circuit. Megohmmeters are commonly used to check for the breakdown of insulation on coils. If the insulation breaks down, there is a loss of electricity to the ground. This occurs most frequently in high-voltage systems. A *synchroscope* is a meter that shows the phase relationship between two voltages. Synchroscopes are used when two AC generators are paralleled or synchronized. Voltmeters provide a terminal voltage reading (AC or DC). A *frequency meter* is a meter that indicates the electrical frequency in the system. Electrical equipment is designed to operate at a specific frequency and voltage. Any variation in frequency affects all electric motors, electric clocks, and other inductive electrical devices in the plant.

SAFETY PRACTICES AND PROCEDURES

Standards and national, state, and local codes are used to protect people and property from electrical dangers. A code is a regulation or legal minimum requirement. A standard is an accepted reference or practice. Codes and standards ensure electrical equipment is built and installed safely and that every effort is made to protect people from electrical shock. Electrical safety has been advanced by the efforts of the Occupational Safety and Health Administration (OSHA), National Fire Protection Association (NFPA), and state safety laws.

Occupational Safety and Health Administration (OSHA)

The *Occupational Safety and Health Administration (OSHA)* is a federal agency that requires all employers to provide a safe environment for their employees. All work areas must be free from hazards likely to cause serious harm. The provisions of this act are enforced by federal inspection.

OSHA has developed color codes to help ensure a safe environment. **See Figure 9-34.** The color codes help quickly identify fire protection equipment, physical hazards, dangerous parts of machines, radiation hazards, and locations of first aid equipment. With few exceptions, OSHA uses the NEC® to help ensure a safe electrical environment.

Color	Examples
Red	Fire protection equipment and apparatus, portable containers of flammable liquids, emergency stop pushbuttons and switches
Yellow	Caution and for marking physical hazards, waste containers for explosive or combustible materials, caution against starting, using, or moving equipment under repair, identification of the starting point or power source of machinery
Orange	Dangerous parts or machines, safety starter buttons, the exposed parts (edges) of pulleys, gears, rollers, cutting devices, power jaws
Purple	Radiation hazards
Green	Safety, location of first aid equipment (other than fire-fighting equipment)

Figure 9-34. OSHA-mandated color coding is used to provide visual information on the location of safety information and on the dangers that are present in an area.

National Fire Protection Association (NFPA)

The *National Fire Protection Association (NFPA)* is a national organization that provides guidance in assessing the hazards of the products of combustion. The NFPA sponsors the development of the National Electrical Code® (NEC®) and the NFPA 70E standard.

National Electrical Code® (NEC®). The *National Electrical Code® (NEC®)* is a standard of practices for the installation of electrical products published by the National Fire Protection Association (NFPA). The NEC® is one of the most widely used and recognized consensus standards in the world. The purpose of the NEC® is to protect people and property from hazards that arise from the use of electricity. Improper procedures when working with electricity can cause injury or death. Many city, county, state, and federal agencies use the NEC® to set requirements for electrical installations. The NEC® is updated every three years.

NFPA 70E. The National Fire Protection Association standard NFPA 70E, *Standard for Electrical Safety in the Workplace,* addresses "electrical safety requirements for employee workplaces that are necessary for the safeguarding of employees in their pursuit of gainful employment." Per NFPA 70E, "Only qualified persons shall perform testing work on or near live parts operating at 50 V or more."

Qualified Persons

To prevent an accident, electrical shock, or damage to equipment, all electrical work must be performed by qualified persons. A *qualified person* is a person who is trained in, and has specific knowledge of, the construction and operation of electrical equipment or a specific task, and is trained to recognize and avoid electrical hazards that might be present with respect to the equipment or specific task. NFPA 70E Part II *Safety-Related Work Practices,* Chapter 1 *General,* provides additional information regarding the definition of a qualified person. A qualified person does the following:
• Determines the voltage of energized electrical parts
• Determines the degree and extent of hazards and uses the proper personal protective equipment and job planning to perform work safely on electrical equipment by following all NFPA, OSHA, equipment manufacturer, state, and company safety procedures and practices
• Performs the appropriate task required during an accident or emergency situation
• Understands electrical principles, follows all manufacturer procedures, and abides by approach distances specified by the NFPA
• Understands the operation of test equipment and follows all manufacturer procedures
• Informs other technicians and operators of tasks being performed and maintains all required records

Arc Flash

In addition to the risk of electrical shock, other hazards are present when work is performed on or near energized electrical equipment or conductors. Two serious hazards are electrical arc flashes and electrical arc blasts. *Arc flash* is a short circuit through air. An *arc blast* is an explosion that occurs when the surrounding air becomes ionized and conductive. When insulation or isolation can no longer withstand the applied voltage, an arc flash occurs. An arc flash may occur from phase to ground or from phase to phase. The temperature of the arc flash may reach thousands of degrees and cause an arc blast. The explosion from the arc blast can spread hot gases and melting metal, damage hearing and vision, and send objects flying. Personal protective equipment is required in areas where arc flash and arc blast may be present. **See Figure 9-35.**

Figure 9-35. Personal protective equipment is required when working on electrical equipment where arc flash or arc blast may be present.

OSHA Part 1910.333 gives general requirements for safety-related work practices. OSHA inspectors also carry a copy of NFPA 70E and use it to enforce safety procedures related to arc flash. NFPA 70E gives specific guidelines on actions to be taken to comply with the general OSHA statements. The NEC® requires several types of equipment to be marked to warn qualified persons that a hazard exists. **See Figure 9-36.**

Figure 9-36. An arc-flash warning label must be placed on any electrical equipment that may remain energized during repair or maintenance.

The NFPA requires facility owners to perform a flash hazard analysis. A flash protection boundary must be established around electrical devices. This boundary is determined by calculations that estimate the maximum energy released and the distance that energy travels before dissipating to a safe level. Technicians working within the boundary must have appropriate personal protective equipment (PPE). The IEEE Standard 1584-2002, *Guide for Performing Arc-Flash Hazard Calculations,* gives procedures for determining the incident energy exposure, flash protection boundary, and level of PPE required. Incident energy is expressed in calories per centimeter squared (cal/cm^2). This is a measure of the heat energy applied to a certain area of an object. The object may be a person. The flash protection boundary must be established at the point where the incident energy has fallen below 1.2 cal/cm^2. It is not safe to assume that similar equipment located in different locations has the same flash protection boundary.

Personal Protective Equipment (PPE)

Personal protective equipment (PPE) is clothing and/or equipment worn by a technician to reduce the possibility of injury in the work area. The use of PPE is required whenever work may occur on or near energized exposed electrical circuits or when exposure to chemicals is possible. For maximum safety, PPE must be used as specified in NFPA 70E, OSHA Standard Part 1910 *Subpart I – Personal Protective Equipment* (1910.132 through 1910.138), and other applicable safety mandates.

All PPE and tools are selected for at least the operating voltage of the equipment or circuits to be worked on or near. Equipment, devices, tools, or test equipment must be suited for the work to be performed. Personal protective equipment includes protective clothing, head protection, eye protection, ear protection, hand protection, foot protection, back protection, knee protection, and rubber insulated matting. **See Figure 9-37.**

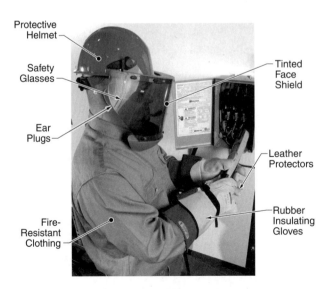

Figure 9-37. Personal protective equipment is used to reduce the possibility of an injury.

Lockout/Tagout

Electrical power and other potential sources of energy must be removed when electrical equipment is inspected, serviced, or repaired. To ensure the safety of personnel working with the equipment, all electrical, pneumatic, and hydraulic power as well as any other potential hazard are removed and the equipment must be locked out and tagged out. *Lockout* is the process of removing the source of electrical power and installing a lock that prevents the power from being turned on. *Tagout* is the process of placing a danger tag on the source of electrical power, which indicates that the equipment may not be operated until the danger tag is removed. Per OSHA standards, equipment is locked out and tagged out before any installation or preventive maintenance is performed. **See Figure 9-38.**

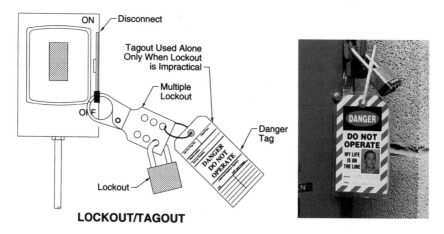

Figure 9-38. For maximum safety, electrical equipment is shut off, locked out, and tagged before work is performed on the circuit.

A danger tag has the same importance and purpose as a lock and is used alone only when a lock does not fit the disconnect device. A danger tag shall be attached at the disconnect device with a tag tie or equivalent and shall have space for the technician's name, craft, and other company-required information. A danger tag must withstand the elements and expected atmosphere for the maximum period of time that exposure is expected. Lockout/tagout is used in the following circumstances:
• power is not required to be ON to a piece of equipment to perform a task
• machine guards or other safety devices are removed or bypassed
• the possibility exists of being injured or caught in moving machinery
• jammed equipment is being cleared
• the danger exists of being injured if equipment power is turned on

General Safety Rules

Many specific rules and procedures are mandated by OSHA, the NEC®, NFPA 70E, and state and local codes. In addition, basic safety practices and procedures include the following:

• Always check all power tools for proper grounding.

• Never use a power tool, extension cord, or droplight with a damaged cord.

• Unplug any electrical cord by pulling the plug, not the cord.

• Never work on electrical equipment while standing in water.

• Always comply with provisions of the NEC® and any applicable state or local codes.

• Shut off, lock out, and tag disconnect switches when working on any electrical circuit or equipment. Use testing equipment to verify all circuits are shut off.

• Always use a fuse puller to remove a fuse.

• Use a portable light or a low-voltage droplight when working on the steam and water side of the boiler.

• Never apply current to any equipment until it has been inspected and checked with the proper meter.

• Never handle capacitors until they have been discharged.

• Protect all electrical equipment from moisture and water.

• Never use water on an electrical fire. Shut off the power and use a Class C rated fire extinguisher. **See Figure 9-39.**

• Never work on a live circuit unless authorized.

• Always use nonconductive ladders when working on or around electrical equipment.

• Use one hand when working on a live circuit to reduce the chance of an electrical shock passing through the heart and lungs.

• Use insulated tools. Check grips for loss of insulation from wear or damage before using.

• Exercise caution in an electrical fire as burning insulation produces toxic fumes.

• Always use recommended protective and safety equipment.

• Learn CPR and first aid.

• Never bypass or disable fuses or circuit breakers.

• Report any injuries to proper authorities.

• Keep rubber runners in front of panelboards and switches.

Figure 9-39. Class C fire extinguishers are used on electrical fires.

CHAPTER 9 LEARNER RESOURCES

ATPeResources.com/QuickLinks
Access Code: 735728

Industry requires steam to produce electricity, petroleum, food, paper, and many other products. Steam can be generated using many different fuels or combustible materials. Fuels most commonly used include coal, fuel oil, and gas. Combustible materials used include industrial waste such as sawdust, pulverized coffee, bagasse, garbage, and methane gas (from sewage disposal plants). In addition, nuclear energy can be used to produce steam. The equipment used to generate steam is very costly. In addition, operation, maintenance, and labor costs required to produce steam must be kept to a minimum. A stationary engineer is responsible for operating the plant and managing steam for maximum safety and efficiency.

BALANCING STEAM AND ELECTRIC LOADS

The steam and the electric loads of a plant must be balanced for maximum efficiency. At times it is more efficient to use electric auxiliaries in place of steam-driven auxiliaries, or steam-driven auxiliaries in place of electric auxiliaries. For optimum efficiency of a steam-driven turbo-generator, it should always operate at approximately 75% of its rated capacity. This is accomplished by selecting either steam-driven or electric auxiliaries. In addition, exhaust steam, bleed steam, and/or extracted steam should be used whenever possible.

HEAT RECOVERY SYSTEMS

Heat recovery systems are implemented to use heat generated by the boiler and auxiliaries that may otherwise be lost. Heat recovery systems commonly include the following:
- *Air preheaters* — used to recover heat from the gases of combustion installed in plants with a relatively constant steam load. Air preheaters are used in larger plants because their cost is prohibitive for smaller plants. **See Figure 10-1.**
- *Economizers,* or fuel savers — reclaim heat from gases of combustion to heat feedwater. Economizers require large breeching and draft fans and are commonly used in large plants with a constant steam load. **See Figure 10-2.**
- *Continuous blowdown* — used to control boiler water total dissolved solids with minimal heat loss.

- *Flash tanks*—used in conjunction with continuous blowdown to recover flash steam from water being removed from the steam and water drum. **See Figure 10-3.**
- *Feedwater heaters* of the open, deaerating, or closed type—the larger the plant is, the more sophisticated the feedwater heater and its controls are. **See Figure 10-4.**
- *Condensate return systems*—for any system that does not have contaminated condensate. Maximizing the amount of condensate returns recovered reduces the amount of makeup water, boiler water treatment, oxygen entering the boiler, and steam required to heat the feedwater.

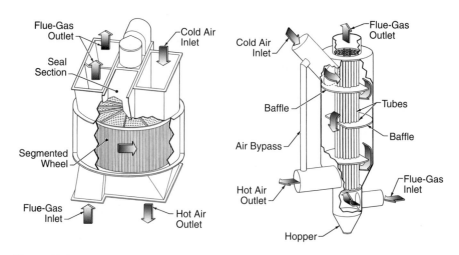

Figure 10-1. An air preheater is a heater used in a heat recovery system that recovers heat from gases of combustion.

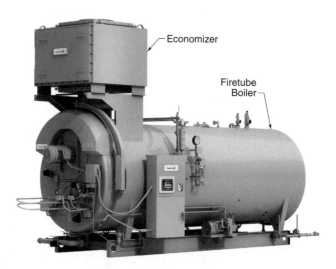

Cleaver-Brooks

Figure 10-2. An economizer uses gases of combustion to heat feedwater.

Cleaver-Brooks
Figure 10-3. A flash tank is used with a continuous blowdown system to recover the flash steam from the water being removed from the steam and water drum.

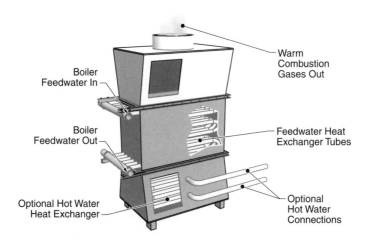

Boiler
Feedwater In

Boiler
Feedwater Out

Optional Hot Water
Heat Exchanger

Warm
Combustion
Gases Out

Feedwater Heat
Exchanger Tubes

Optional
Hot Water
Connections

Figure 10-4. A feedwater heater is an auxiliary component on a boiler that is used to increase plant efficiency by preheating the feedwater with exhaust steam or heat from the gases of combustion leaving the boiler.

Heat recovery systems vary from plant to plant. Heat recovery system equipment must be designed for the specific plant and application. Air preheaters and economizers are used on large boilers operating with a relatively constant load. Constant loads reduce the possibility of corrosion resulting from sweating. Air preheaters and economizers also require additional maintenance, larger draft fans, and breechings. These requirements reduce the cost-effectiveness of air preheaters and economizers on small boilers.

A continuous blowdown and flash tank system can be added inexpensively to small- and medium-sized plants to provide short- and long-term savings. For example, a plant has two boilers. Boiler 1 operates at 400 psig, generating 1,250,000 lb of steam per day. Boiler 2 operates at 200 psig, generating 350,000 lb of steam per day. The heat recovery system consists of a flash tank exhausting to an 18 psi back pressure system, a float-controlled dump valve to the main blowdown tank, and a high water alarm. With this heat recovery system, the following results are produced:

- five percent blow down from the 400 psig boiler generating 1,250,000 lb of steam per day amounts to 62,500 lb
- five percent blow down from the 200 psig boiler generating 350,000 lb of steam per day amounts to 17,500 lb
- heat of liquid boiler pressure 400 psig 428 Btu/lb
- heat of liquid boiler pressure 200 psig 362 Btu/lb
- heat of liquid at flash pressure 18 psig 224 Btu/lb
- latent heat of evaporation flash pressure 18 psig 942.4 Btu/lb
- percent flash steam 400 psig boiler = $\dfrac{428 - 224}{942.4}$ = 21.65%
- percent flash steam 200 psig boiler = $\dfrac{362 - 224}{942.4}$ = 14.64%
- pounds of flash steam 400 psig boiler = 0.2165 × 62,500 = 13,531.25 lb
- pounds of flash steam 200 psig boiler = 0.1464 × 17,500 = 2562 lb
- total of flash steam of 400 psig boiler + 200 psig boiler at 18 psi per day = 16,093.25 lb
- total of flash steam recovered for 200 days per year = 3,218,650 lb

Cogeneration

Cogeneration is using a plant for the generation of electricity and process work at the same time. All power (heat) is used. Cogeneration is not a recent development. Large industrial plants have been cogenerating since the 1930s. Steam turbines used with generators (turbo-generators) were designed to run noncondensing with exhaust steam used to drive steam engines, or for process. As public utilities began to produce electricity inexpensively, plants started to buy electricity rather than cogenerate. This reduced the importance of the turbo-generator used to generate electricity. Older turbo-generators needing work or replacement were retired.

Interest in cogeneration has increased recently as the cost of fuel and electricity has steadily increased. In addition, the development and use of gas turbines, diesel engines, and more compact turbo-generators have made cogeneration more popular. Exhaust gases from gas turbines and diesel engines are used in waste heat boilers. The steam generated is then used for plant process and plant heating. Using cogeneration, it is possible for one plant to supply electricity, process steam, heated water, and air conditioning. **See Figure 10-5.**

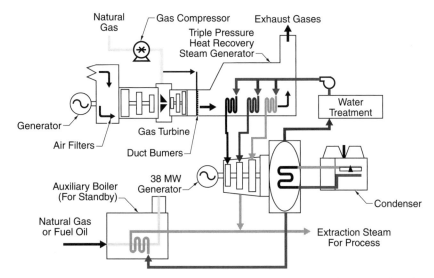

Figure 10-5. Cogeneration is the process of generating electricity and then using the leftover heat from the generation process to heat buildings, to provide process heat, or for further electrical generation.

Steam-to-Water Cycle Generating Plant

A process plant that also generates its own electrical power has a complex steam-to-water cycle. **See Figure 10-6.** A typical process plant has a steam turbine generator (turbo-generator) that has two extraction stages and is operating condensing. At the boiler, water is changed into steam and passes through the superheater, absorbing more heat. From the superheater, steam enters the steam header and divides into branches that go to the steam turbine throttle valve, steam auxiliaries, and pressure-reducing and desuperheating stations.

Steam that enters the steam turbine to generate electricity can be extracted for process at two pressure stages. Condensate from the steam that is used for process is returned to the open feedwater heater by a condensate pump. The steam that is not extracted for process is condensed in the surface condenser and discharged to the hot well. The condenser condensate pump removes water from the hot well and discharges it through the first- and second-stage air-ejector condensers. Water then travels back to the open feedwater heater.

Exhaust steam from auxiliaries is used in a low-pressure steam line for space heating, water heating, or in the open feedwater heater. The pressure-reducing and desuperheating station is used to balance the demand for process steam and electricity on the turbo-generator.

All condensate is returned to the open feedwater heater. In case of condensate loss, a makeup valve opens to compensate for the loss. The feedwater pump removes water from the open feedwater heater and discharges it through the closed feedwater heater where it is heated by extracted steam. The water is then regulated by the feedwater regulator before it passes through the feedwater check and stop valves and enters the boiler.

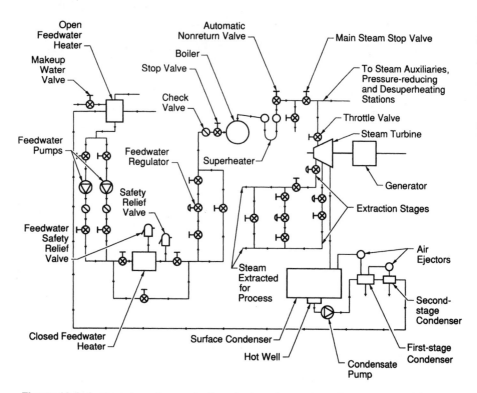

Figure 10-6. A steam-to-water generating plant produces steam for process and for generating electricity.

Waste Heat Boilers

Waste heat boilers serve two functions: to eliminate industrial waste and use the heat to produce steam for plant process or plant heating. Waste heat boilers use heat from gas turbines or diesel engines to produce steam. In addition, waste heat boilers can be designed to burn garbage, acids, pulverized coffee, sawdust, bagasse, tires, industrial by-products, and other combustibles from industrial by-products. Gases of combustion from waste heat boilers can be used to preheat air used in gas turbines and diesel engines. **See Figure 10-7.**

INDUSTRIAL USES OF STEAM

Industrial processes often require large quantities of steam using large boilers at 600 psig to 1000 psig and high temperatures to drive prime movers efficiently. The primary property of steam used in industrial processes is the high heat content produced. The specific application of steam used in industrial processes varies from plant to plant. In addition, the quantity and pressure required also vary depending on the application. Industrial plants are designed for a specific function, with no two industrial plants alike. Industries that commonly use steam for process work include paper, food, brewery, petroleum, and textile industries. **See Figure 10-8.**

Cleaver-Brooks
Figure 10-7. Waste heat boilers are used to eliminate industrial waste. Waste heat boilers use heat to produce steam for plant processes or plant heating.

Figure 10-8. Large industrial plants, such as paper plants and oil refineries, use steam for process work.

A typical large industrial plant includes one boiler operating at a steam pressure of 150 psi and four boilers operating at a steam pressure of 420 psi. Pressure-reducing valves (PRV) are used to reduce steam pressure required to meet all plant requirements. **See Figure 10-9.** Boiler 1 operating at 150 psi supplies steam to a 150 psi process line. At PRV (1), the steam is reduced from 150 psi to 90 psi for another process line. At PRV (2), 90 psi steam is reduced to 5 psi for heating feedwater, building heat and steam for the low-pressure steam turbine, and a low-pressure process line. Boilers 2 and 3 provide 420 psi supply steam to high-pressure turbo-generators. At PRV (3), the 420 psi steam is reduced to 150 psi for the 150 psi process line. Steam pressure from boilers 4 and 5 is reduced to 150 psi at PRV (4). At PRV (5) and (6), 420 psi steam is reduced to 90 psi for plant process. Steam is distributed to turbo-generators straight noncondensing (7) and single extraction noncondensing (8). Extraction steam at 90 psi goes to a plant process line. The 5 psi exhaust steam goes to either the low-pressure condensing steam turbine (9) or the 5 psi heating or process line (10).

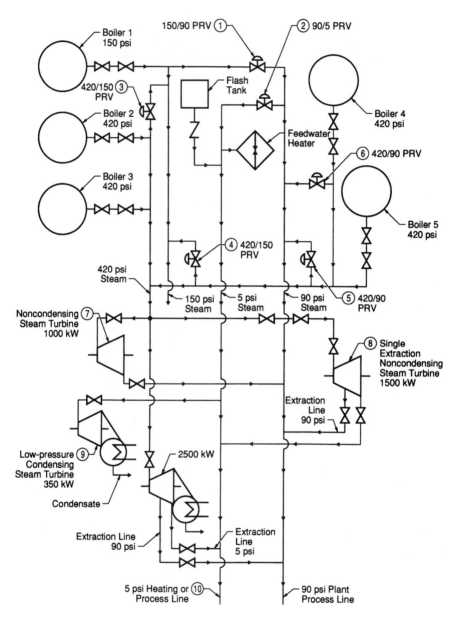

Figure 10-9. Industrial plants use steam at various pressures and temperatures for maximum efficiency.

The steam turbines in this plant act as large reducing valves. Steam pressure is reduced as it passes through the blades or nozzles of the steam turbine. The exhaust from the noncondensing steam turbines (7) and (8) is used for plant process. In addition, electricity can be generated at a very low cost. Plants of this type must have a balanced steam and electric load for optimum efficiency.

Paper Plants

Paper plants are unique in that large quantities of steam are required at relatively low pressures (30 psi to 60 psi) in the manufacturing process. High-pressure steam turbines are used to generate electricity. Exhaust steam from steam turbines is used for plant process. The steam turbine acts as a PRV while electricity is generated. Operating the steam turbine noncondensing is more efficient than operating condensing. This is important in paper plants where cogeneration has been used for many years. Steam lost in the pulp-making process requires a large amount of makeup water. **See Figure 10-10.**

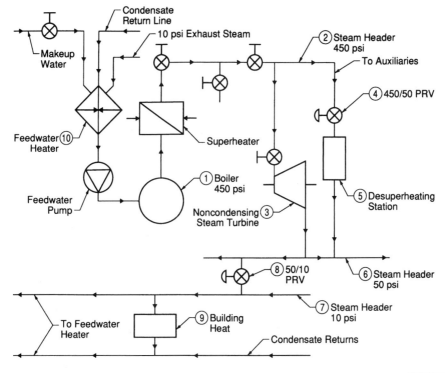

ASI Robicon

Figure 10-10. Paper plants commonly use noncondensing steam turbines to generate electricity.

In a typical paper plant, steam leaves the boiler (1) at 450 psi. From the steam header (2), it enters the noncondensing steam turbine (3) or PRV (4) and desuperheating station (5). Steam exhausts from the steam turbine at 50 psi to the process steam header (6). Steam from the process steam header is used for plant process and for supplying the 10 psi steam header (7) through the PRV (8). The 10 psi steam is used for building heat (9) and for the feedwater heater (10). All condensate collected is returned to the feedwater heater.

Food Processing Plants

Food processing plants use steam for cooking and sterilization. Proper boiler water treatment is necessary to prevent contamination of food or containers in the event of carryover. Food processing plants rarely generate their own electricity. Package boilers producing approximately 250 psi at 400°F are commonly used for food processing. Food processing does not cause rapid steam load fluctuation. Steam in food processing is used in the sterilization, blanching, pasteurization, and dehydration of food. Heat exchangers are commonly used to prevent steam from coming in direct contact with food. Chemicals in the feedwater must be noncaustic to prevent contamination of the food if a leak occurs. **See Figure 10-11.**

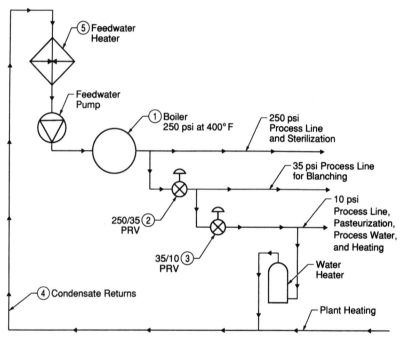

Figure 10-11. Food processing plants require careful design to prevent food contamination.

In a typical food processing plant, steam leaves the boiler (1) at 250 psi at 400°F for plant process and sterilization. The 250 psi steam goes to PRV (2) where it is reduced to a 35 psi process line used for blanching. The 35 psi steam also goes to PRV (3) where it is reduced to 10 psi and then enters the process line used for pasteurization, process water heating, and plant heating. The condensate from process water heating and plant heating (4) returns to the feedwater heater (5). This plant uses one boiler operating at 250 psi and PRVs to produce the required pressure for each process load. The dehydration process uses steam to create a high vacuum in closed chambers. This vacuum is created by steam passing through air ejectors. By maintaining a high vacuum, food can be dried at relatively low temperatures. When all moisture is removed from the food being dried, there is less chance of spoilage when the food is stored.

Breweries

Breweries require large quantities of steam for the brewing process. In addition, breweries commonly cogenerate using extraction steam turbines. Steam is extracted from the steam turbine at a controlled pressure and is used to drive auxiliaries. **See Figure 10-12.** Steam from the boiler (1) passes through the superheater into the main steam header (2). From the main steam header, steam at 450 psi enters the double extraction condensing steam turbine (3). Extraction steam goes into a 150 psi line (4) and into a 50 psi line (5). The remaining steam passes through the steam turbine into a surface condenser. The 150 psi line also is supplied steam when necessary from the 450 psi steam passed through the PRV and desuperheating station (6). A PRV (7) is also used between the 150 psi and 50 psi steam lines (5). The 150 psi and 50 psi steam lines are used for process where the latent heat is reclaimed. The PRV (8) maintains 10 psi on the 10 psi steam line. The 10 psi steam line is used for building heat (9) and heating feedwater in the feedwater heater. Condensate from building heat, process, and the condenser hot well is pumped back to the feedwater heater.

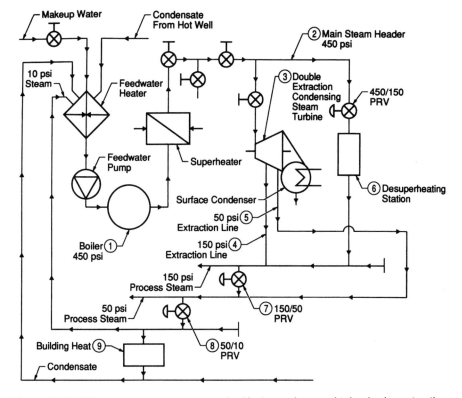

Figure 10-12. Different steam pressures required for breweries are obtained using extraction steam turbines.

Oil Refineries

Oil refineries use large quantities of relatively low-pressure steam (200 psi) for process heating, distillation, and cogeneration, and to keep tank farm oil warm to facilitate pumping. **See Figure 10-13.** Extraction steam turbines and PRVs are used to ensure sufficient quantities of steam for process lines. Steam at 750 psi from the boiler (1) flows to the main steam header (2). Steam flows through the PRV (3) to the 600 psi line, or through the noncondensing steam turbine (4) that exhausts to the 150 psi steam header (7). The 750 psi steam is also supplied to a double extraction condensing steam turbine (5). Extraction lines supply steam to the 600 psi steam header (6) and the 150 psi steam header (7). Steam at 600 psi is used for various processes (8) and to drive a noncondensing steam turbine (9) that exhausts into the 150 psi line (7). The PRV (10) controls the flow of makeup steam to the 150 psi line. A process line (11) uses 150 psi steam and a PRV (12) controls the pressure from the 150 psi to the 15 psi line (13). The 15 psi line is used for heating and the deaerating feedwater heater (14).

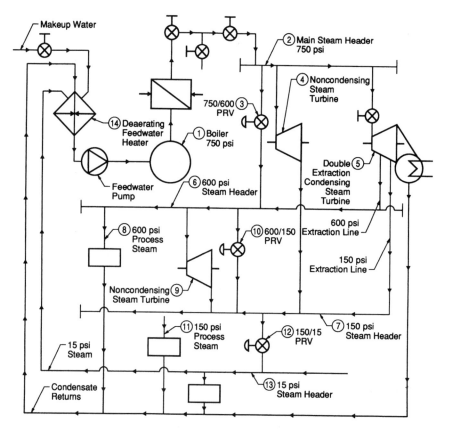

Figure 10-13. Oil refineries commonly cogenerate while operating at high plant efficiencies.

Textile Plants

Textile plants use steam to prepare for the weaving, spinning, dying, and drying processes. The dying process requires the greatest amount of steam. Steam injected into dye vats becomes contaminated and cannot be reclaimed. Approximately 20% to 25% of the steam in the form of condensate is reclaimed in the drying process. Large water softeners are required to provide makeup water to compensate for the quantities of steam lost in the dying and drying processes. **See Figure 10-14.**

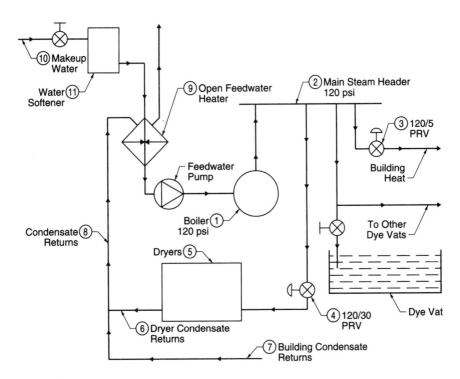

Figure 10-14. Textile plants have very low condensate returns and require large quantities of makeup water.

Steam leaves the boiler (1) at 120 psi and enters the main steam header (2). The steam is then distributed to the plant for use. One line goes to PRV (3), where steam is reduced from 120 psi to 5 psi for building heat. The other line goes to PRV (4), which reduces 120 psi steam to 30 psi steam to supply the dryers (5). Dryer condensate returns (6) and building condensate returns (7) are returned through condensate return lines (8) to the open feedwater heater (9). The large amounts of steam lost in the process are made up by makeup water (10) introduced through the water softener (11). The water softener reduces the amount of chemicals required to treat the boiler water. Improper boiler water treatment can adversely affect the color of the dye and dye consistency between batches.

PLANT EFFICIENCY

Maximum plant efficiency requires minimizing the loss of Btu in the boiler, steam turbine, and generator. Approximately 80% to 90% of the Btu available in fuel is absorbed by the boiler, depending on the heat reclaiming equipment under optimum conditions. **See Figure 10-15.** Of the remaining Btu, 80% is used in the steam turbine to supply power to the generator with approximately 20% lost under optimum conditions. The generator loses approximately 5% of the power supplied under optimum conditions. These amounts do not allow for leakage, such as in packing, piping, gaskets, or faulty traps, in the plant.

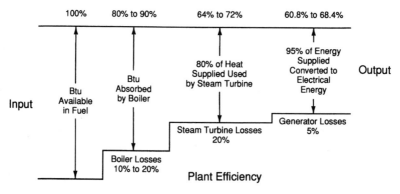

Figure 10-15. Minimizing Btu losses in the boiler, steam turbine, and generator is required for maximum plant efficiency.

Losses in Boilers

Btu losses in boilers occur in the following:
• dry chimney gases
• moisture in air
• water vapor formed by hydrogen in fuel
• moisture in fuel
• incomplete combustion of fuel
• loss of combustibles in refuse
• radiation losses

The greatest controllable Btu loss in the boiler is from loss of heat in gases of combustion. Btu losses also occur from boiler heating surfaces. Air heaters and economizers must be kept clean for maximum heat transfer. Fuel must be free of all moisture. High combustion efficiency and firing rate control are required. Insulation must be properly maintained to prevent heat losses.

Losses in Steam Turbines

A stationary engineer can reduce Btu losses in steam turbines by maintaining the insulation in proper condition, adjusting steam seals to minimize steam leakage, and maintaining the maximum vacuum in the condenser with minimum steam to the air injectors. With extraction steam turbines, high extraction flow rates are necessary for maximum steam turbine efficiency.

Losses in Generators

Generators are designed for maximum efficiency when running at 75% of maximum rating. The primary reason for loss in generators is shutdown resulting from improper maintenance procedures. Bearings and their lubrication must be carefully monitored. The commutator must be dressed and brushes replaced as required. Proper lubricating oil temperature and minimizing heat in the windings will extend life and improve efficiency of the generator.

INCINERATORS

Many million tons of municipal solid waste (MSW) or refuse are generated in the United States each year. Incinerators have been used over the years to reduce the volume of MSW through combustion. In the 1940s, there were approximately 700 municipal incinerators. However, many of these incinerators produced undesirable ash, noxious gases, and odors and were shut down.

The Clean Air Act of 1970 had a great impact on incinerator design and operation. In 1995, the EPA instituted new rules that tightened the regulation of plants using MSW to generate electricity. The new rule required large facilities to use the Maximum Achievable Control Technology (MACT). New incinerators designed for use in the United States use mass burn technology to comply with modern environmental standards. Mass burn plants burn MSW at above 2000°F. In the process, steam is directed to generators to produce electricity. One compacted ton of MSW occupies approximately 1.67 cubic yards. One ton of MSW processed through a mass burn incinerator occupies approximately 0.24 cubic yards, a volume reduction of about 85%.

Refuse-derived fuel (RDF) plants are mass burn plants that process MSW before burning by removing glass, metal, and other inorganic materials. These materials are recycled with the remaining material further processed by shredding, pelletizing, or grinding. The heating value of processed MSW ranges from 4500 to 5000 Btu/lb. Most mass burn incinerators are regulated to transfer specific quantities of heat over time. This requires control and monitoring of fuel content.

In a typical mass burn plant, trucks deliver MSW to an enclosed reception area. **See Figure 10-16.** The MSW is then transferred from the reception area to the feed hopper with an overhead crane. Combustion grates advance the fuel where it is burned at above 2000°F. Air blown from above and below the combustion grates is drawn from the reception area. This maintains a negative pressure to prevent the escape of dust and odor. A waterwall boiler above the combustion grates produces superheated steam. Steam is piped to a turbo-generator, which generates electricity for distribution from the switchyard.

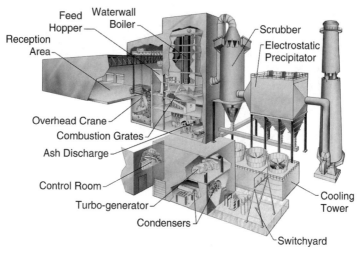

Wheelabrator Spokane Inc.

Figure 10-16. Mass burn plants burn refuse at above 2000°F and produce steam.

The scrubber removes acidic constituents such as sulfur dioxide from the gases of combustion. Condensate is reclaimed at the cooling tower and condensers. Flyash carried by the gases of combustion is trapped and collected by the electro-static precipitator. Bottom ash remaining on the combustion grates is discharged and collected. Approximately 85% of ash produced in mass burn incinerators is bottom ash.

CHAPTER 10 LEARNER RESOURCES
ATPeResources.com/QuickLinks
Access Code: 735728

COMPUTER-INTEGRATED CONTROL SYSTEMS

Computer-integrated control for heating, ventilating, air conditioning (HVAC) systems, and computer-integrated process application systems in a plant has greatly increased in use in recent years. All stationary engineers and maintenance personnel must have a basic understanding of these systems for maximum plant safety and efficiency. Additionally, computer hardware and software capabilities have increased, which requires the acquisition of new knowledge and skills to remain current in the field. Control offers new capabilities but does not eliminate the need for qualified personnel.

COMPUTER-INTEGRATED CONTROL SYSTEMS

A *computer-integrated control system* is a system that uses a computer to control equipment in a building or production plant. Computer-integrated control systems have been used to control boiler and engine room equipment, in the heating, ventilating and air conditioning (HVAC) industry, and for process work since the early 1970s. Computer-integrated control systems can be networked together in a plant-wide system and/or connected to remote locations. Varying in size and sophistication, these electronic devices are rapidly replacing the traditional electromechanical equipment that controls commercial buildings and industrial complexes. In addition, competition among manufacturers has reduced prices and added capabilities with each new product introduced.

Each manufacturer has specific control system strategies. However, any control system strategy using computer-integrated control is not designed to act as the sole equipment safety device. For example, high- and low-temperature limit controls, boiler water high-and low-limit controls, and flame safeguard controls are still required with computer-integrated control systems. In addition, the installation of a computer-integrated control system does not reduce the need for regular preventive maintenance.

Although computer-integrated control systems do not replace the need for a qualified stationary engineer on site, these systems are commonly used to monitor and control functions in commercial and industrial sites. In commercial applications, they are used to monitor and control functions for total control of the "intelligent" buildings being built today. In industrial settings these control systems can be utilized to their full extent to control processes that produce products such as bakery goods, milk products, furniture, and automobiles.

357

When used in a commercial setting, computer-integreated controls are commonly classified as energy management control systems or as direct digital control systems. In an industrial setting, computer-integrated controls are commonly classified as programmable logic controls. An *energy management control (EMC) system* is a solid-state control system that usually supplements other stand-alone control systems such as pneumatic controls. In an EMC system, a computer monitors building conditions and resets pneumatic controls based on what the computer calculates as the desired energy efficiency. The pneumatic controls operate to control the building functions. Because the pneumatic control system is operated by instructions from the computer-integrated system, it is often called a hybrid system.

A *direct digital control (DDC) system* is a solid-state control system in which a building automation system controller is wired directly to controlled devices to turn them ON or OFF. **See Figure 11-1.** Temperature sensors located throughout the system sense the load requirements. Controllers at various locations respond to the load requirements by activating the required equipment. Data in the system is shared through the network communication module. The operator workstation allows system monitoring, direct control, or programming of system functions. A printer is used to print hard copies of the system operation data.

A *programmable logic control (PLC) system* is a solid-state control system that is programmed to automatically control an industrial process or machine. Operations such as transporting parts, positioning parts, and machining are synchronized by the PLC. Most boiler manufacturers integrate PLCs into the boiler control systems of their medium- and large-size boilers. PLCs are also referred to as programmable controllers. A computer-integrated control system commonly includes a central processing unit (CPU), input and output modules, and a programming device.

Central Processing Unit

A *central processing unit (CPU)* is a computer component that interprets and controls all computer program instructions. A *program* is a sequence of coded instructions designed to initiate a desired function. Computer program instructions are designed to perform the functions required to control all controller activities located at the plant equipment. The three major components of a CPU are the processor, the memory, and the power supply.

Processor. A *processor* is a computer device that contains integrated circuits for performing mathematical operations, data interpretation, and diagnostic routines. These circuits interpret input signals, perform necessary calculations, and output signals to control plant equipment. The processor also performs tasks required for interfacing with workstations and other computer-integrated controls.

A CPU may contain more than one processor (multiprocessor) to increase the processing speed. This configuration is normally used in large PLC systems. The main CPU processor is used to store the program and database and execute the program based on the inputs. The computer processor supports the communications between the plant equipment (field device) input and output, the operator, and other PLCs.

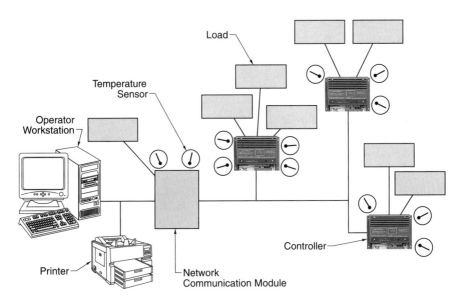

Direct Digital Control (DDC)

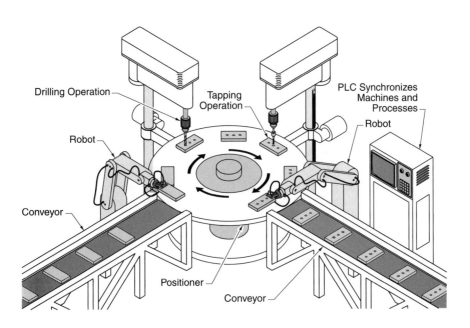

Programmable Logic Control (PLC)

Figure 11-1. Computer-integrated control systems are commonly used to control equipment in HVAC systems and equipment used for manufacturing processes in industrial facilities.

A *proportional-integral-derivative (PID) processor* is a processor that performs independent control tasks such as closed loop control utilizing PID actions. A *proportional-integral-derivative (PID) action* is a control scheme in which proportional action occurs by comparing a sensed condition to a setpoint for determining the output. A PID action is used to eliminate the undesirable "hunting" effect of a controller, where a control decision oscillates and varies from setpoint. The processor is selected for an application such as controlling fast moving equipment or processes. Processors are categorized according to the number of bits they simultaneously use to perform operations. Examples include 8-bit, 16-bit, and 32-bit processors. A 32-bit processor can process data twice as fast as a 16-bit processor.

Memory. *Memory* is computer storage capacity. Computer programs are stored in memory to be executed by the processor as required for the desired control of field devices. These programs can be changed to adapt to plant requirements. The memory system of a PLC consists of the executive section and the application section. The executive section includes supervisory programs that manage all of the system activities, such as execution of the control program and communications with the field devices, other controllers, and the operator. The executive section stores all system instructions, such as math instructions, transfer instructions, and relay instructions. The executive section is not accessible to the user. The application section is where a program for a specific application is stored by the user. Application memory is memory that the user defines to monitor specific inputs and outputs for the control of plant equipment.

Computer memory is classified as volatile memory and nonvolatile memory. *Volatile memory* is computer memory that is lost if power is interrupted. *Nonvolatile memory* is computer memory that is not lost if power is interrupted. Random-access memory (RAM) is volatile memory. Normally RAM is protected by a battery backup in most controllers. Different types of computer memory include read-only memory (ROM), programmable read-only memory (PROM), erasable programmable read-only memory (EPROM), electrically altered read-only memory (EAROM), and electrically erasable programmable read-only memory (EEPROM).

Read-only memory (ROM) is designed to store a fixed program from the manufacturer and cannot be altered. Programmable read-only memory (PROM) cannot be easily altered and is normally used to store a backup copy of RAM. Erasable programmable read-only memory (EPROM) can be erased and reprogrammed by passing the chip (integrated circuit) under an ultraviolet light source, which erases its entire contents, and then reprogramming it if necessary. Electrically altered read-only memory (EAROM) is similar to EPROM but instead of passing the chip under an ultraviolet light to erase it, a voltage is applied to a specific pin to erase it. Electrically erasable programmable read-only memory (EEPROM) is the preferred type of nonvolatile memory for use in controllers because individual characters can be erased and reprogrammed, similar to conventional RAM.

Power Supply. A *power supply* is a computer component that provides DC voltage for the operation of the processors, memory devices, and input/output modules. **See Figure 11-2.** The power supply also monitors and regulates supply voltage to the controller. Most controllers require 120 VAC or 240 VAC, but sometimes can use 24 VDC. The use of an isolation transformer or a constant voltage transformer will prevent interference from other equipment or voltage variation from affecting the power supplied to the controller. An *isolation transformer* is a transformer that eliminates electromagnetic interference (EMI) to other equipment. A *constant voltage transformer* is a transformer that is used where voltage supplied to the controller varies more than the normally acceptable 10% and that compensates for voltage variations at the primary side to maintain a consistent voltage on the secondary side.

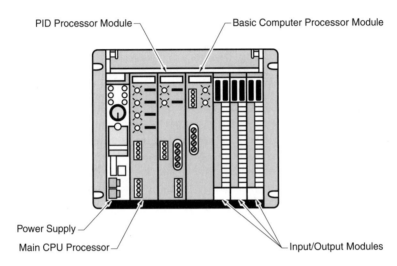

Figure 11-2. A power supply provides DC voltage for the operation of processors and input/output modules.

Input/Output Modules

Input/output (I/O) is the way in which a controller receives and transmits data to and from field devices. Many computer-integrated control systems have individual modules for this purpose. Some systems, usually the smaller, dedicated equipment, have the I/O components built into the controller. These systems have a limited number of inputs and normally can only control one piece of equipment, such as a single air conditioner or a single product-packaging machine.

Field devices transmit and receive signals. Based on the status of an input field device sensed or the process values measured, the CPU transmits signals for the desired control of the output field devices. An *input module* is a device that receives a signal from an input field device classified as a sensor and transmits that signal to the computer. A *sensor* is a device that senses or measures the condition of a variable such as temperature, relative humidity, pressure, flow, level, electrical units, and the position of various mechanical devices. Sensors also monitor relays, switches, and other binary (two position) devices. An *output module* is a device that receives a signal from a computer and transmits that signal to field devices. This signal is used to control devices, such as relays, motor starter coils, and/or solenoids, used to energize system components (dampers, valves, or motors).

Signals. The signal to or from I/O modules can be an analog signal or digital signal. This signal is the data that defines the information stored, transferred, transmitted, and processed. An *analog signal* is data that is sent about the status of pressure, temperature, humidity, airflow, voltage, or resistance. The computer software uses mathematical formulas to calculate the desired condition and processes the analog signals. This information is then transmitted to the equipment to vary the equipment position or speed. A *digital signal* is data that has a distinct predetermined status as one of two states: ON or OFF.

Point Types. All computer-integrated control systems, from digital watches to sophisticated building controls, utilize four possible point types: analog-in (AI) points, analog-out (AO) points, digital-in (DI) points, and digital-out (DO) points. The type of equipment determines the point type, whether it is a signal into or out of the computer, and whether the signal is a continuously varying signal or if it is a two position or staged signal. Equipment status in the system is indicated on a display.

An *analog-in (AI) point* is a remotely located sensor that has a continuously varying signal that bears a known relationship to the value of the measured variable. An example of an AI point type is a thermistor in a stack temperature sensor. **See Figure 11-3.** A *thermistor* is an electrical device that changes electrical resistance with a change in temperature. A thermistor senses temperature and varies a continuous signal based on the surrounding temperature. In a PLC system, this type of signal is sent to an analog input interface module. In a DDC system, this type of signal is sent to an analog input module. This module receives the signal from the remote sensors and translates and transmits it to the CPU.

An *analog-out (AO) point* is a remotely located device that receives a continuously varying signal from the computer and reacts to that signal to control equipment. An example of an AO point type is a modulating damper motor. A signal is received from the computer, which modulates the damper motor to open, close, or move the damper to regulate airflow to a space. In a PLC system, this type of signal is sent to an analog output interface module. In a DDC system, this type of signal is sent to an analog output module. This module receives the signal from the CPU and translates and transmits it to the remote equipment.

A *digital-in (DI) point* is a remotely located sensor that represents a distinct predetermined value. An example of a DI point is a pressure differential switch. The switch closes an electrical contact when water pressure out of a pump is greater than the amount of pressure into the pump as determined by the setpoint. The switch then sends a signal to the computer that represents a predetermined value (FLOW). When the water pressure is less than the setpoint, a distinct separate signal is sent to the computer that represents a second predetermined value (NO FLOW). In a PLC system, this type of signal is sent to a discrete input interface module. In a DDC system, this type of signal is sent to a digital input module. This module receives the signal from the remote sensors and transmits it to the CPU.

A *digital-out (DO) point* is a remotely located device that receives a distinct predetermined value from a computer and reacts to the signal to control a staged procedure. An example of a DO point is a motor starter. The motor starter receives a distinct value from the computer that represents the ON stage to close a contact that starts the motor. A second distinct value is received from the computer that represents a second predetermined value (OFF). In a PLC system, this type of signal is sent to a discrete output interface module. In a DDC system, this type of signal is sent to a digital output module. This module receives the signal from the CPU and transmits it to remote equipment.

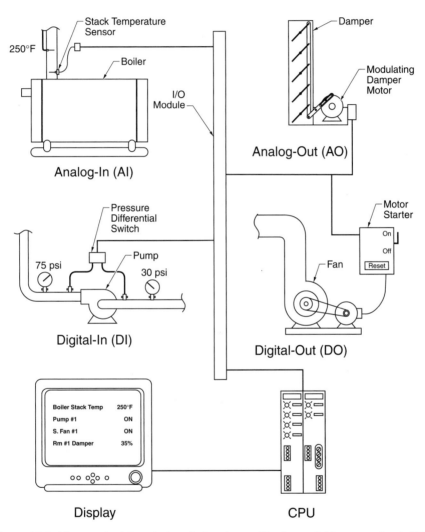

Figure 11-3. All computer-integrated control systems utilize four point types: analog-in (AI) points, analog-out (AO) points, digital-in (DI) points, and digital-out (DO) points.

In a PLC system there are also special I/O interface modules. A *special I/O interface module* is an input/output interface that provides a link between PLCs and field devices by using special types of signals. Special I/O interface modules are categorized as direct-acting interface modules and intelligent interface modules. A *direct-acting interface module* is a special I/O interface module that is capable of processing low-level and fast-input signals. Standard I/O modules cannot read these signals. An *intelligent interface module* is a special I/O interface module that incorporates onboard microprocessors to add intelligence to the interface. An intelligent interface module can perform complete processing tasks independent of the PLC processor and program.

Programming Devices

A *programming device* is a human interaction device used for controlling the actions of a computer. **See Figure 11-4.** Manufacturers have designed programming devices that minimize learning and programming time. These devices allow plant personnel to solve control problems rather than spend valuable time programming. The programming languages are distinctively different for PLCs than for DDC systems but the hardware is similar. Two common types of programming devices are mini programmers and personal computers.

A *mini programmer* is a small handheld device for programming computer-integrated controls used in the past. Mini programmers are still used today in limited applications. Mini programmers resemble handheld calculators but have a larger display, programming instruction keys, and special function keys. Some controllers have built-in mini programmers instead of handheld units. Some built-in programmers are detachable from the controller. Although they are primarily used for editing and inputting control programs, mini programmers can also be used for starting up, changing, and monitoring the control program.

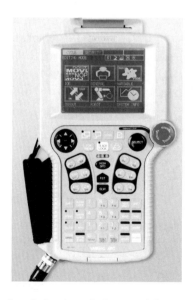

Motoman, Inc.

Figure 11-4. A programming device is a device used for programming and controlling equipment.

Mini programmers are also classified as intelligent or nonintelligent depending on whether they are microprocessor-based or not. A microprocessor-based mini programmer provides the user with many of the features offered by personal computers for off-line programming. A nonintelligent mini programmer can be used to enter and edit a controller program that has limited monitoring and editing capabilities.

A desktop personal computer (PC) is a commonly used programming device that consists of a CPU, a monitor, a keyboard, and usually a printer. Most manufacturers provide the necessary software for entering, editing, documenting, and monitoring of the control program. The large screen of a PC displays more of the control program at a time, making it easier to enter, edit, and monitor the program. A PC can be connected to a local area network (LAN) for continual monitoring, storing files, processing information, and/or acting as a terminal for the plant computer system. Off-site access to the control system is possible using a modem. **See Figure 11-5.** Portable (laptop) computers allow plant personnel to use the computer for normal office tasks and as a programming device to connect directly to the control system.

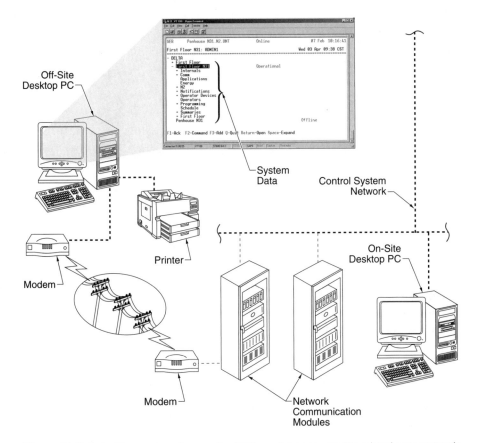

Figure 11-5. A desktop personal computer (PC) can be connected to a local area network (LAN) for continual monitoring, storing files, processing information, and/or acting as a terminal for the plant computer system.

Most manufacturers have a screen graphic interface for computer-integrated control systems. **See Figure 11-6.** The program for the graphic interface resides in the CPU memory of the programming device. Graphic interfaces allow easier programming, system monitoring, and system troubleshooting. They offer greater efficiency when isolating a problem on a system and minimize expensive downtime. In DDC systems, programming devices are called operator-interface consoles. Most DDC systems today do not require an operator-interface console connected to the system, but many large systems have one for convenience.

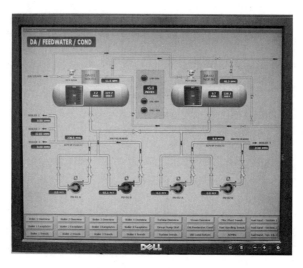

Figure 11-6. System graphics help visualize the association between the system and the boiler equipment.

System Architecture

System architecture is the computer equipment, the strategies used to control the equipment and operations in the plant, and the connections between them that form a complete control system. A *centralized control system* is a process scheme that uses a centralized computer for all information gathering, decision making, and sending of equipment control commands. This scheme has been used over the years, but can result in catastrophic consequences. If the central computer fails, the entire system is inoperable. Most manufacturers have replaced centralized control systems with distributed control systems. **See Figure 11-7.** A *distributed control system* is a control system that uses controllers located at the equipment throughout the plant for information gathering, the decision-making process, and the processing of commands. Control information is transmitted over a network used by other controllers. Typically, a personal computer is connected to the network to monitor and issue commands requiring operator control and to override controller commands of the equipment. Laptop computers allow monitoring at locations throughout the system.

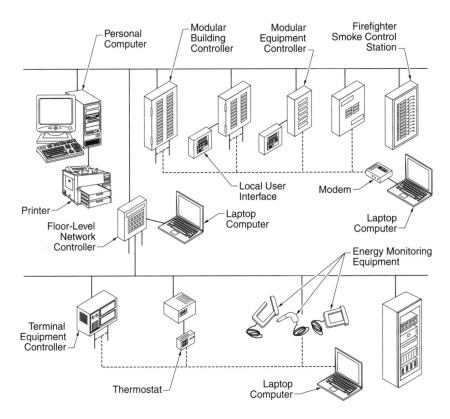

Figure 11-7. A distributed control system uses a network to communicate with specific controllers located at equipment throughout the plant.

The system architecture required varies from plant to plant. Computer hardware and software from different manufacturers is usually proprietary (not compatible with other manufacturers). Most commercial computer-integrated control systems related to boiler operation are direct digital control (DDC) systems. The primary advantages of a computer-integrated control system over a pneumatically-controlled or hybrid system are less maintenance, better control of conditions, easier scheduling and troubleshooting, more information provided about facility conditions, greater efficiency, easier computer control, and remote operation capability. Direct digital control systems can be used to control individual equipment, entire buildings, building complexes, and/or complexes in different geographical locations.

In some industrial applications, such as bakeries, breweries, and candy companies, computer-integrated control systems are used to control a single machine, an application, or a complete process. In these applications, computer-integrated controls are built to withstand harsh industrial plant environments that require the use of PLCs.

Controllers in a system can be networked or can operate as stand-alone controllers. Their capability ranges from very small microcontrollers having as few as 10 I/O points to very large networked systems having several thousand points in the system. Computer-integrated control systems in industrial processes require faster communication between machines and greater accuracy than commercial HVAC computer-integrated control systems.

Networking controllers together in local area networks (LANs) enables each controller to share information with other controllers for total system control. An industrial LAN normally operates at a high speed to accommodate faster processing requirements. A LAN is a medium-distance communication system that can be arranged into different configurations including a star network, common-bus network, or ring network. **See Figure 11-8.** Some large networks may use more than one configuration for increased capabilities.

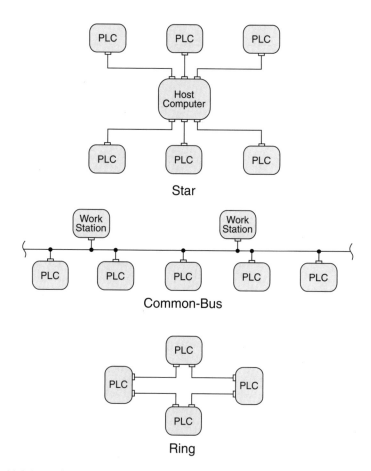

Figure 11-8. Local area network (LAN) configurations used in a computer-integrated control system are selected based on the size of the facility, number of PLCs, speed, and other application requirements.

Star Network. A *star network* is a computer communication network that consists of a multiport host computer with each port connected to the programming port of a controller. Typically this LAN is made up of twisted pair conductors that provide good communication at distances up to 4000'. The main advantage of a star network is that each controller can communicate directly with the host at any time through its connection. A disadvantage of a star network is that it uses central processing rather than distributed processing. The network is totally dependent on the host computer to communicate information from one controller to another controller. If the host computer fails, the system fails. Another disadvantage of a star network is that the wiring cost is high in large installations.

Common-Bus Network. A *common-bus network* is a computer communication network that consists of a main local area network (LAN), or bus, to which individual controllers and workstations are connected. A coaxial cable is typically used in PLC systems for communication up to 18,000'. In a common-bus network, there is communication directly between each controller, and the network can be easily reconfigured to add or remove workstations or controllers. A disadvantage of a common-bus network is determining which controller may transmit at which time. Another disadvantage is that all workstations and controllers are affected by a problem with the bus. **See Figure 11-9.**

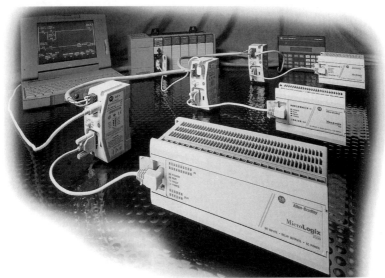

Rockwell Automation, Inc.

Figure 11-9. PLCs connected to a communication network are connected in parallel for improved speed of communication.

Ring Network. A *ring network* is a computer communication network that consists of workstations and controllers linked in series to each other. The ring configuration is not commonly used in industrial environments because failure of any workstation or controller will crash the network. Overcoming this disadvantage requires special wiring methods and combination of configurations to bypass interrupted communication. However, a ring network can be used effectively with fiber-optic LANs and is usually used in industrial environments because of fast communication rates over long distances. Controllers in a network can be classified into two main types: field panel controllers and terminal equipment controllers.

Field Panel Controller. A *field panel controller* is a controller that contains all hardware, firmware (programmed software), and applications software to provide continuous control of connected components. A field panel controller operates on information it collects from other controllers or equipment. Field panel controllers are located in areas of high point densities, such as mechanical rooms, because of the large I/O point capacities. The physical architecture of field panel controllers has evolved from components mounted on large printed circuit boards to a modular design. The modular design greatly improves installation, reliability, and troubleshooting.

Field panel controllers may be called modular building controllers, modular unit controllers, or remote building controllers, depending on the application. Specific components vary depending on the manufacturer, but all require I/O termination, a power supply, and a software interface. An operator interface allows input by the operator. Field panel controllers in an industrial environment are installed in a rack system protected from the plant environment. **See Figure 11-10.**

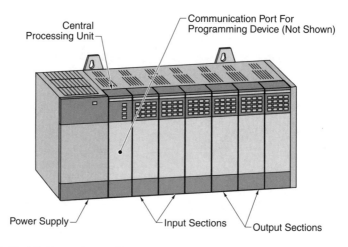

Figure 11-10. A field panel controller with modular design allows easy installation, troubleshooting, and repairs.

A field panel controller can connect directly to 50 points or more, with the capacity for expansion modules or slave I/O modules to increase the point capacity. This increase is limited only by the memory in the controller. The power supply transforms the supplied voltage into several separate voltage ranges to be used internally and externally for I/O points. Protection from line voltage fluctuations (low voltage, high voltage, spikes, and surges) is provided. Power supplies are typically fused for protection from excessive amperages. Most field panel controllers have battery back-up protection against memory loss in the event of a power outage.

The available memory in a field panel controller depends on system sophistication and software features. In computer-integrated control systems, point databases, application programs, and trend logs require a certain amount of memory. Additional memory or memory upgrade may be required to ensure full use of these capabilities. The software (point databases, application programs, and trend logs) allows communications with the I/O points, operator interfaces, and other controllers. Most manufacturers have additional application software programs available that enhance the features of these controllers, such as optimization programs for boilers, chillers and related equipment, and preventative maintenance programs.

The operator interface of a field panel controller typically has several connection points, including a direct connection for a programming device, a modem connection for remote communications with a personal computer utilizing a telephone line, and a trunk connection for remote communications with a personal computer in the same complex. Normally the trunk connection (system trunk) is used for interfacing several controllers to share information. For example, one outside air temperature sensor terminated at one controller may be sufficient, but all the controllers in the system can utilize the temperature information.

Controller technology has improved, allowing the development of a communications controller that can take the place of a field panel controller. A *communications controller* is a microprocessor that controls the communication into and out of a computer device. Most field panel controller features exist in a communications controller except for the individual I/O point support. Communications controllers usually have the capability of utilizing one to three LAN trunks, which can handle many smaller, terminal equipment controllers. Some manufacturers have add-on modules that increase the trunk numbers for increasing the controller point numbers.

Terminal Equipment Controller. A *terminal equipment controller* is a small, stand-alone controller that has a limited number of input and output point connections. Other connections are provided for supply power and LAN trunk connection. With advances in technology, terminal equipment controllers can emulate several field panel controller functions. This offers new capability opportunities to computer-integrated control systems and operators. A terminal equipment controller is used to control equipment in or near the main equipment being controlled, such as heat pumps, variable air volume (VAV) boxes, constant volume boxes, baghouses, and boiler draft fans.

The point capacity of the terminal equipment controllers is limited (typically two to eight inputs and outputs) and usually has no expansion capability. Because terminal equipment controllers were originally designed for controlling small actuators and devices, some come with only digital-out points. Also, most terminal equipment controllers have stand-alone capabilities but are limited to specific applications. Terminal equipment controllers can be networked together on a LAN, or they can directly interface with a computer for control and monitoring purposes. **See Figure 11-11.** In the industrial environment, this type of terminal equipment controller is commonly called a "brick" (small PLC with a limited amount of I/O points).

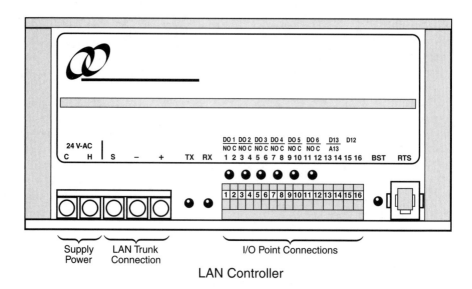

Supply Power
LAN Trunk Connection
I/O Point Connections

LAN Controller

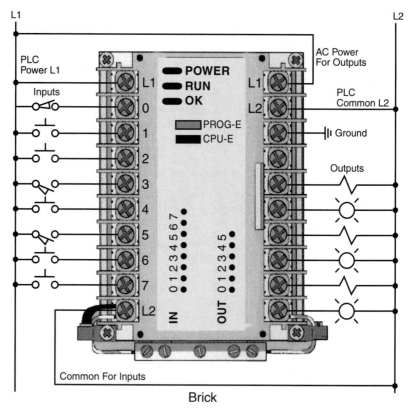

Brick

Figure 11-11. A terminal equipment controller can be connected to a LAN trunk in a total building control system or it can operate as a stand-alone controller.

Computer-Integrated Control Operation

A controller in a computer-integrated control system operates using power to the CPU and I/O modules from the power supply. The power supply contains rectifiers, surge suppressors, and voltage regulators because precise voltage is required for correct operation. An input module receives a signal from a sensor and transmits that signal to the computer. **See Figure 11-12.** Signals from the inputs are converted to low-voltage DC signals used by the CPU. Each input is connected to a specific terminal of the input module. The CPU processes the signals and runs the control program. Based on the processed signals, the program determines the output signals. The output signals are transmitted to the output module, which transfers the low-voltage DC signal into the appropriate signal to be used by the loads. Each input and output receives an address.

An *address* is an identification number assigned to a specific input or output. The identification number corresponds to a software location in the CPU that holds the specific I/O information. The CPU scans the input addresses many times a second, checking for a signal that has changed, arrived at, or been removed from an address. When a signal has changed, arrives or is removed from an input address, the CPU checks its program for the action required in response to the signal change.

Programs are created or modified by typing instructions on the programming device that operates the CPU. Computer-integrated control programs use various computer languages. Most PLCs display the program visually using line diagram symbols similar to electrical circuit line (ladder) diagrams. Most DDC computer programs use a text-based language to explain the operation of the program. Program languages are explained in more detail later in this chapter.

The controller receives inputs and activates loads. Inputs can also activate internal loads inside the controller. These internal loads (virtual points) exist only in the program of the CPU but function like the timers, relays, and counters that the controller replaces. Virtual points take up no space, need no maintenance, are extremely reliable and accurate, and are included in the cost of the program. Their settings can be modified without changing any wiring or replacing components.

Programming Methods

In the past, programming methods for computer-integrated control systems offered few features and limited flexibility. Many systems were supplied with programs written by the manufacturer at the factory using low-level, assembly-type language (decimal, binary, hexadecimal, and Boolean algebra). In some cases manufacturers "burned" their program into a programmable read-only memory (PROM) chip that was supplied with the system. The cost of changing the program was prohibitively high.

The next generation of programming was the EPROM chip. This type of chip makes programming in the field possible. The chip can be erased by passing it under an ultraviolet light and reprogramming. The EEPROM chip allows the user to alter individual words, in much the same way as in ordinary read/write memory. The EPROM chip and EEPROM chip are commonly used in field panel controllers and terminal equipment controllers for application programs. Programming a computer for specific functions requires the use of a programming language.

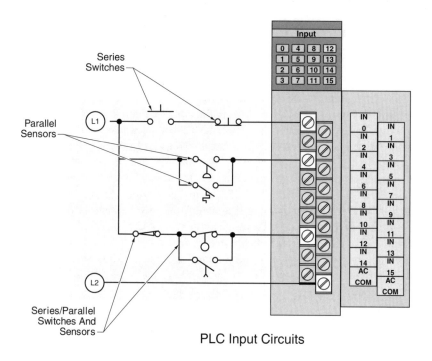

Series Switches

Parallel Sensors

Series/Parallel Switches And Sensors

PLC Input Circuits

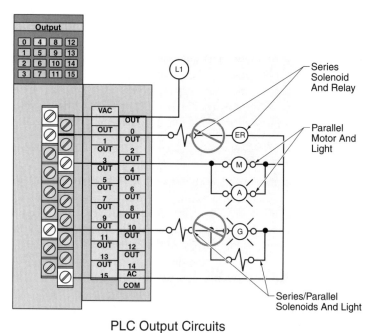

Series Solenoid And Relay

Parallel Motor And Light

Series/Parallel Solenoids And Light

PLC Output Circuits

Figure 11-12. General rules must be followed when installing output devices. Output devices cannot be installed in series from PLC output terminals.

A *programming language* is a standardized communication method used to develop a control program for a controller in a computer-integrated control system. Programming languages commonly used for DDC systems include line programming and function block programming. Programming languages for PLCs have evolved as PLCs have developed over the years into five common languages: line, function block, ladder, Boolean, and Grafcet. The ladder and Boolean programming languages implement operations similarly but differ in the way instructions are entered and represented. The Grafcet programming language implements control instructions based on steps and actions in a graphic-oriented program.

Line Programming. *Line programming* is a method of programming that uses a sequence of computer commands in lines to initiate an operation. Line programming has been used by the computer industry for years. It has a format that is similar to common typical programming, except that specific routines are added to initiate start and stop commands. It is also used to initiate controller output, such as proportional, integral, and derivative (PID) algorithms and tie into occupancy schedules. For example, a program is written to provide chiller compressor control and staging based on supply air temperature to the occupied space. **See Figure 11-13.** The cooling is locked out if the outside air temperature is less than 50°F between machines and the building is not occupied. It also allows the high limit of the cooling loop to ramp up to 100% over a three-minute period after startup. This prevents all chillers and stages of the chillers from starting at the same time.

```
 10 IF(OCC.EQ.OFF.OR.OSA.LT.50.0)THEN HLIMIT=0.0
 20 IF(OCC.EQ.OFF)THEN SECND1=0.0
 30 TABLE(SECND1,HLIMIT,0.0,0.0,180.0,100.0)
 40 TABLE(OSA,SETPT,30.0,65.0,80.0,50.0)
 50 INPUT=0.99*INPUT+0.01*DATEMP
 60 LOOP(0,INPUT,OUTPUT,SETPT,PROPG,INTEGG,DERIVG,SAMPLT,BIAS,
       LLIMIT,HLIMIT,0)
 70 WAIT(1,CHILL1,DELAY1,11)
 80 WAIT(DELAYT,DELAY1,DELAY1,10)
 90 IF(DELAY1.EQ.ON)THEN EMOFF(CHILL1)
100 IF(DELAY1.EQ.OFF.AND.CHILL1.NE.@NONE)THEN RELEAS(CHILL1)
110 WAIT(1,CHILL2,DELAY2,11)
120 WAIT(DELAYT,DELAY2,DELAY2,10)
130 IF(DELAY2.EQ.ON)THEN EMOFF(CHILL2)
140 IF(DELAY2.EQ.OFF.AND.CHILL2.NE.@NONE)THEN RELEAS(CHILL2)
150 WAIT(1,CHILL3,DELAY3,11)
160 WAIT(DELAYT,DELAY3,DELAY3,10)
170 IF(DELAY3.EQ.ON)THEN EMOFF(CHILL3)
180 IF(DELAY3.EQ.OFF.AND.CHILL3.NE.@NONE)THEN RELEAS(CHILL3)
190 DBSWIT(0,OUTPUT,6.0,10.0,CHILL1)
200 DBSWIT(0,OUTPUT,16.0,20.0,CH1S2)
210 DBSWIT(0,OUTPUT,26.0,30.0,CH1S3)
220 DBSWIT(0,OUTPUT,36.0,40.0,CHILL2)
230 DBSWIT(0,OUTPUT,46.0,50.0,CH2S2)
240 DBSWIT(0,OUTPUT,56.0,60.0,CH2S3)
250 DBSWIT(0,OUTPUT,66.0,70.0,CHILL3)
260 DBSWIT(0,OUTPUT,76.0,80.0,CH3S2)
270 DBSWIT(0,OUTPUT,86.0,90.0,CH3S3)
```

Figure 11-13. A line program to control a chiller compressor and staging specifies routines required when supply air temperature changes.

Line programming is flexible and allows control of typical plant functions but can be complex because of the language used and rules that are required. This can be intimidating to plant personnel. However, many of these functions can be simplified by using comments and special formatting. Additionally, most line programming languages now used contain all the features of high-level computer languages and features such as full-screen editors and on-line debugging. This allows the operator to edit, view, change, and debug the control sequence. Line programs are also self-documenting. When programs are updated, a printout documents the change.

Function Block Programming. *Function block programming* is a method of programming that uses a library of preprogrammed routines (blocks) that can be linked together in order to meet job-specific requirements. Function block programming breaks down manufacturer factory-programmed applications into small program blocks. These blocks of information can then be linked together to develop complete program sequences. Each block has areas to fill in for specific functions. Combinations of these blocks can be utilized to provide flexibility of sequences. **See Figure 11-14.**

The primary advantage of function block programming in standard HVAC applications is simplicity. If the control sequence calls for the exact functions provided by the function blocks in the system, the programming effort is minor. However, if the function block sequences required do not match the available function blocks, the programmer must employ custom blocks. This can result in more complicated programs and reduced system performance. Line programming-based systems are usually more effective in building applications that require elaborate control strategies. Computer-integrated control systems with function block programming are generally limited to applications employing simple or traditional pneumatic control strategies.

Ladder Programming. *Ladder programming* is a method of programming that uses ladder diagrams and symbols similar to electrical contact symbols to detail the electrical relationship between controls and loads. A ladder program looks similar to ladder diagrams used by electricians to simplify wiring diagrams illustrating electrical functions in a system. Wiring diagrams show the actual physical location, wire markings, and terminations on the diagram.

Ladder diagrams show the electrical relationship of the controls to the loads. **See Figure 11-15.** Power supplied is depicted as parallel vertical lines with controls and loads between the vertical lines called rungs. The advantage of ladder programming is that plant personnel often have had training in the electrical field and understand ladder diagrams. A minimal amount of training is needed to make the transition to ladder programming. However, as with any other programming, there are many rules that must be followed when ladder programming.

The main function of ladder programming is to control outputs and perform functional operations based on input conditions. When the start pushbutton (20) is pushed, coil C (40) is energized, closing the holding circuit through contact C (40-1). As long as the normally closed overload contacts (30) are closed, the coil remains energized. When the stop pushbutton (10) is pushed, the coil C is de-energized, opening holding contact C. It remains in this de-energized state until the start button is pushed again.

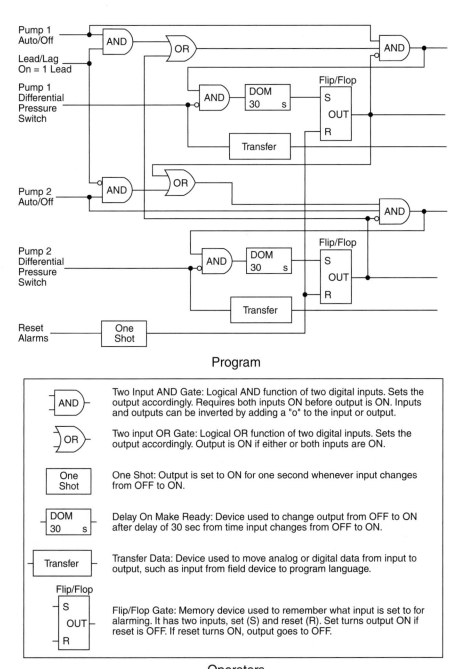

Program

Operators

Figure 11-14. Function block programming uses preprogrammed routines from the manufacturer to create the desired sequence of operation.

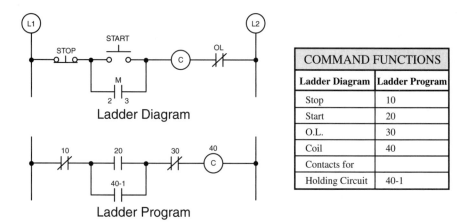

Ladder Diagram

Ladder Program

COMMAND FUNCTIONS	
Ladder Diagram	Ladder Program
Stop	10
Start	20
O.L.	30
Coil	40
Contacts for	
Holding Circuit	40-1

Figure 11-15. A ladder program looks similar to a ladder diagram but uses electrical contact symbols to detail circuit function.

A rung consists of a set of input conditions, represented by contact symbols, and output instructions at the end of the rung, represented by a coil symbol. **See Figure 11-16.** The left side represents L1, the power supply conductor, and the right side represents L2, the neutral conductor. The ladder rung is TRUE when it has continuity. Continuity exists when power flows through the rung from left to right. The contacts are shown in their normal, de-energized state. Circuit continuity is represented on the screen of the programming device as bold lines.

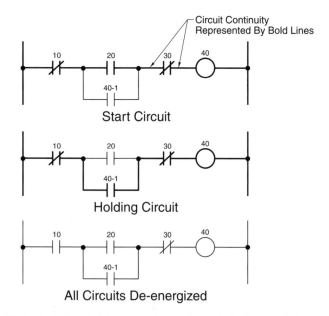

Figure 11-16. Contacts in a ladder program are shown in their normal de-energized state. Circuit continuity is represented on the programming device screen as bold lines.

When the start button is pushed, continuity is straight across the rung. As soon as the coil is energized the holding circuit is established, establishing continuity around the start button through the holding contacts. When the start button is released, the holding circuit is established, keeping the coil energized. When the stop button is pushed the circuit is de-energized.

Boolean Programming. *Boolean programming* is a method of programming that uses Boolean logic operators such as AND, OR, and NOT to control circuit function. The Boolean programming language (Boolean mnemonics) uses these operators (abbreviated using three or four characters) to describe the circuit instruction. **See Figure 11-17.** Boolean programming language can be used to develop and enter a program into PLC memory. However, a PLC may display the entered information in a ladder programming language.

Grafcet Programming. *Grafcet programming* is a method of programming that uses a symbolic graphic language to represent the control program as steps or stages in a process. Grafcet charts provide a flowchart-like representation of the programmed events to take place in each stage of the control program. Few PLCs can be programmed directly using Grafcet. Most manufacturers allow Grafcet programming with the program then being converted into either ladder programming or Boolean programming for downloading into the PLC.

COMPUTER-INTEGRATED CONTROL SYSTEM APPLICATIONS

Computer-integrated control systems are commonly used in industry to monitor and control HVAC functions such as heating and air conditioning, as well as refrigeration, fire safety, and building security. These systems are typically found in office buildings, hospitals, and business complexes. Recent buildings may have control systems for total building control (intelligent buildings). These systems control several additional building functions such as elevators, escalators, and lighting. Computer-integrated control systems are also used to control process systems that produce items such as bakery goods, milk, furniture, and automobiles. These systems are also used to control the operation of specific equipment in large boiler plants, such as electrostatic precipitators, forced and induced draft fans, and fuel-handling equipment.

HVAC Applications

A typical HVAC application of a computer-integrated control system is a building that has several fan powered, variable air volume (VAV) boxes to condition the air supplied into the occupied space. **See Figure 11-18.** A terminal equipment controller, designed specifically to control VAV boxes, controls each box. A room temperature sensor (thermistor) senses the ambient air temperature in each space and varies the signal to the terminal equipment controller. Electrical resistance is created in the thermistor based on the surrounding temperature. The terminal equipment controller determines the required signals to the damper at the inlet of the VAV box and the signals to the valve for a heating coil at the outlet of the VAV box. The damper modulates cold air supplied to the VAV box from a central plant. The valve then modulates hot water to the heating coil. The fan in the VAV box is ON when the building is occupied as determined from the field panel controller.

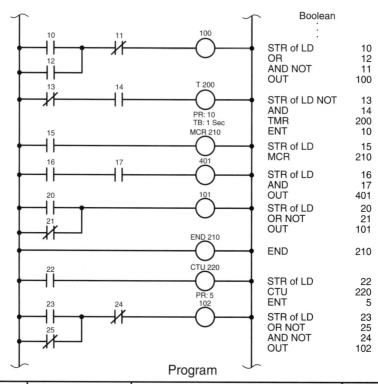

Program

Boolean	Function	Description	Ladder Equivalent
LD/STR	Load/Start	Starts a logic sequence with a NO contact	
LR/STR NOT	Load/Start not	Starts a logic sequence with a NC contact	
AND	And point	Makes a NO contact series connection	
AND NOT	And not point	Makes a NC contact series connection	
OR	Or point	Makes a NO contact parallel connection	
OR NOT	Or not point	Makes a NC contact parallel connection	
OUT	Energize coil	Terminates a sequence with an output coil	
OUT NOT	De-energize coil	Terminates a sequence with a NOT output coil	
OUT CR	Energize internal coil	Terminates a sequence with an internal output	
OUT L	Latch output coil	Terminates a sequence with a latch output	
OUT U	Unlatch output coil	Terminates a sequence with an unlatch output	
TMR	Timer	Terminates a sequence with a timer	
CTU	Up counter	Terminates a sequence with an up counter	
ADD	Addition	Terminates a sequence with an addition function	
SUB	Subtraction	Terminates a sequence with a subtraction function	
MUL	Multiplication	Terminates a sequence with a multiplication function	
DIV	Division	Terminates a sequence with a division function	
CMP	Compare (=, <, >)	Terminates a sequence with a compare function	
JMP	Jump	Terminates a sequence with a jump function	
MCR	Master control relay	Terminates a sequence with an MCR output	
END	End MCR, jump, or program	Terminates a sequence with an end of control flow function	
ENT	Enter value for register	Used to enter preset values of registers	Not required

Operators

Figure 11-17. Boolean programming uses ladder programming and Boolean operators to indicate circuit operation.

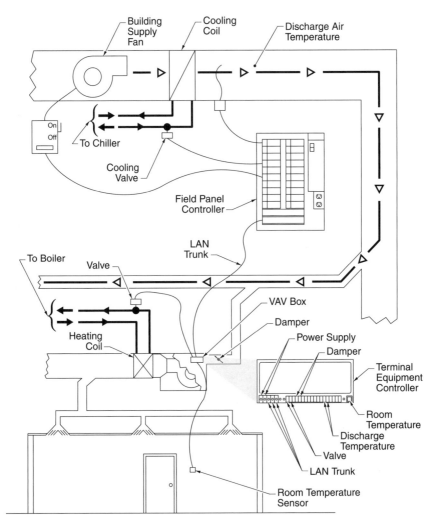

Figure 11-18. A computer-integrated control system for monitoring and controlling HVAC functions uses sensors, terminal equipment controllers, and field panel controllers.

Air supplied to the VAV box from the building supply fan is precooled to satisfy the warmest area in the building. This is accomplished by monitoring all of the room temperature sensors with the field panel controller. Data is communicated via the LAN trunk. The computer processor in the field panel controller selects the warmest room temperature to reset the setpoint of the discharge air temperature sensor, which controls a cooling valve to a common cooling coil for the building. Monitoring the outside air temperature and resetting the setpoint for the temperature sensor monitoring the hot water from a boiler determines the temperature of the hot water supplied to all VAV boxes.

Process Applications

A typical process application of a computer-integrated control system is a PLC system used on a pulverized coal-fired boiler with a forced draft fan. Several inputs are required for communicating data regarding the status of the pulverizer, blower, steam header pressure, superheater differential pressure, furnace draft pressure, gas pressure differential through the boiler, and stack gas analysis (CO, CO_2, and O_2 measurements). The outputs may include control of the coal feeder, pulverizer, blower, forced draft fan, and boiler discharge dampers. **See Figure 11-19.**

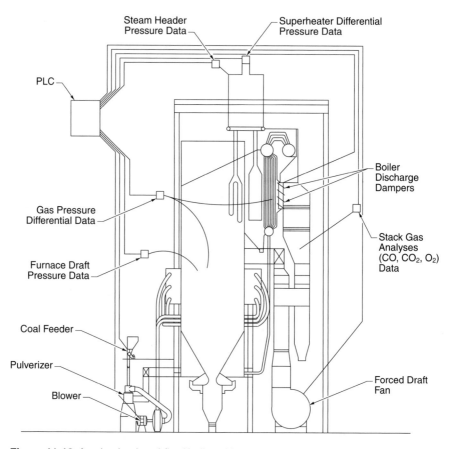

Figure 11-19. A pulverized coal-fired boiler with a computer-integrated control system uses data indicating the status of the pulverizer, blower, steam header pressure, superheater differential pressure, furnace draft pressure, gas pressure differential, and stack gas analysis.

The steam header pressure is monitored by the PLC. As steam header pressure decreases to indicate an increased load, the coal feeder is commanded to open to feed more coal into one of two pulverizers. Depending on which pulverizer the coal is fed into, the PLC signals the pulverizer ON along with the appropriate blower to supply pulverized coal to the burner. As the steam header pressure increases to indicate a drop in load, the PLC signals the coal feeder to close, allowing less coal to be fed to the pulverizer. One pulverizer is operating any time the boiler is operating with a second pulverizer used as a backup.

As the coal feeder modulates open, the PLC signals the forced draft fan (typically with a variable frequency drive) to increase speed to maintain the proper air/fuel ratio. There is a reset available in the PLC program to compensate for the furnace draft pressure and analysis of the stack gases (CO, CO_2, and O_2). The forced draft fan operates any time the boiler is operating.

The gas pressure differential through the boiler is sensed by the PLC and it controls a damper in the breaching of the boiler to maintain the specific gas pressure differential. The signal from this pressure differential can also affect the forced draft fan speed in some installations. Typically, most of these devices have additional control enhancements, depending on the sophistication of the plant, fuel burned, environmental regulations, and age of the equipment.

Computer-Integrated Control System Remote Monitoring

Most computer-integrated control systems are used to monitor system operations remotely. System technicians, maintenance personnel, or building owners can access building operation using an Internet connection, either with a standard web browser or with special software. Some systems even allow authorized users to change system information, such as operating modes or setpoints, remotely. Remote monitoring is particularly helpful when outside consultants are needed to troubleshoot system problems. Some issues may be resolved by monitoring system information via the Internet, reducing the need for expensive on-site visits.

Automation systems can be used to monitor a steam plant remotely, allowing some routine monitoring and control functions to be performed by off-site personnel. **See Figure 11-20.** A boiler's microcomputer burner control system (MBCS) is the local controller that analyzes data from boiler-specific sensors and accessories and makes the necessary data available for other controllers or remote monitoring via the building automation system. Remote monitoring of boilers or other potentially dangerous equipment does not replace the need for qualified personnel in close proximity to the facility. The equipment safety circuitry and manual resets must be maintained to ensure safe operation. However, it does provide additional options for monitoring and control.

COMPUTER-INTEGRATED CONTROL SYSTEM COMPATIBILITY

For years there was no compatibility among computer-integrated control system manufacturers, although most manufacturers maintained compatibility in their own product lines. If an owner of a building purchased another building having a control system from another manufacturer, the two systems could not share information nor could they be controlled from the same operator station. There was also the problem of not being able to interact with large equipment that had its own computer-integrated controls, such as boilers and chillers.

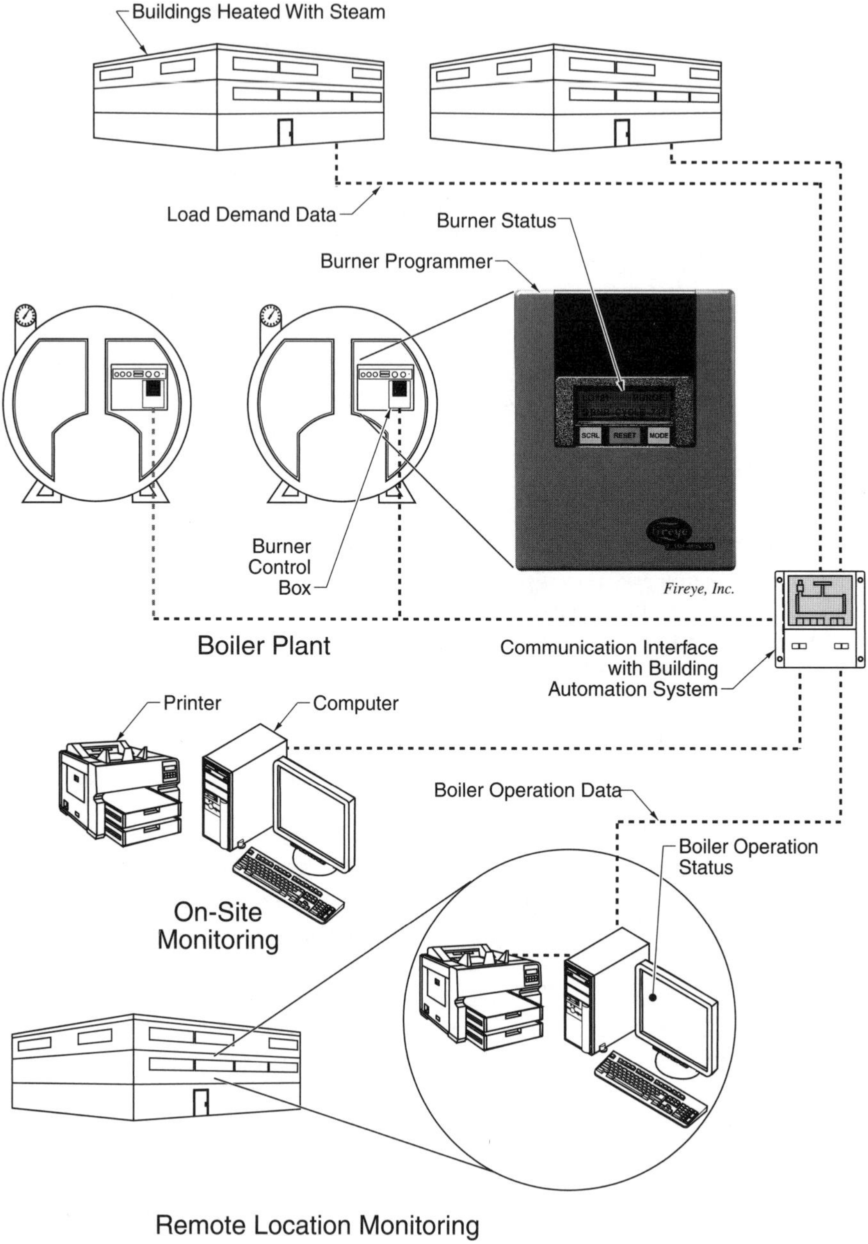

Figure 11-20. Remote access to a building automation system allows off-site personnel to monitor and control a boiler system.

In some cases, manufacturers of building control systems and process control systems worked out arrangements to offer a "black box" solution to the problem of incompatible control systems. A black box interprets communications signals from one control system and passes the information to another control system. These devices (gateways) help building owners use common control systems, but the black box systems are very expensive and not very reliable. A communications protocol is needed for full system integration. The two common standards have emerged in building automation systems are BACnet™ and LonWorks™. An emerging standard called oBIX™ is being developed for communication between building automation systems, intelligent buildings, and enterprise software.

BACnet

Building Automation and Control Networks (BACnet) is an open protocol standard developed and maintained by the American Society of Heating, Refrigerating and Air-Conditioning Engineers (ASHRAE) to standardize communication methods between the building automation systems of different manufacturers. The BACnet protocol was officially published in 1995. In the same year, the American National Standards Institute (ANSI) adopted the BACnet standard.

Standard Objects and Properties. In the past, the building automation system industry used points as a standard. A *point* is an analog input, analog output, digital (binary) input, or digital output that has characteristics determined by a particular manufacturer.

BACnet uses a standard set of objects instead of points. Each object has a standard set of properties that describe the particular object type and its current status to other devices on the BACnet network. Other BACnet devices may communicate with and control the object through these standard object types and properties.

An example of a standard BACnet object is the analog input object. **See Figure 11-21.** The analog input object type represents a sensor input, such as a resistance temperature detector (RTD). This standard analog input object might be seen on a BACnet network through five of its properties. During installation, some of the properties are set, such as "description," "device_type," and "units." Other properties, including "present_value" and "out_of_service," provide information about the status of the sensor input when the sensor is working.

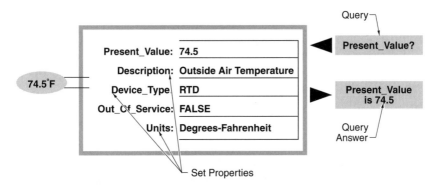

Figure 11-21. The BACnet standard defines standard properties needed for each type of object. The analog input object represents a sensor input.

All standard properties may be read by a BACnet-capable device. For example, if the "present_value" property of an RTD sensor were queried, the analog input object would give a numerical response, such as "74.5."

LONWorks

LONWorks is a protocol for obtaining interoperability in building automation systems using software and hardware. LON technology was developed and introduced by Echelon Corporation. LON and BACnet may coexist on the same network. Both BACnet and LON can provide interoperability, but in different ways. Like BACnet, LON has standard device and object types and the ability to operate over different types of media, and uses public documentation.

Neuron Chips. A key characteristic of LONWorks is the inclusion of a neuron chip on every controller. **See Figure 11-22.** Multiple manufacturers produce neuron chips. Each neuron chip is given a neuron ID, which is a permanent, unique 48-bit code. The neuron chip contains a whole system on the chip, with multiple processors, read-write and read-only memory (RAM and ROM), communications, and input/output (I/O) subsystems.

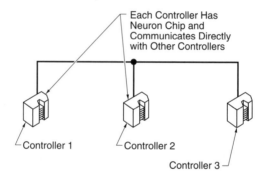

Each Controller Has Neuron Chip and Communicates Directly with Other Controllers

Controller 1   Controller 2

Controller 3

Figure 11-22. Each controller communicates directly with other controllers through the use of a neuron chip.

The read-only memory contains an operating system, communications protocol, and I/O function library. The neuron chip has nonvolatile memory for configuration data and for the application program, both of which are downloaded from a computer over the network. Using the neuron chip provides built-in communication protocol and processors, which removes the need for any development or programming in these areas.

oBIX

oBIX (OASIS Open Building Information eXchange) is a developing technology for using XML-based and Web services-based communications for managing building control systems. The oBIX specification is an open-domain protocol intended to allow communication between the mechanical and electrical services in a building with the control systems for those services.

Most modern control devices include low-cost embedded digital control devices with dedicated wiring for their communications. BACnet and LONTalk are used on these networks. However, these protocols are less efficient and harder to use over a TCP/IP network, such as a plant Ethernet network. oBIX allows resources and devices to be identified with an IP address and URL, similar to an Internet page.

LEADERSHIP IN ENERGY AND ENVIRONMENTAL DESIGN (LEED®)

The Leadership in Energy and Environmental Design (LEED) green building program is a nationally accepted benchmark for the design, construction, and operation of high-performance green buildings. It promotes a whole-building approach to sustainability by recognizing performance in five key areas of human and environmental health: sustainable site development, water savings, energy efficiency, materials selection, and indoor environmental quality.

There are separate rating systems for new construction and for the operation and maintenance of existing buildings. The LEED for Building Operations and Maintenance (O+M) helps building owners and operators measure operations, improvements, and maintenance, with the goal of maximizing operational efficiency while minimizing environmental impacts. LEED O+M addresses whole-building cleaning and maintenance issues, recycling programs, exterior site management, and systems upgrades. Benefits to earning LEED certification may include the following:

• lower operating costs and increased asset value

• reduction in waste sent to landfills

• conservation of energy and water

• healthier and safer environment for occupants

• reduction in harmful greenhouse gas emissions

• tax rebates, zoning allowances, and other incentives in hundreds of cities

• demonstration of owner's commitment to environmental stewardship and social responsibility

To achieve LEED certification, buildings must meet all prerequisites of the rating system plus achieve a minimum of 40 points. **See Figure 11-23.** Certifications Ratings for LEED O+M: Existing Buildings are awarded according to the following point scale:

• Certified 40-49 points

• Silver 50-59 points

• Gold 60-79 points

• Platinum 80-110 points

Proper boiler operation significantly influences the energy efficiency of a building. Points can be awarded toward LEED certification based on building system maintenance and monitoring, water effiency, and emissions reductions. Stationary engineers can help building owners achieve LEED certification through proper operation of a boiler system.

LEED v4 for Operations & Maintenance: Existing Buildings Project Checklist

0	0	0	Energy and Atmosphere		38
Y			Prereq	Energy Efficiency Best Management Practices	Required
Y			Prereq	Minimum Energy Performance	Required
Y			Prereq	Building-Level Energy Metering	Required
Y			Prereq	Fundamental Refrigerant Management	Required
			Credit	Existing Building Commissioning— Analysis	2
			Credit	Existing Building Commissioning—Implementation	2
			Credit	Ongoing Commissioning	3
			Credit	Optimize Energy Performance	20
			Credit	Advanced Energy Metering	2
			Credit	Demand Response	3
			Credit	Renewable Energy and Carbon Offsets	5
			Credit	Enhanced Refrigerant Management	1

U.S. Green Building Council

Figure 11-23. Energy management and control are important components of a LEED certification.

CHAPTER 11 LEARNER RESOURCES
ATPeResources.com/QuickLinks
Access Code: 735728

LICENSING EXAMINATION PREPARATION—THIRD CLASS

Chapter 12, Licensing Examination Preparation – Third Class, consists of questions based on sample questions obtained from licensing agencies across the United Sates and Canada. Third class refers to the lowest level of licensing examinations for stationary engineers. The name of this examination may vary from one licensing agency to another. In addition, licensing agencies have specific requirements governing experience, registration fees, and other examination factors. Chapter 12 consists of ten tests of 25 multiple choice questions and ten tests of 10 essay questions. Essay questions should be answered in detail using proper terminology.

MULTIPLE CHOICE

Test 1

1. ___ are necessary for complete combustion.
   A. Forced draft, turbulence, and heat
   B. Fuel oil, air, time, and mixture
   C. Time, temperature, mixture, and atomization
   D. Time, temperature, and draft

2. If the steam pressure gauge indicates 200 psig, the absolute pressure is ___ psia.
   A. 14.7
   B. 185.3
   C. 214.7
   D. 228.7

3. A boiler is operating at 400 psi, and the safety valve opens at 410 psi and closes at 398 psi. The blowdown is ___ psi.
   A. 10
   B. 12
   C. 12.5
   D. 14.7

4. An oil cooler is commonly used with a(n) ___ when operating in conjunction with a generator as a turbo-generator.
  A. steam turbine
  B. reciprocating steam engine
  C. feedwater heater
  D. air compressor

5. A boiler is producing 40,000 lb of steam/hr at a pressure of 100 psi and a steam temperature of 500°F. The steam produced is ___ steam.
  A. supersaturated
  B. desuperheated
  C. superheated
  D. saturated

6. The economizer on a boiler is used to ___.
  A. heat air for combustion
  B. increase steam temperature
  C. add heat to feedwater
  D. keep tubes clean

7. Soot blowers are normally used with ___ boilers.
  A. scotch marine
  B. cast iron
  C. firetube
  D. watertube

8. Overheating of tubes on a boiler can be caused by ___.
  A. rapid circulation of water
  B. scale on tubes
  C. a high water level
  D. a low firing rate

9. Carbon dioxide in a condensate return system causes ___.
  A. scale
  B. pitting
  C. channeling
  D. plugging of lines

10. A ___ pump is a positive displacement pump.
  A. centrifugal
  B. reciprocating
  C. jet
  D. vacuum

11. A boiler generating 6000 lb of steam/hr produces approximately ___ Btu/hr.
  A. 6000
  B. 60,000
  C. 600,000
  D. 6,000,000

12. One reason a bottom blowdown is required on a boiler is that it ___.
   A. removes surface scum
   B. tests for water level
   C. tests safety valves
   D. discharges sludge and sediment

13. Water columns are connected to the boiler with a minimum of ___″ pipe.
   A. ¾
   B. 1
   C. 1¼
   D. 1½

14. The minimum size of a water column blowdown valve is ___″.
   A. ½
   B. ¾
   C. 1
   D. 1¼

15. The maximum size of bottom blowdown valves on boilers is ___″.
   A. 1
   B. 1½
   C. 2
   D. 2½

16. A ___ allows condensate to pass through, but stops steam.
   A. steam trap
   B. siphon
   C. check valve
   D. bypass

17. When taking over a shift, the first thing to do is check the ___.
   A. water level in each boiler
   B. fuel supply
   C. flame condition
   D. feedwater pumps

18. According to the ASME code, a high-pressure boiler having more than 500 sq ft of heating surface shall have ___.
   A. one safety valve of the proper size
   B. one safety valve with a rupture disc
   C. two lever-type safety valves
   D. two or more safety valves

19. The probable cause of an increase in the temperature of gases of combustion at the boiler outlet is ___.
   A. the burner is on low fire
   B. the steam pressure is too low
   C. dirty heating surfaces
   D. high combustion efficiency

20. Boilers in battery equipped with ___ must have two main steam stop valves and a free-blowing drain between them.
    A. common blowdown lines
    B. manhole openings
    C. handhole openings
    D. continuous blowdown lines

21. In order to be sure the gauge glass is indicating an accurate water level reading, the stationary engineer should ___.
    A. shut the valve to the steam side of the gauge glass
    B. shut the valve to the water side of the gauge glass
    C. blow down the gauge glass
    D. blow down the water column

22. A continuous blowdown line is located ___.
    A. at the NOWL
    B. in the mud drum
    C. in the flash tank
    D. in the steam and water drum well below the NOWL

23. A temperature reading of 80°C is equal to ___°F.
    A. 52
    B. 148
    C. 168
    D. 176

24. A chimney temperature of 250°F above steam temperature indicates ___.
    A. superheated steam
    B. poor efficiency
    C. saturated steam
    D. normal operation

25. A manometer is read in ___.
    A. inches of water
    B. psig
    C. psia
    D. Btu

Test 2

1. The boiling point of water will ___ with an increase in pressure.
    A. not change
    B. be raised
    C. be lowered
    D. raise until 212°F

2. If a lather is not obtained when performing a soap hardness test, the water sample is ___.
    A. alkaline
    B. normal
    C. soft
    D. hard

3. Combustion efficiency can be increased when air is ___.
   A. filtered
   B. mixed with CO_2
   C. preheated
   D. dried

4. ___ draft is produced by a fan located between the boiler and the chimney.
   A. Induced
   B. Forced
   C. Forced-induced
   D. Natural-forced

5. The overspeed trip on a steam turbine should be tested ___.
   A. before putting the steam turbine on-line
   B. daily
   C. weekly
   D. monthly

6. A high CO_2 reading in the gases of combustion is a sign of ___.
   A. incomplete combustion
   B. carryover of oxygen
   C. low combustion efficiency
   D. high combustion efficiency

7. The purpose of chemically treating feedwater is to ___.
   A. prevent priming
   B. turn scale-forming salts to a nonadhering sludge
   C. change scale-forming salts to an adhering sludge
   D. prevent water hammer

8. Water in a deaerating feedwater heater is prevented from entering the steam line by a(n) ___.
   A. check valve
   B. stop valve
   C. loop seal or internal overflow
   D. atmospheric relief

9. A steam turbine operating at 3600 RPM will trip at ___ RPM.
   A. 3650
   B. 3700
   C. 3750
   D. 3960

10. The steam line going to the soot blowers is connected to the main ___.
   A. steam header
   B. steam drum
   C. auxiliary line
   D. pressure-reducing station

11. Before using soot blowers on a boiler, ___.
   A. drain superheater lines
   B. drain soot blower lines
   C. secure all burners
   D. decrease the draft

12. An accumulation of soot on the fire side of boiler tubes ___.
   A. increases boiler efficiency
   B. decreases boiler efficiency
   C. decreases gas temperature
   D. protects tubes from excessive heat

13. A globe valve installed in the feedwater line between the feedwater pump and boiler has pressure from the feedwater pump ___ the valve disc.
   A. under
   B. on top of
   C. under and on top of
   D. check valve over

14. The range of a pressure gauge should be twice the safety valve setting, but not less than ___ time(s) the safety valve setting.
   A. 1
   B. 1¼
   C. 1½
   D. 1¾

15. Hard scale formed on the inside of tubes is best removed by ___.
   A. washing down with water
   B. air- or water-driven turbines
   C. wire brushes
   D. boiling chemical solution

16. To correct a low water condition, ___.
   A. decrease feedwater
   B. increase firing rate
   C. lower firing rate and increase feedwater
   D. secure fires

17. The superheater drain is used to ___.
   A. remove gases from the superheater
   B. remove condensate
   C. relieve pressure
   D. establish steam flow

18. A sudden rise in chimney temperature with no increase in load indicates ___.
   A. the damper is stuck closed
   B. the boiler needs cleaning
   C. there is a broken baffle
   D. there is too much air

19. The Btu content of natural gas is ___.
   A. 100,000 Btu/therm
   B. 100,000 Btu/cu ft
   C. 152,000 Btu/therm
   D. 152,000 Btu/cu ft

20. The safety valve setting on the steam and water drum is ___ the superheater safety valve.
   A. lower than
   B. higher than
   C. the same as
   D. balanced on

21. The Btu content of No. 6 fuel oil is approximately ___.
   A. 18,750 Btu/gal.
   B. 18,750 Btu/lb
   C. 18,750 Btu/barrel
   D. 150,000 Btu/lb

22. Chemicals can be added to a boiler by a ___ pump.
   A. feedwater
   B. vacuum
   C. proportioning
   D. turbine

23. The feedwater stop valve is found between the ___.
   A. boiler and check valve
   B. check valve and feedwater pump
   C. boiler and feedwater pump
   D. boiler and main steam header

24. Sensible heat is ___.
   A. heat that can be measured with a thermometer
   B. 970.3 Btu
   C. heat that cannot be measured
   D. heat that changes a substance's state but not its temperature

25. Balanced draft is ___.
   A. forced draft
   B. induced draft
   C. below atmospheric pressure at the flue outlet
   D. constant furnace pressure slightly below atmospheric pressure

Test 3

1. A flame rod ___.
   A. regulates steam flow
   B. ignites the fuel oil
   C. ensures that the gas pilot is lighted
   D. measures heat in the boiler

2. There are approximately ___ Btu in 1 cu ft of natural gas.
   A. 100
   B. 1000
   C. 3000
   D. 5000

3. The pH of boiler water may be controlled by using ___.
   A. sulphite
   B. sulphate
   C. caustic soda
   D. softened feedwater

4. Boiler water with a pH of 11 is ___.
   A. alkaline
   B. acidic
   C. neutral
   D. hard

5. A scotch marine boiler is easily identified because it ___.
   A. has a brick setting
   B. has a cylindrical internal furnace
   C. is externally fired
   D. has waterwalls

6. The air flow switch at the burner fan housing ___.
   A. ensures the primary air fan is functioning
   B. controls the secondary air damper
   C. controls the boiler operating range
   D. controls the boiler water temperature

7. The ASME code recommends that low water fuel cutoffs be tested by an evaporation test ___.
   A. daily
   B. weekly
   C. monthly
   D. yearly

8. A neutral solution is ___ on the pH scale.
   A. 0
   B. 3.5
   C. 7
   D. 14

9. A dry pipe separator ___.
   A. delivers higher quality steam to the header
   B. distributes feedwater the length of the drum
   C. prevents water hammer
   D. agitates the water in the steam and water drum

10. Draft is carefully controlled to ___.
   A. maintain at least 15% CO in gases of combustion
   B. completely burn fuel supplied to the burner
   C. prevent the boiler from overheating
   D. maintain high levels of excess air in gases of combustion

11. Draft is measured with a ___.
   A. pyrometer
   B. hydrometer
   C. manometer
   D. potentiometer

12. A Ringleman test is used to determine ___.
   A. CO_2 in gases of combustion
   B. smoke density
   C. flame failure
   D. steam quality

13. A flame scanner proves (verifies) ___.
   A. beta rays
   B. heat
   C. pilot and main flame
   D. X rays

14. According to the ASME code, when warming up a boiler the air cock is closed when ___.
   A. steam starts to come out of the boiler
   B. the steam pressure gauge reads approximately 25 psi
   C. the steam pressure gauge reads 100 psi
   D. the boiler goes on-line

15. A motor rated at 25 HP delivering 20 HP is operating at ___% capacity.
   A. 75
   B. 80
   C. 85
   D. 90

16. A deposit of scale or mud in an HRT boiler could cause ___.
   A. a bag to develop
   B. a draft loss
   C. overheating of the breeching
   D. lowering of the chimney temperature

17. Waterwall tubes protect the refractory and ___.
   A. increase boiler heating surface
   B. act as stays for tube sheets
   C. allow the operator to blow down while steaming
   D. help control water circulation

18. Steam at 250 psia and 400°F is considered ___ steam.
   A. supersaturated
   B. saturated
   C. superheated
   D. steam temperature cannot exceed 212°F

19. Sensible heat is heat that ___.
   A. changes a substance's state but not its temperature
   B. changes a substance's temperature and its state
   C. changes a substance's temperature but not its state
   D. is 970.3 Btu/lb of steam

20. Extra-heavy pipe must be used on blowdown lines when the steam pressure exceeds ___ psi.
  A. 15
  B. 100
  C. 125
  D. 200

21. To check the water level on a boiler on-line, it is best to ___.
  A. give the boiler a bottom blowdown
  B. blow down the gauge glass
  C. lower the steam pressure
  D. check the log sheet

22. A steam pressure gauge has a siphon between the boiler and the pressure gauge to ___.
  A. allow for expansion and contraction
  B. prevent water from entering the pressure gauge
  C. protect the pressure gauge from high steam pressure
  D. prevent steam from entering the pressure gauge

23. If a boiler evaporates 50,000 lb of steam/hr and is evaporating 10 lb of steam/sq ft of heating surface, the heating surface of the boiler is ___ sq ft.
  A. 50
  B. 500
  C. 5000
  D. 50,000

24. Oxygen in feedwater causes ___.
  A. priming
  B. foaming
  C. pitting
  D. scale

25. It is permissible to have one blowdown valve on a boiler operated under ___ psi.
  A. 100
  B. 125
  C. 150
  D. 200

Test 4

1. At atmospheric pressure, water boils and turns to steam at ___°F.
  A. 212
  B. 215
  C. 220
  D. 250

2. A steam pressure gauge is protected against excessive temperature by a ___.
  A. steam trap
  B. siphon
  C. separator
  D. Hartford loop

3. The only type of safety valve allowed on a boiler by the ASME code is a ___ safety valve.
  A. ball and lever
  B. deadweight
  C. spring-loaded check
  D. spring-loaded pop

4. The primary function of a steam boiler is to ___.
  A. boil water
  B. burn fuel
  C. generate steam efficiently
  D. generate electricity

5. If the burner is firing and no water is in the gauge glass, the best procedure to follow is to ___.
  A. blow down the low water fuel cutoff
  B. check the water level with the try cocks
  C. shut down the burner and secure the boiler
  D. add water to the boiler

6. A high-pressure boiler operates at a pressure above ___ psi and 6 BHP.
  A. 15
  B. 30
  C. 100
  D. 150

7. A safety valve on a boiler pops ___.
  A. when the boiler reaches its working pressure
  B. between working pressure and MAWP
  C. at 15 psi
  D. at 150 psi

8. In a firetube boiler, the heat and gases of combustion pass ___.
  A. around the tubes
  B. through the waterwalls
  C. through the water legs
  D. through the tubes

9. As the steam pressure in a boiler increases, there is a corresponding increase in the ___.
  A. BHP
  B. water level
  C. volume of steam
  D. temperature of steam and water

10. The safety valve on a boiler ___.
  A. prevents the boiler from exceeding its MAWP
  B. adds makeup water when pressure is high
  C. secures boiler and bypasses steam
  D. releases air and noncondensable gases

11. A steam trap ___.
   A. is a device that traps condensate and steam
   B. acts as a cushion in steam lines
   C. automatically removes air and water without loss of steam
   D. is a device that controls superheat temperatures

12. The amount of excess air in gases of combustion is normally determined by the analysis of ___.
   A. CO_2
   B. O_2
   C. CO
   D. C

13. Before lighting off a boiler during a cold startup, the operator must first ___.
   A. check for the correct water level
   B. close all drains on main steam line
   C. close all damper outlets
   D. close the main steam stop valves

14. Boilers should be warmed up slowly when starting up to ___.
   A. eliminate water hammer
   B. allow uniform expansion of drums and tubes
   C. maintain a steady water level
   D. prevent an explosion

15. Combustion requires fuel, oxygen, and ___.
   A. carbon dioxide
   B. heat
   C. draft
   D. carbon monoxide

16. The three most common fuels used in boilers are ___.
   A. gas, fuel oil, and coal
   B. gas, fuel oil, and wood
   C. gas, wood, and waste
   D. gas, fuel oil, and bagasse

17. Incomplete combustion can be caused by ___.
   A. insufficient air
   B. insufficient fuel
   C. too much excess air
   D. too high a furnace temperature

18. After replacing a gauge glass but before placing it in service, it is necessary to ___.
   A. open the try cock and check the water level
   B. crack open the steam valve and warm the gauge glass
   C. crack open the water valve and warm the gauge glass
   D. tighten the packing gland nuts

19. Two bottom blowdown valves are required on boilers operating at pressures above ____ psi.
    A. 100
    B. 125
    C. 150
    D. 175

20. Extra-heavy steel pipe is required on a boiler bottom blowdown line when the pressure is above ____ psi.
    A. 100
    B. 125
    C. 150
    D. 175

21. Package boilers are programmed to ____.
    A. ignite the burner in low fire
    B. ignite the burner in high fire
    C. ignite the burner before purging the furnace
    D. increase excess air discharged

22. There are ____ sq ft of heating surface in a watertube 3½" in diameter and 20' long.
    A. 1.832
    B. 1.96
    C. 18.32
    D. 19.6

23. To change 1 lb of water from and at 212°F into steam at atmospheric pressure, ____ Btu are necessary.
    A. 144.0
    B. 180.0
    C. 970.3
    D. 1000.0

24. A surface condenser is one that ____.
    A. is open to the atmosphere
    B. is equipped with a tail pipe
    C. has circulating water around the outside of the tubes and exhaust steam flowing through the tubes
    D. has circulating water flowing through the tubes and exhaust steam around the outside of the tubes

25. A boiler is rated at 30,000 lb/hr and has a 5% blowdown loss. The minimum-size feedwater pump to supply this boiler is ____ gal./hr.
    A. 3782
    B. 4000
    C. 30,000
    D. 31,500

Test 5

1. Three types of draft used in power plants are ___.
    A. natural, forced, and induced
    B. natural, forced, and gravity
    C. natural, flow, and induced
    D. natural, induced, and gravity

2. In a watertube boiler, heat and gases of combustion pass ___.
    A. through the casing
    B. through the tubes
    C. around the tubes
    D. through and around the tubes

3. When the burner is in high fire, the ___.
    A. amount of fuel supplied to the burner is decreased
    B. air-fuel ratio is at maximum rate
    C. air-fuel ratio is at minimum rate
    D. burner steam pressure is too high

4. A boiler with a draft gauge showing a positive pressure indicates ___.
    A. the draft requires calibrating
    B. furnace pressure and atmospheric pressure are the same
    C. atmospheric pressure is above furnace pressure
    D. furnace pressure is above atmospheric pressure

5. Downcomer tubes are found ___.
    A. on the superheater drain
    B. in the chimney to sample gases of combustion
    C. in a firetube boiler
    D. in a watertube boiler

6. A boiler connected in battery requires a free-blowing drain between the two stop valves when the ___.
    A. steam line is over 4″ in diameter
    B. boiler has a nonreturn valve
    C. boiler has a manhole opening
    D. boiler operates above 100 psi

7. Most furnace explosions are caused by ___.
    A. failure to purge the furnace at startup
    B. overpurging the furnace at startup
    C. mechanical failures
    D. damaged refractory

8. Baffles in a boiler are designed to direct gases of combustion for close contact with the ___.
    A. boiler drums
    B. boiler heating surfaces
    C. blowdown lines
    D. mud drum

9. The overspeed tripping device on a steam turbine must be tripped when ___.
   A. the turbine is 10% below set speed
   B. the load is on the unit
   C. the steam turbine is 30% over set speed
   D. shutting down the steam turbine

10. One gallon of No. 6 fuel oil weighs approximately ___ lb.
   A. 6.5
   B. 7.6
   C. 8.2
   D. 9.3

11. Three types of boiler feedwater pumps are ___.
   A. gear, rotary, and lobe
   B. reciprocating, centrifugal, and turbine
   C. reciprocating, centrifugal, and gear
   D. centrifugal, gear, and lobe

12. The most important fitting on a boiler is the ___.
   A. bottom blowdown valve
   B. gauge glass
   C. safety valve
   D. water column

13. Viscosity of fuel oil is a ___.
   A. method of finding flash point
   B. measure of internal resistance to flow
   C. method of finding the fire point
   D. method of determining heating value

14. Boiler water treatment is required to ___.
   A. determine the hardness of water
   B. determine the degree of corrosion
   C. prevent scale formation on heating surfaces
   D. increase total solids in water

15. The safety valve ___.
   A. prevents the boiler from exceeding its MAWP
   B. removes air from the boiler
   C. opens whenever the boiler is drained
   D. never fully opens

16. Try cocks are used ___.
   A. only on watertube boilers
   B. only on firetube boilers
   C. to blow down water column
   D. as a secondary means of determining water level

17. A gauge glass is used to ___.
   A. indicate the water level in the boiler
   B. cut off the fuel at low water
   C. take a boiler water sample
   D. test the pH of the boiler water

18. Before using soot blowers, it is necessary to ___.
    A. let the steam line warm up for 1 hr
    B. reduce air flow through the furnace
    C. open the steam valve on the main floor
    D. open the steam line drain on the soot blower

19. Safety valve repairs should be done by the ___.
    A. chief engineer or shift supervisor
    B. boiler inspector
    C. safety valve manufacturer or an authorized representative
    D. boiler operator

20. Steam at a temperature of 350°F and a pressure of 50 psi is ___.
    A. saturated
    B. superheated
    C. dry saturated
    D. wet saturated

21. A duplex double-acting reciprocating feedwater pump has ___.
    A. one steam and one water piston
    B. two water pistons larger than the two steam pistons
    C. one steam piston and two water pistons
    D. two steam pistons 2 to 2½ times larger in area than the water pistons

22. An accumulation test is used to determine ___.
    A. furnace volume
    B. heating surface
    C. safety valve capacity
    D. evaporation rate

23. A watertube boiler has 250 tubes that are 20′ long and 3″ in diameter. If the boiler produces 1 BHP for every 10 sq ft of heating surface, the maximum BHP it will produce is ___ BHP.
    A. 125
    B. 392
    C. 750
    D. 1500

24. During a boiler inspection, the main steam stop valves are ___.
    A. left open
    B. closed, locked, and tagged
    C. removed from the line
    D. equalized

25. When a quick-opening valve is used as a blowdown valve, it is located ___.
    A. before the flash tank
    B. between the slow-opening valve and blowdown tank
    C. between the boiler and slow-opening valve
    D. on the continuous blowdown line

Test 6

1. Lifting levers on safety valves are used to ___.
   A. relieve excess pressure
   B. check the relieving capacity
   C. manually test valves
   D. vent the boiler

2. The frequency of performing a bottom blowdown is determined by ___.
   A. steaming capacity
   B. boiler water analysis
   C. operating routine
   D. operator on shift

3. Normally a water column and gauge glass should be blown down ___.
   A. at the start of a shift and whenever in doubt of the water level
   B. at least once a day
   C. whenever the water level fluctuates
   D. whenever the water level is high

4. A thermostatic trap opens when the ___.
   A. float rises
   B. steam enters the trap body
   C. steam pressure increases
   D. the bellows contract

5. An economizer located in the path of the gases of combustion increases boiler efficiency approximately 1% for each ___°F rise in feedwater temperature.
   A. 5
   B. 10
   C. 15
   D. 20

6. The purpose of a vent on the open feedwater heater is to ___.
   A. prevent flooding
   B. prevent pressure increase
   C. remove oxygen and noncondensable gases
   D. add makeup water

7. Steam pressure gauges are calibrated in ___.
   A. inches of water
   B. inches of mercury
   C. pounds per square foot
   D. pounds per square inch

8. Bottom blowdown lines are located ___.
   A. at the lowest part of the water side of a boiler
   B. at the lowest part of the waterwall header
   C. on the steam and water drum
   D. at the lowest part of the combustion chamber

9. Sewer lines are protected from pressure and temperature by the ___ when blowing down a boiler.
  A. steam trap
  B. expansion tank
  C. blowdown tank
  D. compression tank

10. The upper connection to a gauge glass is closed. If the boiler has a normal water level, the level on the gauge glass will show ___ glass.
  A. full
  B. one-half
  C. one-quarter
  D. no water

11. Combustion controls on a boiler are used to ___.
  A. eliminate the need for an operator
  B. control steam temperature
  C. control the air-fuel ratio
  D. increase boiler capacity

12. Sealing strips that warn of excessive radial tolerance are found on ___.
  A. solenoid valves
  B. steam turbines
  C. steam-driven pumps
  D. a surface condenser

13. Prepurge, air-to-fuel ratio, and postpurge of the burner are functions of the ___.
  A. condenser
  B. fuel interlock
  C. solenoid valve
  D. combustion controls

14. A water column ___.
  A. reduces fluctuations of water in the gauge glass
  B. prevents steam from flashing
  C. prevents water from flashing
  D. is needed on all boilers

15. The low water fuel cutoff can be tested by ___.
  A. lifting the lever or handle
  B. securing the burner
  C. allowing the water level to drop
  D. opening the fuel valve

16. Boilers that are equipped with air heaters and economizers usually have ___ draft fans.
  A. only forced
  B. only induced
  C. natural
  D. forced and induced

17. Baffles are used in the steam and water drum of watertube boilers to ___.
   A. provide better circulation
   B. improve the separation of steam and moisture
   C. prevent fluctuations of the water level
   D. clean the steam

18. A closed feedwater heater is always installed ___.
   A. on the discharge side of the feedwater pump
   B. on the suction side of the feedwater pump
   C. before the open feedwater heater
   D. above the feedwater pump to give a positive head

19. The purpose of a boiler pressure control is to ___.
   A. regulate the furnace pressure
   B. regulate the draft
   C. start and stop the burner firing cycle
   D. start and stop the feedwater pump

20. A gas pressure regulator is used to control the ___.
   A. gases of combustion
   B. gas pressure at the burner
   C. air-fuel ratio
   D. gas flow to the burner

21. If the chimney temperature is 250°F higher than the steam and water temperature, the boiler is ___.
   A. operating efficiently
   B. not operating efficiently
   C. operating at maximum firing rate
   D. operating at minimum firing rate

22. The most controllable heat loss in a boiler is from the loss of heat in ___.
   A. saturated steam
   B. convection of boiler surfaces
   C. gases of combustion
   D. hydrogen in the fuel

23. On the feedwater line, the valve closest to the boiler shell is a ___ valve.
   A. check
   B. stop
   C. blowdown
   D. makeup

24. Four basic systems found on boilers are ___.
   A. steam, feedwater, fuel, and draft
   B. steam, exhaust, fuel, and draft
   C. steam, fuel, condensate, and blowdown
   D. steam, condensate, draft, and exhaust

25. If a low water condition occurs, the operator should first ___.
   A. add water to the boiler
   B. secure the burners
   C. raise firing rate and determine cause of low water
   D. close steam stop valve to conserve water

Test 7

1. Spalling in a boiler occurs in the ___.
   A. main steam line
   B. combustion chamber
   C. bottom blowdown line
   D. soot blower

2. Economizers are used in plants to ___.
   A. control superheater outlet temperature
   B. heat feedwater using exhaust steam
   C. heat feedwater using gases of combustion
   D. remove radiant heat from the furnace

3. Fuel oil with a high viscosity ___.
   A. is hard to pump
   B. is easy to pump
   C. is not used in boilers
   D. limits the firing rate

4. Boiler water is chemically treated to ___.
   A. eliminate foaming
   B. eliminate priming
   C. increase surface tension
   D. prevent scale from forming on the boiler heating surfaces

5. Steam pressure gauges are calibrated in ___.
   A. inches of mercury
   B. pounds per square inch
   C. pounds per square foot
   D. absolute pressure

6. Feedwater temperature raised 10°F from exhaust steam or heat from gases of combustion results in an increase of ___% in plant efficiency.
   A. 0
   B. 1
   C. 5
   D. 10

7. The evaporation of 34.5 lb of water from and at 212°F is a standard for ___.
   A. the number of Btu in 1 lb of coal
   B. performing an evaporation test
   C. 1 BHP
   D. calibrating a calorimeter

8. The tubes in a watertube boiler are fastened by ___.
   A. expanding and welding
   B. rolling and expanding in drums and headers
   C. rolling and beading in drums and headers
   D. ring welding

9. A dry pipe separator in a boiler ___.
   A. superheats steam
   B. maintains high-quality steam
   C. gives the boiler a surface blowdown
   D. prevents priming

10. A ___ valve offers the least resistance to flow.
   A. globe
   B. gate
   C. needle
   D. check

11. If valves are installed between the boiler and water column, they must be ___.
   A. os&y or lever-type, locked or chained closed
   B. os&y or lever-type, locked or chained open
   C. 250 psi standard valves
   D. globe valves

12. Steam withdrawn from a steam turbine under pressure control is ___ steam.
   A. exhaust
   B. extraction
   C. bleed
   D. superheated

13. In a closed feedwater heater, the ___.
   A. steam and water mix
   B. steam and water do not mix
   C. heater is on the suction side of the pump
   D. feedwater pump is between the boiler and feedwater heater

14. A compound gauge registers ___.
   A. vacuum and pressure at the same time
   B. either vacuum or pressure
   C. the total pressure on the boiler
   D. the difference between vacuum and pressure

15. An Orsat analyzer measures ___ in gases of combustion.
   A. carbon dioxide
   B. oxygen
   C. carbon monoxide
   D. hydrogen

16. The Fyrite analyzer can be used to measure ___.
   A. sulfur dioxide
   B. flyash
   C. oxygen
   D. pounds of steam per hour

17. Automatic combustion controls on a boiler control ___.
   A. water and fuel ratio
   B. fuel and air ratio
   C. steam and air ratio
   D. chemical to firing ratio

18. If the firing rate increases on a boiler equipped with a convection-type superheater, the superheater temperature will ___.
   A. increase
   B. decrease
   C. stay constant
   D. remove scale outside of firetubes

19. A thermostatic steam trap can be used ___.
   A. only on low-pressure systems
   B. on high- or low-pressure systems
   C. only on high-pressure systems
   D. on nonreturn condensate systems

20. Heat that changes the temperature of a substance but not its state is ___.
   A. latent heat
   B. heat of evaporation
   C. sensible heat
   D. heat of fusion

21. ___ is carrying over of particles of water with the steam.
   A. Pitting
   B. Flashing
   C. Splashing
   D. Priming

22. If a boiler is equipped with an automatic nonreturn valve, it must be ___.
   A. on the blowdown line
   B. on the feedwater line next to boiler
   C. the valve closest to the boiler shell
   D. between the main steam stop valve and header

23. A broken baffle on a watertube boiler would cause ___.
   A. tubes to leak
   B. an increase in steam pressure
   C. an increase in chimney temperature
   D. carryover

24. A feedwater pump will become steambound if the ___.
   A. steam pressure gets too high
   B. water temperature in the open feedwater heater is too high
   C. water temperature in the closed feedwater heater is too high
   D. water level in the open feedwater heater is too high

25. Overheating of tubes in a watertube boiler results if ___.
   A. scale builds up on the outside of the tubes
   B. scale builds up on the inside of the tubes
   C. the firing rate increases too quickly
   D. soot deposits on tubes

Test 8

1. Complete combustion occurs when ___.
   A. white smoke comes out of the chimney
   B. steam pressure rises quickly
   C. all combustible material is burned with a minimum of excess air
   D. all combustible material is burned with as much air as possible

2. According to the ASME code, safety valve repair should be done by the ___.
   A. boiler operator under the direction of an inspector
   B. ASME inspector
   C. manufacturer or authorized representative
   D. maintenance department

3. In a steam generating plant, the economizer is used to ___.
   A. heat feedwater
   B. preheat fuel oil
   C. superheat steam
   D. desuperheat steam

4. When testing a safety valve by hand, the pressure on the boiler should be at least ___% of the safety valve set pressure.
   A. 30
   B. 50
   C. 75
   D. 100

5. Heat absorbed by water when changing from liquid to steam at the boiling point is ___.
   A. sensible heat
   B. latent heat
   C. specific heat
   D. superheat

6. To raise the temperature of 1 lb of water from 32°F to 212°F, ___ Btu are needed.
   A. 144
   B. 180
   C. 250
   D. 970.3

7. Baffles are placed between the tubes of a boiler to ___.
   A. direct the travel of gases of combustion
   B. brace the tubes
   C. prevent bending of tubes
   D. maintain pressure in the furnace

8. In a pressure atomizing fuel-oil burner, atomization of fuel oil is produced by the ___.
   A. steam pressure and burner tip
   B. oil strainer and rotary cup pressure
   C. burner tip and high fuel-oil pressure
   D. check valve and auxiliary line

9. The steam inlet connections of a condenser should be provided with ___.
   A. a vacuum breaker
   B. an air vent
   C. an atmospheric relief valve
   D. at least 2″ of insulation

10. When a hydrostatic test is performed on a boiler, each safety valve should be ___.
   A. set at its MAWP
   B. blanked or gagged
   C. held to its seat by increasing spring tension
   D. left alone (no adjustment necessary)

11. In a surface condenser, the cooling water ___.
   A. passes around the tubes
   B. mixes with the exhaust steam
   C. passes through the tubes
   D. is only used in the hot well

12. Compared to No. 6 fuel oil, No. 4 fuel oil has ___ Btu per gallon.
   A. more
   B. less
   C. the same number of
   D. more if heated

13. A bypass damper on an air heater ___.
   A. reduces the fan load
   B. prevents moisture from collecting at low firing rates
   C. secures the air preheater
   D. reduces the chimney temperature

14. The minimum-size manhole is ___.
   A. 11″ × 11″
   B. 12″ × 12″
   C. 11″ × 13″
   D. 11″ × 15″

15. The diameter of a chimney may have to be increased ___.
   A. to get rid of excessive soot
   B. when the burner has been converted from coal to natural gas
   C. if the boiler smokes continually
   D. if the maximum amount of flue gases has been increased

16. Staybolts are generally used on ___.
   A. watertube boilers
   B. firetube boilers
   C. feedwater heaters
   D. the main superheater header

17. Before draining a boiler, ___.
   A. allow the boiler to cool
   B. call the inspector
   C. open blowdown valves with 10 psi to 12 psi on the boiler
   D. use surface blowdown first

18. Cast iron water columns may be used for pressures up to ___ psi.
    A. 100
    B. 250
    C. 300
    D. 500

19. All high-pressure boilers must have more than one method to determine water level: a gauge glass and ___.
    A. low water fuel cutoff
    B. whistle valve
    C. water column
    D. try cocks

20. If dry saturated steam at 200 psi is passing through a reducing valve and is reduced to 25 psi, it is ___.
    A. superheated
    B. at its corresponding temperature and pressure
    C. dry saturated
    D. supersaturated

21. Three boilers are connected to a common header and each boiler is carrying 100 psi. The pressure gauge on the header reads ___ psi.
    A. 100
    B. 200
    C. 300
    D. 400

22. ___ is caused by allowing sudden extreme changes of temperature in a furnace.
    A. Spalling
    B. A drop in chimney temperature
    C. Priming
    D. Pitting

23. ___ is always necessary in the combustion process.
    A. Nitrogen
    B. Hydrogen
    C. Oxygen
    D. Methane

24. Waterwalls should be blown down ___.
    A. daily
    B. weekly
    C. when the furnace is cool
    D. once every six months

25. An automatic nonreturn valve is used with a(n) ___.
    A. engine
    B. pump
    C. boiler
    D. condenser

Test 9

1. Feedwater is treated chemically before it enters the boiler drum to ___.
    A. prevent foaming
    B. eliminate blowing down the boiler
    C. change scale-forming salts to an adhering sludge
    D. change scale-forming salts to a nonadhering sludge

2. An operating pressure control on a steam boiler is used to ___.
    A. control high and low fire
    B. start and stop the feedwater pump
    C. control the operating range of the boiler
    D. control the operating temperature

3. Soot blowers are most commonly found ___.
    A. on HRT boilers
    B. on watertube boilers
    C. in the main breeching
    D. on vertical boilers

4. A duplex reciprocating feedwater pump operates on the principle that the ___.
    A. steam side must be 2 to 2½ times larger in diameter than the water side
    B. water piston area must be 2 to 2½ times larger than the steam piston
    C. steam piston area must be 2 to 2½ times larger than the water piston
    D. steam piston area must equal the water piston area

5. When testing the safety valve by hand, it is the accepted practice to have ___.
    A. no pressure on the boiler
    B. a full gauge glass of water
    C. at least 75% popping pressure before testing
    D. safety valves are not tested by hand

6. Oxygen in a boiler drum causes ___.
    A. priming
    B. foaming
    C. pitting
    D. carryover

7. ___ steam turbines allow condensate to be reclaimed for use in the system.
    A. Two-stage
    B. Reciprocating
    C. Noncondensing
    D. Condensing

8. With water at a temperature of 212°F, 1 BHP is equivalent to the evaporation of 34.5 lb of water every ___ hour(s).
    A. 1
    B. 8
    C. 12
    D. 24

9. In a closed feedwater heater, the ___.
    A. steam and water mix
    B. steam and water do not mix
    C. heated feedwater is stored
    D. oxygen is removed from the water

10. Sensible heat is heat that ___.
    A. changes the state of a substance but not its temperature
    B. changes the temperature of a substance but not its state
    C. is released when burning a fuel
    D. is added to feedwater

11. When performing a hydrostatic test on a boiler, the safety valve must be ___.
    A. gagged or removed from the boiler and the opening blanked
    B. plugged at the outlet
    C. left alone so that the popping point can be noted
    D. the safety valves are not affected

12. If repairs are to be made on a safety valve, they must be made by the ___.
    A. mechanical inspector
    B. insurance inspector
    C. chief engineer
    D. manufacturer of the safety valve or authorized representative

13. The proper way to stop a steam turbine when taking it off-line is to ___.
    A. shut the main steam stop valve
    B. shut the steam turbine throttle valve
    C. shut the exhaust steam valve
    D. trip the overspeed device by hand

14. Approximately ___ of air is required to completely burn 1 lb of fuel oil.
    A. 15 cu ft
    B. 15 lb
    C. 30 cu ft
    D. 30 lb

15. In order to reduce the viscosity of a fuel oil, its ___.
    A. temperature must be reduced
    B. temperature must be increased
    C. pressure must be reduced
    D. pressure must be increased

16. Water is prevented from backing up into the steam line on a deaerator by ___.
    A. a check valve
    B. exhaust steam pressure
    C. a float control valve
    D. an internal overflow or a loop seal

17. A steam pressure gauge must have a range of at least ___.
    A. the working pressure of the boiler
    B. 1½ times the working pressure of the boiler
    C. the MAWP
    D. 1½ times the MAWP

18. Electrostatic precipitators are used in power plants to ___.
   A. remove flyash from gases of combustion
   B. measure smoke conditions
   C. collect boiler impurities
   D. demineralize feedwater

19. Valves located between the boiler and water column must be ___.
   A. locked or sealed open
   B. regular globe valves
   C. steel and rated at 300 psi standard
   D. kept closed

20. The boiling point of water ___ with an increase in pressure.
   A. does not change
   B. increases
   C. decreases
   D. increases up to a point and then decreases

21. A hole is drilled in the end of a staybolt ___.
   A. for expansion and contraction
   B. to warn of a corroded stay
   C. to add strength
   D. no holes are ever drilled in a staybolt

22. When taking a boiler off-line, a vacuum is prevented by ___.
   A. opening the air cock
   B. blowing down
   C. dumping the boiler at once
   D. popping the safety valve

23. A valve that requires five 360° turns of the handwheel to change from fully open to fully closed is a ___ valve.
   A. quick-opening
   B. blowdown
   C. slow-opening
   D. ball

24. A thermostatic steam trap opens when the ___.
   A. bucket rises
   B. steam enters the trap
   C. bellows contract
   D. steam pressure increases

25. A combustion control on a boiler ___.
   A. controls steam temperature
   B. controls steam pressure
   C. eliminates the need for an operator
   D. controls the air-fuel ratio

Test 10

1. When coal is classified, ___ refers to hardness.
   A. rank
   B. grade
   C. run of mine
   D. slack

2. ___ gauge glasses are used on boilers with pressures exceeding 250 psi.
   A. Tubular
   B. Flat
   C. Cast iron
   D. Metal

3. Oxygen and noncondensable gases are removed from feedwater by ___.
   A. adding sodium sulfate
   B. adding sodium sulfite
   C. heating the water
   D. routing through the condenser

4. ___ combustion is when all fuel is burned using only the theoretical amount of air.
   A. Theoretical
   B. Secondary
   C. Complete
   D. Perfect

5. Watertube boilers require less water than ___ boilers.
   A. firetube
   B. industrial watertube
   C. bent-tube
   D. steam

6. A boiler equipped with a draft gauge and registering a negative pressure indicates ___.
   A. 0″ of water pressure in the furnace
   B. negative atmospheric pressure
   C. furnace pressure is greater than atmospheric pressure
   D. atmospheric pressure is greater than furnace pressure

7. Pressure atomizing burners atomize fuel oil using ___.
   A. steam pressure
   B. air pressure
   C. a rotary cup
   D. none of the above

8. The ___ on the steam turbine functions as a limit switch.
   A. governor
   B. oil control pump
   C. overspeed trip
   D. sealing strips

9. Circulating water passes through a surface condenser ___.
   A. outside the tubes
   B. inside the tubes
   C. in direct contact with exhaust steam
   D. outside the relief valve

10. The pour point of a fuel oil is ___.
   A. the lowest temperature at which oil will flow
   B. its internal resistance to flow
   C. only used with lubricating oils
   D. the same as its fire point

11. A condensate pump is located 75′ below the feedwater heater and its discharge gauge reads 75 psi. The pressure at the inlet to the heater is ___ psi.
   A. 30.525
   B. 36.525
   C. 42.525
   D. 48.525

12. Higher numbered fuel oils produce ___ lower numbered fuel oils.
   A. more Btu/gal. than
   B. fewer Btu/gal. than
   C. the same amount of Btu/gal. as
   D. one-half the Btu/gal. as

13. A boiler evaporates 48,000 lb of steam/hr. It evaporates 12 lb of steam/sq ft of heating surface per hour. The boiler has ___ sq ft of heating surface.
   A. 4000
   B. 5000
   C. 40,000
   D. 50,000

14. A steam turbine exhaust end and condenser are prevented from excess pressure by ___.
   A. closing in on the throttle valve
   B. the atmospheric relief valve
   C. the overspeed trip
   D. the governor

15. At 0 psi, ___ Btu are necessary to change 1 lb of water from and at 212°F into steam.
   A. 144
   B. 180
   C. 970.3
   D. 1200

16. Automatic boilers are programmed to ignite the burner ___.
   A. in low fire
   B. in high fire
   C. before purging the furnace
   D. with dampers closed

17. ___ boilers can operate at higher pressures than firetube boilers.
   A. HRT
   B. Scotch marine
   C. Vertical firetube
   D. Watertube

18. The first step when using a soot blower is to ___.
   A. open steam line drain
   B. warm up the steam lines slowly
   C. turn soot blower elements slowly
   D. actuate steam turbine blades

19. The purpose of a vent on an open feedwater heater is to ___.
   A. allow air into the heater
   B. remove air and other noncondensable gases
   C. act as an overflow
   D. prevent pressure buildup in the heater

20. Cooling water and steam are not in direct contact with each other in a(n) ___ condenser.
   A. barometric
   B. open
   C. closed
   D. surface

21. A boiler with 3450 sq ft of heating surface designed for 5 sq ft of heating surface per BHP has ___ HP.
   A. 100
   B. 345
   C. 690
   D. 1000

22. A water column ___.
   A. dampens fluctuating water level
   B. must be equipped with ball checks
   C. prevents foaming in the gauge glass
   D. is used on all boilers

23. A closed feedwater heater is always installed ___.
   A. on the suction side of the feedwater pump
   B. on the discharge side of the feedwater pump
   C. at a high elevation to increase head
   D. at the lowest possible position

24. ___ is not required for combustion.
   A. Atomization of fuel
   B. Heat
   C. Oxygen
   D. Nitrogen

25. Heating value of natural gas is expressed in ___.
   A. Btu/lb
   B. therms
   C. Btu/gal.
   D. specific gravity

ESSAY

Test 1

1. A boiler is equipped with a superheater and has an automatic nonreturn valve. Explain how this boiler must be warmed up and cut in on-line.

2. Explain how to allow for expansion and contraction of the main steam line between the boiler and steam header.

3. Explain how to replace a tubular gauge glass with the boiler in service. Include all safety precautions.

4. According to the ASME code, when must a boiler have two main steam stop valves? What type of valves may be used?

5. Why does a boiler safety valve pop open rather than open slowly?

6. Describe how a deaerating feedwater heater functions.

7. What causes a feedwater pump to become steambound and how is this condition corrected?

8. Describe the two basic types of superheaters. Where are they located on a boiler?

9. Describe the operation of a reciprocating feedwater pump.

10. What is the function of an overspeed trip on a steam turbine? When is it tested and how often?

Test 2

1. According to the ASME code, when must a boiler have two main steam stop valves? What type of valves may be used?

2. Describe the operation of a reciprocating feedwater pump.

3. What is an economizer? Where is it located? What are some problems encountered in its use?

4. What could cause a reciprocating feedwater pump to not provide enough water to a boiler?

5. Describe the procedure to follow when performing a hydrostatic test.

6. What is an automatic nonreturn valve? Where is it located?

7. What is the purpose of an air heater? Describe two types.

8. How is a boiler prepared for an annual inspection?

9. Describe the wet method of boiler lay-up.

10. Describe the function and location of a surface condenser.

Test 3

1. What is an automatic nonreturn valve? Where is it located?
2. Describe the operation of a centrifugal pump.
3. List all the parts of a centrifugal pump.
4. Define *running condensing.*
5. What is the heating surface of a boiler?
6. What is caustic embrittlement? How can it be prevented?
7. What is the purpose of an air heater? Describe two types.
8. What are common causes of a low water condition?
9. What causes foaming and priming in a boiler? How is this condition corrected?
10. Which safety valves should be set to pop first, the safety valves on the boiler or the safety valves on the superheater?

Test 4

1. How are the steam valves set on a duplex reciprocating feedwater pump?
2. What is caustic embrittlement? How can it be prevented?
3. How can it be determined if the safety valves on a boiler have enough relieving capacity?
4. Define *boiler efficiency.*
5. Why is fuel oil in boiler water dangerous? How can fuel oil get into a boiler?
6. How does the boiler drum pressure compare to the superheater outlet pressure during normal operation of a boiler equipped with a superheater?
7. Describe the condition that develops in economizers if the feedwater temperature is too low.
8. What happens when baffles deteriorate in watertube boilers?
9. What is the purpose of a surface blowdown? Describe its location.
10. Why is a steam valve under pressure never opened quickly?

Test 5

1. What is the purpose of blowing down the water column, gauge glass, and low water fuel cutoff each shift?
2. What procedure must be followed if the water level in the gauge glass is not visible and the burner is still operating?
3. How is a low water fuel cutoff tested?

4. Why is an approved blowdown tank required on a steam boiler?

5. Describe the procedure followed when taking over a shift.

6. List three commonly used fuels and their Btu content.

7. What is the function of the bottom blowdown line? Where and how is it connected to a boiler?

8. What is a steam separator and steam trap? How is each used?

9. Describe two types of feedwater regulators. State where each is used.

10. Describe the function and location of a surface condenser.

Test 6

1. Describe three ways fuel oil is atomized.

2. What test must be performed on a new or repaired boiler?

3. What is the function of a feedwater injector?

4. What is the function of a grid valve?

5. What are the advantages of a superheater used with a steam turbine?

6. Why would a safety valve fail to open at its set pressure?

7. Why are a steam strainer and separator located before a steam turbine?

8. How is a steam pressure gauge protected against excessive temperature?

9. Describe three methods of heat transfer.

10. Why is feedwater heated before it enters the steam and water drum?

Test 7

1. Describe three ways fuel oil is atomized.

2. Why must the safety valve discharge outlet not be reduced or restricted in any way?

3. Why would a safety valve fail to open at its set pressure?

4. What maintains the speed of an auxiliary steam turbine operating at 3600 RPM?

5. List four reasons for losing vacuum in a plant using a surface condenser. How could each condition be corrected?

6. Why is superheated steam used in a steam turbine plant?

7. Why is steam extracted from a steam turbine?

8. What is the function of a thrust bearing on a steam turbine?

9. Why does the auxiliary lubricating oil pump operate after the steam turbine has been secured?

10. Why must a steam turbine rotor rotate during cool down?

Test 8

1. How does a governor control the speed of a steam turbine?
2. Why is a high vacuum desirable in a condenser for a condensing steam turbine?
3. Should the rotor of a steam turbine be moving or stationary during the warm-up period?
4. How do economizers and air heaters affect draft?
5. What are the advantages of a superheater used with a steam turbine?
6. What is the procedure of cold startup of a boiler equipped with a superheater?
7. What is the procedure followed if popping or blowdown pressure on a safety valve must be adjusted?
8. List the auxiliaries commonly used in the operation of a condensing steam turbine.
9. How is an auxiliary steam turbine started?
10. How is an auxiliary steam turbine removed from service?

Test 9

1. What should be done if there is a possibility that the main boiler steam pressure gauge is not accurate?
2. What is the purpose of a steam separator?
3. List all the parts of a centrifugal pump.
4. Describe the operation of commonly used feedwater heaters.
5. What is superheated steam?
6. Why is superheated steam used in steam turbine plants?
7. What is the procedure followed if popping or blowdown pressure on a safety valve must be adjusted?
8. Why must leaks on the bottom blowdown line be promptly repaired?
9. What are the requirements of the discharge line of a safety valve?
10. What procedure should be followed if a safety valve leaks after testing?

Test 10

1. Describe the procedure required to increase the working pressure of a boiler.
2. Describe the location and purpose of the following:

    a. Nonreturn valve

    b. Quick-opening valve

c. Superheater safety valve

d. Free-blowing drain

e. Expansion bend

3. What steps must be taken if the safety valve does not pop with excessive boiler pressure?

4. What are water temperature considerations from an open feedwater heater?

5. What is the function of a telltale hole?

6. What is the function of a feedwater injector?

7. How is scale in a boiler caused and prevented?

8. Describe three methods for identifying overheated bearings.

9. What procedure should be followed if a leak is found between the boiler and the bottom blowdown valves?

10. It takes approximately 15 lb of air to burn 1 lb of coal. How much air is necessary to burn 1 ton of coal?

CHAPTER 12 LEARNER RESOURCES

ATPeResources.com/QuickLinks
Access Code: 735728

LICENSING EXAMINATION PREPARATION–SECOND CLASS

Chapter 13, Licensing Examination Preparation-Second Class, consists of questions based on sample questions obtained from licensing agencies across the United States and Canada. Second class refers to the intermediate level of licensing examinations for stationary engineers. The name of this examination may vary from one licensing agency to another. At the second class level, the stationary engineer is responsible for larger plants and more equipment than third class stationary engineers. Consequently, examination questions test for the ability to apply mathematics, physics, and engineering principles. Essay questions should be answered in detail using proper terminology.

MULTIPLE CHOICE

Test 1

1. A ___ amp fuse is necessary for a circuit with a load of 21,000 watts and a voltage of 220.
   - A. 100
   - B. 150
   - C. 175
   - D. 200

2. All steam lines should be ___.
   - A. level
   - B. pitched in the direction of flow
   - C. pitched back toward the boiler
   - D. pitched for convenience

3. In a Bourdon tube pressure gauge, ___.
   - A. the Bourdon tube tends to straighten with temperature
   - B. the Bourdon tube tends to straighten with pressure
   - C. hydraulic pressure is applied
   - D. hydraulic stress is applied

4. The piston in a duplex pump is prevented from striking the head by the ___.
    A. pressure on the liquid
    B. lap of the steam valve
    C. cushion formed by the piston covering the exhaust post
    D. lead of the steam valve

5. An HRT boiler has a 72″ shell diameter, 0.5″ thick metal plate, and a tensile strength of 50,000 psi. The efficiency of the boiler seam is 80% and factor of safety is 5. The MAWP is ___.
    A. 51
    B. 111
    C. 151
    D. 176

6. Large amounts of combustible material in the furnace ash are most likely to be caused by ___.
    A. not enough forced draft
    B. low furnace temperatures
    C. high furnace temperatures
    D. defective grates

7. If the discharge pressure of a reciprocating pump drops drastically, the pump will probably ___.
    A. stop
    B. decrease in speed
    C. increase in speed
    D. run normally

8. When the evaporation and factor of evaporation of a boiler are known, the equivalent evaporation can be found by ___.
    A. dividing the evaporation by the factor of evaporation
    B. multiplying the evaporation by the factor of evaporation
    C. dividing the factor of evaporation by the evaporation
    D. adding the evaporation to the factor of evaporation

9. Forty degrees Fahrenheit (40°F) is equal to ___°C.
    A. 4.4
    B. 6
    C. 22.2
    D. 40

10. A power plant is located 1 mile above sea level. The barometric pressure is 26.5″ Hg and the condenser vacuum is 20.5″ Hg. The absolute condenser pressure is ___ psia.
    A. 1.95
    B. 2.95
    C. 3.95
    D. 4.95

11. A saybolt viscosimeter is used for ___.
    A. finding the viscosity of steam
    B. finding the heating value of coal
    C. finding the specific gravity of fluids
    D. testing and grading of oils

12. In an inverted bucket steam trap, the bucket ___ as the trap fills with condensate.
    A. rises
    B. tilts
    C. rotates
    D. sinks

13. Too much excess air through a furnace ___.
    A. lowers boiler efficiency
    B. makes a hotter fire
    C. increases boiler efficiency
    D. has little effect on operation

14. An internal feedwater line in a watertube boiler is placed below water level and extended into the drum ___.
    A. so that cold water entering is warmed
    B. for better distribution and less thermal shock
    C. to disrupt the circulation
    D. because it is the most convenient way to install the feedwater line

15. More than one safety valve should be used on a boiler when the boiler has over ___.
    A. 500 BHP
    B. 500°F steam temperature
    C. 500 sq ft of heating surface
    D. 500 psi

16. A soot buildup on boiler tubes results in ___.
    A. higher steam temperature
    B. lower steam temperature
    C. higher efficiency
    D. lower efficiency

17. A(n) ___ analysis is used to determine the percentage of various elements contained in coal, including carbon, oxygen, nitrogen, hydrogen, sulphur, and ash.
    A. total
    B. ultimate
    C. proximate
    D. complete

18. A(n) ___ is located on the inlet line of a steam turbine condenser.
    A. atmospheric relief valve
    B. vacuum breaker
    C. air vent
    D. check valve

19. A closed feedwater heater is located ___.
    A. before the pump and boiler
    B. above the feedwater pump
    C. between the pump and boiler
    D. above the pump and boiler

20. ___ is used to prevent oxidation of boiler drums and tubes.
  A. Sodium sulfite
  B. Sodium phosphate
  C. Calcium phosphate
  D. Magnesium phosphate

21. ___ can cause a tube in a superheater to overheat.
  A. Poor boiler circulation
  B. Scale buildup in the superheater
  C. High firing rates
  D. Leaving the superheater drain open during startup

22. The bottom blowdown valve is used to ___.
  A. remove steam from the header
  B. raise the boiler water level
  C. drain the boiler for servicing and inspection
  D. remove floating impurities

23. A tube sheet is supported in a firetube boiler with ___.
  A. tubes and stays
  B. welded braces
  C. external rods and straps
  D. it does not need supports

24. ___ results in a high carbon dioxide reading in gases of combustion.
  A. Complete combustion
  B. Incomplete combustion
  C. Proper excess of fuel
  D. Air heater leaks

25. Factor of evaporation is ___.
  A. heat produced by boiler divided by 970.3
  B. heat produced by boiler minus inlet feedwater temperature
  C. 970.3 Btu/lb
  D. 34.5 lb/min

Test 2

1. In an inverted bucket steam trap, the bucket ___ as the trap fills with condensate.
  A. rises
  B. tilts
  C. sinks
  D. rotates

2. A boiler is 85% efficient and uses 875 gal. of fuel oil to produce 100,000 lb of steam/hr. To generate 250,000 lb of steam/hr, ___ are needed.
  A. 21.87 gal./min
  B. 218.7 gal./min
  C. 2187.5 gal./hr
  D. 21,875 gal./hr

3. Pressure guages indicate pressure below atmospheric pressure in ___.
   A. PSIG
   B. PSIA
   C. inches of mercury
   D. inches of water column

4. A 10,000 kW turbo-generator operates at 600 psi with a steam temperature of 700°F and an exhaust pressure of 1 psia. The thermal efficiency is ___%.
   A. 16.23
   B. 19.13
   C. 23.14
   D. 51.82

5. A boiler operates at 100 psi and has 1.5" diameter staybolts that are spaced 6" on centers, horizontally and vertically. The load supported by the staybolts is ___ lb.
   A. 3200
   B. 3400
   C. 3600
   D. 3800

6. An ignition arch is used when using a(n) ___ stoker.
   A. sidefeed
   B. underfeed
   C. spreader
   D. chain grate

7. A 50 HP motor runs at 75% efficiency; ___ watts per HP are being used.
   A. 99.5
   B. 560
   C. 995
   D. 5600

8. A 40 HP motor runs at 75% efficiency. The input at maximum load is ___ watts.
   A. 39,787
   B. 41,878
   C. 43,778
   D. 45,787

9. A globe valve, when installed in the feedwater line, must have ___.
   A. inlet pressure under the disc
   B. inlet pressure over the disc
   C. outlet pressure under the disc
   D. does not matter which way it is installed

10. A duplex pump 4 × 3 × 4 pumps water to a height of 300'. The steam pressure required is ___ psi. (Do not allow for friction losses.)
    A. 43
    B. 53
    C. 63
    D. 73

11. A high carbon dioxide reading is an indication of ___.
   A. incomplete combustion
   B. too much excess air
   C. efficient combustion
   D. the boiler needing cleaning

12. The overspeed trip of a steam turbine should function at ___% above normal speed.
   A. 2
   B. 5
   C. 10
   D. 15

13. During assembly, ___ is/are placed around the outer ends of steam turbine blades.
   A. rivets
   B. a rotor
   C. a shroud ring
   D. a stress ring

14. A steam turbine uses 50,000 lb of steam/hr and exhausts into a condenser at 28″ Hg vacuum. It is connected to a generator with an output of 3000 kW. Its water rate is ___ lb/kW.
   A. 10.7
   B. 12.7
   C. 14.7
   D. 16.7

15. Draft is measured in ___.
   A. pounds per square inch
   B. inches of mercury
   C. feet of water
   D. inches of water

16. Safety valve capacity should be ___ the boiler capacity.
   A. the same as or greater than
   B. the same as or less than
   C. at least two times
   D. at least 1½ times

17. Safety valve set pressures should be ___ the MAWP.
   A. at or above
   B. at or below
   C. 1½ times
   D. two times

18. Tube ends of a firetube boiler are secured to tube sheets by ___.
   A. brazing
   B. riveting
   C. expanding and beading
   D. screwing and threading

19. The minimum size blowdown pipe on a water column is ___".
  - A. ½
  - B. ¾
  - C. 1
  - D. 1¼

20. An economizer ___.
  - A. heats feedwater with chimney gases
  - B. increases chimney gas temperature
  - C. removes flyash and soot
  - D. controls the combustion process

21. Sodium ___ is used to eliminate oxygen from boiler water.
  - A. sulfite
  - B. sulfate
  - C. phosphate
  - D. hydroxide

22. Absolute pressures are used when ___.
  - A. determining boiler efficiency
  - B. measuring feedwater temperature
  - C. measuring draft losses through the boiler
  - D. solving gas law and thermodynamic problems

23. ___ can prevent corrosion of air preheaters.
  - A. Using a low sulfur fuel
  - B. Large amounts of excess air
  - C. Maintaining gases of combustion at a higher temperature than the dew point of gases
  - D. Maintaining gases of combustion at a lower temperature than the dew point of gases

24. Exhaust steam supplied to a feedwater heater increases ___.
  - A. plant and boiler efficiency
  - B. plant efficiency only
  - C. boiler efficiency only
  - D. steam turbine efficiency

25. An equation for the relieving capacity of a safety valve is ___.

  A. $Pounds\ of\ steam/hr = \dfrac{Safety\ valve\ area(MAWP+10)}{0.2}$

  B. $Pounds\ of\ steam/hr = \dfrac{Safety\ valve\ area \times MAWP \times 10}{0.2}$

  C. $Pounds\ of\ steam/hr = \dfrac{0.2 \times Safety\ valve\ area}{MAWP+10}$

  D. $Pounds\ of\ steam/hr = \dfrac{Safety\ valve\ area(MAWP+15)}{0.2}$

Test 3

1. An excessively high gases of combustion temperature leaving the boiler indicates ___.
A. a high level of CO_2
B. a high level of SO_2
C. the wrong type of fuel being used
D. dirty tubes

2. One kW is equal to ___.
A. 0.746 HP
B. 1.34 HP
C. 2425 Btu
D. 3014 Btu

3. Cuprous chloride is used in ___.
A. testing boiler water
B. an Orsat combustion gas analyzer
C. a combustion catalyst
D. removing soot from tubes

4. A steam turbine operating at its critical speed would ___.
A. overheat
B. overspeed
C. vibrate excessively
D. need lubrication

5. A 300 HP motor is running at 1800 RPM and is at full load. The torque on the shaft of the motor is ___ ft-lb.
A. 650
B. 875
C. 1126
D. 1750

6. Using exhaust steam to heat feedwater ___ efficiency.
A. increases plant
B. increases boiler
C. decreases plant and boiler
D. does not change plant or boiler

7. The relieving capacity of a safety valve is determined by the equation ___.

A. $Pounds\ of\ steam/hr = \dfrac{0.2 \times Safety\ valve\ area}{MAWP + 10}$

B. $Pounds\ of\ steam/hr = \dfrac{Safety\ valve\ area(MAWP + 15)}{0.2}$

C. $Pounds\ of\ steam/hr = \dfrac{Safety\ valve\ area(MAWP + 10)}{0.2}$

D. $Pounds\ of\ steam/hr = \dfrac{10 \times Safety\ valve\ area \times MAWP}{0.2}$

8. The pressure and temperature limitations on the discharge from a blowdown tank are 150°F and ___ psi.
  A. 0
  B. 5
  C. 10
  D. 15

9. An equation for determining the frequency of an AC generator is ___.
  A. $Frequency = \dfrac{No.\ of\ poles \times RPM}{120}$

  B. $Frequency = \dfrac{No.\ of\ poles + RPM}{120}$

  C. $Frequency = \dfrac{No.\ of\ poles \times 120}{RPM}$

  D. $Frequency = \dfrac{No.\ of\ poles + 120}{RPM}$

10. Oil contamination in condensate returns is most serious in a(n) ___.
  A. closed feedwater heater
  B. open feedwater heater
  C. economizer
  D. is equally serious in all feedwater heaters

11. Boiler tubes are sized by the ___.
  A. outside diameter
  B. inside diameter thickness
  C. outside diameter minus wall
  D. inside diameter minus wall

12. Boiler water normally has a pH range of ___.
  A. 1 to 3
  B. 4 to 6
  C. 10.5 to 12
  D. 11 to 14

13. A boiler generates 20,000 lb of steam/hr and has a factor of evaporation of 1.05. The DBHP is ___.
  A. 10.2
  B. 60.9
  C. 609
  D. 1020

14. A steam pressure gauge should have a range of at least ___.
  A. the MAWP
  B. 1½ times the safety valve setting
  C. 2 times the safety valve setting
  D. 2½ times the safety valve setting

15. Electrostatic precipitators are used to ___.
  A. remove flyash and soot
  B. remove solids in feedwater
  C. remove impurities from steam
  D. determine smoke density

16. The volatile gases from soft coal can be burned more efficiently by supplying more ___ air.
  A. primary
  B. secondary
  C. overfire
  D. underfire

17. The air chamber on the discharge side of a reciprocating feed water pump provides ___.
  A. water storage
  B. constant flow
  C. pressure buildup
  D. steam expansion in the steam cylinder

18. A step-up transformer increases the ___.
  A. amperage while decreasing voltage
  B. resistance while decreasing amperage
  C. amperage while decreasing resistance
  D. voltage while decreasing the amperage

19. Staybolts are commonly used on ___.
  A. watertube boilers
  B. firetube boilers
  C. steam turbines
  D. reciprocating pumps

20. A stop valve and check valve must be installed in the ___.
  A. main steam line
  B. main condensate line
  C. feedwater line close to the boiler shell
  D. blowdown lines from the boiler

21. A throttling calorimeter is used to determine the ___.
  A. amount of heat in the steam
  B. amount of heat in the condensate
  C. efficiency of the superheater
  D. amount of moisture in the steam

22. Tube sheets are supported in a firetube boiler by ___.
  A. rolling and beading of the tubes
  B. the shell, stays, and tubes
  C. the shell and stays
  D. the shell and tube

23. A boiler using coal with a high ash content is most likely using a(n) ___ stoker.
  A. sidefeed
  B. underfeed
  C. spreader
  D. chain grate

24. An ignition arch is used when using a(n) ___ stoker.
   A. sidefeed
   B. underfeed
   C. spreader
   D. chain grate

25. The output of an impulse steam turbine depends on the ___.
   A. operating steam pressure
   B. number of nozzles opened
   C. number of moving blades
   D. number of stationary blades

Test 4

1. A boiler generates 30,000 lb of steam/hr and has a factor of evaporation of 1.04. The DBHP is ___.
   A. 750
   B. 825
   C. 904
   D. 1004

2. The equation for determining factor of evaporation is ___.

   A. $Factor\ of\ evaporation = \dfrac{Enthalpy\ of\ steam - Enthalpy\ of\ feedwater}{970.3}$

   B. $Factor\ of\ evaporation = \dfrac{Enthalpy\ of\ steam - Enthalpy\ of\ feedwater}{34.5}$

   C. $Factor\ of\ evaporation = Enthalpy\ of\ steam - \dfrac{Enthalpy\ of\ feedwater}{970.3}$

   D. $Factor\ of\ evaporation = \dfrac{Enthalpy\ of\ steam - Enthalpy\ of\ feedwater}{33,000}$

3. A duplex reciprocating feedwater pump has the following dimensions: 12″ stroke, 6″ diameter water piston. It operates at 40 strokes/min at 92% efficiency. This pump will discharge ___ gal./hr.
   A. 648.6
   B. 4860
   C. 6486
   D. 64,860

4. Tubes can have smaller metal wall thicknesses than boiler drums when subjected to the same pressures because ___.
   A. they are made of stronger material
   B. they are of seamless construction
   C. of the expansion and contraction
   D. they are smaller in diameter

5. Tube sheets are supported in a firetube boiler by ___.
   A. rolling and beading of the tubes
   B. the shell, stays, and tubes
   C. the shell and stays
   D. the shell and tube

6. The heating surface area of boiler firetubes is determined using the ___.
   A. outside diameter
   B. inside diameter
   C. wall thickness
   D. wall thickness times 2

7. A high differential pressure across a fuel oil strainer indicates ___.
   A. normal operation
   B. the suction valve of the fuel oil pump is closed
   C. the discharge valve of the fuel oil pump is closed
   D. a dirty strainer

8. A steam pressure gauge can be recalibrated with a ___.
   A. pneumatic gauge
   B. calibrating gauge
   C. deadweight pressure-gauge tester
   D. Fyrite tester

9. A ___ valve is most commonly used to throttle steam flow.
   A. globe
   B. gate
   C. check
   D. ball

10. ___ is an air pollutant resulting from the use of coal and oil in industry.
   A. Carbon dioxide
   B. Sulfur dioxide
   C. Hydrogen sulfide
   D. Nitrous oxide

11. A control that starts and stops a burner on pressure demand is a(n) ___.
   A. air flow switch
   B. aquastat
   C. modulating pressure control
   D. pressure control

12. The steam and water in a closed feedwater heater ___.
   A. mix freely
   B. do not mix
   C. mix in the top section of the heater
   D. mix in the bottom section of the heater

13. Soot that collects on the outside of boiler tubes is normally removed with a ___.
   A. brush
   B. chemical soot powder
   C. soot blower
   D. water hose

14. Overheating of boiler tubes on a watertube boiler can be caused by ___.
  A. using soot blowers too frequently
  B. maintaining a constant firing rate
  C. accumulated soot on the inside of the tubes
  D. scale buildup on the inside of the tubes

15. A chain grate stoker can successfully burn ___.
  A. small sizes of bituminous coal
  B. coal with a low ash content
  C. coal with a high volatile content
  D. caking coals

16. The heating value of coal is determined using a(n) ___.
  A. optic pyrometer
  B. calorimeter
  C. enthalpy meter
  D. Fyrite meter

17. A steam turbine operating condensing is operating at maximum efficiency when the condenser pressure is ___″ Hg.
  A. 24
  B. 25.5
  C. 27
  D. 28.5

18. The core of a rotor on an AC generator is laminated to reduce ___.
  A. arcing across coils
  B. hysteresis losses
  C. eddy current loss
  D. friction losses

19. When burning pulverized coal, a ___ temperature must be maintained.
  A. low furnace
  B. high furnace
  C. low primary air
  D. low secondary air

20. The equation used for determining theoretical water HP of a centrifugal pump is ___.

  A. $HP = \dfrac{Total\ foot\ head \times Pounds\ of\ water\,/\,sec}{3300}$

  B. $HP = \dfrac{Total\ foot\ head \times Pounds\ of\ water\,/\,min}{33,000}$

  C. $HP = \dfrac{Discharge\ head \times Pounds\ of\ water\,/\,min}{3300}$

  D. $HP = \dfrac{Velocity \times Head \times Pounds\ of\ water\,/\,min}{33,000}$

21. In steam tables, the heat in liquid is the ___.
   A. total heat minus latent heat
   B. boiling point minus 32
   C. boiling point plus 32
   D. latent heat plus sensible heat

22. In a pressure atomizing burner, the atomization of the oil is accomplished by ___.
   A. steam pressure
   B. air pressure
   C. a rotating cup
   D. a sprayer plate and tip

23. The best time to blow down a boiler is at ___.
   A. its lightest load
   B. its heaviest load
   C. any convenient time
   D. the end of a shift

24. Shear pins are commonly used in a(n) ___.
   A. oil gear pump
   B. stoker
   C. reciprocating feedwater pump
   D. rotary cup oil burner

25. ___ require the least excess air.
   A. Sidefeed stokers
   B. Sprinkler stokers
   C. Pulverizers
   D. Hand stokers

Test 5

1. Shear pins are commonly used in a(n) ___.
   A. oil gear pump
   B. stoker
   C. reciprocating feedwater pump
   D. rotary cup oil burner

2. A centrifugal pump delivers 50,000 gal. of water/hr against a total head of 350′ and is 75% efficient. The HP of the pump is ___.
   A. 58.165
   B. 78.165
   C. 98.165
   D. 118.165

3. The best time to blow down a boiler is at ___.
   A. its lightest load
   B. its heaviest load
   C. any convenient time
   D. the end of a shift

4. Steam at a pressure of 100 psi and a temperature of 425°F is ___.
   A. low quality steam
   B. supersaturated
   C. saturated
   D. superheated

5. Tubes in a watertube boiler are secured by ___.
   A. welding and beading
   B. expanding and beading
   C. expanding and welding
   D. expanding and flaring

6. Priming occurs when ___.
   A. water is carried over with steam
   B. the water in the gauge glass is unstable
   C. there is a low water level in the gauge glass
   D. a boiler is steaming at a maximum level

7. A condensing steam turbine is losing vacuum and the temperature of the condensate in the hot well is 90°F. This indicates ___.
   A. too much circulating water in the condenser
   B. air or noncondensable gases in the condenser
   C. a low condensate level in the condenser hot well
   D. not enough load on the turbine

8. In a pressure atomizing burner, the atomization of the oil is accomplished by ___.
   A. steam pressure
   B. air pressure
   C. a rotating cup
   D. a sprayer plate and tip

9. Spalling is caused by ___.
   A. temperature changes in the furnace
   B. a consistent firing rate
   C. maintaining a high furnace temperature
   D. maintaining a low furnace temperature

10. A ___ is commonly used to measure draft.
   A. steam pressure gauge
   B. hygrometer
   C. hydrometer
   D. manometer

11. A pressure-reducing valve reduces the steam pressure from 100 psi to 15 psi. Without considering heat losses from the valve, the steam at 15 psi has ___.
   A. less heat than at 100 psi
   B. more heat than at 100 psi
   C. increasing heat
   D. the same heat as at 100 psi

12. A centrifugal pump that has a suction lift of 20′, a discharge head of 100′, and a friction head of 10′ has a total head of ___′.
    A. 70
    B. 100
    C. 110
    D. 130

13. Degrees API refers to ___ of fuel oil.
    A. viscosity
    B. enthalpy
    C. specific gravity
    D. pour point

14. A pump 6 × 4 × 12 operates with 100 psi steam. The discharge pressure on the water side is ___ psi.
    A. 125
    B. 150
    C. 200
    D. 225

15. A mixed-pressure steam turbine ___.
    A. admits steam to the impulse blading then the reaction blading
    B. admits steam at two or more different pressure stages
    C. admits steam at the first stage only
    D. operates at a very low pressure

16. In a jet condenser, ___.
    A. water flows through the tubes while steam surrounds the tubes
    B. steam flows through the tubes while water surrounds the tubes
    C. water and steam mix
    D. condensate can be reclaimed

17. Sealing strips found on ___ alert the operator of excessive radial tolerance.
    A. steam turbines
    B. boilers
    C. reciprocating pumps
    D. steam condensers

18. Superheater safety valves are set at a pressure ___ the boiler drum safety valves.
    A. higher than
    B. lower than
    C. same as
    D. higher or lower than

19. If the diameter of a staybolt is doubled, the holding power will ___.
    A. be the same
    B. double
    C. triple
    D. quadruple

20. One mechanical HP is equivalent to ___ Btu.
  A. 1500
  B. 2000
  C. 2545
  D. 3000

21. The pH scale is from 0 to ___.
  A. 10
  B. 12
  C. 14
  D. 18

22. If the discharge valve on a centrifugal pump is closed during its operation, the power input will ___.
  A. be zero
  B. be at a maximum
  C. be at a minimum
  D. not change

23. Proximate analysis of coal indicates the percentage by weight of ___.
  A. moisture, volatile matter, fixed carbon, and ash
  B. moisture, volatile matter, fixed carbon, and Btu content
  C. nitrogen, oxygen, carbon, ash, sulfur, and hydrogen
  D. carbon, oxygen, hydrogen, and sulfur

24. A boiler operating at 85% efficiency consumes 750 gal. of fuel oil/hr to generate 90,000 lb of steam. To produce 300,000 lb of steam, ___ gal. of fuel oil are required.
  A. 1000
  B. 1750
  C. 2250
  D. 2500

25. The metal of a boiler drum has a tensile strength of 60,000 psi and is 1″ thick. The diameter of the drum is 5′ and the joint efficiency is 90%. The bursting pressure is ___ psi.
  A. 1600
  B. 1800
  C. 2000
  D. 2200

Test 6

1. External feedwater treatment is accomplished by passing water through a ___.
  A. zeolite softener
  B. feedwater flash tank
  C. chemical proportioning pump
  D. condensate return tank

2. A boiler has a capacity of 50,000 lb/hr. After six months, its capacity drops to 45,000 lb/hr possibly because of ___.
  A. excessive back pressure
  B. faulty instrumentation
  C. plugged dry pipe outlet
  D. scale buildup on the heating surface

3. A centrifugal pump delivers 50,000 gal. of water/hr against a total head of 350′ and is 75% efficient. The HP of the pump is ___.
   A. 58.165
   B. 78.165
   C. 98.165
   D. 118.165

4. A(n) ___ is not an auxiliary used with a surface condenser.
   A. circulating pump
   B. condensate pump
   C. atmospheric relief valve
   D. low water fuel cutoff

5. A mixed-pressure steam turbine ___.
   A. admits steam to the impulse blading then the reaction blading
   B. admits steam at two or more different pressure stages
   C. admits steam at the first stage only
   D. operates at a very low pressure

6. Shear pins are commonly used in a(n) ___.
   A. oil gear pump
   B. reciprocating feedwater pump
   C. stoker
   D. rotary cup oil burner

7. A boiler operates at 500 psig with a factor of safety of 5. The boiler shell is 5′ in diameter, the tensile strength of the drum plate is 65,000 psi, and the joint efficiency is 90%. The boiler shell thickness is ___″.
   A. 1.000
   B. 1.182
   C. 1.282
   D. 1.750

8. To raise the temperature of 1 lb of water 1°F, ___ Btu is/are required.
   A. 1
   B. 2
   C. 3
   D. 4

9. One mechanical HP is equivalent to ___.
   A. 3300 ft-lb/min
   B. 33,000 ft-lb/min
   C. 3300 ft-lb/hr
   D. 33,000 ft-lb/hr

10. The main difference between a relief valve and a safety valve is that a ___.
   A. safety valve opens slowly
   B. relief valve opens wide when lifted
   C. safety valve opens fully
   D. relief valve pops open

11. Baffles are arranged in watertube boilers to ___.
   A. keep heat from impinging on the tubes
   B. protect the casing from over-heating
   C. help remove soot from the tubes
   D. direct the flow of gases of combustion

12. Staybolts are most commonly used in ___ boilers.
   A. firetube
   B. watertube
   C. forced circulation
   D. high-pressure

13. Scale deposits on the tubes of a watertube boiler can be removed by ___.
   A. hand-operated steam lances
   B. a mechanical hammer-type cleaning tool
   C. wire-brushing all the tubes
   D. a water turbine cutting tool

14. Intensity of heat is measured by ___.
   A. Btu
   B. sensible heat
   C. a thermometer
   D. latent heat

15. In an open feedwater heater, ___.
   A. steam comes in direct contact with water
   B. steam and water do not come in direct contact
   C. moisture is removed from the steam
   D. carryover is prevented by using an internal overflow

16. Discharge limits of a boiler blowdown tank are ___°F and ___ psi.
   A. 100, 15
   B. 112, 0
   C. 150, 1
   D. 150, 5

17. Two boilers with the same capacities are operating at 200 psi and 400 psi, respectively. The approximate ratio of the areas of the safety valves is ___.
   A. 1:2
   B. 1:3
   C. 2:1
   D. 3:1

18. The factor of evaporation can be expressed as ___.

   A. $Factor\ of\ evaporation = \dfrac{Enthalpy\ of\ steam - Enthalpy\ of\ feedwater}{970.3}$

   B. $Factor\ of\ evaporation = \dfrac{Enthalpy\ of\ steam - Enthalpy\ of\ feedwater}{34.5}$

   C. $Factor\ of\ evaporation = \dfrac{Enthalpy\ of\ steam - Enthalpy\ of\ feedwater}{33,000}$

   D. $Factor\ of\ evaporation =$
   $\dfrac{Enthalpy\ of\ steam - \left(Enthalpy\ of\ feedwater - 32\right)}{970.3}$

19. A turbo-generator is operating at 400 psi with 250°F of superheat and is exhausting to a condenser at 28″ Hg. The thermal efficiency is ___%.
   A. 4.5
   B. 7.1
   C. 10.2
   D. 19.74

20. ___ is introduced into the boiler as an oxygen scavenger.
   A. Sodium sulfite
   B. Caustic soda
   C. Zeolite
   D. Sodium phosphate

21. If the feedwater to a boiler is increased in temperature without a change in the firing rate, the steam ___.
   A. flow will decrease
   B. flow will increase
   C. flow will remain constant
   D. temperature will increase

22. The steam flow from a boiler increases and in turn causes the superheated steam to increase in temperature. This boiler is fitted with a ___ superheater.
   A. radiant
   B. radiant-convection
   C. convection
   D. conduction

23. A 200 HP electrically driven feedwater pump operates at 1800 RPM. At full load, ___ ft-lb of torque is developed on the shaft.
   A. 83.5
   B. 372.5
   C. 483.5
   D. 583.5

24. A throttling calorimeter is used to determine the quality of ___.
   A. steam
   B. water
   C. superheated steam
   D. superheated water

25. The output of two similar centrifugal pumps connected in series is ___.
   A. constant pressure-double volume
   B. no change in pressure or volume
   C. double pressure-constant volume
   D. double pressure-double volume

Test 7

1. The pH scale is from 0 to ___.
   A. 10
   B. 12
   C. 14
   D. 18

2. The part of the pH scale that indicates an alkali is ___.
   A. 1 to 6
   B. 7
   C. 8 to 14
   D. 15

3. Saturated steam is steam ___.
   A. at a temperature that corresponds with its pressure point
   B. at a temperature below its corresponding pressure
   C. at a temperature below its dew point
   D. only found in boilers operating below 250 psi

4. An air cock ___.
   A. vents air and prevents vacuum from forming in the boiler drum
   B. tests for proper combustion air
   C. tests for proper steam flow
   D. tests air quality in the boiler room

5. Although boiler tubes and boiler drums withstand the same pressure, the walls of the boiler tubes are much thinner because ___.
   A. the tubes are much smaller in diameter
   B. the tube metal has higher tensile strength
   C. circulation keeps the tubes cool
   D. there are higher temperatures within the boiler drums

6. A duplex pump has 6″ diameter plungers, an 18″ stroke, and 50 working strokes per minute. At 100% efficiency, ___ gpm are discharged.
   A. 55
   B. 110
   C. 220
   D. 330

7. A 5′ diameter boiler drum has a boiler metal thickness of 1.5″, 60,000 psi tensile strength, and 100% joint efficiency. The bursting pressure is ___ psi.
   A. 1500
   B. 3000
   C. 4500
   D. 6000

8. A thermal expansion feedwater regulator operates by ___.
   A. movement generated by the thermo hydraulic diaphragm
   B. floats rising with the density of water
   C. positive displacement
   D. expansion and contraction of a thermostatic element

9. The minimum setting on the blowdown or blowback ring of a safety valve is ___.
   A. 2 psi
   B. 4 psi
   C. 2% of set pressure
   D. 4% of set pressure

10. The minimum size piping for bottom blowdown lines is ___".
    A. ½
    B. ¾
    C. 1
    D. 2½

11. Boiler efficiency is ___.
    A. pounds of steam/lb of fuel divided by Btu in steam
    B. heat absorbed by boiler/hr divided by heat available in fuel used/hr
    C. based on feedwater consumption
    D. weight of water in boiler divided by weight of fuel

12. A watertube boiler operates with a steam pressure of 150 psi. The steam and water drum is 3.5' in diameter. The total force acting on the head of the drum is ___ lb.
    A. 107,817
    B. 207,817
    C. 307,817
    D. 407,817

13. The amount of heat in Btu needed to raise the temperature of 1 lb of a substance 1°F is ___.
    A. latent heat of evaporation
    B. latent heat of fusion
    C. ambient heat
    D. specific heat

14. A ___ is used to determine the heating value of solid fuels.
    A. hydrometer
    B. hygrometer
    C. calorimeter
    D. pyrometer

15. The steam and water drum safety valve shall be set to pop ___ superheater safety valve.
    A. before the
    B. at the same time as the
    C. after the
    D. before or after the

16. If not enough water is delivered after starting a centrifugal pump, a probable cause is that the ___.
    A. packing glands are too tight
    B. pump RPM is too high
    C. suction valve is not fully open
    D. pump is overprimed

17. The fewest number of valves a duplex double-acting pump can have on the water side is ___.
    A. two
    B. four
    C. six
    D. eight

18. If a steam turbine using steam from a boiler trips without warning and the boiler safety valves pop, the operator should ___.
    A. secure the fires and maintain the water level
    B. reduce pressure by blowing down
    C. reduce the fires gradually and blow off steam
    D. open all steam valves to reduce pressure

19. The open feedwater heater is located on the ___ the level of the pump.
    A. suction side of the feed water pump above
    B. discharge side of the feed water pump above
    C. suction side of the feedwater pump below
    D. discharge side of the feedwater pump below

20. An increase in the steam turbine speed at light loads when the governor valve is closed indicates ___.
    A. worn governor pins
    B. a leaking governor valve
    C. excessive bearing clearances
    D. a worn thrust bearing

21. If a boiler in battery has a tube rupture, the automatic nonreturn valve will ___.
    A. prevent water flow to the ruptured boiler
    B. secure all boilers by disconnecting from the header
    C. prevent steam flow from the steam header into the ruptured boiler
    D. prevent steam flow from the ruptured boiler into the header

22. A(n) ___ feedwater heater is a heat exchanger located on the discharge side of a pump used to heat feedwater.
    A. open
    B. closed
    C. deaerating
    D. discharge heat

23. The height of a water column necessary to produce a static pressure of 120 psi at its base is ___′.
    A. 107
    B. 127
    C. 157
    D. 277

24. A feedwater injector may fail to operate because ___.
    A. superheated steam is being used
    B. the water temperature is too high
    C. the suction line is obstructed
    D. all of the above

25. A steam turbine used to reduce high-pressure steam to a lower pressure for processing work or heating is a ___ turbine.
    A. back pressure
    B. bleed
    C. low-pressure
    D. mixed-pressure

Test 8

1. A 75 HP centrifugal pump discharges against a 100′ head and operates at an efficiency of 92%. At full load, this pump can deliver ___ gpm.
   A. 45
   B. 90
   C. 1367
   D. 2734

2. An equation to determine the approximate heating value per pound of fuel oil is ___.
   A. $Btu/lb = 17,780 + (54 \times °API)$
   B. $Btu/lb = 17,780 \times (54 \times °API)$
   C. $Btu/lb = 17,780 + \dfrac{54}{°API}$
   D. $Btu//b = 17,780 \times (54 + °API)$

3. Corrosion in a boiler is caused by ___.
   A. carbon dioxide
   B. silica
   C. oxygen
   D. caustic soda

4. Caustic embrittlement in a boiler is most likely found in the ___ water level.
   A. riveted joint above
   B. riveted joint below
   C. welded joint above
   D. welded joint below

5. A symbol with the letter ___ is found adjacent to the power boiler stamping.
   A. E
   B. R
   C. A
   D. S

6. Mechanical draft is necessary when a boiler is equipped with ___.
   A. an air heater and economizer
   B. scrubbers
   C. waterwalls
   D. soot blowers

7. A(n) ___ feedwater heater is a heat exchanger located on the discharge side of a pump used to heat feedwater.
   A. open
   B. closed
   C. deaerating
   D. discharge heat

8. A synchronous motor is running with its field overexcited. The result is ___.
   A. no electrical effect
   B. a lagging power factor
   C. a leading power factor
   D. an increase in motor efficiency

9. The minimum drain size on a water column is ___".
  A. ½
  B. ¾
  C. 1
  D. 1¼

10. The minimum blowdown valve size on power boilers is ___".
  A. ½
  B. ¾
  C. 1
  D. 2½

11. A boiler constructed of 0.75" plate with 55,000 psi tensile strength and a shell diameter of 36", joint efficiency of 80%, and a factor of safety of 5 has an MAWP of ___ psi.
  A. 102.4
  B. 175.5
  C. 265.6
  D. 366.6

12. A 5000 kW turbo-generator operates at 600 psia and 700°F, with a condenser pressure of 1 psia. The thermal efficiency is ___%.
  A. 19
  B. 25
  C. 75
  D. 85

13. A dry pipe in the steam and water drum ___.
  A. removes air from the steam
  B. increases the quality of the steam
  C. prevents the boiler water from foaming
  D. prevents carryover

14. Three boilers are in battery. One is removed from service and has been cooled and emptied. Before entering the water side of this boiler, ___.
  A. close, tag, and lock the main
  B. ventilate the drum with exhaust blowers
  C. replace check valve steam, blowdown, and feedwater
  D. remove tube sheets stop valves

15. When steam with 10°F of superheat is heated to 100°F of superheat, it can increase the thermal efficiency of a steam turbine by ___%.
  A. 4.5
  B. 5.4
  C. 9
  D. 12

16. Safety valves are designed with ___ to open quickly.
  A. release vents
  B. venturi openings
  C. try cocks
  D. huddling chambers

17. Too little excess air can cause ___.
  A. furnace explosions
  B. smoky fire
  C. flashbacks from the furnace
  D. spalling of the furnace refractory

18. A boiler is operating at 125 psi. The diameter of the steam and water drum head is 3′. The total force on the drum head is ___ lb.
  A. 127
  B. 1272
  C. 12,723
  D. 127,235

19. Two or more safety valves are required on a boiler when the boiler ___.
  A. is operating at over 15 psi
  B. is operating at over 100 psi surface
  C. has over 250 sq ft of heating
  D. has over 500 sq ft of heating surface

20. The tubes in some watertube boilers are pitched to ___.
  A. increase heating surface
  B. increase circulation
  C. increase separation of steam and water
  D. help support boiler headers

21. A sentinel valve ___.
  A. warns the operator of high back
  B. relieves high back pressure
  C. warns the operator of a loss of pressure vacuum
  D. warns the operator of low back pressure

22. An atmospheric relief valve on a condensing turbine protects the ___.
  A. turbine casing
  B. condenser from excessive vacuum
  C. condenser from excessive pressure
  D. condenser from flooding

23. A boiler operating at 100 psi has a saturated steam temperature of 338°F. The temperature of the water in the steam and water drum is ___°F.
  A. 212
  B. 250
  C. 300
  D. 338

24. A(n) ___ steam turbine has high-velocity steam going through a nozzle and striking a set of moving blades.
  A. impulse
  B. reaction
  C. compound
  D. velocity-reaction

25. A safety relief valve is found on ___ boilers.
   A. electric steam
   B. miniature steam
   C. power
   D. hot water

Test 9

1. When oxygen enters a boiler drum, it can cause ___.
   A. erosion
   B. embrittlement
   C. scale to form
   D. corrosion

2. A 24′ long, 3″ OD watertube with a wall thickness of ⅛″ has a heating surface of ___ sq ft.
   A. 14.75
   B. 16.80
   C. 18.85
   D. 20.95

3. The tubes in a straight-tube watertube boiler are inclined ___.
   A. to prevent soot from coating
   B. to increase circulation at the correct angle
   C. to help when draining the boiler tubes
   D. so that the tubes enter the header

4. The minimum diameter for a blowdown line on a water column is ___.
   A. ¼
   B. ½
   C. ¾
   D. 1

5. A dry pipe located in the steam space of firetube and watertube boilers is used to ___.
   A. remove moisture from the steam
   B. remove moisture from the air
   C. add heat to dry steam
   D. cool steam to remove moisture

6. Not enough excess air for combustion can cause ___.
   A. overheating of the refractory
   B. an unstable flame
   C. white smoke
   D. a smoky fire

7. A(n) ___ analysis is used to determine the percentage, by weight, of moisture, volatile matter, fixed carbon, and ash in a coal sample.
   A. total
   B. ultimate
   C. proximate
   D. complete

8. An economizer ___.
   A. recycles water for the boiler
   B. heats feedwater with the gases of combustion
   C. heats air for combustion
   D. analyzes gases of combustion

9. The equation ___ can be used to determine the number of poles in a three-phase generator that operates at 3600 RPM at a frequency of 60 Hz.

   A. $Poles = \dfrac{RPM \times Frequency}{120}$

   B. $Poles = \dfrac{120 \times Frequency}{RPM}$

   C. $Poles = \dfrac{120 - Frequency}{RPM}$

   D. $Poles = \dfrac{RPM - Frequency}{120}$

10. A steam trap blowing steam into the condensate line ___.
   A. lowers plant efficiency
   B. increases plant efficiency
   C. prevents water hammer
   D. would become steambound

11. ___ results in difficulty maintaining boiler pressure and steam coming from the chimney.
   A. A lifted safety valve
   B. Over-firing of the boiler
   C. A leak in the boiler tube
   D. A dirty burner tip

12. A boiler can be hydrostatically tested using water at ___°F.
   A. 60
   B. 100
   C. 180
   D. 212

13. When entering a boiler room, the ___ is/are checked first.
   A. steam pressure
   B. water level
   C. auxiliaries
   D. log sheet

14. A ___ valve can be used for throttling service.
   A. check
   B. gate
   C. globe
   D. ball

15. Water hammer is dangerous because it can ___.
    A. loosen tubes in the boiler
    B. affect the water level
    C. cause pressure to fluctuate
    D. lead to steam pipe rupture

16. A ___ valve is used to limit flow to one direction.
    A. check
    B. gate
    C. globe
    D. ball

17. To replace a handhole gasket on a boiler, it is not necessary to ___.
    A. clean both metal surfaces
    B. use the proper gasket
    C. assure proper seating
    D. maintain the NOWL

18. Steam coming from a staybolt indicates ___.
    A. a broken or corroded staybolt
    B. pressure is too high
    C. not enough scale formed on the bolt
    D. there is excessive pressure in the boiler

19. Try cocks on a boiler ___.
    A. vent the air from drums
    B. are used to check the steam pressure
    C. are a secondary means of determining water level
    D. are used to remove water samples for chemical testing

20. A deaerator in a boiler feedwater system removes ___.
    A. combustibles
    B. total dissolved solids
    C. dissolved gases
    D. CO_2 only

21. A pressure gauge indicates 2 psig. The equivalent pressure on an absolute scale is ___ psia.
    A. 14.7
    B. 15.7
    C. 16.7
    D. 17.7

22. ___ is carryover of water from a boiler to the main steam line.
    A. Flashing
    B. Foaming
    C. Pitting
    D. Priming

23. A ___ is commonly used to measure draft.
    A. pressure gauge
    B. manometer
    C. pyrometer
    D. hygrometer

24. Alkalinity of boiler water may be raised by adding ___.
   A. caustic soda
   B. sulfite
   C. phosphate
   D. amine

25. If a steaming boiler gives a low water alarm and the gauge glass shows no water, ___.
   A. increase the feedwater pressure
   B. open the bypass on feedwater regulator
   C. secure fires, remove from line, and after cooling, inspect for damage
   D. decrease the firing rate until the water level returns

Test 10

1. The equation for determining the approximate Btu content of a fuel oil is ___.
   A. $17,780 + (54 \times °API)$
   B. $17,780 \times (54 \times °API)$
   C. $17,780 \times (54 + °API)$
   D. $17,780 + (54 + °API)$

2. The torque on a 300 HP motor running at 1800 RPM at maximum load is ___ ft-lb.
   A. 763
   B. 875
   C. 905
   D. 975

3. ___ can cause a hard, brittle scale deposit on the heating surfaces of a boiler.
   A. Sodium sulfate
   B. Soda ash
   C. Calcium carbonate
   D. Sodium carbonate

4. Two identical centrifugal pumps pumping in series ___.
   A. double the discharge pressure
   B. double the discharge pressure and volume
   C. double the volume
   D. would not affect pressure or volume

5. Exhaust steam used for feedwater heating results in an increase in ___.
   A. boiler efficiency
   B. plant efficiency
   C. boiler and plant efficiency
   D. steam temperature

6. The steam and water drum of a watertube boiler is 5′ in diameter and has a bursting pressure of 2500 psi. The drum metal has a tensile strength of 65,000 psi with a joint efficiency of 100%. The thickness of the drum metal is ___″.
   A. 1
   B. 1¼
   C. 1½
   D. 1¾

7. A throttling calorimeter is used to measure the quality of ___.
   A. saturated steam
   B. dry steam
   C. superheated steam
   D. the feedwater before it enters the boiler

8. To burn 1 lb of coal, ___ lb of air are required.
   A. 5
   B. 10
   C. 15
   D. 20

9. Ten cu ft of gas at 70°F is heated to 600°F. The final volume of the gas, if the gas pressure is kept constant, is ___ cu ft.
   A. 12.5
   B. 15.0
   C. 17.5
   D. 20.0

10. Five degrees Fahrenheit is equal to ___°C.
   A. −15
   B. −10
   C. 10
   D. 15

11. A bottom blowdown is used to ___.
   A. remove sludge and sediment
   B. lower the steam pressure
   C. take samples of boiler water
   D. add water to the boiler

12. The ends of boiler tubes are secured to headers by ___.
   A. welding
   B. brazing
   C. expanding
   D. sealing rings

13. A scotch marine boiler can be easily recognized because it ___.
   A. is internally fired
   B. is externally fired
   C. has watertubes
   D. has a brick setting

14. The speed of the cup of a motor-driven rotary cup burner is approximately ___ RPM.
   A. 1200
   B. 1800
   C. 2400
   D. 3450

15. A feedwater injector is used on a high-pressure boiler to ___.
   A. reduce surface tension
   B. add chemicals to the boiler water
   C. add water to the boiler
   D. remove air and noncondensable gases from the boiler

16. Atomization of fuel oil in a rotary cup burner occurs using a(n) ___.
   A. rotating cup and primary air
   B. rotating cup and secondary air
   C. rotating cup
   D. air nozzle

17. Staybolts are most commonly used in the construction of ___ boilers.
   A. watertube
   B. firetube
   C. Stirling
   D. steam

18. Oxygen in boiler feedwater causes ___.
   A. scaling of tubes
   B. foaming on the water surface
   C. priming of the boiler water
   D. pitting of drums and tubes

19. If a steaming boiler gives a low water alarm and the gauge glass shows no water, ___.
   A. increase the feedwater pressure
   B. open the bypass on feedwater regulator
   C. secure fires, remove from line, and after cooling, inspect for damage
   D. decrease the firing rate until the water level returns

20. The life of furnace refractory will be extended if a furnace is equipped with ___.
   A. water legs
   B. waterwalls
   C. baffles
   D. an ignition arch

21. An economizer is used in a steam plant to ___.
   A. increase the temperature of the feedwater
   B. increase the temperature of the steam
   C. increase the quality of the steam
   D. remove impurities in the feedwater

22. An electronic combustion tester is used to measure ___ in gases of combustion.
   A. CO_2
   B. CO
   C. O_2
   D. all of the above

23. The ASME code states that the water column connections to the boiler must be at least ___″ in diameter.
   A. ½
   B. ¾
   C. 1
   D. 1¼

24. Draft produced by placing a fan between the boiler and chimney is ___ draft.
   A. induced
   B. forced
   C. balanced
   D. natural

25. High and low water level alarm floats are found in the ___.
  A. blowdown tank
  B. water column
  C. open feedwater heater
  D. closed feedwater heater

ESSAY

Test 1

1. List three types of feedwater regulators. Describe how they function.

2. List three types of steam traps. Describe how they function.

3. When are two or more safety valves required on a boiler?

4. How is feedwater distributed in a watertube boiler and an HRT boiler?

5. What procedure is taken if the water level in the gauge glass is out of sight on a coal-fired boiler?

6. Define *cavitation* and explain where it could occur.

7. How is the number of poles in an AC generator determined?

8. What information is commonly found on a safety valve data plate?

9. What is a steam separator and where is it located?

10. What is the function and location of an atmospheric valve?

Test 2

1. How many and what type of valves are used on a boiler bottom blowdown line?

2. A simplex reciprocating pump with a 4" diameter water cylinder and 6" stroke operates at 50 strokes/min. How many gpm will the pump discharge?

3. What are possible causes for a surface condenser to gradually lose its vacuum?

4. What is the approximate water rate of a 10,000 kW turbo-generator? What BHP is needed for this turbo-generator?

5. What is an economizer and where is it located?

6. What is the function of baffles on a watertube boiler? What would indicate a broken baffle?

7. How many square feet of boiler heating surface are there in a straight-tube watertube boiler with 350 tubes that are 3" in diameter and 20' long?

8. Explain the procedure to follow if a steam line ruptures going to the steam turbine.

9. How much cooling water is needed on a surface condenser connected to a 10,000 kW, straight, condensing turbo-generator? (Assume all values.)

10. Explain the procedure to follow if the feedwater pump discharge pressure drops to boiler pressure.

Test 3

1. What procedure is followed if a safety valve fails to open at its set pressure?

2. What is the function of a feedwater injector?

3. A 10,000 kW turbo-generator operates on 450 psi steam pressure at 750°F and exhausts into a surface condenser with a 28″ Hg vacuum. What is the thermal efficiency of the turbo-generator?

4. Determine the bursting pressure of a boiler drum with the following: tensile strength of metal—60,000 psi; metal thickness—1¾″; drum diameter—5′; efficiency of joint—90%.

5. How are draft fans driven and controlled?

6. Why can boiler tubes be of thinner metal than the heads or shell while still carrying the same steam pressure?

7. What is the function of steam traps used in a steam system? Where are they located?

8. What automatic system is used to control the steam pressure on a boiler? How does it function and what safety devices are used?

9. Define the following terms:

   a. Btu

   b. Specific heat

   c. Boiling point of water

   d. Absolute pressure

   e. Absolute temperature

   f. Absolute zero temperature

10. Describe in detail the meaning of *blowdown of a safety valve.*

Test 4

1. A duplex reciprocating pump has $10 \times 8 \times 6$ on the data plate. What do the numbers mean and how are they used?

2. What is the approximate Btu content per pound of fuel oil with a specific gravity of 25.2°API at 60°F?

3. A centrifugal pump discharges 100 gpm against a head of 250′. What HP is required to drive this pump? Do not allow for friction losses.

4. Describe the steam and water circulation on various types of boilers.

5. Define boiler efficiency, steam turbine efficiency, and plant efficiency. Create and solve a problem for each.

6. Define the following terms:

   a. Volatile

   b. Critical speed

   c. Ultimate analysis

   d. Synchronization

   e. Atomization

7. A boiler generates 150,000 lb of steam/hr from feedwater at 230°F. What is the DBHP if the enthalpy of steam is 1200 Btu/lb?

8. A 50 HP electric motor is 85% efficient. What is the input in watts per HP? What is the total usable power in watts at full load?

9. Is the superheater outlet header safety valve or the steam and water drum safety valve set lower? Why?

10. What happens to the water level on watertube boilers when the firing rate is increased or decreased rapidly? Explain the cause.

Test 5

1. Describe how to replace a round gauge glass while the boiler is in service.

2. Find the heating surface of a firetube and watertube 1' long, 3" in diameter, with a wall thickness of $\frac{3}{16}''$.

3. Define *factor of evaporation*. Assuming any values necessary, set up and solve for the factor of evaporation.

4. A boiler is rated for 100,000 lb/hr but delivers only 90,000 lb/hr. List the probable causes.

5. Define the following terms:

   a. Radiant heat

   b. Convection

   c. Conduction

   d. Latent heat

   e. Sensible heat

   f. Perfect combustion

   g. Complete combustion

   h. Incomplete combustion

6. When preparing a boiler for inspection, what procedures must be followed?

7. Describe steam and condensate flow and circulating water flow in a surface condenser.

8. What can cause the temperature of gases of combustion to increase?

9. Why is a watertube boiler safer to operate than a firetube boiler?

10. Convert the following to degrees Celsius:

   a. 140°F

   b. 100°F

   c. 40°F

   d. 10°F

   e. −12°F

   f. −32°F

Test 6

1. Two boilers are on-line and are cycling ON and OFF. What effect does this have on boiler efficiency and boiler maintenance?

2. A plant has two 600 HP boilers. One operates at 300 psi and the other at 600 psi. Which boiler needs the larger safety valve area and why?

3. What is an atmospheric relief valve, where is it located, and how is it constructed?

4. On the steam side of a reciprocating pump, is the steam inlet or the exhaust outlet larger?

5. What can cause a high water level in a boiler?

6. Explain how thrust is controlled on a steam turbine.

7. What is the maximum and minimum size of a bottom blowdown valve and surface blowdown valve?

8. Define *power factor*.

9. What is the total force on a steam drum 40″ in diameter, 20′ long, with a 450 psi steam pressure?

10. Describe the procedure for laying up a boiler for a short period and long period.

Test 7

1. Describe the function of a hydraulic governor on a steam turbine.

2. What is the water rate of a steam turbine that operates condensing? What is the water rate if it operates noncondensing?

3. What effect is there in raising the boiler pressure from 250 psi to 300 psi?

4. What maintenance is needed on a centrifugal feedwater pump?

5. What are possible causes of tube ruptures in watertube boilers?

6. If a steam line ruptures in a steam turbine room, what would be the effects on the operating steam turbine?

7. What limits the degrees of superheat on steam turbines?
8. A boiler operates at 400 psia and the temperature of the feedwater entering the boiler is 180°F. It generates 12 lb of steam for each lb of fuel burned, and the fuel has a Btu content of 18,600 Btu/lb. What is the thermal efficiency of the boiler?
9. Define the following terms:
   a. Fire point
   b. Pour point
   c. Flash point
   d. Viscosity
10. What are the most common causes of boiler accidents and explosions?

Test 8

1. What are advantages of using a steam atomizing burner?
2. What is the belt speed on an engine with a 6′ diameter pulley running at 150 RPM?
3. What are the most common causes of boiler accidents and explosions?
4. What reverses the operation of a simplex pump at the end of its stroke?
5. Is a reciprocating or centrifugal pump most likely to prime itself?
6. Describe the test used on a boiler to determine the pressure-relieving capacity of the safety valves.
7. If a steam line ruptures in a steam turbine room, what would be the effects on the operating steam turbine?
8. What is the purpose of preheating No. 6 fuel oil?
9. What is the maximum and minimum size of a bottom blowdown valve and surface blowdown valve?
10. What is the principle of operation of an impulse steam turbine?

Test 9

1. Is a reciprocating or centrifugal pump most likely to prime itself?
2. What is the purpose of preheating No. 6 fuel oil?
3. What are possible causes for a decrease in boiler efficiency?
4. What is the principle of operation of a reaction steam turbine?
5. What type of governor is used on a steam turbine-driven centrifugal pump?
6. A reaction steam turbine and an impulse steam turbine have the same horsepower. Which is the larger machine and why?
7. How does a mechanical governor control the speed of a steam turbine?

8. In a process plant, is it more efficient to run a steam turbine that is generating electricity condensing or noncondensing?

9. What action should be taken if there is a leak between the boiler and the bottom blowdown valves?

10. Define *induced, forced,* and *natural draft.*

Test 10

1. What is the principle of operation of a reaction steam turbine?

2. Describe the test used on a boiler to determine the pressure-relieving capacity of the safety valves.

3. In a process plant, is it more efficient to run a steam turbine that is generating electricity condensing or noncondensing?

4. What is the function of a low water fuel cutoff on an automatic package boiler? How is the low water fuel cutoff tested?

5. What can occur if fuel oil is present in the water side of a boiler?

6. List the auxiliaries connected with the operation of a condensing steam turbine and explain the operation of each.

7. How is a steam turbine generator that is running condensing started?

8. List the parts of a centrifugal pump and explain the purpose of each. Which parts wear the fastest?

9. What are the advantages of superheated steam over saturated steam when used in a steam turbine?

10. What is the specific volume of steam at 200 psi and at 28″ vacuum?

CHAPTER 13 LEARNER RESOURCES

ATPeResources.com/QuickLinks
Access Code: 735728

Chapter 14, Licensing Examination Preparation—First Class, consists of questions based on sample questions obtained from licensing agencies across the United States and Canada. First class refers to the highest level of licensing examination for stationary engineers. The name of this examination may vary from one licensing agency to another. At this level, the stationary engineer is responsible for the largest plants with more sophisticated equipment than second-class stationary engineers. Examination questions emphasize applying variations of mathematical equations and applied physics and engineering principles. This level examination typically contains mostly essay questions.

MULTIPLE CHOICE

Test 1

1. Heating value is determined by ___.
   A. flue gas analysis
   B. finding specific gravity
   C. ultimate analysis and Dulong's formula
   D. proximate analysis and Dulong's formula

2. A boiler has an MAWP of 111 psi. The boiler joint efficiency is 80%, tensile strength of metal is 50,000 psi, factor of safety is 5, and boiler shell diameter is 72″. The thickness of the metal plate is ___″.
   A. 0.25
   B. 0.50
   C. 0.75
   D. 1.00

3. In steam tables, heat in liquid (h) equals ___.
   A. boiling point in °F minus 32
   B. boiling point in °F plus 32
   C. latent heat plus sensible heat
   D. total heat less latent heat

4. A firetube boiler has an MAWP of 222 psi. The boiler joint efficiency is 80%, tensile strength of metal is 50,000 psi, factor of safety is 5, and thickness of metal plate is 0.5″. The boiler shell diameter is ___″.
   A. 36
   B. 54
   C. 72
   D. 90

5. Scale buildup in a superheater causes ___.
   A. superheater tubes to overheat
   B. superheater tubes to cool
   C. superheater discharge temperature to increase
   D. a drop in steam pressure

6. ___ results in a high O_2 reading in gases of combustion.
   A. Complete combustion
   B. Incomplete combustion
   C. Perfect combustion
   D. Too much excess air

7. By keeping gases of combustion passing through the air heater at a temperature higher than the dew point of gases of combustion, it is possible to ___.
   A. prevent air losses
   B. prevent corrosion
   C. decrease fuel consumption
   D. decrease chimney temperatures

8. A(n) ___ is located on the inlet line of a steam turbine condenser.
   A. atmospheric relief valve
   B. check valve
   C. air vent
   D. vacuum breaker

9. Sodium sulfite used in a boiler drum reduces ___.
   A. scale formation
   B. caustic embrittlement
   C. channeling
   D. oxidation of metal

10. A boiler is 85% efficient and uses 875 gal./hr of fuel oil to produce 100,000 lb/hr of steam. To generate 250,000 lb/hr, ___ gal./hr of fuel oil are needed.
    A. 21.87
    B. 218.7
    C. 2187.5
    D. 21,875.0

11. Plant efficiency can be increased when ___ steam is used to heat boiler feedwater.
    A. live
    B. saturated
    C. exhaust
    D. condensed

12. An economizer uses heat from ___ to heat feedwater.
   A. exhaust steam
   B. live steam
   C. bleed steam
   D. gases of combustion

13. Safety valves are rated ___.
   A. in Btu per hour
   B. at 1½ times the MAWP
   C. in pounds of steam per hour at a given pressure
   D. in percent of blowdown

14. The thermal efficiency of a 5000 kW turbo-generator operating at 600 psia with a steam temperature of 700°F and an exhaust pressure of 1 psia is ___%.
   A. 15.2
   B. 19.4
   C. 51.8
   D. 81.7

15. A boiler operates at 120 psi and has 1.25″ diameter staybolts spaced 6″ on centers horizontally and vertically. The load supported by the staybolts is ___ lb.
   A. 150
   B. 720
   C. 2160
   D. 4320

16. The number of poles on an AC generator is determined by using the following equation:

   A. $No.\,of\,poles = \dfrac{Frequency + 120}{RPM}$

   B. $No.\,of\,poles = \dfrac{Frequency \times 120}{RPM}$

   C. $No.\,of\,poles = \dfrac{Frequency \times RPM}{120}$

   D. $No.\,of\,poles = \dfrac{Frequency + RPM}{120}$

17. Flyash and soot are removed from gases of combustion going up the chimney by ___.
   A. hydrostatic precipitators
   B. accumulator precipitators
   C. electrostatic precipitators
   D. flyash and soot traps

18. Bituminous coal requires large amounts of overfire air because it has a ___ content.
   A. high fixed carbon
   B. low fixed carbon
   C. low volatile
   D. high volatile

19. A steam turbine uses 60,000 lb of steam/hr and exhausts into a surface condenser at 28½″ Hg vacuum. It is connected to a generator with an output of 4500 kW. The water rate is ___ lb/kW.
  A. 10.33
  B. 11.33
  C. 12.33
  D. 13.33

20. A straight-tube watertube boiler with five hundred fifty 3″ diameter tubes that are 22′ long has ___ sq ft of heating surface.
  A. 7500.43
  B. 8205.62
  C. 9503.34
  D. 11,212.57

Test 2

1. A(n) ___ is used to reduce pulsation and maintain a constant pressure on the discharge side of a reciprocating pump.
  A. air chamber
  B. accumulator
  C. hydraulic chamber
  D. regulating valve

2. A straight-tube watertube boiler with five hundred fifty 3″ diameter tubes that are 18′ long has ___ sq ft of heating surface.
  A. 648
  B. 2475
  C. 7775
  D. 93,300

3. A chain grate stoker most commonly burns coal with a ___ content.
  A. low ash
  B. high ash
  C. low clinker
  D. high clinker

4. The torque on the shaft of a 500 HP steam turbine running at full load at 3600 RPM is approximately ___ ft-lb.
  A. 650
  B. 730
  C. 825
  D. 930

5. Bituminous coal requires large amounts of overfire air because it has a ___ content.
  A. high fixed carbon
  B. high volatile
  C. low fixed carbon
  D. low volatile

6. The amount of moisture in saturated steam can be determined by using a(n) ___.
   A. throttling calorimeter
   B. Orsat analyzer
   C. pyrometer
   D. Fyrite® analyzer

7. Approximately ___ Btu are required to raise the temperature of 100 lb of water from 60°F to 180°F.
   A. 1200
   B. 12,000
   C. 18,000
   D. 99,960

8. The bursting pressure of a 6′ diameter boiler drum that has a tensile strength of 65,000 psi, metal thickness of 1½″, and joint efficiency of 90% is ___ psi.
   A. 692
   B. 1384
   C. 2438
   D. 4283

9. A boiler has a factor of evaporation of 1.04 and is rated at 904 BHP. This boiler produces approximately ___ lb of steam/hr.
   A. 10,000
   B. 20,000
   C. 25,000
   D. 30,000

10. ___ is the comparison of heat energy (enthalpy) supplied to the steam turbine to heat energy used by the steam turbine.
   A. Heating value
   B. Entropy of the turbine
   C. Mechanical efficiency of the steam turbine
   D. Thermal efficiency of the steam turbine

11. When burning pulverized coal, a ___ temperature must be maintained.
   A. low secondary
   B. low furnace
   C. high furnace
   D. low primary air

12. In steam tables, heat in liquid (h) equals ___.
   A. boiling point in °F minus 32
   B. boiling point in °F plus 32
   C. latent heat plus sensible heat
   D. total heat less latent heat

13. A centrifugal pump discharges 150 gpm against a head of 300′. The amount of HP needed to drive this pump is ___ HP. Do not allow for friction losses.
   A. 11.36
   B. 13.63
   C. 15.98
   D. 20.02

14. A boiler generates 200,000 lb of steam/hr from feedwater at 220°F with a factor of evaporation of 1.04. The DBHP is ___.
  A. 909
  B. 5252
  C. 6029
  D. 7590

15. ___ air supplied to a furnace controls the rate of combustion.
  A. Overfire
  B. Primary
  C. Secondary
  D. Combustion

16. Degrees API refers to ___ of fuel oil.
  A. viscosity
  B. enthalpy
  C. density
  D. specific gravity

17. New boilers and boilers that have had extensive repair work done on their steam or water side should have a(n) ___ test performed.
  A. accumulation
  B. X-ray
  C. hydrostatic
  D. electrostatic

18. A(n) ___ test is performed on a boiler to check the operation of the low water fuel cutoff by allowing the water level in the boiler to drop slowly.
  A. hydrostatic
  B. evaporation
  C. accumulation
  D. low water

19. A ___ governor is used on a steam turbine where close control of RPM is required.
  A. flyball
  B. flywheel
  C. hydroelectric
  D. hydraulic

20. ___ is/are used to prevent leakage between stages of steam turbine blades.
  A. Steam seals
  B. Labyrinth packing
  C. Water seals
  D. Both water and steam seals

Test 3

1. ___ air supplied to a furnace controls the rate of combustion.
  A. Overfire
  B. Primary
  C. Secondary
  D. Combustion

2. The ASME code states that a boiler pressure gauge should have a range of two times the safety valve setting, but not less than ___ the safety valve setting.
   A. 1½ times
   B. 1¼ times
   C. 2% of
   D. 8% of

3. In a jet condenser, ___.
   A. water flows in the tubes while steam surrounds the tubes
   B. steam flows in the tubes while water surrounds the tubes
   C. water and steam mix
   D. condensate can be reclaimed

4. If the staybolt diameter is doubled, the holding power will ___.
   A. be the same
   B. double
   C. triple
   D. quadruple

5. The heating surface of a firetube 1′ long, 3″ in diameter, with a wall thickness of ³⁄₁₆″ is ___ sq ft.
   A. 0.343
   B. 0.687
   C. 6.871
   D. 10.62

6. Seven degrees Fahrenheit is equal to ___°C.
   A. −13.9
   B. −9.1
   C. 3.1
   D. 13.9

7. Approximately ___ Btu are required to raise the temperature of 100 gal. of water from 60°F to 180°F.
   A. 1200
   B. 9996
   C. 12,000
   D. 99,960

8. The intensity of heat is measured ___.
   A. in Btu
   B. as sensible heat
   C. as latent heat
   D. with a thermometer

9. Two boilers with the same capacities are operating at 300 psi and 600 psi, respectively. The approximate ratio of the areas of their safety valves is ___.
   A. 1:2
   B. 1:3
   C. 2:1
   D. 3:1

10. Steam turbines are commonly designed to have steam velocity ___ times as great as blade velocity.
   A. 1½
   B. 2
   C. 2½
   D. 3

11. Axial thrust in a steam turbine is not controlled by a ___.
   A. counterflow turbine
   B. Kingsbury thrust bearing
   C. balance piston
   D. surface blowdown valve

12. A sudden increase in chimney temperature with no increase in firing rate indicates ___.
   A. soot buildup on tubes
   B. a low water condition
   C. scale on the water side
   D. broken baffles on the fire side

13. If the feedwater temperature is increased without a change in the firing rate, the steam ___.
   A. flow will decrease
   B. flow will increase
   C. flow will remain constant
   D. temperature will increase

14. The output of two similar centrifugal pumps connected in parallel is ___.
   A. constant pressure, double volume
   B. no change in pressure or volume
   C. double pressure, constant volume
   D. constant pressure, no change in volume

15. The maximum size bottom blowdown valve on a boiler is ___".
   A. 1.25
   B. 1.50
   C. 2.00
   D. 2.50

16. A thermoexpansion feedwater regulator operates on ___.
   A. movement generated by the thermohydraulic diaphragm
   B. changes in density of water and steam
   C. positive displacement
   D. expansion and contraction of a thermostatic element

17. The steam and water drum of a boiler is 4′ in diameter. The force on the drum head is ___ lb if it operates at 200 psi.
   A. 130,469
   B. 260,542
   C. 361,912
   D. 543,822

18. A boiler explosion is more likely in a(n) ___ boiler.
   A. cross drum
   B. HRT
   C. bent-tube watertube
   D. straight-tube

19. Failure to properly purge a boiler after an ignition failure could lead to ___.
   A. a boiler explosion
   B. spalling of the refractory
   C. a furnace explosion
   D. overheating of tubes in the first pass

20. Scale buildup on boiler heating surfaces is caused by ___.
   A. hard water
   B. soft water
   C. loss of condensate returns
   D. improper feedwater treatment

Test 4

1. The pressure at the base of a vertical mercury column 3′ high is ___ psi.
   A. 14.7
   B. 17.7
   C. 20.5
   D. 23.8

2. The minimum size connection between the gauge glass and water column is ___″.
   A. ¼
   B. ½
   C. ¾
   D. 1

3. A dry pipe in the steam and water drum ___.
   A. removes air from steam
   B. removes the total dissolved solids from the steam
   C. prevents foaming
   D. improves steam quality

4. To blow down a boiler equipped with a quick-opening and slow-opening valve on the bottom blowdown line, the ___.
   A. quick-opening valve is opened first and closed last
   B. quick-opening valve is opened first and closed first
   C. slow-opening valve is opened first and closed first
   D. order of opening valves is rotated to allow even wear

5. The main purpose of the water column on a boiler is to ___.
   A. show the water level in the boiler
   B. provide a mount for attaching the gauge glass and try cocks
   C. reduce turbulence of water in the gauge glass
   D. a water column is found only on firetube boilers

6. A ___ is a device used to remove soot deposits from around tubes and to permit better heat transfer in the boiler.
 A. tube polisher
 B. soot removal agent
 C. soot blower
 D. safety blower

7. Two or more safety valves are required on a boiler ___.
 A. operating over 15 psi
 B. operating over 100 psi
 C. with over 250 sq ft of heating surface
 D. with over 500 sq ft of heating surface

8. A steam trap blowing steam into the condensate line can cause ___.
 A. lower plant efficiency
 B. increased plant efficiency
 C. water hammer in the water column
 D. the condensate pump to become steam bound

9. ___ is carryover of water from a boiler to the main steam line.
 A. Flashing
 B. Foaming
 C. Pitting
 D. Priming

10. Failure to properly purge a boiler after an ignition failure could lead to ___.
 A. a boiler explosion
 B. a high water condition
 C. a furnace explosion
 D. overheating of the superheater

11. The main function of a condenser is to ___.
 A. increase the steam turbine water rate
 B. decrease the steam turbine water rate
 C. raise the cooling water temperature
 D. increase the steam turbine back pressure

12. A ___ is a large auxiliary device where air moves upward to mix with falling water, resulting in cooling of the water.
 A. cooling tower
 B. surface condenser
 C. spray pond
 D. vacuum condenser

13. A ___ is commonly used to measure furnace pressure.
 A. pressure gauge
 B. manometer
 C. pyrometer
 D. hygrometer

14. Furnace refractory life will be extended if the furnace is equipped with ___.
 A. baffles
 B. water legs
 C. waterwalls
 D. an ignition arch

15. Boilers in battery must have two main steam stop valves and a free-blowing drain between them if the boilers are equipped with ___.
    A. manhole openings
    B. handhole openings
    C. common blowdown lines
    D. continuous blowdown lines

16. When the temperature of condenser cooling water increases with no increase in volume, the vacuum ___.
    A. increases
    B. decreases
    C. remains the same
    D. first decreases then increases

17. The maximum velocity of water in feedwater lines is limited to ___.
    A. 500 ft/min
    B. 500 ft/sec
    C. the same velocity as steam leaving the boiler
    D. the pump capacity in gpm

18. ___ is the process where the water level in a boiler momentarily drops with a decrease in steam demand.
    A. Compression
    B. Expansion
    C. Shrink
    D. Swell

19. The continuous blowdown line is located ___.
    A. at the NOWL
    B. in the mud drum
    C. in the flash tank well below the water column
    D. in the steam and water drum well below the NOWL

20. Water in a deaerator feedwater heater is prevented from entering the steam line by a(n) ___.
    A. check valve
    B. stop valve
    C. loop seal or internal overflow
    D. atmospheric relief valve

Test 5

1. An overspeed trip on a steam turbine that operates at 3600 RPM would activate at ___ RPM.
    A. 3650
    B. 3700
    C. 3750
    D. 3960

2. The Btu content of natural gas is ___.
   A. 100,000 Btu/therm
   B. 150,000 Btu/therm
   C. 100,000 Btu/cu ft
   D. 150,000 Btu/cu ft

3. The ASME code recommends that the low water fuel cutoff be tested by an evaporation test ___.
   A. daily
   B. weekly
   C. monthly
   D. yearly

4. A lead sulfite cell detects ___.
   A. burner pilot and main flame
   B. ultraviolet rays
   C. X-rays
   D. steam quality

5. When warming up a boiler, the air cock should be closed when ___.
   A. steam starts to come out of the air cock
   B. the boiler pressure gauge reads about 15 psi to 25 psi
   C. the boiler pressure gauge reads 100 psi
   D. as soon as the boiler is put on-line

6. Waterwall tubes ___.
   A. act as stays on tube sheets
   B. allow blowdown while the boiler is operating
   C. help control water circulation
   D. increase boiler heating surface

7. Extra-heavy pipe must be used on blowdown lines when the steam pressure is over ___ psi.
   A. 15
   B. 100
   C. 125
   D. 200

8. The only type of safety valve allowed by the ASME code on a boiler is a ___ safety valve.
   A. spring-loaded pop-type
   B. spring-loaded check
   C. deadweight
   D. ball and lever

9. As the steam pressure in a boiler increases, there is a corresponding increase in the ___.
   A. boiler horsepower
   B. water level
   C. temperature of steam and water
   D. volume of steam

10. The maximum velocity of water in feedwater lines is limited to ___.
    A. 500 ft/min
    B. 500 ft/sec
    C. the same velocity as steam leaving the boiler
    D. the pump capacity in gpm

11. A steam turbine uses a total of 40,000 lb of steam/hr while generating 2000 kW. The water rate is ___ lb of steam/kWh.
    A. 10
    B. 15
    C. 20
    D. 25

12. To determine the total dissolved solids in suspension in boiler water, a(n) ___ meter is used.
    A. electrostatic
    B. conductivity
    C. thermocouple
    D. hydrostatic

13. Most furnace explosions are caused by ___.
    A. failure to purge the furnace at startup
    B. overpurging the furnace at startup
    C. mechanical failure
    D. damaged refractory

14. The overspeed tripping device on a steam turbine must be tripped when ___.
    A. the steam turbine is 30% over set speed
    B. shutting down the steam turbine
    C. under load
    D. the steam turbine is 10% below set speed

15. An accumulation test is used to determine ___.
    A. heating surface
    B. furnace volume
    C. evaporation rate
    D. safety valve relieving capacity

16. Fuel oil with a high viscosity ___ than fuel oil with a low viscosity.
    A. has a lower firing rate
    B. is easier to pump
    C. is harder to pump
    D. is not as commonly used in boilers

17. Steam expands ___ times when going through a steam turbine from 400 psia to a 29″ vacuum. *Note:* 400 psia is 1.16 cu ft/lb and 29″ vacuum is 633 cu ft/lb.
    A. 246
    B. 346
    C. 546
    D. 646

18. As much as ___ lb of water are required to condense 1 lb of steam, depending on ambient air temperature and cooling water temperature.
    A. 10
    B. 15
    C. 35
    D. 50

19. When heating feedwater with exhaust steam or gases of combustion, there is a 1% increase in plant efficiency for every ___°F increase in feedwater temperature.
    A. 10
    B. 15
    C. 20
    D. 25

20. If valves are installed between the boiler and water column, they must be ___.
    A. globe valves
    B. 250 psi standard valves
    C. os&y or lever-type, locked or chained open
    D. os&y or lever-type, locked or chained closed

ESSAY

Test 1

1. What is the thermal efficiency of a straight condensing steam turbine? Enthalpy of steam at throttle is 1350 Btu/lb, enthalpy of steam at exhaust is 1050 Btu/lb, and enthalpy of condensate is 102 − 32.

2. Explain *ultimate analysis* and *proximate analysis*.

3. Explain how a water column is installed on a boiler.

4. How is an internal and external boiler inspection done?

5. How many pounds of water does 1 gal. of No. 6 fuel oil evaporate? Heat available in 1 gal. of No. 6 fuel oil is 146,000 Btu/gal., heat in steam leaving the boiler is 1200 Btu/lb, temperature of feedwater entering boiler is 210°F, and boiler efficiency is 82%.

6. Prove that there is a point on the Celsius and Fahrenheit scales where they read the same.

7. Describe the purpose and location of the following:

   a. Continuous blowdown heat recovery system

   b. Total dissolved solids

   c. Dry pipe

   d. Scrubbers

8. What are the harmful effects of scale on the heating surface of a boiler? Explain internal and external feedwater treatment for boilers.

9. A 60″ diameter boiler drum has ¾″ thick metal and a tensile strength of 60,000 psi. The joint efficiency is 90% and MAWP is 300 psi. What is the factor of safety?

10. A boiler develops 5000 BHP. The temperature of the feedwater entering the boiler is 225°F and the enthalpy of the steam leaving the boiler is 1200 Btu/lb. How many pounds of steam per hour is this boiler generating?

11. Explain how to install a steam pressure gauge on a boiler.

12. What is the function of an air ejector?

13. How many Btu are equivalent to

    a. 1 mechanical HP

    b. 1 BHP

    c. 1 kW

14. Define *centrifugal* and *centripetal force* in reference to a steam turbine.

15. Define *factor of evaporation* and explain how it is used.

16. What is the purpose of a throttling calorimeter in a steam plant?

17. When two similar centrifugal pumps are connected in series, what happens to the discharge pressure and flow?

18. Is an electrically driven feedwater pump or a steam turbine-driven feedwater pump more economical to operate?

19. What types of boilers are equipped with waterwalls and what purpose do they serve?

20. What is boiler heat balance?

Test 2

1. Explain *ultimate analysis* and *proximate analysis*.

2. Give a brief description of the following valves:

    a. Atmospheric

    b. Nonreturn

    c. Slide

    d. Relief

    e. Pilot

    f. Needle

3. How does the condition of refractory in a boiler furnace affect the boiler efficiency?

4. A flat boiler surface is supported by 1″ diameter staybolts. The staybolts are allowed a maximum stress of 7500 psi. If the boiler is to operate at an MAWP of 200 psi, what would be the maximum pitch or spacing of the staybolts?

5. What routine maintenance is required to keep a boiler in good condition?

6. What is the total force acting on the inside surface of a 60″ diameter boiler drum that is 18′ long and under a pressure of 400 psi?

7. Describe the steps for paralleling two AC generators.

8. Prove that the stress on the longitudinal seam of a boiler drum is twice as great as the stress on the circumferential seam.

9. Explain how a water column is installed on a boiler.

10. How many pounds of water does 1 gal. of No. 6 fuel oil evaporate? Heat available in 1 gal. of No. 6 fuel oil is 146,000 Btu/gal., heat in steam leaving the boiler is 1200 Btu/lb, temperature of feedwater entering the boiler is 180°F, and boiler efficiency is 83%.

11. What types of boilers are equipped with waterwalls and what purpose do they serve?

12. A 60″ diameter boiler drum has a tensile strength of 60,000 psi, boiler seam efficiency of 90%, factor of safety of 5, and an MAWP of 250 psi. Find the thickness of the plate required.

13. How much does steam expand going from 200 psia to 28″ Hg vacuum (change in volume)?

14. Describe how the thermal efficiency of a steam turbine is affected when steam bled from various stages is used to heat feedwater.

15. How much torque is developed on the shaft of a 250 HP steam turbine operating at 1800 RPM at full load?

16. A boiler produces 80,000 lb of steam/hr at 250 psi. The steam has a heat content of 1200 Btu/lb. The feedwater entering the boiler is 210°F. What is the DBHP?

17. Why do firetube boilers require stays?

18. Define the following:

a. Tensile stress

b. Compression stress

c. Shear stress

19. Define the following:

a. Perfect combustion

b. Complete combustion

c. Incomplete combustion

20. Why do staybolts have holes drilled in the ends? What is the diameter and depth of the drilled holes?

Test 3

1. A 60″ diameter boiler drum has a tensile strength of 50,000 psi, boiler seam efficiency of 90%, factor of safety of 5, and an MAWP of 200 psi. Find the thickness of the plate required.

2. A 10′ long boiler has a pressure of 150 psi, 36″ drum diameter, and ⅝″ metal thickness. What is the stress on the circumferential and longitudinal seams?

3. When is the best time to blow down the waterwall on a watertube boiler? Explain in detail.

4. Describe the flow of steam from the boiler to a desuperheating station.

5. Describe the operation of a two-element feedwater regulator.

6. A condensate pump on a surface condenser operates with a 29″ vacuum suction and a 100′ head discharge. It pumps 50,000 lb of water/hr. If the unit is 85% efficient, how much HP is required to drive this unit?

7. A boiler produces 60,000 lb of steam/hr at 250 psi. The steam has a heat content of 1200 Btu/lb. The feedwater entering the boiler is 180°F. What is the DBHP?

8. How many pounds of steam are required to generate 1 kW of electricity?

9. How are expansion and contraction compensated for in main steam lines?

10. If 10 cu ft of air are heated at a constant pressure from 70°F to 335°F, what is the final volume?

11. How are the blowdown line and valves installed on a high-pressure boiler?

12. Three 2500 kW AC turbo-generators are running in parallel. If the governor on one fails toward the closed position, what would happen to each unit?

13. What size steam line is needed for a saturated steam boiler that carries 200 psi pressure and generates 60,000 lb of steam/hr? The velocity in the line must not exceed 5000 fpm. (The specific volume of steam at 200 psi is 2.15 cu ft/lb.)

14. Describe a chain grate stoker.

15. Where is a Kingsbury thrust bearing used?

16. What is the function of shroud rings?

17. How are boiler tubes connected to drums of watertube boilers? How are they connected to the tube sheet of firetube boilers?

18. What effect does an increase in firing rate have on the superheater outlet steam temperature?

19. What is the approximate evaporation rate when firing coal with a Btu content of 14,000 Btu/lb?

20. How many pounds of air are needed to burn 1 lb of fuel? How many cubic feet does that convert to?

Test 4

1. What effect does ON/OFF operation of a boiler have on boiler maintenance and boiler life?

2. In a steam turbine surface condenser, how much circulating water is needed per pound of steam condensed?

3. What is the purpose of staging in a reaction steam turbine and an impulse steam turbine?

4. How many pounds of steam are required to generate 1 kW of electricity?

5. Why are steam lines insulated?

6. How are expansion and contraction compensated for in main steam lines?

7. If 10 cu ft of air are heated at a constant pressure from 70°F to 600°F, what is the final volume?

8. Convert the following temperatures from Fahrenheit to Celsius:

   a. 0°F

   b. 30°F

   c. 75°F

   d. 150°F

9. Explain the function and installation of a pop-type safety valve.

10. A boiler produces 100,000 lb of steam/hr with a factor of evaporation of 1.06. What is the developed boiler horsepower?

11. A boiler is operating at a rate of 2000 DBHP. The boiler operates with an efficiency of 80% and is using a fuel that has a caloric value of 14,000 Btu/lb. The temperature of the feedwater entering the boiler is 230°F. Find the following:

    a. Amount of feedwater needed per hour

    b. Amount of fuel used per hour

    c. Approximate amount of combustion air needed per hour

12. What are the principal advantages of heating feedwater?

13. Two AC generators are running in parallel. What keeps them in step and running at the same frequency?

14. What factors are considered when selecting a drive for a feedwater pump?

15. Explain how a thermohydraulic feedwater regulator functions.

16. What are priming and foaming? How can these conditions be prevented?

17. What are the principal elements that compose coal, No. 6 fuel oil, and natural gas? Explain the combustion process of each type of fuel.

18. A boiler feedwater pump is feeding 18,500 lb of water/hr to the boiler. The discharge line water velocity is 400 fpm. What is the size of the discharge line? (*Note:* Water weighs 62.4 lb/cu ft.)

19. Describe an automatic combustion control system.

20. A boiler generates 30,000 lb of steam/hr at 400 psi. What is the approximate safety valve relieving area needed on this boiler? What area is needed if the pressure is reduced to 250 psi?

Test 5

1. How many adjustments can be made on a safety valve?

2. Can boilers be classified by the steam and water circulation?

3. Describe the principal parts of a coal-pulverizing system.

4. Describe the function of primary air and secondary air in the combustion process.

5. Why must induced draft fans be used with air heaters and economizers?

6. What can be done to improve the efficiency of a steam plant?

7. What routine on-line and off-line maintenance work is required on a boiler?

8. Describe two types of air heaters found on boilers.

9. A boiler feedwater pump is feeding 25,000 lb of water/hr to the boiler. The discharge line water velocity is 400 fpm. What is the size of the discharge line? (*Note:* Water weighs 62.4 lb/cu ft.)

10. How is an ignition arch used on a chain grate stoker?

11. A chain grate stoker 10′ wide is burning anthracite coal. The grate speed is 6″ per minute and the coal is 6″ deep. At what rate, in tons per hour, is coal being used? (*Note:* Anthracite coal weighs 60 lb/cu ft.)

12. What is the Btu content of 1 lb of fuel oil with a specific gravity of 20°API?

13. What problems are caused by oxygen and carbon dioxide in boiler water? How can these problems be eliminated?

14. Define the following terms:

    a. Saturated steam

    b. Superheated steam

    c. Quality of steam

    d. Enthalpy of steam

15. How is the outlet damper controlled on an automatic boiler?

16. When operating a boiler, what heat losses can be controlled?

17. A boiler generates 100,000 lb of steam/hr with an evaporation rate of 12.5 lb of steam/lb of fuel. Two hundred cubic feet of air are needed per pound of fuel, and the forced draft fan has a discharge pressure of 4″ of water. What is the HP needed to drive the forced draft fan?

18. Describe how a zeolite water softener works.

19. What are the advantages of using a continuous blowdown system?

20. List the parts of an AC generator and describe the function of each part.

Test 6

1. An AC generator has a frequency of 60 cycles/sec and is operating at 1800 RPM. How many poles does it have?

2. What is the purpose of a synchroscope?

3. How is load applied to an AC generator that is running in parallel with another unit?

4. Describe a front feed spreader stoker.

5. Describe the construction and function of a regenerative heater.

6. What are losses that can occur in boiler operation?

7. How can heat from the continuous blowdown line be reclaimed?

8. List and describe the functions of the parts of a DC generator.

9. How are two DC generators paralleled?

10. Why do large steam plants use watertube boilers while small steam plants use firetube boilers?

11. Why are feedwater pumps in a plant usually located on the first floor or basement level?

12. What are reasons for lowering the MAWP on a boiler?

13. Is it efficient to use bleed or extraction steam from different stages of steam turbines for heating feedwater and process use?

14. What type of feedwater pump drive would a process plant that generates its own electrical power have?

15. Describe how gases of combustion are analyzed using a Fyrite® analyzer and an Orsat analyzer.

16. How is a surface condenser cleaned (from opening to closing)?

17. What purpose does an air ejector serve?

18. Why must the auxiliary lubricating pump on a steam turbine function properly?

19. When does a boiler require a hydrostatic test? Describe how this test is done.

20. Describe how labyrinth packing prevents leakage on a shaft.

Test 7

1. What data is found on a safety valve? What causes a safety valve to pop open?

2. Explain the procedure for starting a condensing turbo-generator.

3. Explain how two AC generators are paralleled.

4. When does a boiler require a hydrostatic test? Describe how this test is done.

5. How does a hydraulic governor control steam turbine speed?

6. Describe how gases of combustion are analyzed using a Fyrite® analyzer and an Orsat analyzer.

7. How is a battery of boilers connected to a steam header to allow for expansion and contraction?

8. Why can small steam turbines operate at higher speeds than large steam turbines?

9. What is an exciter and how does it function?

10. What percentage of heat in the fuel is carried away by gases of combustion?

11. A reciprocating pump $5 \times 3 \times 6$ delivers water at 200 psi. What steam pressure is needed to operate this pump?

12. What is the primary cause of steam turbine blade deposits? What effect does this have on steam turbine operation?

13. A steam turbine is operating a generator that is generating 6000 kW. The water rate of the steam turbine is 14 lb/kW. How much boiler horsepower is required with a feedwater temperature of 200°F?

14. What is the thermal efficiency of a steam turbine that operates with a throttle pressure of 350 psi at 600°F and exhausts into a surface condenser at 29″ Hg vacuum?

15. How does a jet condenser differ from a surface condenser?

16. How much cooling water is needed to condense 60,000 lb of steam/hr if the temperature of the condensate is 115°F and the steam at the inlet to the condenser has a Btu content of 1000 Btu/lb? The cooling water is 70°F at the inlet and 95°F at the outlet.

17. Why is the induced draft fan larger than the forced draft fan on a given size unit?

18. Describe steam pressure and velocity present in an impulse steam turbine.

19. How much feedwater is flowing through a 4″ feedwater line if the feedwater has a velocity of 400 fpm?

20. How much pressure is at the base of a water column 100″ high? How much pressure is at the base of a column of mercury 100″ high?

Test 8

1. When is the best time to blow down a boiler? In what order should bottom blow-down valves be used?

2. When is the best time to blow down the waterwalls on a boiler?

3. What materials are used in the construction of water columns?

4. Describe the conditions that determine the volume of combustion space in a boiler.

5. Describe a system to chemically treat boiler water. What determines the amount of chemical treatment added?

6. Explain the combustion process in a boiler furnace.

7. What is the purpose of a blowdown tank? What pipeline connections are required?

8. What type of circulating pump is commonly used with a surface condenser?

9. Define *enthalpy*.

10. What can cause a furnace explosion?

11. A 2000 HP boiler is operating at 200 psi. What HP is needed to drive the feedwater pump on this boiler?

12. Does the forced draft fan in question 11 require more or less HP than the feedwater pump?

13. What is the rim speed of a 4′ flywheel turning at 200 RPM?

14. What is the torque on a 10,000 kW steam turbine shaft running at 1800 RPM at full load?

15. Describe the advantages of using a combustion control system.

16. How is thrust of the shaft controlled in reaction and impulse steam turbines?

17. What are the principal heat losses in a boiler?

18. Why is the rotor of a generator made of laminated steel sections?

19. Compare the water rates of a condensing and noncondensing steam turbine of equal capacity.

20. How does a multistage centrifugal pump on a common shaft with the same diameter impellers build up the pressure in each stage?

Test 9

1. Why are bypass dampers used on air preheaters?

2. Describe the advantages of using a combustion control system.

3. What is the purpose of and maintenance procedures for a motor-generator set?

4. How is a positive displacement pump protected from excessive pressure on the discharge side?

5. What factors must be considered when selecting a new boiler feedwater pump?

6. Describe a velocity-pressure diagram of a reaction steam turbine.

7. What are the cold clearances on an average steam turbine?

8. How is steam leakage around fixed and moving blades prevented in a reaction steam turbine?

9. Why is it important to save the condensate from the hot well of a surface condenser?

10. What is the specific volume of steam at the inlet of a steam turbine using saturated steam at 250 psi? What is the specific volume of steam at the exhaust with a pressure of 28″ Hg?

11. What effect does condensing or noncondensing have on the thermal efficiency of a steam turbine?

12. What are steam surface condenser tubes made of? How are they attached to the tube sheet?

13. What is the difference between bleed steam and extraction steam taken from a steam turbine?

14. What is the function of a mechanical governor and an overspeed trip on a steam turbine?

15. What is the function of an atmospheric relief valve?

16. Define the following terms:

    a. Gauge pressure

    b. Absolute pressure

    c. Enthalpy

    d. Latent heat of evaporation

    e. Sensible heat

    f. Specific heat

17. A boiler carrying 200 psig has a safety valve 2½″ in diameter. What is the total upward force on the safety valve disc?

18. When is extra-heavy pipe required on bottom blowdown lines of a boiler?

19. How is the proper location of a water column determined?

20. What is the minimum size for tubing, steel pipe, or wrought iron pipe used in the installation of a steam pressure gauge?

Test 10

1. Describe the velocity-pressure diagram of a velocity compound impulse steam turbine.

2. List the location and function of equipment used in a feedwater system.

3. List three ways to control the amount of air delivered to a furnace using a forced draft fan.

4. A boiler carrying 150 psig has a safety valve 2″ in diameter. What is the total upward force on the safety valve disc?

5. Can plant efficiency be increased if a closed feedwater heater is used?

6. A simplex reciprocating pump has a water piston diameter of 12″ and a stroke of 14″. It operates at 100 strokes/min. What is the discharge in gallons per minute if the pump is 80% efficient?

7. A boiler operates at 100 psi steam pressure and is supplied with feedwater from a feedwater heater that has an internal pressure of 10 psi. The boiler water level is located 30′ above the pump and the heater is 45′ above the pump. What is the total head on the pump? Do not allow for friction losses.

8. How should horizontal steam lines be pitched?

9. What would the amount of expansion in a 175′ long steel steam pipe be if its temperature changed from 85°F to 355°F?

10. Steam passing to a steam separator has a quality of 93%. If 5600 lb of steam/hr pass through and the separator collects 285 lb of water, what is the efficiency of the separator?

11. Explain how superheated steam can be desuperheated.

12. Define the following:

   a. Viscosity

   b. Flash point

   c. Fire point

   d. Pour point

13. A coal sample has the following ultimate analysis:
   Nitrogen 2%
   Oxygen 5%
   Carbon 85%
   Ash 2%
   Sulfur 1%
   Hydrogen 5%
   Using Dulong's formula, find the heating value of the coal.

$$Btu/lb = 14,450C + 62,000\left(H - \frac{O}{8}\right) + 4050S$$

14. Describe an underfeed stoker and explain how it functions.

15. What are the disadvantages of using pulverized coal as a fuel?

16. Explain what takes place in a furnace after fuel is introduced.

17. Discuss the routine duties and responsibilities of a boiler operator.

18. What is the danger of high water and low water in a boiler?

19. What are the circumferential and longitudinal stresses on a boiler shell 48″ in diameter, 12′ long, and with a plate thickness of 1″ if the steam pressure is 400 psi?

20. On a steam turbine, how does a mechanical governor differ from a hydraulic governor?

CHAPTER 14 LEARNER RESOURCES

ATPeResources.com/QuickLinks
Access Code: 735728

Test 1

1. *A boiler is equipped with a superheater and has an automatic nonreturn valve. Explain how this boiler must be warmed up and cut in on-line.*
   Answer:
   1. Check for the following:
   • Furnace is free and clear of all tools.
   • Boiler has proper water level.
   • Superheater drain is open.
   • Free-blowing drain between automatic nonreturn and main steam stop valve is open.
   • Air cock is open.
   2. Follow manufacturer's recommendation. Light off the boiler and let it warm up slowly.
   3. Monitor the water level in the boiler drum.
   4. When the boiler reaches approximately 25 psi, close the air cock.
   5. When the boiler reaches approximately 85% of the line pressure, open the equalizing line around the main steam stop valve. Warm up the line and open the main steam stop valve slowly. Open the valve fully, then back it off slightly to keep it from jamming in the open position.
   6. Open the automatic nonreturn valve. Bring up pressure on the incoming boiler slowly. When the steam pressure on the incoming boiler is approximately 1 lb higher than the line pressure, the automatic nonreturn valve will cut the boiler in on-line.
   7. When 20% of the boiler's steaming capacity is reached, the superheater drain valve may be closed.
   8. Check the water level, condition of fire, and all operating temperatures and pressures.
   9. Put the boiler on automatic control. Check to make sure the automatic controls are functioning properly.
   10. Resume normal shift.

2. *Explain how to allow for expansion and contraction of the main steam line between the boiler and steam header.*
   Answer: To allow for the expansion and contraction of the main steam line between the boiler and steam header, expansion bends are used.

487

3. *Explain how to replace a tubular gauge glass with the boiler in service. Include all safety precautions.*

Answer: Assuming the gauge glass is broken, the following procedure is performed:

1. Using the chain-operated shut-off valves, secure steam and water to the gauge glass.
2. Open the gauge glass blowdown valve.
3. Remove the gauge glass nuts and washers.
4. Check to make sure no broken glass has fallen into the line going to the water side of the gauge glass.
5. Obtain new gauge glass and gauge glass washers. *Note:* If gauge glass must be cut to size, it must be cut ¼″ shorter than the inside measurements (distance between top gauge glass connection and bottom gauge glass connection). This is to allow for expansion of the gauge glass when it warms up.
6. Install the new gauge glass using new gauge glass washers.
7. Tighten the gauge glass nuts hand-tight, then about one-quarter turn with a wrench.
8. Crack the steam valve to the gauge glass to warm up the glass.
9. When the glass is warmed up, open the steam and water valves wide, and close the gauge glass blowdown valve. Then check for leaks. *Note:* When working on the gauge glass, use gloves and a full face mask for eye protection. Keep checking for the proper water level using try cocks or other secondary means of determining boiler water level.

4. *According to the ASME code, when must a boiler have two main steam stop valves? What type of valves may be used?*

Answer: The ASME code states that high-pressure boilers in battery that have a manhole opening must have two main steam stop valves and a free-blowing drain between them. These valves should be os&y gate valves that show whether they are open or closed by the position of their spindle. They should always be either wide open or closed. The ASME code further states that an automatic nonreturn valve may be used in place of one stop valve. If an automatic nonreturn valve is used, it must be located as close to the boiler as practical.

5. *Why does a boiler safety valve pop open rather than open slowly?*

Answer: A safety valve prevents the boiler from exceeding its MAWP. It must open fully and remain open until there is a definite drop in pressure (blowdown or blowback). It then must close tightly without chattering. The safety valve is designed to pop open. This is accomplished by the huddling chamber, which provides steam pressure with a larger area to work on. This increases the total upward force, which overcomes the valve spring force, which keeps the valve in a closed position.

As an example, a boiler is carrying 100 psi with a 3″ diameter safety valve. The total upward force acting on the safety valve disc is determined as follows:

Total force = Pressure × Area

$TF = P \times A$

$TF = 100 \times 0.7854D^2$

$Area = 0.7854D^2$

$TF = 100 \times 0.7854 \times 3 \times 3$

$TF = 100 \times 7.0686$

$TF = \mathbf{707\ lb}$

The spring must exert a downward force of 707 lb to keep the valve closed. As soon as the steam pressure starts to overcome the spring force, it allows the steam to enter the huddling chamber. This increases to a 3½″ diameter area exposed to the steam.

$TF = P \times A$
$Area = 0.7854D^2$
$TF = 100 \times 0.7854 \times 3.5 \times 3.5$
$TF = 100 \times 9.62115$
$TF = \textbf{962 lb}$

The total upward force increases from 707 lb to 962 lb, an increase of 255 lb, which causes the safety valve to pop open.

6. *Describe how a deaerating feedwater heater functions.*
Answer: A deaerating feedwater heater functions by having steam and water mix together in the shell. There is a vent condenser on top of the shell. Condensate and makeup water enter, passing through tubes before entering the shell below. Air and noncondensable gases separate from the steam and pass up through the vent condenser and are discharged or vented to the atmosphere. The heater is equipped with an internal overflow that discharges to a float trap that goes to waste. Oil and floating impurities are therefore removed from the surface of the water.

A deaerating feedwater heater is an open feedwater heater. It is located on the suction side of the feedwater pump but at a higher elevation. Its purpose is to heat feedwater, remove air and noncondensable gases, act as a reservoir for feedwater, remove some hardness from water, separate oil when using exhaust steam, and add makeup water to the system.

7. *What causes a feedwater pump to become steam bound and how is this condition corrected?*
Answer: A feedwater pump becomes steambound when the feedwater temperature gets too hot. Water entering the suction side of the feedwater pump flashes into steam. To correct this condition, cold water must be added to the feedwater heater to lower its temperature. The operator must then determine the cause of the increase in feedwater temperature and take necessary corrective action.

8. *Describe the two basic types of superheaters. Where are they located on a boiler?*
Answer: The two basic types of superheaters are radiant and convection superheaters. The radiant superheater is located among the generating tubes and receives most of its heat by radiation. The convection superheater, located in the second to last pass of the gases of combustion leaving the boiler, receives most of its heat by convection. The radiant superheater temperature decreases with an increase in boiler load. The convection superheater temperature increases with an increase in boiler load. To maintain a fairly constant superheater temperature, boilers have been designed to have convection superheaters in series with a radiant superheater.

9. *Describe the operation of a reciprocating feedwater pump.*
Answer: The reciprocating feedwater pump operates on the principle that the steam piston must be 2 to 2½ times larger in area than the water piston. This is why a reciprocating feedwater pump using steam at 100 psi can develop 250 psi of water pressure on the pump discharge. When the steam piston moves through each stroke, a specific amount of water is discharged.

10. *What is the function of an overspeed trip on a steam turbine? When is it tested and how often?*

Answer: The overspeed trip on a steam turbine shuts the steam turbine down to prevent it from destroying itself from excessive speed. The overspeed trip secures steam to the steam turbine when it exceeds its designed speed by 10%. The overspeed trip should be tested when warming up a steam turbine before putting it in service and should also be tripped when taking a steam turbine out of service.

Test 2

1. *According to the ASME code, when must a boiler have two main steam stop valves? What type of valves may be used?*

Answer: The ASME code states that high-pressure boilers in battery that have a manhole opening must have two main steam stop valves and a free-blowing drain between them. These valves should be os&y gate valves that show whether they are open or closed by the position of their spindle. They should always be either wide open or closed. The ASME code further states that an automatic nonreturn valve may be used in place of one stop valve. If an automatic nonreturn valve is used, it must be located as close to the boiler as practical.

2. *Describe the operation of a reciprocating feedwater pump.*

Answer: The reciprocating feedwater pump operates on the principle that the steam piston must be 2 to 2½ times larger in area than the water piston. This is why a reciprocating feedwater pump using steam at 100 psi can develop 250 psi of water pressure on the pump discharge. When the steam piston moves through each stroke, a specific amount of water is discharged.

3. *What is an economizer? Where is it located? What are some problems encountered in its use?*

Answer: An economizer is a heat exchanger used to heat feedwater before it enters the boiler drum. An economizer reclaims heat from gases of combustion that would otherwise be lost to the atmosphere. Economizers are used mostly on large watertube boilers. They may be a unit type outside the boiler casing or an integral type inside the boiler casing.

Economizers are located in the direct path of gases of combustion. Water passes through tubes that are surrounded by gases of combustion. To overcome the restriction to the gases of combustion, induced draft fans are required. Economizer tubes must be kept clean and free of soot. Soot restricts heat transfer. The temperature of the gases of combustion must be high enough to prevent sweating of tubes, which causes corrosion. In addition, economizers must be inspected for leaks. Leaking water combined with soot also causes corrosion.

4. *What could cause a reciprocating feedwater pump to not provide enough water to a boiler?*

Answer: The following could cause a reciprocating feedwater pump to not provide enough water to a boiler:
- No water in the open feedwater heater.
- Pump suction or discharge valves closed.
- Suction or pump discharge valves worn, leaking, or not seating properly.
- Worn rings or liner in the water cylinder.

- Steam admission valves worn or not set properly.
- Steam piston rings worn or broken.
- Steam cylinder worn.
- Pump has become steambound.

5. *Describe the procedure to follow when performing a hydrostatic test.*
Answer: The following procedure is followed to perform a hydrostatic test:
1. Secure the main steam stop valves.
2. Open the free-blowing drain between the main steam stop valves.
3. Remove high-low water level whistle alarm and plug the opening.
4. Remove and blank-flange or gag the safety valve.
5. Fill the boiler with 100°F water until it comes out of the air cock.
6. Close the air cock.
7. Use a hand pump to bring the pressure up to 1½ times the safety valve setting.
8. Look for signs of pressure dropping.
9. Lower the pressure to the safety valve setting.
10. Carefully examine the boiler for leaks or bulging of metal.
11. After testing, reduce the pressure to 0 psi.
12. Replace the high-low water level whistle alarm and replace the safety valve or remove gags.
13. Check the fire side for any tools, lights, ladders, etc.

6. *What is an automatic nonreturn valve? Where is it located?*
Answer: An automatic nonreturn valve cuts a boiler in on-line automatically, takes a boiler off-line automatically, and prevents the system from being drained of steam in case of a failure on the steam side of a boiler. It can be closed manually but can only open by a difference in pressure across the valve. It is located on the main boiler steam line as close to the boiler as practical, replacing one of the boiler's main steam stop valves.

7. *What is the purpose of an air heater? Describe two types.*
Answer: An air heater increases combustion efficiency by supplying preheated primary and secondary air to the furnace. It also increases the life of the firebrick and furnace refractory. Air heaters are located in the breeching and reclaim heat from gases of combustion that would otherwise be lost to the atmosphere.

The two basic types of air heaters are the convection and regenerative air heaters. The convection air heater can be of tubular or plate design. In the tubular convection air heater, gases of combustion pass through the tubes while air being heated flows around the tubes. The air being heated makes several passes over tubes to increase transfer of heat. The plate convection air heater uses plates instead of tubes. They both use the counterflow principle where gases of combustion and air to be heated flow in opposite directions.

The regenerative air heater is found in larger boiler installations, and because of the many moving parts, they require more care and maintenance. They consist of a large round casing divided into three zones: the gases of combustion zone, sealing zone, and air zone. A slow-moving rotor consisting of honeycomb plates rotates through the three zones. The air for combustion is forced through the air heater to the furnace by the forced draft fan.

Air heaters, because of their location, are continually bombarded with soot and flyash. They must be kept clean for maximum efficiency. This is accomplished by soot blowers.

8. *How is a boiler prepared for an annual inspection?*
Answer: A boiler is prepared for annual inspection using specific procedures to prevent injury to plant personnel or damage to the boiler. The procedure is as follows:
1. Take the boiler off-line.
2. Allow the boiler to cool slowly.
3. Make sure the air cock is open and there is no vacuum on the boiler.
4. Secure all main steam stop valves and feedwater valves to the boiler by locking and tagging out.
5. Make sure the free-blowing drain between the stop valves is open. This proves that the stop valve closest to main steam header is not leaking.
6. Never dump a hot boiler. Sludge will bake on the heating surface.
7. Never dump the boiler until it is ready to be flushed.
8. After dumping and flushing out the boiler, secure, lock, and tag out bottom blowdown valves, or remove valves and blank-flange the line to the bottom blowdown tank.
9. Overhaul the bottom blowdown valves.
10. Clean the fire side and water side of the boiler.
11. Open up feedwater regulators and low water fuel cutoffs and clean them.
12. Remove all plugs from cross Ts.
   Note: Never enter the steam and water drum with a conventional droplight. Use a spotlight or low-voltage droplight. Never allow anyone into a boiler until you, personally, have checked to see that all valves are secured, locked, and tagged out.
13. After checking to make sure the boiler is clean on the fire side and water side, notify the boiler inspector that the boiler is ready. Be prepared to assist the boiler inspector during inspection, and point out any problems that are cause for concern. All defects should be reported to the boiler inspector.
14. After the inspection is complete, close up the boiler and check the fire side and water side for tools.
15. Use new gaskets when closing up the boiler.

9. *Describe the wet method of boiler lay-up.*
Answer: The wet method of boiler lay-up is as follows:
1. Thoroughly clean the fire side and water side of the boiler to prevent corrosion.
2. Inspect the fire side and water side for tools or rags.
3. Close the steam and water side using new gaskets.
4. Fill the boiler using chemically treated water at about 100°F until water comes out of the air cock. Close the air cock.
5. Maintain a pressure in the boiler slightly above atmospheric pressure.

10. *Describe the function and location of a surface condenser:*
Answer: A surface condenser is used to create a vacuum on the exhaust end of a steam turbine or reciprocating steam engine. This reduces the back pressure, lowering the water rate, and increases the efficiency of the steam turbine or reciprocating steam engine.
   The exhaust steam and cooling water of a surface condenser do not mix. Water flows through the tubes that are surrounded by exhaust steam. Surface condensers are classified as single-pass or multipass, depending on whether the cooling water makes one or more passes before leaving the condenser.

Test 3

1. *What is an automatic nonreturn valve? Where is it located?*
Answer: An automatic nonreturn valve cuts a boiler in on-line automatically, takes a boiler off-line automatically, and prevents the system from being drained of steam in case of a failure on the steam side of a boiler. It can be closed manually but can only open by a difference in pressure across the valve. It is located on the main boiler steam line as close to the boiler as practical, replacing one of the boiler's main steam stop valves.

2. *Describe the operation of a centrifugal pump.*
Answer: A centrifugal pump uses centrifugal force to develop a pressure rise for moving a liquid. The centrifugal pump imparts kinetic energy to the liquid, which overcomes potential energy. As the liquid passes through the rotating impeller, the kinetic energy overcomes potential energy and gives it centrifugal force. Once the liquid is thrown from the impeller, the casing directs the liquid to the proper destination.

3. *List all the parts of a centrifugal pump.*
Answer: The basic parts of a centrifugal pump and their function are
- Casing: Outer part of the pump that directs water flow from the pump.
- Impeller: Rotating element that imparts centrifugal force to the liquid.
- Impeller shaft: Supports the impeller and transmits rotational force to the impeller from the drive.
- Shaft bearings: Support the impeller and shaft in a fixed axial and radial position.
- Packing glands or mechanical seals: Prevent leakage of liquid between casing and shaft.

4. *Define running condensing.*
Answer: A steam turbine or reciprocating steam engine is running condensing when it exhausts into a condenser. The condenser forms a vacuum as the exhaust steam condenses. This reduces the back pressure and water rate to increase efficiency.

5. *What is the heating surface of a boiler?*
Answer: The heating surface of the boiler is any part of the boiler that has water on one side and heat from gases of combustion on the other.

6. *What is caustic embrittlement? How can it be prevented?*
Answer: Caustic embrittlement is the internal breakdown or cracking of the boiler metal as a result of high alkalinity in the boiler water. The high alkaline material tends to build up along seams, tube ends, and riveted joints. Caustic embrittlement can be prevented by maintaining proper alkalinity of boiler water at all times. This is accomplished by using caustic soda or soda ash and careful control of boiler blowdowns. A continuous blowdown system is most helpful in controlling proper boiler water treatment.

7. *What is the purpose of an air heater? Describe two types.*
Answer: An air heater increases combustion efficiency by supplying preheated primary and secondary air to the furnace. It also increases the life of the firebrick and furnace refractory. Air heaters are located in the breeching and reclaim heat from gases of combustion that would otherwise be lost to the atmosphere.

The two basic types of air heaters are the convection and regenerative air heaters. The convection air heater can be of tubular or plate design. In the tubular convection air heater, gases of combustion pass through the tubes while air being heated flows around the tubes. The air being heated makes several passes over tubes to increase transfer of heat. The plate convection air heater uses plates instead of tubes. They both use the counterflow principle where gases of combustion and air to be heated flow in opposite directions.

The regenerative air heater is found in larger boiler installations, and because of the many moving parts, they require more care and maintenance. They consist of a large round casing divided into three zones: the gases of combustion zone, sealing zone, and air zone. A slow-moving rotor consisting of honeycomb plates rotates through the three zones. The air for combustion is forced through the air heater to the furnace by the forced draft fan.

Air heaters, because of their location, are continually bombarded with soot and flyash. They must be kept clean for maximum efficiency. This is accomplished by soot blowers.

8. *What are common causes of a low water condition?*
Answer: A low water condition can be caused by feedwater pump failure, no water in feedwater heater, failure of makeup water valve, malfunction of feedwater regulator, leaky bottom blowdown valves, and/or boiler tube failure.

9. *What causes foaming and priming in a boiler? How is this condition corrected?*
Answer: Foaming in a boiler results from high surface tension caused by impurities on the surface of the water or high total dissolved solids. It can lead to priming or carryover of water into the main steam line. Correcting this condition requires the boiler surface to be blown down to remove surface impurities, then a bottom blowdown to reduce total dissolved solids in the boiler water.

10. *Which safety valves should be set to pop first, the safety valves on the boiler or the safety valves on the superheater?*
Answer: The safety valves on the superheater must be set to pop before the safety valves on the boiler. This ensures steam flow through the superheater tubes to prevent them from overheating.

Test 4

1. *How are the steam valves set on a duplex reciprocating feedwater pump?*
Answer: Before the steam valves can be set, the pump must be placed in mid-position, which is done as follows:
- Secure steam admission and steam exhaust valves and feedwater suction and feedwater discharge valves. (There are four valves total: two on the steam side and two on the water side.)
- Open drains on the steam and water sides.

- Bar the pump toward the steam side as far as it will go.
- Scribe a mark on the steam piston rod flush with the packing gland.
- Bar the pump in the opposite direction as far as it will go. Mark the steam rod again, flush with the packing gland.
- With dividers, find the center between the two marks. Mark the steam piston rod.
- Bar the pump toward the steam side, lining the last mark flush with the packing gland.
- The rocker arm should be perpendicular to the steam piston rod. One side of pump is now in mid-position.

Follow the same procedure on the other side of the pump. The pump is now ready to have the valves set. *Note:* When barring the pump over, bar it against the crosshead or spool piece, never against the rocker arm.

With the pump in mid-position, proceed to set the valves.

- Mark and remove the valve chest cover.
- Move the slide valve so that it covers all ports.
- Move the valve in one direction as far as it will go. Measure the steam port opening.
- Move the valve in the opposite direction as far as it will go. Measure the steam port opening.

Note: The steam port openings should be equal, with about one half of a port open on each side when the D slide is at its end position. This distance is the lost motion. The lost motion allows the pump to make a full stroke, and it also prevents the pump from striking the head. After setting the lost motion nut, move one valve to uncover a port, otherwise the pump cannot start. Once in motion, the duplex reciprocating feedwater pump can never stop with all the ports covered.

2. *What is caustic embrittlement? How can it be prevented?*

Answer: Caustic embrittlement is the internal breakdown or cracking of the boiler metal as a result of high alkalinity in the boiler water. The high alkaline material tends to build up along seams, tube ends, and riveted joints. Caustic embrittlement can be prevented by maintaining proper alkalinity of boiler water at all times. This is accomplished by using caustic soda or soda ash and careful control of boiler blowdowns. A continuous blowdown system is most helpful in controlling proper boiler water treatment.

3. *How can it be determined if the safety valves on a boiler have enough relieving capacity?*

Answer: Safety valves are rated in number of pounds of steam per hour they are capable of relieving at a given pressure. Three equations can be used.

Pounds of steam/hour =

$$\frac{Quantity\ of\ fuel\ burned/hr \times Heating\ value/unit\ of\ fuel \times 0.75}{1100}$$

$$W = \frac{c \times H \times 0.75}{1100}$$

Note: The assumed boiler efficiency is 0.75; 1100 is the assumed Btu value in 1 lb of steam.

According to ASME code,

$$W = \frac{c \times H \times 0.75}{h}$$

W = pounds of steam generated per hour
c = quantity of fuel burned per hour
H = heating value per unit of fuel
h = constant, the latent heat of the steam (in Btu/lb)

The second method that can be used is determined by the square feet of heating surface in the boiler.

For firetube boilers:

$$W = (8 \times HS_b) + (14 \times HS_{ww})$$

For watertube boilers:

$$W = (10 \times HS_b) + (16 \times HS_{ww})$$

W = pounds of steam generated per hour
HS_b = boiler heating surface (in sq ft)
HS_{ww} = waterwall heating surface in furnace (in sq ft)

The values obtained from these equations are then compared to the total relieving capacity of the safety valves.

The third method is to perform an accumulation test on the boiler.

4. *Define boiler efficiency.*
   Answer: Thermal efficiency and boiler efficiency are synonymous. It is the ratio of heat output to heat input, or the amount of fuel needed to produce a certain amount of heat. It is expressed mathematically as follows:

   Thermal efficiency =

   $$\frac{Pounds\ of\ steam/hr\left[Btu\ content\ of\ steam - \left(Feedwater\ temperature - 32\right)\right]}{Units\ of\ fuel/hr \times Btu\ content\ per\ unit\ of\ fuel}$$

   $$TE = \frac{W_S\left[H_S - \left(T_{fw} - 32\right)\right]}{W_f - C}$$

   TE = thermal efficiency
   W_S = pounds of steam per hour
   H_S = enthalpy of steam (in Btu/lb)
   T_{fw} = feedwater temperature
   32 = base temperature from which total heat is calculated
   W_f = units of fuel per hour
   C = Btu content per unit of fuel

5. *Why is fuel oil in boiler water dangerous? How can fuel oil get into a boiler?*
   Answer: Fuel oil in boiler water can settle out on boiler heating surfaces and lead to the formation of blisters, bags, or burned out tubes. Fuel oil also increases surface tension in the steam and water drum, leading to foaming, priming, or carryover. Fuel oil can get into boiler water by leaking heating coils in fuel oil tanks.

6. *How does the boiler drum pressure compare to the superheater outlet pressure during normal operation of a boiler equipped with a superheater?*
   Answer: The superheater outlet pressure is always slightly lower (approximately 5 lb) than the boiler drum pressure. Unless there is a difference in pressure, there could be no steam flow. In order for any fluid to flow, there must be a pressure differential and the flow is always from the high-pressure area to the low-pressure area.

7. *Describe the condition that develops in economizers if the feedwater temperature is too low.*
Answer: The economizer is located in the direct path of the gases of combustion. Its purpose is to heat the feedwater before it enters the steam and water drum. If the feedwater temperature is too low, it will create sweating of the economizer tubes. This causes corrosion of the economizer tubes. Soot and water will form sulfuric acid, which corrodes the economizer metal.

8. *What happens when baffles deteriorate in watertube boilers?*
Answer: Baffles direct the flow of gases of combustion around the boiler heating surfaces so that the maximum amount of heat can be absorbed by the boiler. If baffles were leaking, the gases of combustion would take the path of least resistance. There would be an increase in the chimney temperature and the boiler would have trouble maintaining its proper steam pressure.

9. *What is the purpose of a surface blowdown? Describe its location.*
Answer: A surface blowdown removes impurities on the surface of the boiler water. This reduces surface tension and prevents foaming, priming, and carryover. The surface blowdown discharges into the blowdown tank. A surface blowdown is located on the steam and water drum at the NOWL.

10. *Why is a steam valve under pressure never opened quickly?*
Answer: When steam flows from a high-pressure area to a low-pressure area, it drops in pressure and increases in velocity. This leads to water hammer and could lead to a pipe rupture. If the steam valve is connected to the boiler, it can lift water from the boiler into the steam line. It could cause a reciprocating pump or reciprocating steam engine to break its cylinder head. In a steam turbine, the water impinging on the steam turbine blades would cause severe erosion of the blades.

Test 5

1. *What is the purpose of blowing down the water column, gauge glass, and low water fuel cutoff each shift?*
Answer: Blowing down the water column and gauge glass removes sludge or sediment that could result in an inaccurate indication of the boiler water level. Blowing down the low water fuel cutoff also removes sludge and sediment and verifies that the control is operative. Blowing down the low water fuel cutoff does not replace the evaporation test that should be performed once a month as recommended by the ASME code.

2. *What procedure must be followed if the water level in the gauge glass is not visible and the burner is still operating?*
Answer: The burner must be secured immediately. The boiler must be taken off-line and allowed to cool slowly, then it must be inspected for damage that may have resulted from overheating. Before the boiler is put back into service, the cause of the low-water condition must be determined and corrected.

3. *How is a low water fuel cutoff tested?*
Answer: To test a low water fuel cutoff, it should be blown down once a shift. When it is blown down, the burner should shut down. The low water fuel cutoff should also be tested once a month using an evaporation test. This is done by securing all feedwater pumps to that boiler and allowing the water level to drop slowly. The boiler operator must watch the water level in the gauge glass to make sure the burner shuts off properly.

4. *Why is an approved blowdown tank required on a steam boiler?*
Answer: The blowdown tank must be built according to the ASME Unfired Pressure Vessel Code. When a steam boiler is blown down, the blowdown tank is subjected to the steam boiler pressure and temperature even though it is vented to the atmosphere. In addition, the blowdown tank is also subjected to thermal shock.

5. *Describe the procedure followed when taking over a shift.*
Answer: The shift operator should report for work at least 30 minutes before the shift begins. This should give the shift operator enough time to check all the equipment for proper operating conditions before relieving the operator on duty. The procedure for taking over a shift is as follows:
- Check the water level in all boilers that are on-line.
- Blow down the water column, gauge glass, and low water fuel cutoff on all boilers that are on-line.
- Check the boiler steam pressure.
- Check the condition of fires.
- Check all running auxiliaries for proper temperature, pressure, and lubrication.
- Ask the operator being relieved if any difficulties occurred during that shift.
- Check the boiler log for any orders left by the chief engineer.
- If everything is in order, relieve the operator and take over the shift.

Note: Once the boiler operator takes over a shift, that boiler operator is held responsible for any problems that occur. If a boiler drops a tube or a pump burns out a bearing 5 minutes into the shift, the boiler operator on duty will be held responsible.

6. *List three commonly used fuels and their Btu content.*
Answer: The three most commonly used fuels in boilers are coal, fuel oil, and natural gas. The following Btu contents are approximate values and vary depending on where the fuels come from.
- Bituminous coal: 12,000 Btu/lb
  Anthracite coal: 15,000 Btu/lb
- No. 6 fuel oil: 150,000 Btu/gal.
  No. 4 fuel oil: 146,000 Btu/gal.
- Natural gas: 100,000 Btu/therm

7. *What is the function of the bottom blowdown line? Where and how is it connected to a boiler?*
Answer: The bottom blowdown line is used to control high water, remove sludge and sediment, control chemical concentration, and dump the boiler for cleaning and inspection.

The bottom blowdown lines are connected to the lowest part of the water side of a firetube boiler and the mud drum of a watertube boiler. The lines must run full size with no reducers or bushings. If they are exposed to direct furnace heat, they must be protected by a refractory pier or wrapped with protective insulation. Piping may not be exposed to freezing or corrosive conditions. The opening in the boiler setting should be arranged to provide for free expansion and contraction, and blowdown lines must be exposed for easy inspection.

8. *What is a steam separator and steam trap? How is each used?*

Answer: A steam separator is a device used to remove moisture from steam and allows steam to pass through. It is located in the steam and water drum to remove moisture trapped in the steam before the steam enters the main steam line. It is also located before the throttle valve of prime movers. In plants having reciprocating steam engines it is called a receiver separator.

A steam trap is an automatic device that removes air and water from a steam line without loss of steam. It is located on main steam headers, on the end of steam lines, and where steam lines make a 90° turn going up. Steam traps are found on all heat exchangers where steam condenses when giving up its heat.

9. *Describe two types of feedwater regulators. State where each is used.*

Answer: Thermoexpansion feedwater regulators are found mostly on large boilers in which the feedwater pumps run continuously. A thermoexpansion feedwater regulator consists of an inclined tube called a thermostat that is located at the NOWL of the boiler. The top of the thermostat is connected to the steam side of the boiler, and the bottom to the water side of the boiler. The inclined tube then registers the amount of water in the boiler. The top of the inclined tube is free to expand and contract and is connected by linkage to a valve in the feedwater line. The bottom of the inclined tube is fixed and cannot move.

As the water level in the boiler drops, it also drops in the inclined tube and steam takes its place. This causes the tube to expand slightly, and the linkage moves the feedwater valve toward an open position. When the water level in the boiler rises, it also rises in the inclined tube, replacing the steam. This causes the tube to contract, and the linkage causes the valve in the feedwater line to move toward a closed position. By using an inclined tube rather than a vertical tube, there is a larger surface of steam and water, making the regulator more sensitive. The inclined tube feedwater regulator is mechanical and works on the principle of expansion and contraction.

Float type feedwater regulators are used on package firetube and watertube boilers where the feedwater pumps run intermittently. They are also located at the NOWL of the boiler. They consist of a float in a float chamber, which connects to mercury tubes or micros witches. There are two mercury tubes or micros witches because the regulator is often a combination feedwater regulator and low water fuel cutoff. The top of the float chamber is connected to the uppermost part of the steam space and the bottom of the chamber is connected to the water side of the boiler, well below the NOWL. The float chamber will have the same water level as the boiler.

As the water level drops in the boiler, the float in the float chamber starts to move down and the linkage causes the mercury tube to tilt. This completes an electrical circuit, and the feedwater pump starts up and pumps water into the boiler. With a rise in water level, the float through the same linkage moves up, causing the electrical circuit to break, shutting down the feedwater pump.

If for some reason the feedwater pump fails to pump water to the boiler, the float drops lower and the second mercury tube tilts and breaks the electrical circuit to the burner, shutting the burner OFF on a low water condition. The float chamber is also fitted with a blowdown valve to keep the chamber free of sludge and sediment and also to test the control for proper operation.

10. *Describe the function and location of a surface condenser:*
Answer: A surface condenser is used to create a vacuum on the exhaust end of a steam turbine or reciprocating steam engine. This reduces the back pressure, lowering the water rate, and increases the efficiency of the steam turbine or reciprocating steam engine.

The exhaust steam and cooling water of a surface condenser do not mix. Water flows through the tubes that are surrounded by exhaust steam. Surface condensers are classified as single-pass or multipass, depending on whether the cooling water makes one or more passes before leaving the condenser.

Test 6

1. *Describe three ways fuel oil is atomized.*
Answer: To atomize fuel oil means to break it up into fine mist-like particles that can come into more intimate contact with oxygen in the air for complete combustion. This can be accomplished by mechanical atomization, air or steam atomization, or rotary cup atomization.

Pressure atomizing burners, also called mechanical atomizing burners, require fuel oil at the proper temperature and pressure to atomize the fuel oil. They are either plug-and-tip or sprayer plate atomizing burners. In the plug-and-tip burner, fuel oil enters the burner tube and is forced through the plug channel where atomization takes place. The fuel oil leaves the burner tip rotating at a high velocity. In the sprayer plate burner, fuel oil enters the burner tube where it is forced through the sprayer plate channel where atomization takes place. The atomized fuel oil leaves the burner tip spinning at a high velocity.

Air or steam atomizing burners use either air or steam to atomize fuel oil. Air atomizing burners require an air compressor to atomize the fuel oil. They are inside-mixing burners and are used primarily in small installations. Steam atomizing burners are either outside-mixing or inside-mixing burners. In the outside-mixing burner, fuel oil and steam mix outside the burner. In the inside-mixing burner, fuel oil and steam mix inside the burner. Steam is used in both cases to atomize the fuel oil. Fuel oil is supplied at pressures and temperatures lower than that in the pressure atomizing burners. The fuel oil pressure is approximately 50 lb, and the steam for atomization must be 20 lb higher than the fuel oil pressure. The steam and fuel oil do not mix until the steam is directed across the path of the fuel oil, flowing outside or inside the burner nozzle, depending on the type used.

The rotary cup atomizing burner can burn a wide range of fuel oil at low pressures and temperatures. It is ideal for automatic operation. Atomization is accomplished by a fuel oil discharged from a spinning cup and high-velocity air supplied by the primary air fan.

2. *What test must be performed on a new or repaired boiler?*
Answer: All new boilers and boilers that have had extensive repair work on their steam and water side or have had a low water condition develop should be given a hydrostatic test. This test requires that the safety valves be removed and blank-flanged or gagged. High-low water level whistle alarms are removed and plugged. The boiler is filled with 100°F water until it comes out of the air cock. The air cock is closed and

pressure is built up to 1½ times the safety valve setting. The pressure is watched to make sure it is holding. The pressure is then dropped to safety valve popping pressure and the boiler is carefully checked for signs of leaks or bulging of metal.

3. *What is the function of a feedwater injector?*
   Answer: A feedwater injector is a secondary means of supplying feedwater to a boiler. It is not an efficient means of feeding water to the boiler. It can only lift water about 20′ under ideal conditions. The maximum temperature of the water a feedwater injector can handle is 130°F to 150°F.

   The feedwater injector works on the principle that kinetic energy is greater than potential energy. Steam passing through a nozzle drops in pressure, expands, and increases in velocity. It picks up the water, and velocity or kinetic energy forces open the check valve on the feedwater line, discharging the water into the boiler.

4. *What is the function of a grid valve?*
   Answer: A grid valve is used on steam turbines. It extracts steam from one or more of its stages for plant process. Steam flow is regulated by a positioning control that actuates a servo-motor connected by linkage to a grid valve and governor system that controls the steam turbine speed. The grid valve is used to maintain pressure in the extraction line but will close or throttle back if the load requires more steam. If not enough steam is supplied from the extraction line, a steam regulator will open and steam will be supplied from an auxiliary steam line.

5. *What are the advantages of a superheater used with a steam turbine?*
   Answer: A superheater is a heat exchanger located in either the radiant or convection zone of a boiler. It increases the temperature of the steam but not the pressure. By increasing the heat of the steam before it enters the steam turbine, the superheater cuts down condensation losses and makes it possible to extract more heat from the steam to increase the thermal efficiency of the steam turbine.

6. *Why would a safety valve fail to open at its set pressure?*
   Answer: A safety may fail to open at its set pressure because of corrosion. The operator should lower the boiler pressure below its set pressure and pop the safety by hand. If the safety valve pops, the operator should bring the pressure back up and make sure the safety valve is operating properly. If the safety valve does not pop by hand at its set pressure, a new boiler should be warmed up and cut in on-line. The other boiler should be taken out of service and the faulty safety valve replaced.

7. *Why are a steam strainer and separator located before a steam turbine?*
   Answer: If there is no steam strainer separator in the line before the steam turbine throttle valve, the steam turbine blades traveling at high velocity would be seriously damaged by loose particles of matter or water carried along with the steam.

8. *How is a steam pressure gauge protected against excessive temperature?*
   Answer: A steam pressure gauge is protected from the temperature of live steam by a siphon installed between the boiler and the steam pressure gauge. The siphon forms a water leg so that live steam cannot enter the Bourdon tube in the steam pressure gauge.

9. *Describe three methods of heat transfer.*
Answer:
1. Radiation: Method of heat transfer that occurs when heat is trans- ferred without a material carrier.
2. Convection: Method of heat transfer that occurs as heat is transferred by currents through a fluid.
3. Conduction: Method of heat transfer in which molecules come in direct contact with each other, and energy is passed from one molecule to another.

Heat always travels from a hot area to a cold area. There must be a temperature difference for heat to flow. The greater the difference in temperature is, the greater the movement will be.

10. *Why is feedwater heated before it enters the steam and water drum?*
Answer: Feedwater is heated before it enters the steam and water drum for the following reasons:
- Fuel consumption is reduced. For every 10°F rise in feedwater temperature using exhaust steam or heat from gases of combustion, there is approximately a 1% savings in fuel.
- Thermal shock to the boiler is reduced, which extends the life of the boiler.
- The boiler capacity is increased slightly.
- Boiler and plant efficiency are increased resulting from fuel savings.

Test 7

1. *Describe three ways fuel oil is atomized.*
Answer: To atomize fuel oil means to break it up into fine mist-like particles that can come into more intimate contact with oxygen in the air for complete combustion. This can be accomplished by mechanical atomization, air or steam atomization, or rotary cup atomization.

Pressure atomizing burners, also called mechanical atomizing burners, require fuel oil at the proper temperature and pressure to atomize the fuel oil. They are either plug-and-tip or sprayer plate atomizing burners. In the plug-and-tip burner, fuel oil enters the burner tube and is forced through the plug channel where atomization takes place. The fuel oil leaves the burner tip rotating at a high velocity. In the sprayer plate burner, fuel oil enters the burner tube where it is forced through the sprayer plate channel where atomization takes place. The atomized fuel oil leaves the burner tip spinning at a high velocity.

Air or steam atomizing burners use either air or steam to atomize fuel oil. Air atomizing burners require an air compressor to atomize the fuel oil. They are inside-mixing burners and are used primarily in small installations. Steam atomizing burners are either outside-mixing or inside-mixing burners. In the outside-mixing burner, fuel oil and steam mix outside the burner. In the inside-mixing burner, fuel oil and steam mix inside the burner. Steam is used in both cases to atomize the fuel oil. Fuel oil is supplied at pressures and temperatures lower than that in the pressure atomizing burners. The fuel oil pressure is approximately 50 lb, and the steam for atomization must be 20 lb higher than the fuel oil pressure. The steam and fuel oil do not mix until the steam is directed across the path of the fuel oil, flowing outside or inside the burner nozzle, depending on the type used.

The rotary cup atomizing burner can burn a wide range of fuel oil at low pressures and temperatures. It is ideal for automatic operation. Atomization is accomplished by a fuel oil discharged from a spinning cup and high-velocity air supplied by the primary air fan.

2. *Why must the safety valve discharge outlet not be reduced or restricted in any way?*
Answer: The steam discharged when the safety valve pops open drops in pressure, increases in velocity, and expands (increases in volume). Restriction or reduction of discharge piping would cause back pressure, preventing the safety valve from relieving the required pounds of steam it is rated for.

3. *Why would a safety valve fail to open at its set pressure?*
Answer: A safety may fail to open at its set pressure because of corrosion. The operator should lower the boiler pressure below its set pressure and pop the safety by hand. If the safety valve pops, the operator should bring the pressure back up and make sure the safety valve is operating properly. If the safety valve does not pop by hand at its set pressure, a new boiler should be warmed up and cut in on-line. The other boiler should be taken out of service and the faulty safety valve replaced.

4. *What maintains the speed of an auxiliary steam turbine operating at 3600 RPM?*
Answer: The speed of auxiliary turbines are maintained by a mechanical governor. The speed of a generating turbine is maintained by a hydraulic governor. Both turbines are equipped with overspeed trips.

5. *List four reasons for losing vacuum in a plant using a surface condenser. How could each condition be corrected?*
Answer: The following are reasons vacuum could be lost in a plant using a surface condenser and remedies to correct the conditions.
   • Insufficient cooling water: Increase water flow and check intake water line. If a cooling tower is the water source, check makeup water supply.
   • Fouled air ejectors: Change over air ejectors and clean faulty ejector.
   • Dirty condenser: Remove from service as soon as possible for cleaning.
   • Loss of water seal in hot well: Reestablish water seal.
   • Waterlogged condenser: Check condenser condensate pump and return to service.

6. *Why is superheated steam used in a steam turbine plant?*
Answer: Superheated steam causes less wear on steam turbine blading. Superheated steam has a higher heat content, which produces more work. Using superheated steam gives a higher thermal efficiency. There is also less condensation loss.

7. *Why is steam extracted from a steam turbine?*
Answer: Steam is extracted from a steam turbine for plant process. This is reclaiming latent heat that would otherwise be lost in the condenser to the cooling water, which also increases steam turbine and plant efficiency.

8. *What is the function of a thrust bearing on a steam turbine?*
Answer: The axial clearance between the moving and stationary blades of the steam turbine and steam nozzles and moving blades is very critical. The thrust bearing keeps the rotor in its proper fixed position to prevent damage to the steam turbine.

9. *Why does the auxiliary lubricating oil pump operate after the steam turbine has been secured?*
Answer: The auxiliary lubricating oil pump is kept running to keep the bearings cool and help the rotor cool slowly.

10. *Why must a steam turbine rotor rotate during cool down?*
Answer: Small steam turbines are allowed to cool while they are at idle. Large steam turbines are equipped with a turning gear to keep the rotor moving slowly to prevent warping during cooling.

Test 8

1. *How does a governor control the speed of a steam turbine?*
Answer: The governor used on steam turbines driving auxiliaries is a mechanical governor. The mechanical governor consists of a speed-sensing element that through linkage controls the speed of the steam turbine shaft.

The speed-sensing element consists of weights mounted on the steam turbine shaft. The centrifugal force tends to move the weights out as the speed of the turbine increases. Through linkage, this tends to close down the steam to the steam turbine. As the steam turbine speed decreases, the weights move in and the linkage opens the steam valve to the steam turbine. There is a separate overspeed trip. It consists of a pin mounted in the steam turbine shaft and is held in by spring tension. If the steam turbine exceeds its safe operating speed by 10%, centrifugal force causes the pin to pop out and hit an emergency trip that secures the steam to the steam turbine.

2. *Why is a high vacuum desirable in a condenser for a condensing steam turbine?*
Answer: The higher the vacuum in the condenser is, the lower the back pressure will be on the steam turbine. This allows the steam to expand further, extracting more heat from the steam. It also reduces the steam turbine water rate to increase its efficiency.

3. *Should the rotor of a steam turbine be moving or stationary during the warm-up period?*
Answer: The rotor and blades of a steam turbine must be slowly moving as soon as steam is admitted to the steam turbine to prevent the rotor from bowing or warping. The rotor is only supported by two bearings. One is at the throttle and one is at the exhaust end. By having the rotor in motion, it is heated up evenly.

4. *How do economizers and air heaters affect draft?*
Answer: Economizers and air heaters are located in the direct path of the gases of combustion and restrict their flow. Both forced and induced draft fans are used to overcome this restriction.

5. *What are the advantages of a superheater used with a steam turbine?*
Answer: A superheater is a heat exchanger located in either the radiant or convection zone of a boiler. It increases the temperature of the steam but not the pressure. By increasing the heat of the steam before it enters the steam turbine, the superheater cuts down condensation losses and makes it possible to extract more heat from the steam to increase the thermal efficiency of the steam turbine.

6. *What is the procedure of cold startup of a boiler equipped with a superheater?*
Answer: Boiler startup procedures vary depending on boiler type and type of fuel used. Manufacturers' recommended startup procedures should be consulted, then proceed as follows:
   1. Inspect the fire side of the boiler to make sure it is clear of tools, rags, etc.
   2. Check the water level of the boiler.
   3. Make sure the boiler vent, superheater drain, and free-blowing drain between the main boiler stop valve and automatic nonreturn valve are open.
   4. Light off the boiler and warm up slowly.
   5. Open the equalizing line around the main steam stop valve to equalize pressure and warm up the line. Then open the main steam stop valve.
   6. Keep a close watch on the boiler water level.
   7. When steam pressure reaches 25 psi, close the boiler vent.
   8. Open the automatic nonreturn valve when steam pressure on the incoming boiler reaches 75% to 85% of line pressure.
   9. Test the flame failure control.
   10. Bring the boiler pressure up slowly and allow the automatic nonreturn valve to cut the boiler in on-line.
   11. Close the free-blowing drain.
   12. Close the superheater drain when adequate flow has been established (approximately 20% of boiler capacity).
   13. Put all boiler controls on automatic and make sure they take over.

7. *What is the procedure followed if popping or blowdown pressure on a safety valve must be adjusted?*
Answer: Before making any changes affecting popping or blowback pressure on a boiler, the boiler inspector must be notified. In order to make any adjustments, the seals must be broken. All safety valves have a nameplate that shows the popping pressure and blowback. Any changes require a new nameplate showing new popping pressure and blowdown. Obtain one from the safety valve manufacturer and reattach it to the safety valve. New seals must also be installed.

8. *List the auxiliaries commonly used in the operation of a condensing steam turbine.*
Answer: The auxiliaries commonly used in the operation of a condensing steam turbine are as follows:
   • Condenser: Condenses steam from steam turbine exhaust, forming a vacuum. Lowers back pressure on steam turbine.
   • Circulating water pump: Provides condenser with cooling water needed to condense steam and maintain vacuum.
   • Condensate pump: Removes condensate from the condenser hot well and discharges it to the feedwater heater to be reused.
   • Air ejectors: Single- or two-stage ejectors used to remove air and noncondensable gases from the condenser, which increases the vacuum.

9. *How is an auxiliary steam turbine started?*
Answer: The procedure for starting an auxiliary steam turbine that drives a feedwater pump is as follows:
   1. Open the steam turbine exhaust valves and casing drains to allow the steam turbine to warm up.
   2. Check the steam turbine to make sure it is clear of tools, rags, etc.

3. Check the oil level on all bearings.
4. Open the feedwater suction and discharge valves on the pump.
5. Crack the throttle valve enough to cause the steam turbine to spin. Secure the throttle and listen for rubbing noises.
6. If no rubbing occurs, open the throttle and bring the steam turbine up to speed, close the cylinder drains, and make sure the governor takes over.
7. Test the overspeed trip manually, then reset the overspeed trip and bring the steam turbine back up to speed.
8. Check all bearings to ensure they are getting proper lubrication and cooling.
9. Remove the other feedwater pump from service if necessary.

10. *How is an auxiliary steam turbine removed from service?*
Answer: An auxiliary steam turbine is removed from service as follows:
1. Use the manual overspeed trip to secure steam to the steam turbine.
2. When the steam turbine stops spinning, secure the exhaust valve and open all cylinder drains.
3. If the steam turbine is driving a feedwater pump, secure the suction and discharge valves.
4. Reset the overspeed trip for the next startup.

Test 9

1. *What should be done if there is a possibility that the main boiler steam pressure gauge is not accurate?*
Answer: Obtain a steam pressure gauge that is known to be accurate. Install it on the inspector's test connection that is alongside the boiler steam pressure gauge and compare readings. If the boiler steam pressure gauge is not accurate, it must be removed and recalibrated using a deadweight pressure- gauge tester if one is available. If one is not available the boiler steam pressure gauge must be sent out for repairs or replaced with a new boiler steam pressure gauge. As soon as a steam pressure gauge has been installed, remove and carefully store the one that was used for testing so it will always be on hand. *Note:* Do not lose the water in the siphon while working on the boiler steam pressure gauge or the new steam pressure gauge may be damaged by live steam.

2. *What is the purpose of a steam separator?*
Answer: A steam separator removes moisture and impurities from the steam before the steam enters a reciprocating steam engine or a steam turbine. This prevents damage to the cylinder head or blades of the steam turbine.

3. *List all the parts of a centrifugal pump.*
Answer: The basic parts of a centrifugal pump and their function are
• Casing: Outer part of the pump that directs water flow from the pump.
• Impeller: Rotating element that imparts centrifugal force to the liquid.
• Impeller shaft: Supports the impeller and transmits rotational force to the impeller from the drive.
• Shaft bearings: Support the impeller and shaft in a fixed axial and radial position.
• Packing glands or mechanical seals: Prevent leakage of liquid between casing and shaft.

4. *Describe the operation of commonly used feedwater heaters.*
Answer: Two commonly used feedwater heaters are
• Open feedwater heater: It is located on the suction side of the feedwater pump but at a higher elevation. Exhaust steam and condensate returns come into intimate

contact with each other and oxygen, and other noncondensable gases are driven off and vented to the atmosphere. The open feedwater heater is equipped with an internal overflow to prevent flooding the system and an oil separator to remove oil present in the returns. The feedwater pump takes its suction from the feedwater heater and delivers water to the boiler under pressure.

- Economizer: It is located between the feedwater pump and boiler in the direct path of the gases of combustion. Water passes through the tubes that pick up heat that would otherwise be lost to the atmosphere. Economizers must be kept free of soot deposits or flyash, which would act as insulation and retard heat transfer. This is accomplished by using soot blowers. Economizers are used mostly in larger boilers that have a fairly constant load to prevent sweating, which would lead to corrosion.

Feedwater must be heated to drive off oxygen and other noncondensable gases, reduce thermal shock to the boiler, increase boiler life, reduce fuel consumption, and increase plant thermal efficiency if using exhaust steam or gases of combustion.

5. *What is superheated steam?*
   Answer: Superheated steam is steam at a higher temperature than its corresponding pressure. Saturated steam is steam at its corresponding temperature and pressure. When more heat is added at the same pressure it becomes superheated steam. For example, steam at 100 psi has a corresponding temperature of 338°F. If it is heated to 538°F it has 200°F of superheat.

6. *Why is superheated steam used in steam turbine plants?*
   Answer: Superheated steam is used in steam turbine plants to cut down on condensation losses. It also allows the steam turbine to extract more heat from the steam, which increases its thermal efficiency. When condensate is eliminated from steam, the steam turbine blading is protected from steam erosion.

7. *What is the procedure followed if popping or blowdown pressure on a safety valve must be adjusted?*
   Answer: Before making any changes affecting popping or blowback pressure on a boiler, the boiler inspector must be notified. In order to make any adjustments, the seals must be broken. All safety valves have a nameplate that shows the popping pressure and blowback. Any changes require a new nameplate showing new popping pressure and blowdown. Obtain one from the safety valve manufacturer and reattach it to the safety valve. New seals must also be installed.

8. *Why must leaks on the bottom blowdown line be promptly repaired?*
   Answer: A leaky bottom blowdown line is costly and also very dangerous. Leaky bottom blowdown lines cause a loss of chemically treated boiler water. This requires more makeup water to be added to the boiler, which requires more chemicals. If the leak in the bottom blowdown line is not repaired, the hole can enlarge and result in an overheated boiler if the feedwater pump cannot supply enough water to the boiler. If the line ruptures, serious boiler damage would result. The ASME code requires any repair to the bottom blowdown line to be supervised by a boiler inspector.

9. *What are the requirements of the discharge line of a safety valve?*
   Answer: The discharge line of a safety valve must be free to expand and contract and not put strain on the body of a safety valve to ensure its proper operation and prevent distortion to the valve.

10. *What procedure should be followed if a safety valve leaks after testing?*
Answer: If a safety valve leaks after it is tested, it should be popped by hand to try to clear the seat of any rust or scale particles. If after two or three attempts it still leaks, another boiler should be warmed up and cut in on-line. The boiler with the leaky safety valve should be taken out of service and cooled slowly. When pressure is off the boiler, the safety valve should be removed and sent back to the manufacturer for an overhaul or replaced with a new safety valve. If there is only one boiler in the plant, the boiler should be taken out of service as soon as possible. Under no circumstances should the boiler be left unattended until the safety valve has been repaired or replaced.

Test 10

1. *Describe the procedure required to increase the working pressure of a boiler.*
Answer: If increasing the working pressure of the boiler does not exceed its MAWP and does not require resetting the safety valves, make the necessary adjustments to the operating controls based on the type in the plant. If, however, the increase in pressure exceeds the MAWP of the boiler, nothing can be done to increase the pressure. If the safety valves must be reset, the boiler inspector must be notified. Changing the popping pressure requires a new safety valve nameplate. This must come from the safety valve manufacturer. The safety valve manufacturer should also make the changes and reseal the valves or, if necessary, replace the safety valves with new valves of the proper capacity.

2. *Describe the location and purpose of the following:*
Answer:
  a. Nonreturn valve: Found on main boiler steam line as close to the boiler as possible. Cuts boiler in on-line automatically, cuts boiler off-line automatically, and protects system from draining in event of a failure on steam side of anyone boiler.
  b. Quick-opening valve: Valve located on blowdown line closest to the boiler shell on boilers operating at 100 psi or higher. The quick-opening valve functions as a sealing valve.
  c. Superheater safety valve: Located on outlet header of superheater. Superheater safety valve is set to pop before main safety valves to establish steam flow through superheater to prevent warping or burning out tubes.
  d. Free-blowing drain: Located on boiler main steam line between the two boiler stop valves. Used to remove condensate from steam line when warming up the line before boiler is cut in on-line. Also used to make sure automatic nonreturn valve is not leaking when performing a hydrostatic test on the boiler.
  e. Expansion bend: Found on boiler main steam line. Allows for expansion and contraction. Also used on long runs of steam lines to allow for expansion and contraction in place of expansion joints.

3. *What steps must be taken if the safety valve does not pop with excessive boiler pressure?*
Answer: Secure all burners to the boiler and allow pressure in the boiler to drop below the safety valve popping pressure, then pop all safety valves manually. Relight the boiler and bring pressure up to safety valve popping pressure. If the safety valves pop, proceed with normal operation. When the boiler is down for

the annual inspection, have the safety valves removed sent to the manufacturer for a complete overhaul. When the safety valves are put back on the boiler and before the boiler is put back in service, establish a safety valve testing procedure. The safety valves will stick to their seats because of corrosion buildup if there is a lack of testing.

4. *What are water temperature considerations from an open feedwater heater?*
Answer: If the water temperature in an open feedwater heater gets too high, the feedwater pump could become steambound. This could cause a low water condition to develop in the boiler, which could lead to overheating or burned out tubes. If the water is allowed to get too cold, oxygen would not be vented from the feedwater heater and would enter the boiler. Oxygen leads to pitting and corrosion in the boiler. In addition, cold water can cause thermal shock to the boiler heating surface. Finally, if the oil separator malfunctions, oil could get into the boiler, causing foaming to develop that would lead to priming and carryover. Oil on the heating surfaces of the boiler could also cause overheating of the boiler heating surface.

5. *What is the function of a telltale hole?*
Answer: A telltale hole alerts the boiler operator of a broken staybolt. A telltale hole is a $3/16''$ hole drilled into a boiler staybolt. This hole extends $1''$ into the boiler water. The staybolt is used on firetube boilers to hold the inner and outer wrapper sheets of the boiler together to keep the flat surfaces from bulging. If a staybolt shears (breaks), water would come through the telltale hole. The operator must replace the staybolt as soon as possible. Staybolts are subjected to shear stress in addition to severe corrosion, which results in leaks.

6. *What is the function of a feedwater injector?*
Answer: A feedwater injector functions as a secondary means of getting water into a boiler. The feedwater injector works on the principle that kinetic energy overcomes potential energy. Steam passing through a nozzle drops in pressure, expands, and increases in velocity. It picks up the water and velocity or kinetic energy and forces open the check valve on the feedwater line, discharging water into the boiler.

7. *How is scale in a boiler caused and prevented?*
Answer: Scale in a boiler is caused by improper feedwater treatment. All raw water contains some amount of scale-forming salts. If the boiler water is tested daily and proper chemicals are added in the correct amounts, the scale-forming salts change to a nonadhering sludge. This nonadhering sludge is removed by the bottom blowdown valves. By proper testing, chemical control, surface blowdown, and bottom blowdown, scale should not form on any boiler heating surface.

8. *Describe three methods for identifying overheated bearings.*
Answer: Three methods to identify overheated bearings are
   - Smell: When the oil in journal bearings starts to overheat, it gives off a distinct odor and/or changes color.
   - Feel: If the operator touches a journal bearing and cannot keep a hand on it, it is running hot.
   - Noise: Journal bearings squeal when they are overheated.

9. *What procedure should be followed if a leak is found between the boiler and the bottom blowdown valves?*

Answer: If the leak makes it difficult to maintain boiler water level, remove the boiler from service immediately and continue feeding the boiler with water. If at any time it puts the other boilers in danger, secure water to the leaking boiler and save the others that are on-line.

If it is a minor leak, warm up another boiler and cut it in on-line. Remove the leaking boiler from service, let it cool slowly, and notify the boiler inspector. The ASME code states that the boiler inspector has jurisdiction concerning how and what type of repair is allowed.

10. *It takes approximately 15 lb of air to burn 1 lb of coal. How much air is necessary to burn 1 ton of coal?*

Answer:

1 ton of coal = 2000 lb

15 × 2000 = 30,000 lb of air

Test 1

1. *List three types of feedwater regulators. Describe how they function.*

Answer: A *thermoexpansion feedwater regulator* is connected to the boiler and main feedwater line. The thermostat is located at the NOWL and is connected to the steam and water side of the boiler. If the water level drops, the steam space within the thermostat increases. This increases the temperature in the thermostat, causing it to expand. The increase in the length of the thermostat moves the mechanical linkage, which opens the regulator valve. As the regulator valve on the main feedwater line opens, water enters the boiler and raises the water level. The opposite occurs if the boiler water level is high. The thermostat contracts and the linkage moves the regulator valve toward the closed position.

A *thermohydraulic feedwater regulator* consists of a regulating valve, bellows, generator, and stop valves. The control element is the generator, which is a tube within a tube. The generator is located at the NOWL and the inner tube is connected to the steam and water side of the boiler. Stop valves are needed to secure the steam and water to the feedwater regulator generator. The outer tube of the generator is connected to the bellows with copper tubing and is filled with condensate. When the boiler water level drops, the inner tube of the generator has a larger steam space. The heat from the steam is released to the condensate in the outer tube. The condensate within the bellows expands, which moves the regulating valve toward the open position. When the water level in the boiler rises, the reverse action takes place.

A *float feedwater regulator* consists of a float chamber, float, and mercury switch or microswitch. The float chamber is connected to the steam and water side of the boiler. It is located at the NOWL. When the boiler water level drops, the float in the float chamber drops and mechanically moves the mercury switch. The mercury switch is connected electrically to a feedwater pump starter relay that energizes the pump motor. The pump will continue to run until the float rises and moves the mercury switch in the opposite direction to shut down the feedwater pump.

2. *List three types of steam traps. Describe how they function.*

Answer: *Thermostatic steam traps* consist of a body, valve, valve seat, and bellows. When steam comes in contact with the bellows, it causes it to expand and the valve to close. As condensate builds up in the trap and surrounds the bellows, it contracts and the valve opens to discharge the condensate and any air that may be in the line.

Float thermostatic steam traps consist of a body and a float that controls a valve and a thermostatic element. When condensate enters the body of the trap, the float is raised and the valve opens to allow condensate to flow into the condensate return line. If there is no condensate in the body, the valve will close. The thermostatic element opens only if there is a temperature drop around the element caused by air or noncondensable gases in the body. When steam enters the body, the thermostatic element expands and closes.

Inverted bucket steam traps consist of a body, an inverted bucket, a valve, and linkage that connects the valve and bucket. Steam and condensate enter the trap from the bottom of the inverted bucket. If only steam enters the trap, it lifts the bucket and the valve will stay closed. When condensate enters the trap, the bucket sinks and the discharge valve opens. A small hole at the top of the bucket releases air and noncondensable gases that would otherwise be trapped in the bucket. Air, noncondensable gases, and condensate leave the body of the trap when the valve opens.

3. *When are two or more safety valves required on a boiler?*
Answer: The ASME code states that boilers having more than 500 sq ft of heating surface must have two or more safety valves.

4. *How is feedwater distributed in a watertube boiler and an HRT boiler?*
Answer: In watertube boilers, feedwater enters the steam and water drum through an internal feed line that extends approximately 80% of the drum length. They are installed so that incoming feedwater is discharged below the surface of the water and over the entire length. This is accomplished using a perforated pipe or water trough.

In HRT boilers over 40″ in diameter, the feed line enters above the center row of tubes. Incoming feedwater is discharged over approximately three-fifths the length of the boiler drum. Either a perforated line or water trough must be used.

5. *What procedure is taken if the water level in the gauge glass is out of sight on a coal-fired boiler?*
Answer: First, it must be determined whether the gauge glass is full or empty. This can be done by using the try cocks. A full gauge glass and an empty gauge glass look the same. If water comes out of the top try cock, blow down the boiler until water is at its proper level. If, however, steam comes out of the bottom try cock, the fires must be secured. Draft to the furnace must be cut off. Depending on the method of coal feed being used, secure the fire. With a hand-fired boiler, pull the fire. With a stoker-fired boiler, secure coal feed and speed up stoker to run fire into ash pit. When pressure has dropped, allow the boiler to cool and have the boiler inspected for signs of damage from overheating. Also, find out why the low water condition happened and repair it.

6. *Define cavitation and explain where it could occur.*
Answer: Cavitation occurs in centrifugal pumps. It can be caused by carrying too high a temperature, causing the liquid to flash and resulting in a partial vacuum. This results in pitting and wearing away of the pump casing and impeller. It can also cause damage to the pump shaft, bearings, and pump seals from the vibration caused by cavitation.

7. *How is the number of poles in an AC generator determined?*
Answer: The number of poles in an AC generator is a function of its frequency and speed and is determined by using the following equation:

$$Frequency = \frac{No.\ of\ poles \times RPM}{120}$$

or

$$No.\ of\ poles = \frac{120 \times Frequency}{RPM}$$

Assuming the following values,
Frequency = 60 Hz
RPM = 3600

$$No.\ of\ poles = \frac{120 \times 60}{3600}$$

No. of poles = **2**

The number of poles of any AC generator can be determined if the frequency and RPM are known.

8. *What information is commonly found on a safety valve data plate?*
   Answer: The information found on a safety valve data plate may vary slightly depending on the manufacturer. However, the data plate should provide the following information:
   • manufacturer's name or trademark
   • manufacturer's design or type number
   • size of valve in inches (seat diameter)
   • popping pressure setting (in psig)
   • blowdown (in psi)
   • capacity (in lb of steam/hr)
   • lift of valve (in inches)
   • year built or code mark
   • ASME symbol
   • serial number

9. *What is a steam separator and where is it located?*
   Answer: A steam separator is a device used to remove moisture while allowing the steam to pass through. When the steam separator changes the direction of steam flow, heavier water droplets will separate. Steam separators are found in steam and water drums, in the steam line before a steam turbine throttle valve, and in the steam line before the throttle valve of a reciprocating steam engine.
   A steam separator helps prevent water hammer in steam lines, conserves the energy of the steam, and protects the steam turbine blading from erosive action of wet steam. It also protects the valves, pistons, and cylinders of reciprocating steam engines from damage caused by water in the steam.

10. *What is the function and location of an atmospheric valve?*
    Answer: An atmospheric valve prevents pressure buildup on the exhaust side of reciprocating steam engines or steam turbines. In noncondensing plants, they are called back pressure valves. In condensing plants, they are called atmospheric relief valves. Atmospheric valves can be located on the main exhaust line or directly at the top of a condenser as close to the exhaust steam line as possible.

Test 2

1. *How many and what type of valves are used on a boiler bottom blowdown line?*
   Answer: According to the ASME code, boilers operating at 100 psi or higher must have two bottom blowdown valves. The two valves may be two slow-opening valves that

require five full turns of the handwheel to open or close fully, or one quick-opening and one slow-opening valve. Both shall be extra-heavy, 250 psi, ASME-approved valves. When the MAWP exceeds 125 psi, the valves must be made of extra-heavy bronze, brass, or malleable iron suitable for the temperature and pressures involved.

2. *A simplex reciprocating pump with a 4" diameter water cylinder and 6" stroke operates at 50 strokes/min. How many gpm will the pump discharge?*
Answer:

$$gpm = \frac{LANE}{231}$$

gpm = gallons per minute
L = length of stroke of piston found on data plate
A = area of water cylinder (in sq in.)
N = number of strokes per minute
E = pump efficiency (90%)
231 = cubic inches per gallon
$A = 0.7854D^2$
$A = 0.7854 \times 4 \times 4$
$A = 12.5664$ sq in.

$$gpm = \frac{6 \times 12.5664 \times 50 \times 0.90}{231}$$

$gpm =$ **14.688**

3. *What are possible causes for a surface condenser to gradually lose its vacuum?*
Answer: Reasons for loss of vacuum in a surface condenser are the condensate pump is not removing water from the condenser hot well fast enough, air is leaking into the condenser, there is a faulty air ejector, the temperature of the cooling water is rising, the condenser tubes are plugging up, and/or there is insufficient cooling water.

4. *What is the approximate water rate of a 10,000 kW turbo-generator? What BHP is needed for this turbo-generator?*
Answer: Water rate varies with the type of steam turbine used. A straight noncondensing steam turbine has a water rate of 24 lb/kW, whereas a condensing steam turbine has a rate of 14 lb/kW. The BHP required to operate a straight noncondensing steam turbine is

$$BHP = \frac{WR \times kW}{34.5}$$

BHP = boiler horsepower
WR = 24 lb/kW
kW = 10,000
34.5 = lb of steam/BHP

$$BHP = \frac{24 \times 10,000}{34.5}$$

$$BHP = \frac{WR \times kW}{34.5}$$

$BHP =$ **6956.5**

The BHP required for a condensing turbine is

$$BHP = \frac{WR \times kW}{34.5}$$

$$BHP = \frac{14 \times 10,000}{34.5}$$

$$BHP = \mathbf{4058}$$

5. *What is an economizer and where is it located?*
Answer: An economizer is used to heat feedwater before it enters the boiler steam and water drum by reclaiming heat from the gases of combustion before they enter the chimney. The feedwater passes through the economizer tubes that are surrounded by gases of combustion. An economizer is located in the direct path of the gases of combusion.

6. *What is the function of baffles on a watertube boiler? What would indicate a broken baffle?*
Answer: Baffles on a watertube boiler direct the flow of gases of combustion so they come into close contact with the boiler heating surfaces. The operator can tell if the baffles break by a sudden increase of the chimney temperature and difficulty maintaining steam pressure in the boiler.

7. *How many square feet of boiler heating surface are there in a straight-tube water-tube boiler with 350 tubes that are 3″ in diameter and 20′ long?*
Answer:

$HS = C \times L \times N$
HS = heating surface (in sq ft)
C = circumference of tube (in ft)
L = length of tube (in ft)
N = number of tubes
$Circumference = \pi \times Diameter\ of\ tube$
$C = 3.1416D$
$HS = 3.1416 \times D \times L \times N$

Note: Change D to feet: $\dfrac{3}{12}$

$HS = 3.1416 \times \dfrac{3}{12} \times 20 \times 350$

$HS = 0.7854 \times 20 \times 350$
$HS = \mathbf{5497.8\ sq\ ft}$

8. *Explain the procedure to follow if a steam line ruptures going to the steam turbine.*
Answer: A ruptured steam line going to the steam turbine is a very serious problem. If at all possible, get to the boiler room and secure the main steam line. If that is not possible, secure the fuel to the boiler. In some plants, there is a triple-acting nonreturn valve that will take the boiler out of service. The steam turbine will most likely trip out and auxiliary lighting should come on. It will be necessary to evacuate the turbine room.

9. *How much cooling water is needed on a surface condenser connected to a 10,000 kW, straight, condensing turbo-generator? (Assume all values.)*

Answer: To find the pounds of cooling water needed per pound of steam on a surface condenser, the following equation is used:

$$Lb\ water\ /\ Lb\ steam = \frac{H_e - (T_c - 32)}{T_1 - T_2}$$

H_e = enthalpy of exhaust steam

$T_c - 32$ = enthalpy of condensate in hot well

T_1 = temperature of cooling water leaving condenser

T_2 = temperature of cooling water entering condenser

Assume the following values: H_e = 1050 Btu/lb; T_c = 98°F; T_1 = 90°F; T_2 = 70°F.

$$Lb\ water\ /\ Lb\ steam = \frac{1050 - (98 - 32)}{90 - 70}$$

Lb water/Lb steam = **49.2**

Assuming a water rate of 14 lb of steam/kW, 140,000 lb (14 × 10,000) of steam/kW must be condensed:

Total lb water/hr = $WR \times kW \times Lb\ water/Lb\ steam$

Total lb water/hr = 14 × 10,000 × 49.2

Total lb water/hr = **6,888,000**

10. *Explain the procedure to follow if the feedwater pump discharge pressure drops to boiler pressure.*

Answer: If the feedwater pump discharge pressure drops to boiler pressure, no water can get into the boiler. The following procedure must be taken:

1. Reduce boiler firing rate.
2. Start up a new feedwater pump immediately.
3. If water level in the boiler is still dropping, take the boiler off-line.
4. Determine the cause of the low water condition.

Test 3

1. *What procedure is followed if a safety valve fails to open at its set pressure?*

Answer: If a safety valve fails to open at its set pressure, proceed as follows:

1. Reduce the firing rate on the boiler to reduce steam pressure.
2. Test all safety valves by hand.
3. Bring the boiler up to its popping pressure.
4. If the safety valves do not open at their set pressure, check the pressure gauge to make sure it is accurate.
5. If the pressure gauge proves to be accurate, do not leave the boiler unattended.
6. Remove the boiler from service as soon as possible and have the safety valves removed and returned to the manufacturer for an overhaul.

2. *What is the function of a feedwater injector?*

Answer: A feedwater injector acts as an auxiliary or a secondary means of feeding water to a boiler. It is not an efficient means of feeding water to a boiler because it cannot handle hot water without the danger of becoming steambound. Under ideal conditions, it has a maximum lift of 20′ and has a tendency to kick out under a fluctuating steam load.

The feedwater injector works on the principle that kinetic energy is greater than potential energy. Steam passing through a nozzle drops in pressure, expands, and increases in velocity. Steam picks up the water and velocity or kinetic energy, forces open the check valve, discharging the water into the boiler.

3. *A 10,000 kW turbo-generator operates on 450 psia steam pressure at 750°F and exhausts into a surface condenser with a 28" Hg vacuum. What is the thermal efficiency of the turbo-generator?*
Answer:

$$TE = \frac{H_a - H_e}{H_a - (T_c - 32)}$$

TE = thermal efficiency
H_a = enthalpy of steam at throttle
H_e = enthalpy of steam at exhaust
$T_c - 32$ = enthalpy of condensate

$$TE = \frac{1387.1 - 1105.2}{1387.1 - (100 - 32)}$$

$$TE = \frac{281.9}{1319.1}$$

TE = **0.2137** or **21.37%**
Note: The enthalpy of the condensate is based on a 5% to 10% condensation of steam at the exhaust.

4. *Determine the bursting pressure of a boiler drum with the following: tensile strength of metal—60,000 psi; metal thickness—1¾"; drum diameter—5'; efficiency of joint—90%.*
Answer:

$$BP = \frac{S \times T \times E}{R}$$

BP = bursting pressure
S = tensile strength of metal (in psi)
T = thickness of metal (in inches)
E = efficiency of joint
R = radius of drum (in inches)

$$BP = \frac{60,000 \times 1.75 \times 0.90}{30}$$

BP = **3150 psi**

5. *How are draft fans driven and controlled?*
Answer: Draft fans may be driven by electric motors or steam turbines. Some plants require one electric and one steam-driven fan for plant flexibility and balancing the steam and electric loads. Draft can be controlled by regulating the speed of the fan, dampers, inlet vanes on the fan, and a hydraulic coupling between the drive and fan.

6. *Why can boiler tubes be of thinner metal than the heads or shell while still carrying the same steam pressure?*
Answer: The stress set in the metal of a boiler drum or tube is determined by the boiler pressure, diameter of the boiler drum or tube, and thickness of the boiler drum or tube

metal. The following equation is used to find longitudinal stress:

$$S_l = \frac{P \times D}{2T}$$

S_l = longitudinal stress
P = pressure in vessel, in psi
D = diameter of drum or tube, in inches
T = thickness of drum or tube, in inches

The pressure is the same in the drum and tube; therefore, the ratio $D \div T$ will affect the amount of stress within the drum or tube. The tube is much smaller in diameter than the drum and can therefore be of thinner metal.

Example:

Using the following values find the longitudinal stress on the drum and the tube: $D = 3''$ (tube); $T = 0.125''$ (tube); $D = 36''$ (shell); $T = 1.5''$ (shell); *Pressure on boiler* = 100 psi

Tube:

$$S_l = \frac{P \times D}{2T}$$

$$S_l = \frac{100 \times 3}{2 \times 0.125}$$

$$S_l = \textbf{1200 psi}$$

Shell:

$$S_l = \frac{P \times D}{2T}$$

$$S_l = \frac{100 \times 36}{2 \times 1.5}$$

$$S_l = \textbf{1200 psi}$$

7. *What is the function of steam traps used in a steam system? Where are they located?*
Answer: Steam traps remove condensate and noncondensable gases from the steam lines without losing steam. Steam traps are used in the steam system wherever steam releases its heat and condenses.

8. *What automatic system is used to control the steam pressure on a boiler? How does it function and what safety devices are used?*
Answer: An automatic system consists of automatic combustion controls that regulate fuel supply, air supply, air-fuel ratio, and removal of gases of combustion. The amount of fuel supplied to the burner must be in proportion to the steam pressure required. Three basic types of automatic combustion control systems are ON/OFF, positioning, and metering.

The ON/OFF combustion control system consists of the following:
• Pressure control: Regulates the operating range of the boiler by starting and stopping the burner on boiler steam pressure demand.
• Modulating pressure control: Regulates high and low fire. The burner must always start off in low fire and shut down in low fire.
• Programmer: Controls the starting sequence and firing cycle of the burner. It controls the operating sequence of the blower, burner motor, ignition system, fuel valve, and all other components in the ON/OFF control system. It provides

a purge cycle before burner ignition and after burner shutdown to cut down the danger of a furnace explosion. In conjunction with the flame scanner, it is designed to de-energize the fuel valve in four seconds after loss of flame signal. It also locks out and must be reset manually after a flame failure. ON/OFF combustion controls can also use solid-state or microprocessor-based circuitry to control burner functions.

Positioning combustion controls function using steam header pressure to actuate a master control unit. Fuel and air supply to the furnace is adjusted to meet steam demands.

Metering combustion controls function using a pressure differential of steam across an orifice. This pressure differential is used to identify steam outputs and regulate air and fuel supplied to the burner.

Automatic combustion controls using solid-state and microprocessor circuitry have provisions for flame-out and burner failure conditions for maximum safety. Another safety device is a sail switch that will interrupt the firing cycle if there is no primary air at the burner. There is also a safety device, called a high-pressure cutout with manual reset, that is used to secure the burner in an event of overpressure.

9. *Define the following terms:*
   Answer:
   a. Btu: British thermal unit-measurement of quantity of heat. It is the unit of heat necessary to raise the temperature of 1 lb of water 1°F.
   b. Specific heat: The amount of heat required to raise the temperature of 1 lb of a substance 1°F.
   c. Boiling point of water: Temperature at which water boils. Changing state from liquid to vapor varies depending on pressure applied to surface of water. At atmospheric pressure (sea level), water boils at 212°F.
   d. Absolute pressure: The sum of gauge pressure and atmospheric pressure.
   e. Absolute temperature: Temperature on the Fahrenheit scale plus 460.
   f. Absolute zero temperature: The temperature 460° below zero Fahrenheit. There is no molecular motion and there is complete absence of heat at that temperature.

10. *Describe in detail the meaning of blowdown of a safety valve.*
    Answer: The blowdown of a safety valve is the drop in pressure from the point of popping until the safety valve reseats. It can be 2 lb to 8 lb but never more than 4% of its set pressure. For pressures between 200 psi and 300 psi, the blowdown shall not be less than 1% of the set pressure.

Test 4

1. *A duplex reciprocating pump has 10 × 8 × 6 on the data plate. What do the numbers mean and how are they used?*
   Answer: The first number (10) refers to the steam cylinder diameter, the second number (8) refers to the liquid cylinder diameter, and the third number (6) refers to the length of the pump stroke. All are given in inches.

From the data given, the pressure that can be developed on the liquid side for any given steam pressure used can be found. The ratio of the pressures is inversely proportional to the diameter of the cylinders squared.

$$\frac{P_s}{P_w} = \frac{D_w^{\,2}}{D_s^{\,2}}$$

P_S = pressure on steam cylinder
P_W = pressure on liquid cylinder
D_S^2 = diameter of steam cylinder squared
D_W^2 = diameter of liquid cylinder squared

The capacity (in gpm) of any reciprocating pump is found by using the following equation:

$$gpm = \frac{LANE}{231}$$

gpm = gallons per minute
L = length of stroke of cylinder (in inches)
A = area of liquid cylinder (in sq in.)
N = number of strokes per minute
E = pump efficiency

There are 231 cubic inches in 1 gallon. Using this information, cubic inches per minute can be converted into gallons per minute.

2. *What is the approximate Btu content per pound of fuel oil with a specific gravity of 25.2°API at 60°F?*
Answer: If the °API of a fuel oil is known, the approximate Btu content per pound of fuel oil is found by using the following equation:
Btu/lb of fuel oil = 17,780 + (54 × °API)
Btu/lb = 17,780 + (54 × 25.2)
Btu/lb = 17,780 + 1360.8
Btu/lb = **19,140.8**

3. *A centrifugal pump discharges 100 gpm against a head of 250′. What HP is required to drive this pump? Do not allow for friction losses.*
Answer:

$$HP = \frac{ft\text{-}lb / min}{33,000\ ft\text{-}lb / min \times E}$$

ft = foot head on pump
lb/min = pounds per minute from discharge
E = efficiency of pump

$$HP = \frac{250 \times (100 \times 8.33)}{33,000 \times 0.8}$$

$HP = $ **7.89**
Note: One gal. of water weighs 8.33 lb.

4. *Describe the steam and water circulation on various types of boilers.*
Answer: In externally fired firetube boilers, the cooler water drops down the center and is heated. As steam bubbles form, they rise up the sides of the shell and break through the surface of the water into steam space.

In internally fired firetube boilers, cooler water moves down the sides of the boiler shell and is heated. As steam bubbles form, they rise up the middle and break through surface of the water into the steam space.

In firetube boilers, the circulation is somewhat erratic because of a large volume of water and the location of the heating surface.

In inclined tube watertube boilers, cooler water leaves the bottom of the steam and water drum through downcomer nipples to the front header. As water is heated, it rises, moving toward the rear header. Steam bubbles form in the tubes and move into the rear header, then to the steam space.

In bent-tube watertube boilers, the cooler water leaves the steam and water drum through downcomer tubes to the mud drum. As it is heated, it rises and steam bubbles form, passing up to the rear drum and through the top tubes into the steam and water drum.

5. *Define boiler efficiency, steam turbine efficiency, and plant efficiency. Create and solve a problem for each.*
Answer: Boiler efficiency is the ratio of heat available in the fuel as fired and heat that the water and steam absorb.

$$TE = \frac{W_s \left[H_s - \left(T_{fw} - 32 \right) \right]}{W_f \times C}$$

TE = thermal efficiency (boiler efficiency)
W_S = pounds of steam per hour
H_S = enthalpy of steam
T_{fw} = feedwater temperature
32 = base temperature for determining enthalpy (heat in substance)
W_f = units of fuel per hour
C = Btu content per unit of fuel
Assume the following values: W_S = 12.8 lb; H_S = 1200 Btu/lb; T_{fw} = 182°F; W_f = 1; C = 17,000 Btu/lb.

$$TE = \frac{12.8 \times \left[1200 - \left(182 - 32 \right) \right]}{1 \times 17,000}$$

TE = **0.79 or 79%**
The thermal efficiency of a steam turbine is the ratio between heat supplied to the steam turbine and heat used by the steam turbine.

$$TE = \frac{H_a - H_e}{H_a - \left(T_c - 32 \right)}$$

TE = thermal efficiency of steam turbine
H_a = enthalpy of steam at throttle
H_e = enthalpy of exhaust steam
T_C = temperature of condensate
Assume the following values: H_a = 1400 Btu/lb; H_e = 1050 Btu/lb; T_C = 112°F.

$$TE = \frac{1400 - 1050}{1400 - \left(112 - 32 \right)}$$

TE = **0.265 or 26.5%**

Plant efficiency is the ratio of heat energy available in the fuel to the energy produced from this fuel.

$$E = \frac{Btu/kW}{C \times Btu/lb \ of \ fuel}$$

E = plant efficiency
Btu/kW = Btu in 1 kW
C = pounds of fuel
$Btu/lb \ of \ fuel$ = Btu in 1 lb of fuel

Assume the following values: $Btu/kW = 3413$; $C = 1$; $Btu/lb \ of \ fuel = 17,000$.

$$E = \frac{3413}{1 \times 17,000}$$

$E = \textbf{0.20}$ or $\textbf{20\%}$

Plant efficiency can also be found using the following equation, which includes the generator with an assumed efficiency of 95%.

$E_p = E_b \times E_t \times E_g$
$E_p = 0.79 \times 0.265 \times 0.95$
$E_p = \textbf{0.19888}$ or $\textbf{20\%}$

6. *Define the following terms.*
   Answer:
   a. Volatile: Quality of a substance that vaporizes, giving off a gas at relatively low temperatures.
   b. Critical speed: Speed at which a rotating element vibrates most violently.
   c. Ultimate analysis: Chemical process used to determine nitrogen, oxygen, carbon, ash, sulfur, and hydrogen in a coal specimen.
   d. Synchronization: Process of putting two pieces of equipment in step. When putting an AC generator on the board, both generators must be in parallel and in phase or synchronized.
   e. Atomization: Breaking up of a liquid into a fine mist, as in the atomization of fuel oil for combustion.

7. *A boiler generates 150,000 lb of steam/hr from feedwater at 230°F. What is the DBHP if the enthalpy of steam is 1200 Btu/lb?*
   Answer: If the factor of evaporation is unknown, the DBHP may be solved by using the following equation:

$$DBHP = \frac{W_s\left[H_s - (T_w - 32)\right]}{33,475}$$

$DBHP$ = developed boiler horsepower
W_S = pounds of steam per hour
H_S = enthalpy of steam
T_{fw} = feedwater temperature
32 = base temperature of determining enthalpy (heat in substance)
33,475 = Btu in 1 BHP

$$DBHP = \frac{W_s\left[H_s - (T_w - 32)\right]}{33,475}$$

$$DBHP = \frac{150{,}000\left[1200-(230-32)\right]}{33{,}475}$$

$$DBHP = \frac{150{,}000 \times 1002}{33{,}475}$$

$DBHP = \textbf{4489.92}$

8. *A 50 HP electric motor is 85% efficient. What is the input in watts per HP? What is the total usable power in watts at full load?*

Answer: The total usable power is found by using the following equation:

$$Watts\ input = \frac{HP \times 746\ watts\,/\,HP}{Efficiency\ of\ motor}$$

$$Watts\ input = \frac{1 \times 746\ watts\,/\,HP}{0.85}$$

Watts input = **877.65 watts/HP**
Usable power in watts = HP output × Watts/HP
Usable power in watts = 50 × 746
Usable power in watts = **37,300 W**

9. *Is the superheater outlet header safety valve or the steam and water drum safety valve set lower? Why?*

Answer: The superheater safety valve is always set to pop first. This ensures a flow of steam through the superheater tubes. If the main safety valves on the boiler drum were to pop first, this could cause the superheater tubes to overheat because of an interruption of steam flow through the superheater.

10. *What happens to the water level on watertube boilers when the firing rate is increased or decreased rapidly? Explain the cause.*

Answer: Any sudden increase in the boiler load would cause the water level in the gauge glass to first go up then rapidly drop. This is caused by the sudden flow of steam leaving the steam and water drum. The drop in steam pressure causes a certain amount of internal flashing to take place, forcing the water level to go up. When there is a sudden drop in load, the water level first will drop rapidly because of the increase in steam pressure and then slowly come up.

Test 5

1. *Describe how to replace a round gauge glass while the boiler is in service.*

Answer: The following procedure should be followed when replacing a round gauge glass:

1. Secure the water and steam gauge glass stop valves.
2. Open the gauge glass blowdown valve.
3. Check the water level by a secondary means.
4. Remove the gauge glass guards.
5. Remove the gauge glass nuts.
6. Remove all broken glass and old washers.

7. Cut a new glass ¼″ shorter than inside measurements to allow for installation and expansion.
8. Place nuts and new washers on gauge glass.
9. Center gauge glass to allow for expansion.
10. Tighten nuts by hand, then one-quarter turn with a wrench.
11. Crack steam to the gauge glass to warm it up.
12. Close the gauge glass blowdown valve and open the water gauge glass stop valve.
13. Open the steam gauge glass valve fully.
14. Check for leaks. Tighten nuts if necessary.
15. Replace gauge glass guards.
Note: Never work on a gauge glass without using eye protection.

2. *Find the heating surface of a firetube and watertube 1′ long, 3″ in diameter, with a wall thickness of ³⁄₁₆″.*
Answer:

$$HS = C \times L \times N$$
$$C = \pi \times D \times L \times N$$

HS = heating surface (in sq ft)
C = circumference of tubes
L = length of tubes
N = number of tubes
D = diameter of tubes

Heating surface is measured in sq ft and the diameter of the tubes is measured in ft.
Note: For firetube boilers, use the inside diameter (ID) of the tube to determine the circumference of the tube. The ID is determined by taking the outside diameter (OD) minus two times the tube wall thickness. For watertube boilers, use the OD to determine the circumference of the tube.

Firetube:

$$HS = 3.1416 \times D \times L \times N$$
$$D = 3 - (³⁄₁₆ + ³⁄₁₆) = 2⅝″$$

Note: Change D to feet: $\dfrac{2.625}{12}$

$$\frac{V_1}{T_1} = \frac{V_2}{T_2} \ or \ V_2 = \frac{V_1 T_2}{T_1}$$

$$HS = \textbf{0.6872 sq ft}$$

Watertube:

$$HS = 3.1416 \times D \times L \times N$$

Note: Change D to feet: $\dfrac{3}{12}$

$$HS = 3.1416 \times \frac{3}{12} \times 1 \times 1$$

$$HS = \textbf{0.7854 sq ft}$$

3. *Define factor of evaporation. Assuming any values necessary, set up and solve for the factor of evaporation.*

   Answer: Factor of evaporation is a correction factor that is needed to find the DBHP of a boiler.

   $$FE = \frac{H_s - \left(T_{fw} - 32\right)}{970.3}$$

   FE = factor of evaporation
   H_S = enthalpy of steam
   T_{fw} = feedwater temperature
   32 = base temperature for determining enthalpy (heat in substance)
   970.3 = latent heat of evaporation of water at 212°F
   Note: T_{fw} − 32 is the enthalpy of the feedwater entering the boiler drum.
   Assume the following values: H_S = 1200 Btu/lb; T_{fw} = 182°F.

   $$FE = \frac{1200 - \left(182 - 32\right)}{970.3}$$

   $$FE = \frac{1200 - 150}{970.3}$$

   $$FE = \frac{1050}{970.3}$$

   $$FE = \mathbf{1.082}$$

4. *A boiler is rated for 100,000 lb/hr but delivers only 90,000 lb/hr. List the probable causes.*

   Answer:
   - The baffles directing the flow of gases of combustion are broken or deteriorated.
   - The boiler tubes are coated with soot, resulting in a high chimney temperature.
   - The water side of the boiler tubes are scaled, retarding heat transfer.
   - There are improper combustion conditions, mixture, atomization temperature, and time.
   - The boiler feedwater temperature is lower than that specified by the manufacturer.

5. *Define the following terms:*

   Answer:
   a. Radiant heat: Heat that travels in waves similar to light waves.
   b. Convection: Heat transfer by currents in a fluid.
   c. Conduction: Heat transfer by molecules coming into direct contact with each other and energy is passed from one to another.
   d. Latent heat: Hidden heat; heat added to a substance that will change its state without a change in temperature.
   e. Sensible heat: Heat that changes the temperature of a substance but not its state.
   f. Perfect combustion: Burning of all the fuel with the theoretical amount of air.
   g. Complete combustion: Burning of all the fuel supplied using minimal amount of excess air.
   h. Incomplete combustion: Occurs when all the fuel supplied is not burned, resulting in smoke and soot.

6. *When preparing a boiler for inspection, what procedures must be followed?*

Answer: The following valves must be closed, locked, and tagged out when preparing a boiler for inspection:
- main steam stop valves
- all feedwater valves going to the boiler
- bottom blowdown valves after the boiler has been dumped and flushed

The following precautions are necessary to prevent injury to plant personnel:
- Never dump a hot boiler. Sludge will bake on heating surfaces.
- Never dump a boiler unless it is ready to be opened and flushed out.
- Always vent the boiler drum to prevent a vacuum from forming.
- Never enter the steam and water side of a boiler unless personally checking to make sure steam stop valves, feedwater valves, and bottom blowdown valves are properly secured, locked, and tagged out.
- Make sure there is no vacuum in the boiler before attempting to remove the manhole cover or handholes.
- To avoid the possibility of electrocution, never use conventional droplights in the steam and water side of boiler.

7. *Describe steam and condensate flow and circulating water flow in a surface condenser.*

Answer: Steam enters the surface condenser at the top and flows around the condenser tubes. The condensate falls to the bottom of the condenser to the hot well where it is removed by a condensate pump. The circulating water enters the condenser on the side, passing through the water box and into the condenser tubes. The cooling water makes two or more passes before the water leaves the condenser. The flow is designed so that steam coming in will come in contact with the condenser tubing and condenser as quickly as possible to create a vacuum. The water passes through the tubes to prevent them from collapsing.

8. *What can cause the temperature of gases of combustion to increase?*

Answer: The temperature of gases of combustion can increase because of poor heat transfer because of soot on the tubes, poor heat transfer because of scale on the tubes, and broken or deteriorated baffles in gas passes, causing the gases of combustion to short-circuit.

9. *Why is a watertube boiler safer to operate than a firetube boiler?*

Answer: A firetube boiler having the same BHP as a watertube boiler has a larger volume of water. A failure on its pressure side makes a boiler explosion more likely. If there is a failure in a watertube boiler, a tube will rupture but not explode.

In a firetube boiler, gases of combustion are directly impinged on the drum or internal furnace, whereas in a watertube boiler gases of combustion are passed only around the tubes. There is less chance of overheating the metal. Firetube boilers have extensive flat surfaces requiring elaborate use of stays that is not necessary on watertube boilers.

10. *Convert the following to degrees Celsius:*
Answer:

$$°Celsius = \frac{°Fahrenheit - 32}{1.8}$$

$$°C = \frac{°F - 32}{1.8}$$

a. 140°F

$$°C = \frac{140 - 32}{1.8}$$

$$°C = \frac{108}{1.8}$$

$$°C = \mathbf{60}$$

b. 100°F

$$°C = \frac{100 - 32}{1.8}$$

$$°C = \frac{68}{1.8}$$

$$°C = \mathbf{37.78}$$

c. 40°F

$$°C = \frac{40 - 32}{1.8}$$

$$°C = \frac{8}{1.8}$$

$$°C = \mathbf{4.44}$$

d. 10°F

$$°C = \frac{10 - 32}{1.8}$$

$$°C = \frac{-22}{1.8}$$

$$°C = \mathbf{-12.22}$$

e. −12°F

$$°C = \frac{-12 - 32}{1.8}$$

$$°C = \frac{-44}{1.8}$$

$$°C = \mathbf{-24.44}$$

f. −32°F

$$°C = \frac{-32 - 32}{1.8}$$

$$°C = \frac{-64}{1.8}$$

$$°C = -35.56$$

Test 6

1. *Two boilers are on-line and are cycling ON and OFF. What effect does this have on boiler efficiency and boiler maintenance?*
Answer: Combustion and boiler efficiency are lower when a boiler cycles ON and OFF. The reason for this is that the furnace cools during the OFF cycle and requires time to reach a good combustion temperature when starting. Heat is also lost as air passes through the furnace during the OFF cycle. There is a greater chance of the boiler heating surfaces getting coated with soot. Maintenance work in the furnace area will increase because of the heating and cooling of the refractory. The refractory will break up under those conditions. Cleaning of boiler tubes must be done more frequently.

2. *A plant has two 600 HP boilers. One operates at 300 psi and the other at 600 psi. Which boiler needs the larger safety valve area and why?*
Answer: The 300 psi boiler will need the larger safety valve area than the 600 psi boiler because of the lower differential across the valves. The lower the pressure is, the larger the valve needed. This can be shown using the following equation and making a comparison of the valve areas.

300 psi	600 psi
$A = \dfrac{0.2W}{P + 10}$	$A = \dfrac{0.2W}{P + 10}$

A = area of safety valve seat
W = pounds of steam generated per hour
P = pressure on boiler

Assuming a factor of evaporation of 1, both cases will be 600×34.5, or 20,700 lb/hr (*Note: W = HP × 34.5*).

300 psi	600 psi
$A = \dfrac{0.2 \times 20,700}{300 + 10}$	$A = \dfrac{0.2 \times 20,700}{300 + 10}$
$A = \dfrac{4140}{310}$	$A = \dfrac{4140}{610}$
$A = \mathbf{13.35}$ **sq in.**	$A = \mathbf{6.79}$ **sq in.**

The 300 psi boiler would require an opening area of 13.35 sq in., whereas the 600 psi boiler would require only a 6.79 sq in. opening.

3. *What is an atmospheric relief valve, where is it located, and how is it constructed?*
Answer: An atmospheric relief valve is used to prevent pressure buildup on the exhaust side of a reciprocating steam engine or steam turbine. They can be located on the main exhaust line or attached to the condenser as close to the exhaust line as possible. In plants running condensing, they are called atmospheric relief valves.

In plants running noncondensing, they are called back pressure valves. The atmospheric relief valve is designed to stay closed as long as the condenser is under a vacuum. If the vacuum is lost, it will open, allowing steam to exhaust to the atmosphere. This will protect the condenser from excessive steam pressure, which could cause tubes to collapse. The atmospheric relief valve has a water seal to prevent air from leaking into the condenser. The water seal must be adjusted so there is a slight drip from the overflow to ensure that the valve is sealed.

4. *On the steam side of a reciprocating pump, is the steam inlet or the exhaust outlet larger? Explain.*
Answer: The exhaust outlet is larger than the steam inlet. To illustrate this, the specific volume at the specified pressure must be known.
Example:
    100 psi inlet = 3.89 cu ft/lb
    0 psi outlet = 26.8 cu ft/lb
As shown, the steam outlet requires six times the area because of the increase in volume.

5. *What can cause a high water level in a boiler?*
Answer: The following are reasons for a high water level in a boiler:
  • The top line to the gauge glass is closed or plugged, giving a false reading.
  • There is a faulty feedwater regulator.
  • There is a sudden demand for steam.

6. *Explain how thrust is controlled on a steam turbine.*
Answer: End thrust or axial thrust on a steam turbine must be carefully controlled because of the very close clearances. It can be taken care of using a Kingsbury thrust bearing, double-flow steam turbine, or dummy pistons.

7. *What is the maximum and minimum size of a bottom blowdown valve and surface blowdown valve?*
Answer: The maximum size of a bottom blowdown and surface blowdown valve is 2½". The minimum size of a bottom blowdown valve is 1" except for boilers with less than 100 sq ft of heating surface. They may use a ¾" valve. There is no minimum size for surface blowdown valves.

8. *Define power factor:*
Answer: Power factor is (1) that portion of energy in an AC circuit that can be applied to useful work or (2) the phase relationship between current and voltage.

9. *What is the total force on a steam drum 40" in diameter, 20' long, with a 450 psi steam pressure?*
Answer: The total force on the drum is the force on the shell plus the force on the two heads.
$$TF_S = P \times A_S$$
$$A_S = C \times L$$
$$A_S = 3.1416 \times D \times L$$
TF_S = total force on shell
P = steam pressure, in psi
A_S = area of boiler shell, in sq in.
C = circumference of shell

L = length of shell
$TF_S = 450 \times 3.1416 \times 40 \times (20 \times 12)$ (convert 20′ to inches)
$TF_S =$ **13,571,712 lb**
TF on the two heads:
$TF_h = 2 \times P \times A_h$
TF_h = total force on head, in lb
P = pressure on head, in psi
A_h = area of front and rear heads, in sq in.
$A_h = 0.7854D^2$
$TF_h = 2 \times P \times 0.7854D^2$
$TF_h = 2 \times 450 \times 0.7854 \times 40 \times 40$
$TF_h =$ **1,130,976 lb**
TF on drum:
$TF_d = TF_S + TF_h$
$TF_d = 13,571,712 + 1,130,976$
$TF_d =$ **14,702,688 lb**

10. *Describe the procedure for laying up a boiler for a short period and long period.*
Answer: A boiler can be laid up wet or dry. The method of lay-up used depends on the length of lay-up time and plant conditions. If the boiler is needed on short notice, the wet method is recommended. If the boiler is to be laid up for an extended period or if there is danger of freezing, the dry method is recommended.
Wet lay-up:
   1. The boiler must be cleaned on both the steam and water side and fire side.
   2. Inspect the steam and water side for tools, rags, etc. Then, using new gaskets, close it up.
   3. Using warm chemically treated water (about 100°F), fill the boiler until water comes out of the air cock.
   4. Secure the air cock; maintain boiler pressure slightly above atmospheric pressure.
Dry lay-up:
   1. The boiler must be thoroughly cleaned on both the steam and water side and fire side.
   2. The boiler must be completely dry.
   3. Secure the steam and water lines so that no moisture can enter the boiler.
   4. Place a tray of moisture-absorbing chemicals in both fire and water sides of the boiler.
   5. Using new gaskets, close up the steam and water side of the boiler.
   6. Check chemicals periodically and replace them when necessary. The ASME code recommends using either quick lime or silica gel.

Test 7

1. *Describe the function of a hydraulic governor on a steam turbine.*
Answer: Hydraulic governors are used when the speed of the steam turbine must remain constant, as is the case of generating electricity. The basic Westinghouse oil-relay governor is a completely enclosed hydraulic system using a single source of oil for governing and steam turbine lubrication. The four essential elements are
   • oil pump for supplying lubricating oil for governing and lubrication
   • pair of governor weights secured to a spring steel strap mounted across the governor hub and driven from the steam turbine rotor shaft

- governing oil valve that transforms changes in force received from the governor weights into oil pressure changes. It acts as an oil relief valve, the pressure setting of which depends on the turbine speed
- servo-motor that actuates the steam admission valve in response to governing oil pressure changes

In the hydraulic governor, the main oil pump supplies oil at a constant high pressure (approximately 135 psi) for operation of the governor. The small amount of oil used by the speed-sensing element enters below the governor cup valve through a fixed orifice. An increase in speed causes an increase of centrifugal force on the governor weights, which causes a decrease in governor oil pressure signal.

The decreased governor oil pressure signal, which acts on the lowest diaphragm chamber in the speed changer, indicates a decreased oil pressure signal in the servo-motor pilot valve. The decreased oil pressure signal causes a decrease in the upward force on the servo-motor pilot valve. This causes a downward movement of the pilot valve, which positions the ports, admitting high-pressure oil to the chamber beneath the servo-motor piston and simultaneously connecting the chamber above the piston to the drain. As the piston moves upward, it closes the steam admission valves. The compensating lever and spring return the pilot valve to mid-position. The steam turbine speed can be adjusted manually using the knurl knob on the top of the speed changer or automatically from a remote location.

2. *What is the water rate of a steam turbine that operates condensing? What is the water rate if it operates noncondensing?*

   Answer: The average water rate for a condensing steam turbine is 14 lb of steam/kW. For a noncondensing steam turbine, it is 24 lb of steam/kW. The water rate of a steam turbine can be found using the following equation:

   $$WR = \frac{Btu / kW}{H_a - H_e}$$

   WR = water rate, in lb steam/kW
   Btu/kW = 3413 Btu, heat in 1 kW
   H_a = enthalpy of steam at throttle
   H_e = enthalpy of steam at exhaust

   Assume the following values for a condensing steam turbine: Btu/kW = 3413 Btu; H_a = 1300 Btu/lb; H_e = 1050 Btu/lb.

   $$WR = \frac{3413}{1300 - 1050}$$

   $$WR = \frac{3413}{250}$$

   WR = **13.65 lb/kW**

   Assume the following values for a noncondensing steam turbine: Btu/kW = 3413 Btu; H_a = 1300 Btu/lb; H_e = 1150 Btu/lb.

   $$WR = \frac{3413}{1300 - 1150}$$

   $$WR = \frac{3413}{150}$$

   WR = **22.75 lb/kW**

3. *What effect is there in raising the boiler pressure from 250 psi to 300 psi?*
Answer: Raising the boiler pressure would decrease heat transfer between the gases of combustion and water in the boiler because of the increase of water temperature from 400°F to 417°F. If additional heating surface such as an air heater or economizer were added, it would be possible to increase boiler efficiency.

4. *What maintenance is needed on a centrifugal feedwater pump?*
Answer: The following maintenance procedures are needed on a centrifugal feedwater pump.
1. Packing should be tightened or replaced when needed. Shaft seals should be replaced when the pump is completely overhauled or when it is leaking.
2. Check alignment and condition of shaft coupling. Realign or replace.
3. During a complete overhaul, replace worn rings, bearings, and shaft if damaged from gland packing. Clean and replace oil to bearings. Inspect impeller for wear. Replace all gaskets.

5. *What are possible causes of tube ruptures in watertube boilers?*
Answer: The following are possible causes of tube ruptures in watertube boilers:
• Scale on the water side of tubes causes overheating and blistering of tubes, which eventually leads to tube failure.
• Low water condition can lead to overheated tubes and possible rupture.
• Blocked tube that lacks water circulation will eventually blister and rupture.
• Oil on boiler heating surfaces prevents heat transfer and can cause a rupture.
• Soot blower elements out of alignment can cut tube with steam while blowing soot. The leaking tube will eventually rupture.

6. *If a steam line ruptures in a steam turbine room, what would be the effects on the operating steam turbine?*
Answer: The steam turbine that the steam line was feeding would immediately start to slow down. The generator would be out of phase, and if there were another generator on-line running in parallel, it would try to pick up the full load. If the operator could not dump some electric load, the generator would probably trip on overload and cause a plant blackout.

7. *What limits the degrees of superheat on steam turbines?*
Answer: The degrees of superheat in a steam turbine is limited by the metals used. Temperatures up to 1100°F are frequently used on large steam turbines.

8. *A boiler operates at 400 psia and the temperature of the feedwater entering the boiler is 180°F. It generates 12 lb of steam for each lb of fuel burned, and the fuel has a Btu content of 18,600 Btu/lb. What is the thermal efficiency of the boiler?*
Answer: The thermal efficiency of a boiler is determined by using the following equation:

$$TE = \frac{W_s\left[H_s - \left(T_{fw} - 32\right)\right]}{W_f \times C}$$

TE = thermal efficiency
W_S = pounds of steam per hour
H_S = enthalpy of steam
T_{fw} = feedwater temperature
32 = base temperature for determining enthalpy (heat in substance)

W_f = pounds of fuel/hr
H_f = heat available in fuel/lb
W_s = 12 lb/hr
H_s = 1204.5 Btu/lb
T_{fw} = 180°F
W_f = 1 lb/hr
H_f = 18,600 Btu/lb

$$TE = \frac{12\left[1204.5 - (180 - 32)\right]}{18,600 \times 1}$$

$$TE = \frac{12 \times 1056.5}{18,600}$$

$$TE = \frac{12,678}{18,600}$$

$TE = \mathbf{0.6816}$ or $\mathbf{68.16\%}$

9. *Define the following terms:*
   Answer:
   a. Fire point: The temperature at which fuel oil burns continuously when exposed to an open flame.
   b. Pour point: The lowest temperature at which fuel oil will flow as a liquid.
   c. Flash point: The temperature at which heated fuel oil produces a vapor that flashes when exposed to an open flame.
   d. Viscosity: Measurement of the internal resistance of a fluid to flow. By heating a fluid, its viscosity is lowered.

10. *What are the most common causes of boiler accidents and explosions?*
    Answer: The following are the most common causes of boiler accidents and explosions:
    • low water, which could cause tube failure
    • improper purging of furnace
    • failure to test low water fuel cutoff, flame scanner, and safety valves
    • improper feedwater treatment that could lead to scale buildup, corrosion, oxygen pitting, or caustic embrittlement

Test 8

1. *What are advantages of using a steam atomizing burner?*
   Answer: Advantages of using a steam atomizing burner are less preheating of fuel oil, lower fuel oil pressure needed, ability to burn dirtier fuel oil, and lower initial cost.

2. *What is the belt speed on an engine with a 6′ diameter pulley running at 150 RPM?*
   Answer: The belt speed can be found by using the following equation. (*Note:* Belt speed is the same as rim speed and is measured in feet per minute, or fpm.)
   Belt speed = Pulley circumference × RPM
   Circumference = π × Diameter of pulley, in ft
   Belt speed = 3.1416 × 6 × 150
   Belt speed = **2827.44 fpm**

3. *What are the most common causes of boiler accidents and explosions?*
Answer: The following are the most common causes of boiler accidents and explosions:
- low water, which could cause tube failure
- improper purging of furnace
- failure to test low water fuel cutoff, flame scanner, and safety valves
- improper feedwater treatment that could lead to scale buildup, corrosion, oxygen pitting, or caustic embrittlement

4. *What reverses the operation of a simplex pump at the end of its stroke?*
Answer: To reverse the operation of a simplex pump at the end of its stroke, an auxiliary piston is used. This is a steam-thrown steam valve and is attached to and drives the main steam valve. This admits steam to the steam cylinder.

5. *Is a reciprocating or centrifugal pump most likely to prime itself?*
Answer: A reciprocating pump is more likely to prime itself because it is a positive displacement pump. A centrifugal pump cannot lift water and must be fitted with a foot valve to keep the suction line filled with water or it must be primed or supplied with positive head pressure.

6. *Describe the test used on a boiler to determine the pressure-relieving capacity of the safety valves.*
Answer: An accumulation test is performed to determine the pressure-relieving capacity of safety valves by shutting off all steam outlets from the boiler and increasing the firing rate to a maximum. The safety valves should relieve all the steam without the pressure increasing more than 6% above the MAWP. During this test, the boiler operator must maintain an NOWL.

7. *If a steam line ruptures in a steam turbine room, what would be the effects on the operating steam turbine?*
Answer: The steam turbine that the steam line was feeding would immediately start to slow down. The generator would be out of phase, and if there were another generator on-line running in parallel, it would try to pick up the full load. If the operator could not dump some electric load, the generator would probably trip on overload and cause a plant blackout.

8. *What is the purpose of preheating No. 6 fuel oil?*
Answer: No. 6 fuel oil is preheated to lower its viscosity so it can be pumped through pipelines, increase the fuel oil temperature so it can be properly atomized by the burner, and prevent the fuel oil from solidifying in fuel oil storage tanks.

9. *What is the maximum and minimum size of a bottom blowdown valve and surface blowdown valve?*
Answer: The maximum size of a bottom blowdown and surface blowdown valve is 2½″. The minimum size of a bottom blowdown valve is 1″ except for boilers with less than 100 sq ft of heating surface. They may use a ¾″ valve. There is no minimum size for surface blowdown valves.

10. *What is the principle of operation of an impulse steam turbine?*
Answer: An impulse steam turbine uses steam velocity as a force acting in a forward direction on a blade or bucket mounted on a wheel. To produce the force, steam is routed through a nozzle and gains velocity before striking the blades on the wheel. This causes the shaft to rotate. After striking the blades, the steam exits through an exhaust opening.

Test 9

1. *Is a reciprocating or centrifugal pump most likely to prime itself?*
Answer: A reciprocating pump is more likely to prime itself because it is a positive displacement pump. A centrifugal pump cannot lift water and must be fitted with a foot valve to keep the suction line filled with water or it must be primed or supplied with positive head pressure.

2. *What is the purpose of preheating No. 6 fuel oil?*
Answer: No. 6 fuel oil is preheated to lower its viscosity so it can be pumped through pipelines, increase the fuel oil temperature so it can be properly atomized by the burner, and prevent the fuel oil from solidifying in fuel oil storage tanks.

3. *What are possible causes for a decrease in boiler efficiency?*
Answer: The following can cause a decrease in boiler efficiency:
- Soot accumulated on the boiler heating surfaces acts as an insulator.
- Scale buildup on the water side of heating surfaces reduces heat transfer.
- Oil on the water side of heating surfaces reduces heat transfer.
- Air-fuel ratio, improper atomization or temperature, or insufficient time to complete the combustion process can cause improper combustion.
- Baffles that direct the path of gases of combustion break, causing a short-circuiting of the gases.

4. *What is the principle of operation of a reaction steam turbine?*
Answer: A reaction steam turbine uses fixed blades designed so that each pair of blades acts as a single nozzle. Steam expands between each pair of blades and the steam gains velocity while losing pressure. The steam strikes the moving blades, giving up energy and losing velocity. As steam leaves, it gives a reactive force to the moving blades and a loss of pressure occurs.

5. *What type of governor is used on a steam turbine-driven centrifugal pump?*
Answer: A mechanical steam turbine governor with an overspeed trip device is used. This governor is not a constant-speed governor. It tends to vary in RPM slightly with changes in load. This is not critical when driving a pump, but is if connected to a generator.

6. *A reaction steam turbine and an impulse steam turbine have the same horsepower. Which is the larger machine and why?*
Answer: The reaction steam turbine is larger. The reaction steam turbine is a low-pressure steam turbine that is inefficient when used with high steam pressures because of steam losses around the blades. Steam pressure drops per element must be small. Therefore, the steam turbine is relatively large.
  The impulse steam turbine is primarily a high-speed machine that uses large pressure drops. Thermal efficiency is high since there is no steam loss around the blades. This makes the steam turbine small per unit of power output.

7. *How does a mechanical governor control the speed of a steam turbine?*
Answer: A mechanical governor with an overspeed trip is used to control speed of a steam turbine. Mechanical governors consist of a speed-sensing element and linkage that control the governor valve through the force acting on the speed-sensing

element. The speed-sensing element consists of weights mounted on the steam turbine shaft. The centrifugal force of the rotating weights is opposed by a spring that moves the linkage of the governor valve toward a closed position. The overspeed trip consists of a bolt mounted in the shaft and opposed by an adjustable spring. It remains in one position until the shaft speed reaches the point where the centrifugal force on the bolt overcomes the spring tension. It then flies out and trips a latch device that releases a butterfly valve, closing off the steam supply.

8. *In a process plant, is it more efficient to run a steam turbine that is generating electricity condensing or noncondensing?*
Answer: A condensing or noncondensing steam turbine generating electricity in a process plant can be equally efficient and would depend upon the type of operation. Noncondensing steam turbines used in steam plants that require large amounts of low-pressure process or heating steam can be run very efficiently as long as exhaust steam is used.

Condensing steam turbines operating on high exhaust vacuum can be run efficiently. The exhaust pressure allows steam to expand to a greater volume and give up more heat to perform more work. Some advantages and disadvantages of running condensing and noncondensing are as follows:
Condensing steam turbine:
   • Efficiency of the unit increases with increase in vacuum.
   • Condensate can be reclaimed.
   • Heat in condensate can be reclaimed.
Noncondensing steam turbine:
   • There is high efficiency if all of the exhaust steam is used.
   • There is low efficiency if the exhaust is allowed to go to the atmosphere.
   • There is no condensate return.
   Plant conditions determine the type of steam turbine best suited for a plant.

9. *What action should be taken if there is a leak between the boiler and the bottom blowdown valves?*
Answer: With a small leak, the boiler can still be operated if the water level can be easily maintained. However, the boiler should be removed from service for repairs as soon as possible. If maintaining a water level is difficult, the boiler should be secured immediately. All leaks on the blowdown line are considered dangerous. The boiler must be removed from service and cooled slowly. The repairs must be made in accordance with the ASME code.

10. *Define induced, forced, and natural draft.*
   Answer:
   • Induced draft is produced when the fan or blower located between the boiler and chimney pulls gases of combustion from the boiler and discharges them into the chimney.
   • Forced draft is produced when the fan or blower located in front of the boiler forces air into the furnace.
   • Natural draft is produced by a difference in the weight of hot gases of combustion inside a chimney and cold air outside the chimney.

Test 10

1. *What is the principle of operation of a reaction steam turbine?*
Answer: A reaction steam turbine uses fixed blades designed so that each pair of blades acts as a single nozzle. Steam expands between each pair of blades and the steam gains velocity while losing pressure. The steam strikes the moving blades, giving up energy and losing velocity. As steam leaves, it gives a reactive force to the moving blades and a loss of pressure occurs.

2. *Describe the test used on a boiler to determine the pressure-relieving capacity of the safety valves.*
Answer: An accumulation test is performed to determine the pressure-relieving capacity of safety valves by shutting off all steam outlets from the boiler and increasing the firing rate to a maximum. The safety valves should relieve all the steam without the pressure increasing more than 6% above the MAWP. During this test, the boiler operator must maintain an NOWL.

3. *In a process plant, is it more efficient to run a steam turbine that is generating electricity condensing or noncondensing?*
Answer: A condensing or noncondensing steam turbine generating electricity in a process plant can be equally efficient and would depend upon the type of operation. Noncondensing steam turbines used in steam plants that require large amounts of low-pressure process or heating steam can be run very efficiently as long as exhaust steam is used.

Condensing steam turbines operating on high exhaust vacuum can be run efficiently. The exhaust pressure allows steam to expand to a greater volume and give up more heat to perform more work. Some advantages and disadvantages of running condensing and noncondensing are as follows:
Condensing steam turbine:
• Efficiency of the unit increases with increase in vacuum.
• Condensate can be reclaimed.
• Heat in condensate can be reclaimed.
Noncondensing steam turbine:
• There is high efficiency if all of the exhaust steam is used.
• There is low efficiency if the exhaust is allowed to go to the atmosphere.
• There is no condensate return.
Plant conditions determine the type of steam turbine best suited for a plant.

4. *What is the function of a low water fuel cutoff on an automatic package boiler? How is the low water fuel cutoff tested?*
Answer: If a low water condition occurs in the boiler, the low water fuel cutoff will secure the fuel at the burner, shutting the burner down. An indicating light or alarm can also be used with the low water fuel cutoff.

The low water fuel cutoff can be tested by blowing down the float chamber of the control until the burner shuts down. The low water fuel cutoff can also be tested by shutting down the feedwater pump and letting the water level drop slowly until the burner shuts down. Whenever testing the low water cutoff, the burner must be firing.

5. *What can occur if fuel oil is present in the water side of a boiler?*
Answer: Fuel oil in the water side of a boiler can result in overheating of boiler tubes by retarding the transfer of heat. This could result in the formation of bags or blisters on

the tubes and lead to tube rupture. Fuel oil will also gather on the surface of the water level in the steam and water drum, increasing surface tension. This leads to foaming, priming, and carryover.

6. *List the auxiliaries connected with the operation of a condensing steam turbine and explain the operation of each.*
Answer:
- Condenser: Found on the exhaust side of the condensing steam turbine. It creates a vacuum, reducing the back pressure on the steam turbine and decreasing its water rate.
- Condensate pump: Removes water from the condenser hot well and discharges it to the open feedwater heater.
- Circulating water pump: Supplies cooling water to the condenser.
- Air ejectors: Connected to the condenser to remove air and other noncondensable gases to maintain a high vacuum.

7. *How is a steam turbine generator that is running condensing started?*
Answer: Once the boiler room is ready, the main steam line to the turbo-generator should be warmed and all condensate removed through a free-blowing drain. The warming up can be done by cracking the bypass around the main steam stop valve going to the steam turbine. After warming up the line, proceed as follows:
1. Start the auxiliary oil pump and check the oil pressure.
2. Check the level in the oil reservoir.
3. Check the cylinder drain valves. They should be in the open position.
4. Open the gland leak-off valves.
5. Drain the condensate from and admit steam to the main steam header up to the throttle valve.
6. Establish circulating water flow through the condenser.
7. Establish seals on the high- and low-pressure ends for starting.
8. Admit enough steam to start the rotor and then shut it off.
9. Listen for rubbing sounds.
10. If no rubbing is heard, admit enough steam to maintain 200 RPM for ½ hour to warm the casing and rotor evenly.
11. Turn on cooling water to the oil cooler to maintain 100°F outlet oil temperature.
12. Close the cylinder drains.
13. Adjust high- and low-pressure seals for normal operating conditions.
14. Start the condensate pump.
15. Seal the atmospheric valve.
16. Start the air ejectors.
17. Bring the steam turbine slowly up to rated speed until the governor takes over.
18. Trip the steam turbine emergency control to check operation.
19. Close the throttle valve and reset the trip. Bring the steam turbine up to rated speed. Make sure the governor takes over.
20. Check the oil temperature and the temperature of the cooling water in the air ejector. Check the temperature of the generator cooling medium.
21. Place the unit on-line as quickly as possible with at least 20% of the rated load.

8. *List the parts of a centrifugal pump and explain the purpose of each. Which parts wear the fastest?*
Answer: The parts of a centrifugal pump are
- Casing: Housing surrounding the impeller. The suction line and discharge line attach to the casing. The upper casing is removable and the lower casing is stationary.

- Impeller: Rotating element through which liquid passes and imparts energy to the liquid.
- Impeller shaft: Connected at one end to the motor or steam turbine. The impeller is attached to the shaft.
- Shaft bearings: Support the impeller shaft.
- Coupling: Attaches pump shaft, drive motor, or steam turbine.
- Packing or mechanical seals: Prevents leakage between pump casing and shaft.
- Wearing rings: Attaches to casing and are fixed. Impeller rotates inside wearing rings. Made to wear before the impeller.

The parts that wear the fastest are the mechanical seals, packing and shaft, wearing rings, coupling, and bearings.

9. *What are the advantages of superheated steam over saturated steam when used in a steam turbine?*
   Answer: Advantages of superheated steam over saturated steam are as follows:
   - higher thermal efficiency
   - more heat energy and therefore can do more work
   - less erosion of steam turbine blades
   - lowers steam turbine water rate
   - smaller steam lines can be used
   - no condensation in steam lines

10. *What is the specific volume of steam at 200 psi and at 28″ vacuum?*
    Answer: Steam at 200 psi has a specific volume of 2.13315 cu ft/lb. The specific volume of steam at 28″ vacuum is 333.6 cu ft/lb. Both amounts are listed in the steam tables.

Test 1

1. *What is the thermal efficiency of a straight condensing steam turbine? Enthalpy of steam at throttle is 1350 Btu/lb, enthalpy of steam at exhaust is 1050 Btu/lb, and enthalpy of condensate is 102 – 32.*

   Answer:

   $$TE = \frac{H_a - H_e}{H_a - (T_c - 32)}$$

   TE = thermal efficiency
   H_a = enthalpy of steam at throttle
   H_e = enthalpy of steam at exhaust
   T_c = enthalpy of condensate

   $$TE = \frac{1350 - 1050}{1350 - (102 - 32)}$$

   $TE = \mathbf{23.4\%}$

2. *Explain ultimate analysis and proximate analysis.*

   Answer: Ultimate analysis is used to determine the percentage of the various elements present in a coal specimen by weight. The NO CASH method is helpful in determining this.

   N - nitrogen
   O - oxygen
   C - carbon
   A - ash
   S - sulfur
   H - hydrogen

   The amount of each element present in the coal specimen determines the heating value, which is expressed in Btu/lb.

   Proximate analysis is used to determine the amount of moisture, volatile matter, fixed carbon, and ash in a coal specimen.

3. *Explain how a water column is installed on a boiler.*
Answer: A water column is located at the NOWL so that the lowest visible part of the gauge glass is 2″ to 3″ above the boiler's highest heating surface on a firetube boiler, and 2″ above the lowest permissible water level according to the boiler manufacturer's specifications in a watertube boiler. The top line is connected to the steam side of the boiler and must be above the top of the gauge glass. The bottom line is connected to the water side of the boiler and must be below the bottom of the gauge glass. The minimum size of water column piping is 1″ and there must be a cross T at any 90° bend for inspection. If there are valves on the steam or water connection line, they must be os&y gate valves or lever-type valves that show by their position if they are open or closed, and they should be locked or sealed open.

4. *How is an internal and external boiler inspection done?*
Answer: An internal inspection is made with the boiler off-line. After the boiler has been thoroughly cleaned on both fire and steam and water sides, proceed as follows:
1. Carefully check the fire side for signs of overheating of the boiler heating surface.
2. Check for proper alignment of all soot blower elements.
3. Check brickwork for signs of flame impingement.
4. Check all refractory for spalling.
5. Check the condition of burner diffusers.
6. Check the condition of all baffles for signs of deterioration.

After inspecting the fire side, inspect the steam and water side as follows:
1. Check for signs of oil, scale, pitting, and corrosion.
2. Check internal feedwater lines.
3. Check baffles, dry pipe, and any braces or stays.
4. Check all cross Ts on the water column, feedwater regulator, and low water fuel cutoffs to make sure they are all clear.
5. Check the bowl of the float type low water fuel cutoffs for sludge or scale buildup.
6. Check the operation of floats in the low water fuel cutoffs to ensure free operation.

An external boiler inspection should be made with the boiler under pressure as follows:
1. Test the popping pressure of the safety valves.
2. Check to see that the safety valves are not leaking.
3. Test the low water fuel cutoff.
4. Check for the proper firing cycle.
5. Test the flame failure control.
6. Check the bottom blowdown line and valves for leaks.
7. Check for leaks in the packing on all boiler valves.

5. *How many pounds of water does 1 gal. of No. 6 fuel oil evaporate? Heat available in 1 gal. of No. 6 fuel oil is 146,000 Btu/gal., heat in steam leaving the boiler is 1200 Btu/lb, temperature of feedwater entering boiler is 210°F, and boiler efficiency is 82%.*
Answer:

$$E = \frac{W\left[H_s - \left(T_{fw} - 32\right)\right]}{Gal. \text{ of } oil / hr \times Btu / gal.}$$

$$W = \frac{E \times Gal. \text{ of } oil/hr \times Btu/gal.}{H_s - \left(T_{fw} - 32\right)}$$

W = pounds of steam/hr
E = efficiency of unit
H_s = heat in steam leaving boiler
T_{fw} = feedwater temperature
$Btu/gal.$ = heat available in the fuel

$$W = \frac{0.82 \times 1 \times 146,000}{1200 - \left(210 - 32\right)}$$

W = **117.14 lb/hr** or **117.14 lb/gal. of fuel oil**

6. *Prove that there is a point on the Celsius and Fahrenheit scales where they read the same.*
   Answer: To solve this problem, two equations are necessary to find two unknowns.

   $$^\circ C = {}^\circ F$$
   $$^\circ C = \frac{F - 32}{1.8}$$
   $$^\circ C = \frac{C - 32}{1.8}$$

   Substitute C for F because $^\circ C = {}^\circ F$

   $$1.8C = C - 32$$
   $$1.8C - C = -32$$
   $$0.8C = -32$$
   $$C = \frac{-32}{0.8}$$
   $$C = -40$$
   $$^\circ C = {}^\circ F \text{ at } -40^\circ$$

7. *Describe the purpose and location of the following:*
   Answer:
   a. Continuous blowdown heat recovery system: Prevents heat losses through the blowdown of solids from the boiler water. Heat is reclaimed first by dropping the pressure on the boiler water so that the flash steam can be reclaimed, then heating makeup water with the blowdown water. The continuous blowdown line is connected to the steam and water drum well below the low water level.
   b. Total dissolved solids: Solids that circulate with the water in a boiler. The solids are formed by the combining of chemicals used in the boiler water treatment process and help remove foreign matter from the boiler.
   c. Dry pipe: Main function is to improve the quality of outgoing steam. It is found in the steam and water drum of some boilers.
   d. Scrubbers: Remove moisture and solids from the steam. Scrubbers are essential on boilers equipped with superheaters. They are located in the steam and water drum of watertube boilers.

8. *What are the harmful effects of scale on the heating surface of a boiler? Explain internal and external feedwater treatment for boilers.*
   Answer: Scale on the boiler heating surface may lead to overheating of boiler metal, resulting in blisters, bags, and possible tube rupture or boiler explosion.

Internal feedwater treatment is accomplished by adding chemicals to the boiler drum to control scale, corrosion, and pitting. External feedwater treatment is the treating of feedwater before it enters the boiler. This is accomplished using water softeners and open or deaerating feedwater heaters.

9. *A 60″ diameter boiler drum has ¾″ thick metal and a tensile strength of 60,000 psi. The joint efficiency is 90% and MAWP is 300 psi. What is the factor of safety?*
Answer:

$$MAWP = \frac{STE}{FR}$$

$$F = \frac{STE}{MAWP \times R}$$

MAWP = maximum allowable working pressure
S = tensile strength of metal (in psi)
T = metal thickness (in inches)
E = joint efficiency
F = factor of safety
R = radius of drum

$$F = \frac{60,000 \times 0.75 \times 0.90}{300 \times 30}$$

$$F = \mathbf{4.5}$$

10. *A boiler develops 5000 BHP. The temperature of the feedwater entering the boiler is 225°F and the enthalpy of the steam leaving the boiler is 1200 Btu/lb. How many pounds of steam per hour is this boiler generating?*
Answer:

$$W = \frac{BHP \times Btu / BHP}{H_s - \left(T_{fw} - 32\right)}$$

W = pounds of steam/hr
BHP = boiler horsepower
H_s = heat per pound
T_{fw} = feedwater temperature
Btu/BHP = 33,475

$$W = \frac{5000 \times 33,475}{1200 - \left(225 - 32\right)}$$

$$W = \mathbf{166,211.5 \ lb/hr}$$

11. *Explain how to install a steam pressure gauge on a boiler.*
Answer: The steam pressure gauge must have a range of 1½ to 2 times the safety valve popping pressure. It must be visible and protected from extreme heat or cold. A siphon must be installed to protect the Bourdon tube from live steam. One-quarter inch brass or copper pipe may be used for temperatures of up to 405°F. When steel or wrought iron pipe is used, they must be at least ½″ and the minimum size siphon is ¼″. Shut-off valves should be lever-handle cock or os&y gate valves that show by their position whether they are open or closed. The pressure gauge must be connected to the highest part of the steam space of the boiler.

12. *What is the function of an air ejector?*
Answer: An air ejector removes air and noncondensable gases from a surface condenser to help maintain a high vacuum.

13. *How many Btu are equivalent to 1 mechanical HP, 1 BHP, and 1 kW?*
Answer:
  a. 1 mechanical HP

$$1 \text{ mechanical HP: } Btu \, / \, hr = \frac{ft\text{-}lb \, / \, min \times 60 \text{ min/hr}}{Mechanical \ equivalent \ of \ heat}$$

$$Btu \, / \, hr = \frac{33,000 \times 60}{778}$$

Btu/hr = **2545**
  b. One BHP is also equivalent to 33,475 Btu/hr. This results from multiplying 970.3 Btu (the Btu needed for latent heat of evaporation) by 34.5 lb (pounds of water evaporated per hour from and at 212°F for 1 HP).
  Btu/hr = 970.3 Btu × 34.5 lb
  Btu/hr = **33,475**
  c. 1 kW: There are 1.341 mechanical HP in 1 kWh; therefore,
  Btu/kWh = 1.341 × 2545
  Btu/kWh = **3413** when taken to the third decimal place

14. *Define centrifugal and centripetal force in reference to a steam turbine.*
Answer: Centrifugal force and centripetal force can be seen in a steam turbine. Centrifugal force is the force that pulls blades or buckets away from the rotating shaft. Centripetal force is the force exerted inward toward the rotating shaft opposite centrifugal force.

15. *Define factor of evaporation and explain how it is used.*
Answer: Factor of evaporation is a correction factor for developed boiler horsepower (DBHP).

$$FE = \frac{H_s - \left(T_{fw} - 32 \right)}{970.3}$$

$$DBHP = \frac{W \times FE}{34.5}$$

FE = factor of evaporation
H_s = enthalpy of steam
T_{fw} = feedwater temperature
970.3 = latent heat of evaporation of water at 212°F
W = lbs of steam/hr
34.5 = pounds of water evaporated per hour from and at 212°F for 1 HP

16. *What is the purpose of a throttling calorimeter in a steam plant?*
Answer: A throttling calorimeter is used to measure the amount of moisture in steam. Once the quality of steam has been determined, the total heat per pound of steam can be found. With this figure, boiler thermal efficiency can be found.

17. *When two similar centrifugal pumps are connected in series, what happens to the discharge pressure and flow?*
Answer: The pressure doubles. This arrangement is the same as having a two-stage pumping unit. The volume through each pump remains unchanged.

18. *Is an electrically driven feedwater pump or a steam turbine-driven feedwater pump more economical to operate?*
Answer: There are many factors that must be considered in order to select one drive over the other. If it is possible to use all the steam turbine exhaust for heating or process, it is usually more economical to use the steam turbine-driven feedwater pump. However, if the electric load is extremely low, it may be necessary to use the electrically driven pump in order to balance the plant's steam electric load.

19. *What types of boilers are equipped with waterwalls and what purpose do they serve?*
Answer: Watertube boilers are normally equipped with waterwalls. Watertube boilers that burn pulverized coal or use stokers usually have waterwalls because of the high heat release in the furnace. Waterwalls prolong the life of the furnace refractory and also add to the heating surface of the boiler. One square foot of waterwall surface will generate as much as 16 lb of steam/hr.

20. *What is boiler heat balance?*
Answer: Boiler heat balance involves accounting for all the heat available in a fuel. It is called a balance because the sum of all the heat units consumed, either usefully or wastefully, must equal (balance) the number of heat units available in the fuel.
Heat in fuel = Heat absorbed by boiler + Losses in boiler
Losses can result from
- gases of combustion to atmosphere
- incomplete combustion
- moisture in fuel
- moisture in air
- burning hydrogen producing water vapor
- unburned combustibles
- radiation

Test 2

1. *Explain ultimate analysis and proximate analysis.*
Answer: Ultimate analysis is used to determine the percentage of the various elements present in a coal specimen by weight. The NO CASH method is helpful in determining this.
N - nitrogen
O - oxygen
C - carbon
A - ash
S - sulfur
H - hydrogen
The amount of each element present in the coal specimen determines the heating value, which is expressed in Btu/lb.
Proximate analysis is used to determine the amount of moisture, volatile matter, fixed carbon, and ash in a coal specimen.

2. *Give a brief description of the following valves:*
Answer:
   a. Atmospheric (relief) valve: Used to prevent a pressure buildup on the exhaust side of a reciprocating steam engine or steam turbine. It is located on the main

exhaust line or attached to the top of a surface condenser as close to the exhaust line as possible.

b. Nonreturn valve: Found on boiler main steam line. Located as close to the boiler shell as practical. Cuts the boiler in on-line automatically, cuts the boiler off-line automatically, and protects the system from draining in case one boiler fails.

c. Slide valve: Sometimes called a simple D type slide valve. Used on reciprocating steam pumps and engines. They are outside-admission inside-exhaust valves. One valve that controls the admission of steam to and the exhaust from both sides of the steam piston.

d. Relief valve: Spring-loaded valve used to relieve pressure on the discharge side of pumps. Also used on feedwater heaters, outlet side of fuel oil heaters, hot water tanks, and any other equipment needing protection from excessive pressure.

e. Pilot valve: Used to control another valve. It receives a small signal pressure and amplifies it to position another valve.

f. Needle valve: Micrometer type valve used to meter the flow of liquid.

3. *How does the condition of refractory in a boiler furnace affect the boiler efficiency?*
Answer: Furnace refractory in poor condition can affect boiler efficiency in the following manner:

• Infiltration of air to the furnace, giving excess air to the combustion process and resulting in loss of heat to the gases of combustion.

• Radiation loss through casing around the furnace area.

• Reduction of furnace temperature because of radiation loss from the refractory to the combustion process.

4. *A flat boiler surface is supported by 1" diameter staybolts. The staybolts are allowed a maximum stress of 7500 psi. If the boiler is to operate at an MAWP of 200 psi, what would be the maximum pitch or spacing of the staybolts?*
Answer:

$$MAWP = \frac{a \times C}{A}$$

$$A = \frac{a \times C}{MAWP}$$

$$A = \frac{0.7854 \times D^2 \times C}{MAWP}$$

$MAWP$ = maximum allowable working pressure
a = cross sectional area of staybolt
C = stress constant
A = area supported by one staybolt
D = diameter of staybolt

$$A = \frac{0.7854 \times 1 \times 1 \times 7500}{200}$$

$A = $ **29.4525 sq in.** or **5.43" × 5.43"**

5. *What routine maintenance is required to keep a boiler in good condition?*
Answer: The following is required to keep a boiler in good condition:

• Keep the boiler setting tight to prevent air leaks.

• Maintain proper feedwater temperature to reduce thermal shock.

- Set operating controls so that the burner runs for longer periods than it is off to protect the refractory from spalling as a result of cooling. Continuous operation is always best.
- Test the boiler feedwater daily and carefully control chemicals to prevent scale formation, oxygen pitting, and corrosion.
- Use bottom blowdown to remove sludge and sediment from the water side of the boiler.
- Use a soot blower to keep tubes clean.
- All valve packing should be replaced as necessary to prevent leakage.
- Lubricate valve spindles with a mixture of graphite and oil for easy opening and closing.

6. *What is the total force acting on the inside surface of a 60″ diameter boiler drum that is 18′ long and under a pressure of 400 psi?*
Answer:

   Total force = Pressure × Area
   The total force on the drum will be the force on the shell and the force on the two heads.
   TF on shell
   $$TF_s = P \times A_s$$
   $$TF_s = P \times Circumference \times L$$
   $$TF_s = P \times 3.1416 \times D \times L$$
   $$TF_s = 400 \times 3.1416 \times 60 \times 18 \times 12$$
   $$TF_s = \mathbf{16,286,054\ lb}$$
   TF on the two heads
   $$TF_h = P \times 2 \times A_h$$
   $$TF_h = P \times 2 \times 0.7854 \times D^2$$
   $$TF_h = 400 \times 2 \times 0.7854 \times 60 \times 60$$
   $$TF_h = \mathbf{2,261,952\ lb}$$
   TF on drum
   $$TF_d = TF_s + TF_h$$
   $$TF_d = 16,286,054 + 2,261,952$$
   $$TF_d = \mathbf{18,548,006\ lb}$$

7. *Describe the steps for paralleling two AC generators.*
   Answer: Assuming that the incoming turbo-generator is up to its rated speed, proceed as follows:
   1. Adjust voltage to line voltage with field rheostat.
   2. Frequency of both turbo-generators should be the same.
   3. With the use of a synchroscope, synchronize the turbo-generators. This is necessary to bring the turbo-generators in step.
   4. When the hand stops at the mark on the dial, indicating synchronization, close the breaker switch.
   5. Increase the load on the unit that just came on-line by slowly increasing the speed. Balance loads on machines.

8. *Prove that the stress on the longitudinal seam of a boiler drum is twice as great as the stress on the circumferential seam.*
   Answer: Stress is the result of force acting on an object. The material of the object resists this force. Stress is found by using the following equation:
   $$Stress = \frac{Force \text{ (acting on object)}}{Area \text{ (material resisting the force)}}$$

On the longitudinal seam,

$$S_l = \frac{Total\ force}{Area} = \frac{P \times A}{A_1}$$

$S_l = longitudinal\ stress$
P = pressure on the boiler drum
A = area the pressure is acting on
A_1 = area of material resisting the force

The area the pressure is acting on is the projected area A-B-C-D.

A (A-B-C-D) = $Diameter \times Length\ of\ drum$
$A = D \times L$

$$S_l = \frac{P \times D \times L}{A_1}$$

$A_1 = 2 \times L \times T$

$$S_l = \frac{P \times D \times L}{2 \times L \times T}$$

Simplifying

$$S_l = \frac{P \times D}{2T}$$

On the circumferential seam,

$$S_c = \frac{P \times A}{A_1}$$

S_c = circumferential stress
P = pressure on the boiler drum
A = area pressure is acting on
A_1 = area resisting circumferential break

The area on which the pressure is acting is the drum head. The area of the drum head is found by using the following equation:

$A = 0.7854D^2$

$$S_c = \frac{P \times 0.7854D^2}{A_1}$$

The area resisting the break is the product of the circumference of the drum and the thickness of the material.

A_1 = area of material resisting the force
$A_1 = Circumference \times Thickness$
$A_1 = \pi \times D \times T$

$$S_c = \frac{P \times 0.7854D^2}{3.1416 \times D \times T}$$

$$S_c = \frac{P \times D}{4T}$$

Comparing the two formulas shows that the stress on longitudinal seam is twice as great as the stress on the circumferential seam.

$$S_l = \frac{P \times D}{2T} \qquad S_c = \frac{P \times D}{4T}$$

9. *Explain how a water column is installed on a boiler.*

   Answer: A water column is located at the NOWL so that the lowest visible part of the gauge glass is 2″ to 3″ above the boiler's highest heating surface on a firetube boiler, and 2″ above the lowest permissible water level according to the boiler manufacturer's specifications in a watertube boiler. The top line is connected to the steam side of the boiler and must be above the top of the gauge glass. The bottom line is connected to the water side of the boiler and must be below the bottom of the gauge glass. The minimum size of water column piping is 1″ and there must be a cross T at any 90° bend for inspection. If there are valves on the steam or water connection line, they must be os&y gate valves or lever-type valves that show by their position if they are open or closed, and they should be locked or sealed open.

10. *How many pounds of water does 1 gal. of No. 6 fuel oil evaporate? Heat available in 1 gal. of No. 6 fuel oil is 146,000 Btu/gal., heat in steam leaving the boiler is 1200 Btu/lb, temperature of feedwater entering boiler is 180°F, and boiler efficiency is 83%.*

   Answer:

$$E = \frac{W\left[H_s - \left(T_{fw} - 32\right)\right]}{Gal.\ of\ oil\,/\,hr \times Btu\,/\,gal.}$$

$$W = \frac{E \times Gal.\ of\ oil\,/\,hr \times Btu\,/\,gal.}{H_s - \left(T_{fw} - 32\right)}$$

   W = pounds of steam/hr
   E = efficiency of unit
   H_s = heat in steam leaving boiler
   T_{fw} = feedwater temperature
   $Btu/gal.$ = heat available in the fuel

$$W = \frac{0.83 \times 1 \times 146,000}{1200 - \left(180 - 32\right)}$$

   W = **115 lb/hr** or **115 lb/gal. of fuel oil**

11. *What types of boilers are equipped with waterwalls and what purpose do they serve?*

    Answer: Watertube boilers are normally equipped with waterwalls. Watertube boilers that burn pulverized coal or use stokers usually have waterwalls because of the high heat release in the furnace. Waterwalls prolong the life of the furnace refractory and also add to the heating surface of the boiler. One square foot of waterwall surface will generate as much as 16 lb of steam/hr.

12. *A 60″ diameter boiler drum has a tensile strength of 60,000 psi, boiler seam efficiency of 90%, factor of safety of 5, and an MAWP of 250 psi. Find the thickness of the plate required.*

    Answer:

$$MAWP = \frac{STE}{FR}$$

$$T = \frac{MAWP \times F \times R}{SE}$$

$MAWP$ = maximum allowable working pressure
S = tensile strength of material, in psi
T = thickness of metal, in inches
E = efficiency of joint
F = factor of safety
R = radius of the boiler drum, in inches

$$T = \frac{250 \times 5 \times 60}{60,000 \times 0.90}$$

T = **1.39″**

13. *How much does steam expand going from 200 psia to 28″ Hg vacuum (change in volume)?*
Answer: From the steam tables, the specific volume of steam at 200 psia is 2.288 cu ft/lb, and at 28″ vacuum it is 333.6 cu ft/lb. The amount of expansion is from 2.288 cu ft/lb to 333.6 cu ft/lb. The change in volume is 333.6 divided by 2.288, or 145.8 times.

14. *Describe how the thermal efficiency of a steam turbine is affected when steam bled from various stages is used to heat feedwater.*
Answer: The thermal efficiency of the steam turbine is raised. When steam is bled from various stages it can no longer be charged to the steam turbine because the heat has been reclaimed. The main increase comes from the saving of latent heat, which is used to heat the feedwater. Normally the latent heat in the steam is lost in the condenser.

15. *How much torque is developed on the shaft of a 250 HP steam turbine operating at 1800 RPM at full load?*
Answer: With turbo-generators, torque on the shaft is found by using the following equation:

$$T_q = \frac{5252 \times HP}{RPM}$$

Note: 5252 is a constant.

$$T_q = \frac{5252 \times 250}{1800}$$

T_q = **729.4 ft-lb**

16. *A boiler produces 80,000 lb of steam/hr at 250 psi. The steam has a heat content of 1200 Btu/lb. The feedwater entering the boiler is 210°F. What is the DBHP?*
Answer:

$$DBHP = \frac{W\left[H_s - \left(T_{fw} - 32\right)\right]}{33,475}$$

W = pounds of steam/hr
H_s = enthalpy of steam
T_{fw} = feedwater temperature
33,475 = heat in 1 BHP

Note: Total heat output (in Btu) by the boiler per hour is determined.

$$DBHP = \frac{80,000\left[1200-(210-32)\right]}{33,475}$$

$$DBHP = \frac{80,000(1200-178)}{33,475}$$

$$DBHP = \frac{80,000 \times 1022}{33,475}$$

$$DBHP = \mathbf{2442.42}$$

17. *Why do firetube boilers require stays?*
Answer: Firetube boilers have extensive flat surfaces that require stays. Tube sheets, crown sheets, water legs, and wrapper sheets all need stays. The forces acting on a flat surface tend to act in one direction whereas forces acting on a curved surface do not.

18. *Define the following:*
Answer:
   a. Tensile stress: Occurs when two forces of equal intensity act on an object, pulling in opposite directions. An example is the pressure within a boiler drum trying to pull the drum apart.
   b. Compression stress: Occurs when two forces of equal intensity act from opposite directions, pushing toward the center of an object. Support beams on a watertube boiler are subject to compression stress as are the firetubes in a firetube boiler.
   c. Shear stress: Occurs when two forces of equal intensity act parallel to each other but in opposite directions. Rivets are subjected to shear stress.

19. *Define the following:*
Answer:
   a. Perfect combustion: The burning of all the fuel supplied to the boiler using only the theoretical amount of air required. Only can be accomplished in a laboratory, never in actual plant operation.
   b. Complete combustion: The burning of all the fuel supplied to the boiler using the minimum amount of excess air.
   c. Incomplete combustion: Occurs when all the fuel supplied to the boiler is not burned, resulting in soot and smoke.

20. *Why do staybolts have holes drilled in the ends? What is the diameter and depth of the drilled holes?*
Answer: Staybolts have holes drilled in the ends to warn the operator if the bolts have sheared or corroded. Water will come out of the telltale hole of the staybolt, indicating that it has failed. The hole drilled in the end is 3/16″ in diameter and must extend at least 1″ into the water side of the boiler. Leaky staybolts must be replaced as soon as possible.

Test 3

1. *A 60″ diameter boiler drum has a tensile strength of 50,000 psi, boiler seam effi-ciency of 90%, factor of safety of 5, and an MAWP of 200 psi. Find the thickness of the plate required.*
Answer:

$$MAWP = \frac{STE}{FR}$$

$$T = \frac{MAWP \times F \times R}{SE}$$

$MAWP$ = maximum allowable working pressure
S = tensile strength of material, in psi
T = thickness of metal, in inches
E = efficiency of joint
F = factor of safety
R = radius of the boiler drum, in inches

$$T = \frac{200 \times 5 \times 30}{50,000 \times 0.90}$$

$$T = \mathbf{0.67''}$$

2. *A 10′ long boiler has a pressure of 150 psi, 36″ drum diameter, and ⅝″ metal thickness. What is the stress on the circumferential and longitudinal seams?*
Answer:

$$S_c = \frac{P \times D}{4T}$$

$$S_l = \frac{P \times D}{2T}$$

S_c = stress on circumferential seam
S_l = stress on longitudinal seam
P = pressure drum
D = diameter of drum
T = thickness of drum

$$S_c = \frac{150 \times 36}{4 \times 0.625}$$

$$S_c = \mathbf{2160\ psi}$$

$$S_l = \frac{150 \times 36}{2 \times 0.625}$$

$$S_l = \mathbf{4320\ psi}$$

3. *When is the best time to blow down the waterwall on a watertube boiler? Explain in detail.*
Answer: Waterwalls are located in the furnace and are subjected to extremely high temperatures (up to 2500°F). They are never blown down until the furnace has cooled down. The blowing down is usually done when the boiler is being tak-en off-line, after the furnace has cooled but while the boiler still has some steam pressure. The waterwall blowdown valves should be kept locked at all other times.

4. *Describe the flow of steam from the boiler to a desuperheating station.*
Answer: At the boiler, water is changed into steam and passes through the super-heater, absorbing more heat. From the superheater, steam enters the steam header and divides into branches that go to the steam turbine throttle valve, steam auxil-iaries, and pressure-reducing and desuperheating stations.

5. *Describe the operation of a two-element feedwater regulator.*
Answer: The two-element feedwater regulator is found on medium to large boilers that are subject to sudden load changes, such as those in process plants. They are controlled by two elements. One is the rise and fall of the water level in the boiler. The other is the steam flow from the boiler measured by taking the difference in pressure across the superheater, if the boiler is so equipped, or between two points between an orifice in the main steam line. This pressure difference is applied across a diaphragm and gives motion to a valve mechanism on the feedwater regulator.

6. *A condensate pump on a surface condenser operates with a 29" vacuum suction and a 100' discharge head. It pumps 50,000 lb of water/hr. If the unit is 85% efficient, how much HP is required to drive this unit?*
Answer:

$$HP_d = \frac{ft\text{-}lb\,/\,min}{33,000 \times E}$$

HP_d = drive horsepower
$ft\text{-}lb/min$ = ft-lb/min of power needed

$$33,000 = \frac{ft\text{-}lb\,/\,min}{HP}$$

E = efficiency
$H_t = H_d + H_f - H_s$
$H_t = H_d + H_f + SL$
H_t = total head, in ft
H_d = discharge head, in ft
H_f = friction head, in ft
H_s = suction head, in ft
SL = suction lift, in ft
$H_t = 100 + 0 + 32.8$
$H_t = \mathbf{132.8'}$

The pounds of water per minute will be 50,000 divided by 60, or 833.3 lb/min.

$$HP_d = \frac{ft\text{-}lb\,/\,min}{33,000 \times E}$$

$$HP_d = \frac{132.8 \times 833.3}{33,000 \times 0.85}$$

$HP_d = \mathbf{3.95}$

7. *A boiler produces 60,000 lb of steam/hr at 250 psi. The steam has a heat content of 1200 Btu/lb. The feedwater entering the boiler is 180°F. What is the DBHP?*
Answer:

$$DBHP = \frac{W\left[H_s - \left(T_{fw} - 32\right)\right]}{33,475}$$

W = pounds of steam/hr
H_s = enthalpy of steam
T_{fw} = feedwater temperature
33,475 = heat in 1 BHP
Note: Total heat output (in Btu) by the boiler per hour is determined.

$$DBHP = \frac{60,000\left[1200-(180-32)\right]}{33,475}$$

$$DBHP = \frac{60,000\left(1200-148\right)}{33,475}$$

$$DBHP = \frac{60,000 \times 1052}{33,475}$$

$$DBHP = \mathbf{1886}$$

8. *How many pounds of steam are required to generate 1 kW of electricity?*
Answer: It depends on whether the steam turbine is operated condensing or noncondensing.

$$WR = \frac{Btu\,/\,kW}{H_a - H_e}$$

WR = water rate, in lb steam/kW
Btu/kW = 3413 Btu, heat in 1 kW
H_a = enthalpy of steam at throttle
H_e = enthalpy of steam at exhaust

Assume the following values for a condensing steam turbine: H_a = 1300 Btu/lb; H_e = 1050 Btu/lb.

$$WR = \frac{3413}{1300 - 1050}$$

$$WR = \frac{3413}{250}$$

$$WR = \mathbf{13.65\ lb\ of\ steam/kW}$$

Assume the following values for a noncondensing steam turbine: H_a = 1300 Btu/lb; H_e = 1150 Btu/lb.

$$WR = \frac{3413}{1300 - 1150}$$

$$WR = \frac{3413}{150}$$

$$WR = \mathbf{22.75\ lb\ of\ steam/kW}$$

9. *How are expansion and contraction compensated for in main steam lines?*
Answer: Expansion and contraction of the main steam line going from the boiler to the main steam header are compensated for by expansion bends. Properly designed expansion bends allow for considerable expansion and contraction without putting a strain on steam line flanges. Expansion joints are also used to allow for expansion and contraction of main steam lines in the plant. Expansion joints require some maintenance whereas expansion bends are maintenance-free.

10. *If 10 cu ft of air are heated at a constant pressure from 70°F to 335°F, what is the final volume?*

Answer:

$$\frac{V_1}{T_1} = \frac{V_2}{T_2} \;\; or \;\; V_2 = \frac{V_1 T_2}{T_1}$$

V_1 = initial volume, in cu ft
V_2 = final volume, in cu ft
T_1 = initial temperature, in °R
T_2 = final temperature, in °R

Assume the following values: V_1 = 10 cu ft; T_1 = (70 + 460); T_2 = (335 + 460).

$$V_2 = \frac{V_1 T_2}{T_1}$$

$$V_2 = \frac{10 \times 795}{530}$$

$$V_2 = \textbf{15 cu ft}$$

11. *How are the blowdown line and valves installed on a high-pressure boiler?*
Answer: The bottom blowdown line and valves are under the jurisdiction of the ASME code. All boilers must have a blowdown line fitted with a valve or cock. The blowdown line must be located at the lowest part of the water side of a firetube boiler or the mud drum of a watertube boiler. Boilers carrying 100 psi or more must have two bottom blowdown valves. They may be two slow-opening valves that require five full turns of the handwheel to open or close fully, or one quick-opening and one slow-opening valve. If a quick-opening valve is used, it must be the valve closest to the boiler shell.

The ASME code further states that when the MAWP exceeds 125 psi, the blowdown line and fitting between the boiler and valves must be extra-heavy bronze, brass, or malleable iron, depending on the temperatures and pressure involved. The bottom blowdown lines must run full size and have no reducers or bushings. The two bottom blowdown valves shall be at least 250 psi standard valves and be extra-heavy. If the bottom blowdown line is exposed to direct furnace heat, it must be protected by a refractory pier or wrapped with a protective insulation. The bottom blowdown line must be installed so that it can be inspected and the opening in the boiler setting arranged to provide for free expansion and contraction of the bottom blowdown lines.

12. *Three 2500 kW AC turbo-generators are running in parallel. If the governor on one fails toward the closed position, what would happen to each unit?*
Answer: The steam turbine with the governor that failed would run at a constant load while the other two units would share the remaining load equally. All units would be in phase, providing the steam is not cut off completely by the failed governor, in which case the unit would trip electrically from the board. The remaining two units would carry the remaining load, provided the load is not above their capacity. If it is too high, the generators could overheat, causing a thermal overload, and disconnect from the board.

13. *What size steam line is needed for a saturated steam boiler that carries 200 psi pressure and generates 60,000 lb of steam/hr? The velocity in the line must not exceed 5000 fpm. (The specific volume of steam at 200 psi is 2.15 cu ft/lb.)*

Answer:

$$V = \frac{Q}{A}$$

V = velocity of fluid (in fpm)
Q = quantity of fluid [in cubic feet per minute (cfm)]
A = cross-sectional area of pipe (in sq ft)
V = 5000 feet per minute

$$Q = \frac{Pounds\ of\ steam\,/\,hr}{60\ min/hr} \times Specific\ volume$$

$$Q = \frac{60,000}{60\ min/hr} \times 2.15$$

Q = 2150 cfm

$$A = \frac{Q}{V}$$

$$A = \frac{2150}{5000}$$

A = 0.43 sq ft
A = 0.43 x 144 (to convert to sq in.)
A = 61.92 sq in.
$A = 0.7854D^2$

$$D = \sqrt{\frac{A}{0.7854}}$$

$$D = \sqrt{\frac{61.92}{0.7854}}$$

$D = \mathbf{8.88''}$

14. *Describe a chain grate stoker.*
   Answer: A chain grate stoker's parts include a coal conveyor that supplies coal to a coal hopper, regulating coal gate that controls the depth of the coal on the grate, and a traveling grate driven by an electric motor or a steam engine. Coal is deposited on the front of the grate and travels into the furnace. By raising or lowering the coal gate, the depth of coal can be varied. An ignition arch made of refractory reflects heat downward to the green coal, igniting it. The ignition arch also helps burn the volatile gases as they are distilled off the coal.
   Coal best suited for chain grate stokers is noncaking bituminous coal with a relatively high ash content. The ash content helps protect the grate from overheating. Air for combustion is introduced through windboxes that are separately zoned and individually adjusted to keep blowing the coal to the front of the furnace to ensure it is burned completely before it drops into the ash hopper at the end and below the grates. Secondary air is introduced into the front of the furnace through overfire air nozzles to ensure complete combustion, thus eliminating smoke. The amount of coal burned on the traveling grate is regulated by the thickness of the fuel bed, rate of grate travel, and amount of air supplied.

15. *Where is a Kingsbury thrust bearing used?*
   Answer: A Kingsbury thrust bearing is found on the low-pressure end of the turbine shaft. It is used to take up the thrust formed as steam passes through the steam

turbine blades to maintain the proper axial clearance between the stationary and moving blades.

16. *What is the function of shroud rings?*
Answer: Shroud rings are found at the top of the moving blades of a steam turbine. They are made up of sections of metal strips that have holes that the tips of the steam turbine blades pass through and are peened over. Shroud rings minimize steam leakage over the tops of the steam turbine blades.

17. *How are boiler tubes connected to drums of watertube boilers? How are they connected to the tube sheet of firetube boilers?*
Answer: Boiler tubes are rolled and expanded in watertube boilers. In firetube boilers, boiler tubes are expanded in the tube sheet and then the ends are beaded over to prevent the tube ends from burning.

18. *What effect does an increase in firing rate have on the superheater outlet steam temperature?*
Answer: The effect depends on the type of superheater the boiler has. With a convection superheater, an increase in firing rate causes an increase in the temperature of the superheated steam leaving the superheater outlet. With a radiant superheater, the superheated steam leaving the superheater outlet decreases in temperature with an increase in firing rate. In order to maintain a uniform temperature, boilers have a combination of radiant and convection superheaters.

19. *What is the approximate evaporation rate when firing coal with a Btu content of 14,000 Btu/lb?*
Answer:

$$Evaporation\ rate = \frac{Btu/lb\ of\ fuel \times TE}{H_s - \left(T_{fw} - 32\right)}$$

TE = boiler efficiency (thermal efficiency)
H_s = heat of steam
T_{fw} = temperature of feedwater
Assume the following values: TE = 80%; H_s = 1200 Btu/lb; T_{fw} = 210°F.

$$Evaporation\ rate = \frac{14,000 \times 0.80}{1200 - \left(210 - 32\right)}$$

$$Evaporation\ rate = \frac{11,200}{1022}$$

Evaporation rate = **10.96 lb of steam/lb of fuel**

20. *How many pounds of air are needed to burn 1 lb of fuel? How many cubic feet does that convert to?*
Answer: This value varies, depending on the amount of excess air needed for complete combustion. An average value that is used is 15 lb of air per lb of fuel. To convert the pounds of air to cubic feet of air, use the conversion factor 0.075 lb per cu ft of air.

$$Cubic\ feet\ of\ air = \frac{Pounds\ of\ air}{0.075\ lb/cu\ ft}$$

$$Cu\ ft\ of\ air = \frac{15}{0.075}$$

$Cu\ ft\ of\ air =$ **200 cu ft**

Test 4

1. *What effect does ON/OFF operation of a boiler have on boiler maintenance and boiler life?*
   Answer: ON/OFF operation of a boiler results in increased maintenance on furnace refractory because of the heating up during the firing cycle and cooling off during the OFF cycle. The starting and stopping of the boiler feedwater pump can cause some degree of thermal shock to the steam and water side of the boiler. These conditions can be minimized by setting the ON/OFF controls so that the burner runs continuously, modulating between high and low fire and always running for longer periods than it is OFF. Also, feedwater temperature should be carried as high as possible without causing the feed water pump to become steam bound.

2. *In a steam turbine surface condenser, how much circulating water is needed per pound of steam condensed?*
   Answer:

$$Lb\ water\ /\ Lb\ steam = \frac{H_e - (T_c - 32)}{T_1 - T_2}$$

   Lb water/Lb steam = pounds of cooling water needed per pound of steam
   H_e = enthalpy of exhaust steam
   T_c = temperature of condensate
   $T_c - 32$ = enthalpy of condensate
   T_1 = temperature of cooling water leaving
   T_2 = temperature of cooling water entering
   Assume the following values: H_e = 1050 Btu/lb; T_c = 102°F; T_1 = 90°F; T_2 = 70°F.

$$Lb\ water\ /\ Lb\ steam = \frac{1050 - (102 - 32)}{90 - 70}$$

$$Lb\ water\ /\ Lb\ steam = \frac{980}{20}$$

   Lb water/Lb steam = **49**

3. *What is the purpose of staging in a reaction steam turbine and an impulse steam turbine?*
   Answer: In a reaction steam turbine, staging is necessary to maintain small pressure drops across each stage to prevent excessive steam leakage around blades and around the drum and stationary blades. A reaction steam turbine usually has many stages with small pressure drops across each stage. In an impulse steam turbine, there are large pressure drops across each stage. Before each stage, the steam passes through a nozzle to gain velocity. This velocity drops as the steam passes through the blades. At the end of a stage, the steam passes through a nozzle again to gain velocity before entering the next stage.

4. *How many pounds of steam are required to generate 1 kW of electricity?*
Answer: It depends on whether the steam turbine is operated condensing or noncondensing.

$$WR = \frac{Btu/kW}{H_a - H_e}$$

WR = water rate, in lb steam/kW
Btu/kW = 3413 Btu, heat in 1 kW
H_a = enthalpy of steam at throttle
H_e = enthalpy of steam at exhaust
Assume the following values for a condensing steam turbine: H_a = 1300 Btu/lb; H_e = 1050 Btu/lb.

$$WR = \frac{3413}{1300 - 1050}$$

$$WR = \frac{3413}{250}$$

WR = 13.65 lb of steam/kW
Assume the following values for a noncondensing steam turbine: H_a = 1300 Btu/lb; H_e = 1150 Btu/lb.

$$WR = \frac{3413}{1300 - 1150}$$

$$WR = \frac{3413}{150}$$

WR = 22.75 lb of steam/kW

5. *Why are steam lines insulated?*
Answer: Steam lines are insulated to prevent radiant heat losses that would cause the steam to condense. Water (condensate) in steam lines can cause water hammer and rupturing of steam lines. Condensate could also damage reciprocating pumps, reciprocating steam engines, and steam turbines.

6. *How are expansion and contraction compensated for in main steam lines?*
Answer: Expansion and contraction of the main steam line going from the boiler to the main steam header are compensated for by expansion bends. Properly designed expansion bends allow for considerable expansion and contraction without putting a strain on steam line flanges. Expansion joints are also used to allow for expansion and contraction of main steam lines in the plant. Expansion joints require some maintenance whereas expansion bends are maintenance-free.

7. *If 10 cu ft of air are heated at a constant pressure from 70°F to 600°F, what is the final volume?*
Answer:

$$\frac{V_1}{T_1} = \frac{V_2}{T_2} \text{ or } V_2 = \frac{V_1 T_2}{T_1}$$

V_1 = initial volume, in cu ft
V_2 = final volume, in cu ft
T_1 = initial temperature, in °R
T_2 = final temperature, in °R

Assume the following values: $V_1 = 10$ cu ft; $T_1 = (70 + 460)$; $T_2 = (600 + 460)$.

$$V_2 = \frac{V_1 T_2}{T_1}$$

$$V_2 = \frac{10 \times 1060}{530}$$

$$V_2 = \textbf{20 cu ft}$$

8. *Convert the following temperatures from Fahrenheit to Celsius:*
   Answer:

   $$^\circ Celsius = \frac{^\circ Fahrenheit - 32}{1.8}$$

   $$^\circ C = \frac{^\circ F - 32}{1.8}$$

   a. 150°F

   $$^\circ C = \frac{150 - 32}{1.8}$$

   $$^\circ C = \frac{118}{1.8}$$

   $$^\circ C = \textbf{65.56}$$

   b. 75°F

   $$^\circ C = \frac{75 - 32}{1.8}$$

   $$^\circ C = \frac{43}{1.8}$$

   $$^\circ C = \textbf{23.89}$$

   c. 30°F

   $$^\circ C = \frac{30 - 32}{1.8}$$

   $$^\circ C = \frac{0 - 32}{1.8}$$

   $$^\circ C = \textbf{-1.1}$$

   d. 0°F

   $$^\circ C = \frac{0 - 32}{1.8}$$

   $$^\circ C = \frac{-32}{1.8}$$

   $$^\circ C = \textbf{-17.78}$$

9. *Explain the function and installation of a pop-type safety valve.*
   Answer: Safety valves are located at the highest part of the steam side of the boiler with no valves between the boiler and safety valve. They are bolted to the flange connection coming from the boiler.

The pop-type safety valve has a calibrated spring holding the valve firmly to its seat. When steam pressure acting on the seat of the safety valve starts to overcome spring pressure, the valve slowly starts to lift (feather). This allows steam to enter the huddling chamber, giving the steam a larger area to work on. This increased area increases the total upward force, overcoming the downward force of the spring pressure and causing the safety valve to pop open.

10. *A boiler produces 100,000 lb of steam/hr with a factor of evaporation of 1.06. What is the developed boiler horsepower?*
Answer:

$$DBHP = \frac{W_s \times FE}{34.5}$$

$DBHP$ = developed boiler horsepower
W_S = pounds of steam per hour
FE = factor of evaporation
34.5 = pounds of steam per hour from and at 212°F per horsepower

$$DBHP = \frac{100,000 \times 1.06}{34.5}$$

$$DBHP = \textbf{3072.5}$$

11. *A boiler is operating at a rate of 2000 DBHP. The boiler operates with an efficiency of 80% and is using a fuel that has a caloric value of 14,000 Btu/lb. The temperature of the feedwater entering the boiler is 230°F. Find the following:*
Answer:
a. Amount of feedwater needed per hour

$$DBHP = \frac{W_s\left[H_s - \left(T_{fw} - 32\right)\right]}{33,475}$$

$$W_s = \frac{DBHP \times 33,475}{H_s - \left(T_{fw} - 32\right)}$$

$DBHP$ = developed boiler horsepower
W_S = pounds of steam per hour
H_S = enthalpy of steam
T_{fw} = feedwater temperature
32 = base temperature for determining enthalpy (heat in substance)
33,475 = number of Btu in 1 BHP
Assume the following value: $H_S = 1200$.

$$W_s = \frac{2000 \times 33,475}{1200 - \left(230 - 32\right)}$$

$$W_s = \frac{2000 \times 33,475}{1002}$$

$$W_S = \textbf{66,816 lb/hr}$$

The boiler generates 66,816 lb of steam/hr and therefore requires 66,816 lb of feedwater to replace the steam.

b. Amount of fuel used per hour

$$Lb \ of \ fuel \ / \ hr = \frac{DBHP \times 33,475}{Btu \ / \ lb \ fuel \times E}$$

$$Lb \ of \ fuel \ / \ hr = \frac{2000 \times 33,475}{14,000 \times 0.80}$$

Lb fuel/hr = **5977.7**

c. Approximate amount of combustion air needed per hour

Lb air/hr = Lb fuel/hr × Lb air/Lb fuel

Assume the following value: *Lb air/Lb fuel* = 15

Lb air/hr = 5977.7 × 15

Lb air/hr = **89,665.5**

If cubic feet of air is needed, divide by the factor 0.075 lb/cu ft.

$$Cu \ ft \ air \ / \ hr = \frac{89,665.5}{0.075}$$

Cu ft air/hr = **1,195,540**

12. *What are the principal advantages of heating feedwater?*

Answer: The principal advantages of heating feedwater are

- By heating feedwater in an open feedwater heater, oxygen and other noncondensable gases are vented to the atmosphere before they can get into the steam and water drum. This prevents oxygen pitting and corrosion on the water side of the boiler.
- When water in the feedwater heater is heated by exhaust steam from auxiliaries or bleed steam from the steam turbine, there is a 1% fuel savings for every 10° rise of feedwater temperature.
- When using economizers, the temperature of the gases of combustion going to the chimney is reduced, the feedwater temperature is increased, and every 10° rise results in a 1% fuel saving.
- The higher the temperature of the feedwater is, the less thermal shock there is to the boiler, which increases boiler life.

13. *Two AC generators are running in parallel. What keeps them in step and running at the same frequency?*

Answer: AC generators in parallel tend to remain synchronized. If the speed of one increases, its load increases and slows down while the other machine sheds load and increases speed so the speed of the machines becomes equal again.

Generator speed determines the frequency. A slow speed decreases frequency while a high speed increases the frequency. Minor speed control changes maintain the desired frequency.

14. *What factors are considered when selecting a drive for a feedwater pump?*

Answer: A drive for a feedwater pump is selected based on the following:

- It must have sufficient power to drive the feedwater pump at full load.
- The type of plant, whether it is a power generating or process plant, is considered.
- A steam turbine drive would be very efficient if all of the exhaust steam is used for feedwater or plant heating.
- For flexibility, an advantage is to have both an electric motor-driven pump and a steam turbine-driven pump.

- Electric and steam loads can be balanced when there are steam and electrically driven pumps.
- In a process plant that does not generate electricity, a motor drive would be simple and is started or stopped.

15. *Explain how a thermohydraulic feedwater regulator functions.*
Answer: The control element (generator) of a thermohydraulic feedwater regulator is an inclined tube located at the NOWL of the boiler. The control element is a tube within a tube. The inside tube is connected to the boiler with the top to the steam side and the bottom to the water side. The outer tube is full of condensate and is connected with copper tubing to the bellows on top of the regulating valve in the feedwater line. When the water level in the boiler starts to drop, it increases the steam space in the inner tube of the generator. The heat from the steam causes the condensate to expand, causing a pressure on top of the bellows, which opens the regulating valve in the feed water line, allowing feedwater to enter the steam and water drum. When the water level in the boiler rises, the reverse action occurs.

16. *What are priming and foaming? How can these conditions be prevented?*
Answer: Priming (carryover) occurs when large slugs of water leave the boiler with the steam and enter the main steam line. Priming can be caused by foaming, carrying too high a water level, or a sudden increase in steam flow from the boiler.

Foaming is the formation of layers of steam bubbles trapped below the water surface in the steam and water drum. Foaming is caused by a high surface tension caused by impurities floating on the surface of the water, oil contamination, or high total dissolved solids (TDS) in the boiler water.

Priming can be prevented by the following:
- controlling TDS by proper feed water treatment in conjunction with bottom blowdowns and continuous blowdowns
- maintaining the oil separator in the open feedwater heater to ensure proper operation
- monitoring condensate returns from fuel oil tanks or plant process returns to ensure no oil is getting into the boiler feedwater
- testing boiler water for TDS and maintaining TDS readings in their accepted range

Foaming can be prevented by removing impurities that cause surface tension.

17. *What are the principal elements that compose coal, No. 6 fuel oil, and natural gas? Explain the combustion process of each type of fuel.*
Answer: The principal elements in coal are nitrogen, oxygen, carbon, ash, sulfur, and hydrogen. The combustibles are carbon, sulfur, and hydrogen. They combine with oxygen as follows:

$$C + O_2 \rightarrow CO_2 + \text{Heat}$$
$$S + O_2 \rightarrow SO_2 + \text{Heat}$$
$$2H_2 + O_2 \rightarrow 2H_2O + \text{Heat}$$

No. 6 fuel oil contains the same elements as coal but also has traces of various metals. The carbon content in fuel oil is greater than that in coal. The combustion process is the same in each case.

Natural gas is mostly methane with traces of other gases such as ethane. Methane is CH_4. It contains one part carbon to four parts hydrogen. Each of these gases is combustible as follows:

$$C + O_2 \rightarrow CO_2 + \text{Heat}$$
$$2H_2 + O_2 \rightarrow 2H_2O + \text{Heat}$$

18. *A boiler feedwater pump is feeding 18,500 lb of water/hr to the boiler. The discharge line water velocity is 400 fpm. What is the size of the discharge line? (Note: Water weighs 62.4 lb/cu ft.)*

    Answer:

    $$A = \frac{Q}{V}$$

    A = area of discharge line, in sq ft
    Q = quantity of water, in cubic feet per minute (cfm)
    V = velocity of water, in fpm

    $$Q = \frac{Lb/hr}{60 \text{ min/hr} \times Lb/cu\ ft}$$

    $$Q = \frac{18,500}{60 \times 62.4}$$

    $Q = 4.941$ cfm
    $V = 400$ fpm

    $$A = \frac{4.941}{400}$$

    $A = 0.0124$ sq ft
    $A = 0.0124$ sq ft $\times$ 144 sq in./sq ft
    $A = 1.779$ sq in.
    $A = 0.7854D^2$

    $$D = \sqrt{\frac{A}{0.7854}}$$

    $$D = \sqrt{\frac{1.779}{0.7854}}$$

    $D = \mathbf{1.50''}$

19. *Describe an automatic combustion control system.*

    Answer: An automatic combustion control system must control the fuel supply to the boiler furnace, air supply to the boiler furnace, air-fuel ratio, and gases of combustion leaving the boiler.

    Three basic types of automatic combustion control systems are ON/OFF, positioning, and metering.

    ON/OFF combustion control systems control the operating range of the boiler by starting and stopping the burner on steam pressure demand. Actuation of the burner can be controlled by mechanical linkage, solid state circuitry, or microprocessor-based circuitry.

The positioning control system operates from the steam header pressure, which activates a master control unit that relays pressure signals to a slave unit that controls the air and fuel supplied to the furnace. The master control is supplied on one side with dry filtered air at a controlled pressure. The other side of the master control is supplied with steam pressure from the main steam header. The air pressure is balanced with the steam pressure. A decrease in steam pressure causes a signal to be relayed to a secondary controller to increase the air and fuel supplied to the boiler furnace. A separate control unit is used to control the outlet damper, which receives its signal from the furnace pressure.

The whole system must be adjusted so that any change in the steam pressure will provide a proportional change in the air-fuel ratio supplied to the furnace.

Metering combustion control systems are commonly used on large, high-capacity boilers. Metering combustion controls are more sensitive to load demands and changes. This is accomplished by measuring the difference of pressure across an orifice, which then signals for more or less fuel and air.

20. *A boiler generates 30,000 lb of steam/hr at 400 psi. What is the approximate safety valve relieving area needed on this boiler? What area is needed if the pressure is reduced to 250 psi?*
Answer:

$$A = \frac{0.2W}{MAWP + 10}$$

A = safety valve area, in sq in.
W = pounds of steam generated per hour
$MAWP$ = maximum allowable working pressure, in psi

$$A = \frac{0.2 \times 30,000}{400 + 10}$$

$$A = \frac{6000}{410}$$

A = **14.63 sq in.** (at 400 psi)

$$A = \frac{0.2 \times 30,000}{250 + 10}$$

$$A = \frac{6000}{260}$$

A = **23.08 sq in.** (at 250 psi)

Test 5

1. *How many adjustments can be made on a safety valve?*
Answer: Two adjustments can be made on a safety valve. One is its popping pressure and the other is its blowback. In order to change the popping pressure or blowback on a safety valve, the two seals installed by the safety valve manufacturer must be broken, and any changes would require a new safety valve nameplate from the manufacturer.

In an emergency, a person thoroughly familiar with the construction and operation of the safety valve can make these changes, then notify either the state boiler inspector and/or the insurance company inspector. At the first opportunity, the safety valve should be sent back to the manufacturer for a factory test and new seals installed on the pressure-adjusting nut and blowback pin, and a new nameplate installed to indicate the new popping and blowback pressures.

2. *Can boilers be classified by the steam and water circulation?*
Answer: Three basic steam and water circulation classifications are natural, forced, and once-through. Natural circulation is caused by a difference in density of the steam and water in the boiler. In the hot zones such as the furnace area, the steam rises through risers. In the cooler zones such as rear tubes, the steam and water drop down through downcomers. This is a continual process and the steam and water circulate rapidly. As the pressure on a boiler increases, its natural circulation decreases. The reason for this is that the density of steam increases while the density of water remains the same.

In forced circulation, pumps are used to increase the natural circulation in the boiler. At higher pressures once-through circulation is necessary. In once-through circulation, pumps force water through a continuous path and it leaves the boiler as high-temperature, high-pressure steam.

3. *Describe the principal parts of a coal-pulverizing system.*
Answer:
- Bunker (coal)
- Conveyor (belt)
- Scale
- Feeder
- Pulverizer
- Warm air duct
- Exhauster
- Burner tube
- Forced draft fan

Coal is stored in a bunker until needed. The conveyor carries the coal from the bunker to the coal scale where it is weighed and its weight recorded. The coal scale dumps the coal toward the feeder that controls the amount of coal entering the pulverizer. Once in the pulverizer, the coal is broken up into a powder. To prevent wet coal from packing together in the pulverizer, a warm air duct conveys air into the pulverizer. An exhauster removes the dry powdered coal from the top of the pulverizer and discharges it through the burner tube to the furnace. Forced draft (air) is introduced through a register to blend with the coal as it enters the furnace.

4. *Describe the function of primary air and secondary air in the combustion process.*
Answer: Primary air is air that controls the combustion rate, which is the amount of fuel that can be burned. Secondary air is air that controls the combustion efficiency, that is, how well the fuel is burned.

How primary and secondary air are admitted to the furnace depends on the type of fuel burned and the type of burner used. For example, a chain grate stoker burning bituminous coal has primary air supplied through windboxes under the grates passing up through the fuel bed. Secondary air is supplied through air jets over the fire to help burn the volatile gases forming over the fuel bed.

In pulverizers, primary air carries the fine coal from the pulverizer mill to the burner into the furnace. Secondary air passes through air registers around the burner.

In a rotary cup fuel oil burner, primary air is supplied by a fan driven by the same motor that drives the rotary cup. Secondary air is supplied around the burner and also through a secondary air damper at the bottom of the burner.

5. *Why must induced draft fans be used with air heaters and economizers?*
Answer: Air heaters and economizers are installed on boilers to extract as much heat from gases of combustion as possible before they enter the chimney. Their location causes a restriction, slowing the flow of gases of combustion leaving the boiler. By installing an induced draft fan, the gases of combustion are pulled from the boiler then discharged into the chimney, which relieves the back pressure that would otherwise be formed.

6. *What can be done to improve the efficiency of a steam plant?*
Answer: The water side and fire side of the boiler should be cleaned and the baffles in the paths of gases of combustion checked to make sure they are in good condition. The combustion process should be checked with a combustion analyzer to determine if adjustments are needed for the proper air-fuel ratio, fuel atomization, or fuel temperature. Excess air should be reduced to a minimum.

Steam traps throughout the plant should be checked for leaks or to make sure they are functioning. Insulation on the boiler and all steam lines should be kept in good condition to prevent heat loss. Adjustments should be made to have a steady steam demand on the boiler rather than fluctuating loads. All air leaks into the fire side of the boiler must be sealed. When and if possible, feed water should be heated with exhaust steam or with gases of combustion. A heat recovery system from the continuous blowdown should be installed.

7. *What routine on-line and off-line maintenance work is required on a boiler?*
Answer: Routine maintenance work while a boiler is on-line is as follows:
1. Test the boiler water, add the proper amounts of chemicals, and control the total dissolved solids using continuous blowdowns and bottom blowdowns.
2. Use soot blowers once a shift.
3. Clean burner tips once a shift.

Routine maintenance work when boiler is off-line is as follows:
1. Check furnace refractory and repair if necessary.
2. Check boiler settings for air leaks and repair if necessary.
3. Repack boiler valves as needed.
4. Clean boiler gauge glasses and replace as needed.
5. Check manhole and handhole gaskets and replace as needed.

8. *Describe two types of air heaters found on boilers.*
Answer: The two basic types of air heaters are the convection and regenerative air heaters. The convection air heater can be either a tubular air heater where gases of combustion pass through the tubes and the air for combustion passes around the tubes, or a plate heater. The plates are assembled and spaced evenly in a steel frame. The plates form air and gas chambers. The heat transfer takes place as the air and gases of combustion move past each other in a counterflow direction on opposite sides of the plates. Bypass dampers are sometimes used to control the temperature of the air for combustion and the temperature of gases of combustion going to the chimney. This is important to prevent sweating from occurring, which could lead to sulfuric acid damage.

The regenerative heater consists of a round casing divided into three zones: the air zone, sealing zone, and gases of combustion zone. A slow-moving rotor made up of honeycomb plate elements rotates through the three zones. The plate element is heated by the gases of combustion and gives up its heat to the air for combustion as the rotor slowly turns. The sealing chamber prevents the air for combustion from mixing with the gases of combustion. Counterflow operation also occurs in the regenerative heater. The forced draft fan forces air through the heater while the induced draft fan pulls out gases of combustion, discharging them to the chimney.

9. *A boiler feedwater pump is feeding 25,000 lb of water/hr to the boiler. The discharge line water velocity is 400 fpm. What is the size of the discharge line? (Note: Water weighs 62.4 lb/cu ft.)*
Answer:

$$A = \frac{Q}{V}$$

A = area of discharge line, in sq ft
Q = quantity of water, in cubic feet per minute (cfm)
V = velocity of water, in fpm

$$Q = \frac{Lb/hr}{60 \text{ min/hr} \times Lb/cu\, ft}$$

$$Q = \frac{25,000}{60 \times 62.4}$$

$Q = 6.677$ cfm
$V = 400$ fpm

$$A = \frac{6.677}{400}$$

$A = 0.0167$ sq ft
$A = 0.0167$ sq ft $\times$ 144 sq in./sq ft
$A = 2.405$ sq in.
$A = 0.7854 D^2$

$$D = \sqrt{\frac{A}{0.7854}}$$

$$D = \sqrt{\frac{2.405}{0.7854}}$$

$D = 1.75''$

10. *How is an ignition arch used on a chain grate stoker?*
Answer: An ignition arch is found in the furnace of a chain grate stoker. It is made of refractory material which, when heated, reflects the heat back to the fuel bed to help ignite the green coal coming into the furnace. When burning soft coal, it also burns the volatile gases distilled from the coal bed so that combustion is completed before these gases hit the boiler heating surface. This prevents the formation of soot and smoke.

11. *A chain grate stoker 10' wide is burning anthracite coal. The grate speed is 6" per minute and the coal is 6" deep. At what rate, in tons per hour, is coal being used? (Note: Anthracite coal weighs 60 lb/cu ft.)*
Answer:
Cu ft/min = $W \times D \times S$
W = width of stoker, in ft
D = depth of coal bed, in ft
S = speed of stoker, in fpm

$$Cu\ ft/min = W \times D \times S$$

$$Cu\ ft\,/\,min = 10 \times \frac{6}{12} \times \frac{6}{12}$$

$$Cu\ ft/min = 2.5$$

$$Tons\,/\,hr = \frac{Cu\ ft\,/\,min \times 60\ \text{min/hr} \times lb\,/\,cu\ ft}{2000\ \text{lb/ton}}$$

$$Tons\,/\,hr = \frac{25 \times 60 \times 60}{2000}$$

$$Tons/hr = \mathbf{4.5}$$

12. *What is the Btu content of 1 lb of fuel oil with a specific gravity of 20°API?*
   Answer:
   $Btu/lb = 17{,}780 + (54 \times °API)$
   $Btu/lb = 17{,}780 + (54 \times 20)$
   $Btu/lb = 17{,}780 + 1080$
   $Btu/lb = \mathbf{18{,}860}$

13. *What problems are caused by oxygen and carbon dioxide in boiler water? How can these problems be eliminated?*
   Answer: Oxygen in boiler water can cause pitting (oxidation of metal) in the steam and water side of the boiler. If left unchecked, this will lead to weakening of the boiler metal and leaks. This condition can be eliminated by treating the boiler water with an oxygen scavenger, such as sodium sulfite.

   Carbon dioxide can cause channeling. Channeling is a form of corrosion that most often is found in condensate return lines. Boilers that use soda ash as a boiler water treatment have carbon dioxide leaving the boiler with steam, and it eventually causes channeling and leaks in the return lines. Return line corrosion can be controlled by adding catalyzed hydrazine in the return lines.

   Oxygen and carbon dioxide should be removed from the feedwater before they leave the open feedwater heater. This is done by heating the water and removing any gases.

14. *Define the following terms:*
   Answer:
   a. Saturated steam: Steam at its corresponding temperature and pressure. Any drop in temperature without a corresponding drop in pressure will cause steam to condense.
   b. Superheated steam: Steam at a temperature higher than its corresponding pressure.
   c. Quality of steam: An indication of the amount of moisture in the steam. Steam containing 5% moisture has a quality of 95%.
   d. Enthalpy of steam: The total heat of steam. It is the sum of the sensible heat and latent heat of evaporation.

15. *How is the outlet damper controlled on an automatic boiler?*
   Answer: The outlet damper on an automatic boiler is controlled electrically or mechanically, but in either case the signal is the pressure in the furnace. With ON and OFF controls, it operates electrically as follows:
   1. On startup, the damper opens wide.
   2. The burner goes through a starting cycle.

3. At the end of the cycle, the damper modulates, depending on furnace pressure. The pressure stays constant and changes are made through a relay.

When a boiler operates continuously, it is usually equipped with pneumatic controls. The pressure signal comes from the furnace and activates a positioning type of controller. The damper moves to maintain a constant pressure in the furnace.

16. *When operating a boiler, what heat losses can be controlled?*

Answer: Three heat losses that the operator has some control over are incomplete combustion of fuel; in coal-burning plants, coal going over into the ashpit; and high chimney temperatures.

17. *A boiler generates 100,000 lb of steam/hr with an evaporation rate of 12.5 lb of steam/lb of fuel. Two hundred cubic feet of air are needed per pound of fuel, and the forced draft fan has a discharge pressure of 4″ water. What is the HP needed to drive the forced draft fan?*

Answer:

$$Lb\ fuel\ /\ min = \frac{Lb\ steam\ /\ hr}{60\ min/hr \times lb\ steam\ /\ lb\ fuel}$$

$$Lb\ fuel\ /\ min = \frac{100,000}{60 \times 12.5}$$

$Lb\ fuel/min = 133.33$

$Cu\ ft\ air/min = Lb\ fuel/min \times Cu\ ft\ air/Lb\ fuel$

$Cu\ ft\ air/min = 133.33 \times 200$

$Cu\ ft\ air/min = 26,666$

$$Air\ HP = \frac{QP}{6356}$$

$Air\ HP$ = horsepower (output)

Q = quantity of air, in cu ft/min

P = inches of water pressure

$$Air\ HP = \frac{26,666 \times 4}{6356}$$

$Air\ HP =$ **16.78**

Assume an efficiency of 55% for the driving unit and fan.

$$Drive\ HP = \frac{Output}{Efficiency}$$

$$Drive\ HP = \frac{16.78}{0.55}$$

$Drive\ HP =$ **30.5**

18. *Describe how a zeolite water softener works.*

Answer: A zeolite water softener works through ion exchange. Zeolite is a granular material that is mined or that can be manufactured. It can change calcium carbonate and magnesium carbonate, both of which cause hard scale on the boiler heating surfaces, into sodium carbonate and magnesium sulfate. Sodium carbonate and magnesium sulfate are nonadhering sludges that can be removed from the boiler by continuous and bottom blowdowns.

In the zeolite water softener, raw water enters the softener, passing through diffusers that evenly distribute the water over the zeolite bed. As the water passes through the layers of zeolite, the ion exchange takes place and changes the scale-forming salts to a nonadhering sludge. After the zeolite has softened a number of gallons of water, its supply of sodium will be depleted. The softener must be removed from service and regenerated. In the process of regeneration, the softener is back washed to remove dirt deposits left behind, then a solution of brine is added to replace the sodium in the zeolite. The softener is flushed with clean water to remove excess brine solution. A soap solution test is taken periodically during the flushing stage. As soon as the soap test indicates soft water is present, the softener can be put back into service.

19. *What are the advantages of using a continuous blowdown system?*
    Answer: The advantages of a continuous blowdown system are
    • close control of total dissolved solids in boiler water
    • reduction of bottom blowdowns, which wastes heat, chemicals, and boiler water
    • heat reclaimed by allowing continuous blowdown water to flash; this flash steam is used in the feedwater heater
    • boiler makeup water heated in the flash tank by passing through coils before entering the open feed water heater

20. *List the parts of an AC generator and describe the function of each part.*
    Answer:
    • Frame: Circular in shape and holds the stator (stationary armature) in place.
    • Stator: Made up of a core with coils of insulated wire spaced evenly around the core.
    • Rotor: Revolving field consisting of a laminated core with coils of insulated wire that are wound around the edges.
    • Slip rings: Fastened to one end of the rotor. They are connected to the ends of the field coils and are used to excite or energize the field coils.
    • Bearings: Support rotating element.
    The rotating field poles are energized through the slip rings by an outside DC source. This produces a rotating electromagnetic field. The electromagnetic field passes over the stationary stator windings, inducing an alternating current. The current that is produced is removed by way of the stator leads.

Test 6

1. *An AC generator has a frequency of 60 Hz and is operating at 1800 RPM. How many poles does it have?*
   Answer:
   $$F = \frac{P \times RPM}{120}$$
   $$P = \frac{120 \times F}{RPM}$$
   F = frequency, in Hz
   P = number of field poles in the generator
   RPM = RPM of generator
   120 = constant

$$P = \frac{120 \times 60}{1800}$$

$$P = 4$$

2. *What is the purpose of a synchroscope?*
Answer: A synchroscope is a device used to put an oncoming generator in parallel with the generator in service before the new generator can be put on-line. AC generators produce current with a 60-cycle frequency. Unless the oncoming generator is in step with the machine that is in service, both generators could trip out when the breaker tying in the oncoming generator is thrown.

3. *How is load applied to an AC generator that is running in parallel with another unit?*
Answer: When running AC generators in parallel, the load on one can be increased by a slight increase of speed on the generator. If the speed is decreased, the load will swing back to the other unit.

4. *Describe a front feed spreader stoker.*
Answer: Spreader stokers are designed to burn small sizes of bituminous coal. The parts of the spreader stoker are as follows:
- Coal conveyor: Moves coal from the coal bunker to the coal scale.
- Coal scale: Weighs out correct amounts of coal and dumps it into a coal hopper.
- Coal hopper: Supplies coal to the coal feeder.
- Coal feeder (reciprocating or conveyor type): Delivers coal to the overthrow unit.
- Overthrow unit: Throws coal out in sprinkled form, some of which burn in suspension, and the larger pieces fall and burn on the grates.

Air for combustion is supplied by forced draft fans. Air is distributed under the grates (primary air) and also over fire air (secondary air) to help the combustion process of the coal burned in suspension. The rate of combustion is controlled by the drive speed on the coal feeder and the amount of over fire and underfire air supplied.

5. *Describe the construction and function of a regenerative heater.*
Answer: The regenerative air heater consists of a round casing divided into three zones: the gases of combustion zone, sealing zone, and air zone. A slow-moving rotor, consisting of a honeycomb plate element, rotates through the three zones of the air heater casing.

When the plate element rotates through the gases of combustion zone, it absorbs heat and reduces the temperature of the gases of combustion. After passing through the plate element, the gases of combustion go through the induced draft fan and out the chimney.

The plate element continues to rotate through the sealing zone. In the sealing zone of the air heater casing, gases of combustion are prevented from mixing with the air for combustion. When the plate element rotates through the air zone, the plate element releases heat and is cooled as the air for combustion is heated. The counterflow principle applies in this heater and the plate element is continuously in motion.

6. *What are losses that can occur in boiler operation?*
Answer: The losses that occur in a boiler include
- heat carried away to the atmosphere in gases of combustion (approximately 11%)
- incomplete combustion of fuel (approximately 1.5%)
- moisture in fuel that must be evaporated (less than 1%)

- moisture formed by burning hydrogen that must be evaporated (about 3%)
- coal fired, unburned combustibles in the ash (about 3%)
- radiation and other small losses (approximately 4%)
- heating moisture in the air used for combustion (less than 1%)

Heat losses total 24.5%.

7. *How can heat from the continuous blowdown line be reclaimed?*

Answer: There are two ways that heat from the continuous blowdown is reclaimed. The continuous blowdown discharges to a flash tank where some of the water flashes into steam. The steam goes to the open feed water heater to heat the boiler feedwater. The balance of the water collects at the bottom of the flash tank where the makeup water passing through coils picks up heat before being discharged into the feedwater heater. The flash tank has an internal overflow that keeps the tank about half full of water at all times.

8. *List and describe the functions of the parts of a DC generator:*

Answer: A DC generator consists of the following parts:

- Frame
- Field poles
- Armature
- Commutator
- Brushes
- Bearings

   The frame is circular and is used to hold the field poles in place. There may be two, four, six, eight, or more field poles. They consist of a laminated steel core wound with insulated copper wire. The armature consists of a soft iron laminated core with coils wrapped around the core. The coil ends are connected to the segments of the commutator. The commutator is made of copper segments and is fastened to the armature by clamping rings. The commutator brushes, in sets, carry electrical energy from the commutator to the external circuit.

   The armature is supported by bearings. The generator functions as follows. The field poles establish an electromagnetic field. The armature moves through the magnetic field, inducing an alternating current in the windings that are connected to the commutator. The brushes and commutator rectify the alternating current to direct current.

9. *How are two DC generators paralleled?*

Answer: DC generators are paralleled by

- bringing the incoming generator up to normal operating speed
- closing the negative and equalizer switches on the electric panelboard for the incoming generator
- using the field rheostat, adjusting the voltage of the incoming generator so that it is equal to or slightly below the generator on-line, then closing the positive switch
- balancing the load between both generators with the field rheostat

10. *Why do large steam plants use watertube boilers while small steam plants use firetube boilers?*

Answer: Watertube boilers are capable of producing steam at much higher pressures and temperatures than firetube boilers. They carry less water than a firetube boiler of equal horsepower, are quicker steamers, and can respond to load changes faster than firetube boilers. The design of watertube boilers eliminates the need for braces and stays and allows the use of much thinner metals in their construction.

11. *Why are feedwater pumps in a plant usually located on the first floor or basement level?*
    Answer: In order for the feedwater pump to handle the hottest water possible without becoming steam bound, the feedwater pump is located below the feedwater heater as low as practical. This supplies the feedwater pump with a head pressure. For every vertical foot the heater is above the feedwater pump, a pressure of 0.433 lb is produced at the suction side of the feedwater pump. The higher the pressure is, the higher the boiling point of water will be.

12. *What are reasons for lowering the MAWP on a boiler?*
    Answer:
    • Possible metal fatigue due to age of boiler.
    • Corrosion of boiler drum metal.
    • Thinning of boiler drum metal by erosion.
    • Replacement of drum or internal furnace metal with a welded patch.

13. *Is it efficient to use bleed or extraction steam from different stages of steam turbines for heating feedwater and process use?*
    Answer: Yes, it is very efficient to use bleed steam from a steam turbine for heating and extraction steam at a controlled pressure for plant process. Steam turbines used for generating electricity are designed to use steam at high pressure and superheated for maximum efficiency. Once the steam leaves the steam turbine and enters the condenser, many Btu are lost to the cooling water. It could be compared to a boiler that is equipped with air heaters and economizers to use heat from gases of combustion before they enter the chimney and are lost to the atmosphere.

14. *What type of feedwater pump drive would a process plant that generates its own electrical power have?*
    Answer: A process plant that generates its own electrical power would most likely have an electrical motor and a steam turbine drive. The reason for this is that it helps balance the steam versus electric load. The motor drive is used when the electric load is light or the steam load is high.
    The steam turbine is used when there is a demand for exhaust steam (10 psi) for heating feedwater or domestic water or for building heating. The steam turbine drive will be very efficient if all of the exhaust steam can be used.

15. *Describe how gases of combustion are analyzed using a Fyrite analyzer and an Orsat analyzer.*
    Answer: Fyrite analyzers are used to measure the percentage of oxygen or carbon dioxide in gases of combustion. When using the Fyrite analyzer to measure the percentage of carbon dioxide in gases of combustion, the following procedure is used:
    1. Vent the Fyrite analyzer to the atmosphere by depressing the top of the Fyrite analyzer, and set the indicator on 0%.
    2. Draw a sample of gases of combustion into the Fyrite analyzer. *Note:* The sample of gases of combustion should be taken as close to the boiler outlet as possible so the sample is not diluted with air.

3. The Fyrite analyzer is inverted a couple of times to thoroughly mix with the carbon dioxide absorbing agent.
4. The absorbing agent increases in volume, and a direct reading in percentage of carbon dioxide can be taken.
5. Take the temperature of the boiler room and the temperature of the gases of combustion. Follow the directions on the slide rule calculator supplied with the Fyrite analyzer to determine combustion efficiency and percentage of chimney loss.

Orsat analyzers are used for more complete analysis of gases of combustion. The Orsat analyzer can be used to measure the percentage of carbon dioxide, oxygen, and carbon monoxide in the gases of combustion. The absorbing agent in the carbon dioxide burette is a solution of caustic potash. The absorbing agent in the oxygen burette is pyrogallic acid. The absorbing agent for carbon monoxide is cuprous (copper) chloride. Using the data from the Orsat analyzer, the amount of excess air needed in the combustion process can be determined.

The percentage of excess air over the theoretical requirements is found by using the following equation:

$$\% \ of \ excess \ air = \frac{O_2 - \frac{1}{2}\,CO}{0.263N_2 + \frac{1}{2}\,CO - O_2} \times 100$$

O_2 = oxygen
CO = carbon monoxide
N_2 = nitrogen

16. *How is a surface condenser cleaned (from opening to closing)?*
Answer: This surface condenser is used in conjunction with a steam turbine and is cleaned as follows:
1. With the turbine off-line and the condenser cool, secure the cooling water pump, condensate pump, and air ejectors.
2. Remove both ends of the condenser (referred to as water boxes).
3. With the end covers removed, expose the tube sheet into which the tubes are fastened.
4. Clean the tubes and tube sheets thoroughly by pushing a small brush through each tube or by using compressed air to shoot a rubber plug through each tube. This removes any impurities deposited by the cooling water.
5. Inspect the water side for signs of oil or leaking tubes.
6. After the condenser has been thoroughly cleaned, the water box and covers are replaced using new gaskets.
7. Test the condenser for leaks.
8. Check the circulating water pump and condensate pump for packing, lubrication, and general operating conditions before the turbine is put back in service.

17. *What purpose does an air ejector serve?*
Answer: Air ejectors are used to remove air and noncondensable gases from a surface condenser to help maintain a high vacuum. The higher the vacuum in a surface condenser in a turbine plant the lower the turbine water rate, more heat can be extracted from the steam and a higher efficiency is achieved.

18. *Why must the auxiliary lubricating pump on a steam turbine function properly?*
Answer: Steam turbines are designed to operate at high speeds and high steam temperatures and pressures. The turbine rotor is supported by a thin film of oil.

The auxiliary oil pump is started during the turbine warm-up period and remains in operation until the main lubricating pump takes over. When the turbine is being taken off-line, the auxiliary oil pump automatically starts up and will be kept running until the turbine has been properly cooled. Any failure of the lubricating oil would result in serious damage to the turbine rotor and blading.

19. *When does a boiler require a hydrostatic test? Describe how this test is done.*
   Answer: A hydrostatic test is performed on a boiler when it is first installed and after any extensive repairs to its steam and water side. A boiler inspector may also require a hydrostatic test on a boiler if it develops a low water condition.
   The following hydrostatic test procedure is followed:
   1. Secure boiler main steam stop valves.
   2. Open free-blowing drain between stop valves.
   3. Gag or remove and blank flange boiler safety valves.
   4. If boiler has whistle valve on the water column, remove and plug it.
   5. Open air vent on the boiler.
   6. Then fill the boiler with water that is as close as possible to the ambient temperature of the boiler room (to prevent the boiler from sweating) until water comes out of boiler vent.
   7. Close the boiler vent and bring the pressure up to 1½ times the MAWP of the boiler.
   8. The boiler inspector then checks the boiler for any leaks on its pressure side. If the boiler has a low water condition, the inspector looks for weakness or bulging of the heating surface of the boiler.
      Note: The free-blowing drain should be checked to make sure the boiler main steam stop valve closest to the boiler shell is not leaking. If it were leaking water, the second boiler stop valve with steam pressure on the other side of the valve could cause damage to the valve.
   9. After the test is completed, drop the boiler water level to it NOWL, put back the whistle valve, remove the safety valve gag or blank flanges, and replace the safety valves.

20. *Describe how labyrinth packing prevents leakage on a shaft.*
   Answer: As high-pressure steam enters the labyrinth packing, it drops in pressure and increases in velocity and expands. The packing is designed so that an eddy current is established. The steam then continues to move through the labyrinth packing, repeating the procedure. If the labyrinth packing is on the high-pressure side of the turbine, the steam exiting from the shaft is only slightly higher than the atmospheric pressure, thus preventing high-pressure steam from leaving the steam turbine. The labyrinth packing on the low-pressure side of a steam turbine running condensing prevents air from leaking into the turbine, which would affect the vacuum.

Test 7

1. *What data is found on a safety valve? What causes a safety valve to pop open?*
   Answer: The data plate on a safety valve includes the following information:
   • manufacturer's name or trademark
   • manufacturer's design or type number
   • size of safety valve in inches, seat diameter

- popping pressure setting
- blowdown in pounds of steam per hour
- lift of safety valve in inches
- year built or code mark
- ASME symbol
- serial number

When steam pressure starts to overcome the force of the safety valve spring, the valve starts to open (feather). This allows the steam to enter the huddling chamber and exposes a larger area for the steam to act on. This increases the force, which causes the safety valve to pop open.

2. *Explain the procedure for starting a condensing turbo-generator.*
   Answer: The turbine manufacturer's startup procedures should always be adhered to. The following is a suggested startup procedure:
   1. Warm up steam line to the turbine and drain all condensate from the line.
   2. Start the auxiliary oil pump and check oil pressure.
   3. Check the lubricating oil level in the reservoir.
   4. Open all cylinder drain valves.
   5. Open gland leak-off valves.
   6. Start the condenser circulating water pump.
   7. Establish seals on the turbine high-pressure and low-pressure ends for starting.
   8. Admit steam to start the rotor turning, then shut it off.
   9. Listen for a rubbing noise in the casing or seal locations.
   10. If no noise is heard, admit enough steam to maintain approximately 200 RPM for ½ hour to warm the casing and rotor evenly.
   11. Adjust the cooling water to the oil cooler to maintain 100°F outlet oil temperature.
   12. Close all cylinder drains.
   13. Adjust the high-pressure and low-pressure seals for operating conditions.
   14. Start the condensate pump.
   15. Seal the atmospheric valve.
   16. Start the air ejectors.
   17. Bring the turbine up to rated speed and let the governor take over.
   18. Trip the turbine governor to test operation.
   19. Reset the turbine governor, bring the turbine back to rated speed, and make sure the governor takes over.
   20. Check the oil temperature and air ejector cooling water temperature.
   21. Check the generator cooling medium temperature.
   22. Place the unit on-line with at least 20% of the rated load.
   23. Put the bleed and extraction lines in operation.
      Note: The turbo-generator should be checked frequently during the first few hours of operation to ensure proper temperatures and lubrication.

3. *Explain how two AC generators are paralleled.*
   Answer: To parallel two AC generators, proceed as follows:
   1. Warm up the turbine following the manufacturer's startup procedures, making all safety checks.
   2. Bring the turbine up to normal speed (frequency).
   3. Adjust the voltage of the incoming generator to the line voltage.
   4. Using a synchroscope, synchronize the incoming generator by increasing or decreasing its speed.

  5. When at the point of synchronization, close the circuit breaker.
  6. Apply 20% of the generator capacity by using the governor speed changer.

4. *When does a boiler require a hydrostatic test? Describe how this test is done.*
Answer: A hydrostatic test is performed on a boiler when it is first installed and after any extensive repairs to its steam and water side. A boiler inspector may also require a hydrostatic test on a boiler if it develops a low water condition.
The following hydrostatic test procedure is followed:
  1. Secure boiler main steam stop valves.
  2. Open free-blowing drain between stop valves.
  3. Gag or remove and blank flange boiler safety valves.
  4. If boiler has whistle valve on the water column, remove and plug it.
  5. Open air vent on the boiler.
  6. Then fill the boiler with water that is as close as possible to the ambient temperature of the boiler room (to prevent the boiler from sweating) until water comes out of boiler vent.
  7. Close the boiler vent and bring the pressure up to 1½ times the MAWP of the boiler.
  8. The boiler inspector then checks the boiler for any leaks on its pressure side. If the boiler has a low water condition, the inspector looks for weakness or bulging of the heating surface of the boiler.
      Note: The free-blowing drain should be checked to make sure the boiler main steam stop valve closest to the boiler shell is not leaking. If it were leaking water, the second boiler stop valve with steam pressure on the other side of the valve could cause damage to the valve.
  9. After the test is completed, drop the boiler water level to it NOWL, put back the whistle valve, remove the safety valve gag or blank flanges, and replace the safety valves.

5. *How does a hydraulic governor control steam turbine speed?*
Answer: An oil relay or hydraulic governor functions using a speed-sensing element, speed changer, and a servo-motor, which moves the steam admission valves. An oil pressure signal is controlled by a set of governor weights moving a cup valve. The signal acts against a diaphragm within the speed changer and produces a signal that activates a pilot valve on the servo-motor. The servo-motor repositions itself, changing the position of steam admission valves.

6. *Describe how gases of combustion are analyzed using a Fyrite analyzer and an Orsat analyzer.*
Answer: Fyrite analyzers are used to measure the percentage of oxygen or carbon dioxide in gases of combustion. When using the Fyrite analyzer to measure the percentage of carbon dioxide in gases of combustion, the following procedure is used:
  1. Vent the Fyrite analyzer to the atmosphere by depressing the top of the Fyrite analyzer, and set the indicator on 0%.
  2. Draw a sample of gases of combustion into the Fyrite analyzer. *Note:* The sample of gases of combustion should be taken as close to the boiler outlet as possible so the sample is not diluted with air.
  3. The Fyrite analyzer is inverted a couple of times to thoroughly mix with the carbon dioxide absorbing agent.
  4. The absorbing agent increases in volume, and a direct reading in percentage of carbon dioxide can be taken.

5. Take the temperature of the boiler room and the temperature of the gases of combustion. Follow the directions on the slide rule calculator supplied with the Fyrite analyzer to determine combustion efficiency and percentage of chimney loss.

Orsat analyzers are used for more complete analysis of gases of combustion. The Orsat analyzer can be used to measure the percentage of carbon dioxide, oxygen, and carbon monoxide in the gases of combustion. The absorbing agent in the carbon dioxide burette is a solution of caustic potash. The absorbing agent in the oxygen burette is pyrogallic acid. The absorbing agent for carbon monoxide is cuprous (copper) chloride. Using the data from the Orsat analyzer, the amount of excess air needed in the combustion process can be determined.

The percentage of excess air over the theoretical requirements is found by using the following equation:

$$\% \text{ of excess air} = \frac{O_2 - \frac{1}{2}CO}{0.263N_2 + \frac{1}{2}CO - O_2} \times 100$$

O_2 = oxygen
CO = carbon monoxide
N_2 = nitrogen

7. *How is a battery of boilers connected to a steam header to allow for expansion and contraction?*
Answer: Each boiler's main steam line has its own expansion bend. The main steam header is supported on brackets from the ceiling of the boiler room. These brackets are equipped with rollers that the main header rests on. These rollers allow the main header to move freely to allow for expansion and contraction.

8. *Why can small steam turbines operate at higher speeds than large steam turbines?*
Answer: Turbine speed is governed by the rim speed of its largest rotating element. The larger the element is, the greater the centrifugal force developed will be. As the speed increases, the chance will be greater that the centrifugal force will overcome the centripetal force holding the element together.

9. *What is an exciter and how does it function?*
Answer: An exciter is a DC generator used to excite the field windings on AC generators, DC generators, or synchronous motors. It functions in the following manner:
1. A magnetic field is established around its field windings.
2. An armature is wound with insulated copper wire and connected to a commutator.
3. The armature rotates, inducing a current in the windings of the armature.
4. The current passes through the commutator and is removed by the carbon brushes. The current then passes through wires to a main board.

10. *What percentage of heat in the fuel is carried away by gases of combustion?*
Answer: Approximately 10% of heat in the fuel is carried away in gases of combustion. This can be greatly reduced if an air heater or economizer is used. The following equation can be used to determine heat loss in gases of combustion:

$$Loss = \frac{W \times Sph \times (T_1 - T_2)}{Btu \text{ in fuel}}$$

W = weight of gases
Sph = specific heat of gases
T_1 = temperature of gases entering chimney
T_2 = temperature of air entering furnace

Assume the following values: 15 lb of air/lb of fuel; $Sph = 0.24$ same as air; $T_1 = 550°F$; $T_2 = 100°F$; *Btu in fuel* = 16,000 Btu/lb.

$$Loss = \frac{(15+1) \times 0.24 \times (550-100)}{16,000}$$

Note: W = 16 (15 air + 1 fuel)

$$Loss = \frac{16 \times 0.24 \times 450}{16,000}$$

$$Loss = \mathbf{10.8\%}$$

11. *A reciprocating pump 5 × 3 × 6 delivers water at 200 psi. What steam pressure is needed to operate this pump?*

    Answer:

    $$\frac{P_s}{P_w} = \frac{D_w^{\,2}}{D_s^{\,2}}$$

    P_s = pressure on steam piston (in psi)
    P_w = pressure on water piston (in psi)
    $D_w^{\,2}$ = diameter squared of water cylinder
    $D_s^{\,2}$ = diameter squared of steam cylinder

    $$P_s = \frac{P_w \times D_w^{\,2}}{D_s^{\,2}}$$

    $$P_s = \frac{200 \times 3 \times 3}{5 \times 5}$$

    $$P_s = \mathbf{72\ psi}$$

12. *What is the primary cause of steam turbine blade deposits? What effect does this have on steam turbine operation?*

    Answer: The main cause of turbine blade deposits is boiler water that is carried over with the steam which, when evaporated, leaves behind the impurities in the boiler water. The impurities are chemicals and solids from the boiler water treatment used to prevent scaling of boiler metal. Carryover of boiler water can be prevented by maintaining the NOWL, preventing fluctuating loads, and maintaining proper boiler water treatment to prevent foaming.

    The effect of turbine blade deposits is an unbalanced rotor. This in turn leads to excessive vibration to a point where the turbine cannot run. The rotor must be removed for cleaning (sandblasting) and rebalanced.

13. *A steam turbine is operating a generator that is generating 6000 kW. The water rate of the steam turbine is 14 lb/kW: How much boiler horsepower is required with a feedwater temperature of 200°F?*

    Answer:

    $$BHP = \frac{W\left[H_s - \left(T_{fw} - 32\right)\right]}{33,475}$$

    BHP = boiler horsepower
    W = pounds of steam per hour
    H_s = heat per pound of steam (enthalpy) assumed to be 1200 Btu/lb
    T_{fw} = feedwater temperature

33,475 = number of Btu in 1 BHP
$W = kW \times Water\ rate$
$W = 6000 \times 14$
$W = 84,000$ lb/hr

$$BHP = \frac{84,000\left[1200 - (200 - 32)\right]}{33,475}$$

$$BHP = \frac{84,000 \times 1032}{33,475}$$

$$BHP = \mathbf{2590}$$

14. *What is the thermal efficiency of a steam turbine that operates with a throttle pressure of 350 psi at 600°F and exhausts into a surface condenser at 29" Hg vacuum?*
Answer:

$$TE = \frac{H_a - H_e}{H_a - H_c}$$

TE = thermal efficiency
H_a = enthalpy of steam at throttle
H_e = enthalpy of steam at exhaust
H_c = enthalpy of condensate

Assume the following values: H_a = 1309 Btu/lb; H_e = 1050 Btu/lb; H_c = 100 – 32 *Btu/lb.*

$$TE = \frac{1309 - 1050}{1309 - (100 - 32)}$$

$$TE = \frac{259}{1241}$$

$$TE = \mathbf{20.9\%}$$

15. *How does a jet condenser differ from a surface condenser?*
Answer: A jet condenser can be compared to an open feedwater heater in which steam and water mix together. In a surface condenser, steam and water do not mix. The cooling water passes through tubes surrounded by steam that condenses, creating the vacuum needed.

16. *How much cooling water is needed to condense 60,000 lb of steam/hr if the temperature of the condensate is 115°F and the steam at the inlet to the condenser has a Btu content of 1000 Btu/lb? The cooling water is 70°F at the inlet and 95°F at the outlet.*
Answer:

$$Lb\ water / Lb\ steam = \frac{H_e - H_c}{T_1 - T_2}$$

H_e = enthalpy of exhaust steam
H_c = enthalpy of condensate
T_1 = temperature of water leaving (in °F)
T_2 = temperature of water entering (in °F)

$$Lb\ water\ /\ Lb\ steam = \frac{1000 - (115 - 32)}{95 - 70}$$

$$Lb\ water\ /\ Lb\ steam = \frac{917}{25}$$

Lb water/Lb steam = 36.68

To condense 60,000 lb per hour

Lb water/hr = Lb steam/hr × Lb water/Lb steam

Lb water/hr = 60,000 × 36.68

Lb water/hr = **2,200,800**

17. *Why is the induced draft fan larger than the forced draft fan on a given size unit?*
Answer: The induced draft fan is located between the boiler and chimney. It must handle gases of combustion leaving the boiler. The volume of the gases of combustion is much greater than the volume of air supplied by the forced draft fan for combustion.

18. *Describe steam pressure and velocity present in an impulse steam turbine.*
Answer: Nozzles located before each stage cause steam to expand and increase in velocity. With the increase in velocity, there is a corresponding decrease in steam pressure. Steam velocity returns to its initial value after passing through the blades before the next stage.

19. *How much feedwater is flowing through a 4" feedwater line if the feedwater has a velocity of 400 fpm?*
Answer:

$$V = \frac{Q}{A}\ or\ Q = V \times A$$

V = velocity of water, in fpm

Q = quantity of water, in cu ft/min

A = cross-sectional area of pipe, in sq ft

$area = 0.7854D^2$

$$area = 0.7854 \times \frac{4}{12} \times \frac{4}{12}$$

$area$ = 0.0872666 sq ft

$Q = V \times A$

Q = 400 × 0.0873

Q = 34.92 cu ft/min

To change cu ft/min to lb/hr:

Lb/hr = Cu ft/min × 60 × 62.4

Lb/hr = 34.92 × 60 × 62.4

Lb/hr = **130,740.5**

Note: There are 62.4 lb in 1 cu ft.

20. *How much pressure is at the base of a water column 100' high? How much pressure is at the base of a column of mercury 100" high?*
Answer: Every vertical foot of water exerts a pressure of 0.433 psi. A column of water 100' high would therefore exert a pressure of 100 × 0.433, or 43.3 psi.

Every vertical inch of mercury exerts a pressure of 0.491 psi. A column of mercury 100" high would therefore exert a pressure of 100 × 0.491 or 49.1 psi.

Test 8

1. *When is the best time to blow down a boiler? In what order should bottom blowdown valves be used?*
Answer: The best time to blow down a boiler is when it is at its lightest load and water circulation is at its minimum. At this time, sludge and sediment tend to settle to the bottom of the boiler and can be removed through the bottom blowdown valves. The proper blowdown procedure depends on the type of blowdown valves on the boiler.

Boilers that have a quick-opening valve and slow-opening valve should be blown down as follows: The quick-opening valve, which is located as close to the boiler as practical, is opened first and closed last. The quick-opening valve is used as a sealing valve. All the wear is on the slow-opening valve.

Boilers with tandem slow-opening valves should be blown down as follows: The valve located closest to the boiler is a sleeve-type (nonwearing) valve and is used as the blowing valve. It is opened last and closed first. The valve farthest from the boiler is the sealing valve.

2. *When is the best time to blow down the waterwalls on a boiler?*
Answer: The waterwalls on a boiler, because of their locations, should only be blown down after the furnace has cooled down. There is still pressure on the boiler, but with the brickwork in the furnace cooled, there is no danger of overheating the waterwall tubes.

3. *What materials are used in the construction of water columns?*
Answer: Water columns are made of cast iron, malleable iron, or steel. According to the ASME code, cast iron can be used for pressures up to 250 psi, malleable iron is used for pressures up to 350 psi, and steel is used for pressures above 350 psi.

4. *Describe the conditions that determine the volume of combustion space in a boiler.*
Answer: The volume of the combustion space in a boiler is governed by the type of fuel used. The volume must be large enough so that combustion is completed before gases of combustion come in contact with the boiler heating surface. Boilers that use bituminous coal must have larger furnaces than those burning anthracite coal. Bituminous coal has a high volatile content, and these gases burn above the grates, thereby requiring more space to complete combustion. Pulverized coal-fired burners also require large furnace volumes to complete the combustion process.

5. *Describe a system to chemically treat boiler water. What determines the amount of chemical treatment added?*
Answer: A continuous treatment system is the most effective system to chemically treat boiler water. The chemicals needed are put into a tank with water (condensate preferred) and mixed with an agitator. The solution is then added to the boiler over a 24-hour period. A proportioning pump that can be finely adjusted is used to add the chemicals. The amount of chemical treatment needed depends on the amount of makeup water added to the boiler and the residual levels required in the boiler water as determined by a chemical analysis.

6. *Explain the combustion process in a boiler furnace.*
Answer: The combustion process within a furnace requires the following;
   1. proper air-fuel ratio,
   2. proper fuel atomization,

3. proper combustion chamber temperature, and
4. time to complete the combustion process.

The reactions that occur in the furnace result from combustibles in the fuel, which are usually carbon, sulfur, and hydrogen.

carbon + oxygen → carbon dioxide + heat
$C + O_2 → CO_2 + heat$
sulfur + oxygen → sulfur dioxide + heat
$S + O_2 → SO_2 + heat$
hydrogen + oxygen → water + heat
$2H_2 + O_2 → 2H_2O + heat$

7. *What is the purpose of a blowdown tank? What pipeline connections are required?*
   Answer: A blowdown tank prevents high temperature water and steam from entering sewer lines when boilers are being blown down. The blowdown tank must be vented to the atmosphere. It is also fitted with internal overflow lines so there is always water in the tank. The internal overflow that goes to the sewer is also fitted with a siphon breaker to prevent the tank from becoming dry. There is also a line on the bottom of the tank so the tank can be dumped for cleaning and inspection. Some tanks also have a water line to cool the discharge if necessary.

8. *What type of circulating pump is commonly used with a surface condenser?*
   Answer: Circulating pumps used with surface condensers are large volume, low-pressure centrifugal pumps. They can be steam-driven or electrically driven. Surface condensers need large amounts of cooling water to maintain the required vacuum. High-pressure water pumps could cause a tube in the condenser to rupture.

9. *Define enthalpy.*
   Answer: Enthalpy is a term used in steam tables in place of heat. Sensible heat becomes enthalpy of liquid. Latent heat becomes enthalpy of evaporation. Enthalpy is measured in Btu/lb.
   Total heat, or enthalpy of steam, is the sum of sensible heat and latent heat of evaporation (starting at a base of 32°F). Enthalpy can be found in the steam tables.

10. *What can cause a furnace explosion?*
    Answer: Furnace explosions result from a buildup of combustible gases in a boiler furnace. Furnace explosions are the result of delayed ignition or improper purging of the furnace. Leaking fuel valves can cause a buildup of fuel in a hot furnace, causing an explosion. Furnace explosions in burners using pulverized coal can be caused when there is an interruption of the flow of pulverized coal to the furnace.

11. *A 2000 HP boiler is operating at 200 psi. What HP is needed to drive the feedwater pump on this boiler?*
    Answer:

$$HP_d = \frac{FH \times Lb / min}{33,000 \text{ ft-lb/min} \times E}$$

HP_d = drive horsepower
FH = foot head on pump
Lb/min = pounds of water pumped per minute
33,000 = ft-lb per minute per horsepower
E = efficiency (assume 80%)

$$FH = \frac{200 \text{ psi}}{0.433 \text{ psi/ft}}$$

$$FH = 462$$

$$Lb / min = \frac{200 \times 34.5}{60}$$

$$Lb/min = 1150$$

$$HP_d = \frac{462 \times 1150}{33,000 \times 0.8}$$

$$HP_d = \textbf{20.125}$$

12. *Does the forced draft fan in question 11 require more or less HP than the feed-water pump?*

Answer: Assume the following values: evaporation rate = 12 lb of steam/lb of fuel; 200 cu ft of air/lb of fuel; discharge pressure = 4″ H_2O; fan efficiency = 55%.

$$HP_d = \frac{QP}{6356 \times E}$$

HP_d = drive horsepower
Q = quantity of air, in cu ft/min
P = pressure, in inches of H_2O
6356 = constant

$$Lb \text{ of fuel} / min = \frac{Lb \text{ of steam} / min}{Lb \text{ of steam} / Lb \text{ of fuel}}$$

$$Lb \text{ of fuel} / min = \frac{1150}{12}$$

Lb of fuel/min = 95.83
$Q = Lb$ of fuel/min $\times$ Cu ft of air/lb
$Q = 95.83 \times 200$
$Q = 19,166$ cu ft/min

$$HP_d = \frac{QP}{6356 \times E}$$

$$HP_d = \frac{19,166 \times 4}{6356 \times 0.55}$$

$$HP_d = \textbf{21.93}$$

The forced draft fan requires approximately the same HP drive as the feed water pump.

13. *What is the rim speed of a 4′ flywheel turning at 200 RPM?*

Answer: Rim speed of a flywheel can be determined using the following equation:

Rim speed = Flywheel circumference $\times$ RPM
Circumference = $\pi \times$ Diameter
$RS = 3.1416 \times 4 \times 200$
$RS = \textbf{2513.28 fpm}$

14. *What is the torque on a 10,000 kW steam turbine shaft running at 1800 RPM at full load?*

Answer:

$$T_q = \frac{5252 \times HP}{RPM}$$

T_q = torque, in ft-lb
HP = horsepower developed
RPM = revolutions per minute
$HP = kW \times 1.34$
$HP = 10,000 \times 1.34$
$HP = 13,400$

$$T_q = \frac{5252 \times 13,400}{1800}$$

T_q = **39,098.2 ft-lb**

15. *Describe the advantages of using a combustion control system.*

Answer: Combustion controls regulate fuel supply, air supply, air-fuel ratio, and removal of gases of combustion to achieve maximum boiler efficiency. The amount of fuel and air supplied must be in proportion to the steam pressure required. A drop in steam pressure requires an increase in air and fuel supplied to the burner. This increase must be controlled so the air-fuel ratio is correct to prevent smoke or flame failure. By maintaining a consistent firing rate, combustion controls improve regulation of feedwater and superheat temperature, reduce fluctuation of water level, and reduce thermal shock, thereby increasing the life of the boiler drum and tubes. Combustion controls also reduce furnace refractory maintenance by maintaining a uniform furnace temperature.

16. *How is thrust of the shaft controlled in reaction and impulse steam turbines?*

Answer: Modern reaction steam turbines use one balance piston to balance all end thrust at full load on the machine. Inequalities of thrust at other loads are compensated for by a thrust bearing. Impulse steam turbines only need a thrust bearing to compensate at all loads.

17. *What are the principal heat losses in a boiler?*

Answer: The principal losses are
- Btu losses resulting from dry gases of combustion to the atmosphere
- incomplete combustion
- moisture in the fuel
- water vapor from burning hydrogen in the fuel
- moisture in the air used for combustion
- unburned combustibles
- radiation

18. *Why is the rotor of a generator made of laminated steel sections?*

Answer: Using laminated steel reduces eddy currents in the rotor and therefore reduces heat that is generated. It is still necessary to use some form of forced circulating and cooling to remove heat from the rotor.

19. *Compare the water rates of a condensing and noncondensing steam turbine of equal capacity.*
Answer:

$$WR = \frac{Btu / kW}{H_a - H_e}$$

WR = water rate
Btu/kW = Btu equivalent to 1 kW
H_a = enthalpy of steam at throttle
H_e = enthalpy of steam at exhaust
Assume the following values: H_a = 1300 Btu/lb; H_e = 1050 Btu/lb (condensing); H_e = 1150 Btu/lb (noncondensing).

$$WR = \frac{3413}{1300 - 1050} \text{ (condensing)}$$

WR = 13.65 lb of steam/kW

$$WR = \frac{3413}{1300 - 1150} \text{ (noncondensing)}$$

WR = 22.75 lb of steam/kW
Condensing = **13.65 lb of steam/kW**
Noncondensing = **22.75 lb of steam/kW**

20. *How does a multistage centrifugal pump on a common shaft with the same diameter impellers build up the pressure in each stage?*
Answer: It is the same as having four separate pumps in series. The water enters the suction of the first impeller and is discharged at a higher pressure into the suction of the second impeller. The second impeller discharges at an even higher pressure into the suction of the third impeller and so on through the fourth impeller. The pressure buildup across each impeller is the same. Care must be taken to balance thrust with this type of pump.

Test 9

1. *Why are bypass dampers used on air preheaters?*
Answer: Bypass dampers are used to control the temperature of the preheated air. The preheated air must be prevented from getting too hot when stokers are used. Air temperatures over 300°F could damage the stoker grates by warping the grates. Also, a low air inlet or air outlet temperature must be prevented during startups and fluctuating boiler loads. This could result in moisture buildup from sweating, causing corrosion of the air heaters.

2. *Describe the advantages of using a combustion control system.*
Answer: Combustion controls regulate fuel supply, air supply, air-fuel ratio, and removal of gases of combustion to achieve maximum boiler efficiency. The amount of fuel and air supplied must be in proportion to the steam pressure required. A drop in steam pressure requires an increase in air and fuel supplied to the burner. This increase must be controlled so the air-fuel ratio is correct to prevent smoke or flame failure. By maintaining a consistent firing rate, combustion controls improve regulation of feedwater and superheat temperature, reduce fluctuation of water level,

and reduce thermal shock, thereby increasing the life of the boiler drum and tubes. Combustion controls also reduce furnace refractory maintenance by maintaining a uniform furnace temperature.

3. *What is the purpose of and maintenance procedures for a motor-generator set?*
Answer: A motor-generator set consists of an AC motor that runs off line voltage and turns a DC generator. There is enough DC power to operate elevators, excite synchronous motors, charge batteries, and supply a DC circuit for control switches. The motor-generator set requires the following attention:
   1. Maintain proper lubrication of the bearings.
   2. Check for arcing between the commutator and brushes.
   3. Adjust the field excitation to maintain correct terminal voltage.
   4. Replace commutator brushes.
   5. Dress commutator.
   6. Remove dust and dirt from windings.

4. *How is a positive displacement pump protected from excessive pressure on the discharge side?*
Answer: To protect a positive displacement pump from excessive pressure on its discharge side, a pressure relief valve is installed between the pump and its discharge valve. The relief valve may discharge back to the suction side or, in the case of fuel oil pumps, back to the return line going to the fuel oil tank.

5. *What factors must be considered when selecting a new boiler feedwater pump?*
Answer: A boiler feedwater pump must meet conditions for pressure and volume. A boiler feedwater pump should be able to deliver two times the maximum steaming capacity of the boiler at the MAWP of the boiler. It can be used effectively with in-line feedwater regulators, and variable drives should be considered. The pump must be reliable and maintain a fairly high efficiency over long running periods.

6. *Describe a velocity-pressure diagram of a reaction steam turbine.*
Answer: Instead of nozzles, the reaction steam turbine uses fixed blades for its first stage. Fixed blades are designed so each pair acts as a single nozzle. Steam expands in both fixed and moving blades. Steam entering the fixed blades drops in pressure but increases in velocity. The steam then enters the blades of the moving element where its velocity drops as it imparts energy to the moving blades. The steam then enters the next row of fixed blades, repeating the process, dropping in pressure and increasing in velocity.
   In leaving the turbine blades at a high velocity, a backward kick is given to the blades, hence the term *reaction*. The diagram shows the pressure drop as the steam passes through the fixed and moving blades as well as the increase in velocity passing through the fixed blades and the drop in velocity passing through the moving blades.

7. *What are the cold clearances on an average steam turbine?*
Answer: Steam turbine clearances are comparatively small. Radial clearances are 0.180″ to 0.250″, which is the distance between the top moving blades and casing. Axial clearances are 0.100″ to 0.200″, which is the distance between the nozzle exits and the leading edge of the blade. Diaphragm gland clearances are 0.002″, which is the distance between the shaft and bottom of the diaphragm. Bearing clearances are 0.001″ per inch of diameter of shaft, with a minimum of 0.005″. A 4″ shaft would rotate in a bearing with an inner diameter of 4.005″, whereas a 10″ shaft would have an inner diameter of 10.010″.

8. *How is steam leakage around fixed and moving blades prevented in a reaction steam turbine?*
Answer: To prevent steam leakage around fixed and moving blades, the clearance between the tips of the fixed blades and shaft must be very close as is the clearance between the moving blades and casing. There are a larger number of elements in the reaction steam turbine that permit a small pressure drop per row of blades. Labyrinth packing and sealing strips are also used to minimize steam leakage.

9. *Why is it important to save the condensate from the hot well of a surface condenser?*
Answer: The condensate from the hot well of a surface condenser is the steam that exhausts into the surface condenser after leaving the steam turbine. It can be compared to distilled water. It is clean hot water and is returned to the feedwater system for reuse.

10. *What is the specific volume of steam at the inlet of a steam turbine using saturated steam at 250 psi? What is the specific volume of steam at the exhaust with a pressure of 28" Hg?*
Answer: From the saturated steam tables, steam at 265 psia has a volume of 1.754 cu ft/lb. From the saturated steam tables, steam at 28" Hg vacuum has a volume of 333.6 cu ft/lb. The volume would change if the 250 psi steam were superheated or if the exhaust steam were partially condensed.

11. *What effect does condensing or noncondensing have on the thermal efficiency of a steam turbine?*
Answer: Condensing steam turbines have a higher thermal efficiency because condensate is reclaimed for use in the system. The condensate retains some heat, which requires less heat from the open feedwater heater. Noncondensing steam turbines that exhaust steam to the atmosphere have a very low thermal efficiency.

12. *What are steam surface condenser tubes made of? How are they attached to the tube sheet?*
Answer: Steam surface condenser tubes are made of brass or copper and are attached to the tube sheet by rolling or expanding the tube into the tube sheet.

13. *What is the difference between bleed steam and extraction steam taken from a steam turbine?*
Answer: Bleed steam is steam removed from various stages of a steam turbine and used for heating feedwater. Extraction steam is removed from the steam turbine at various stages at a controlled pressure for plant process.

14. *What is the function of a mechanical governor and an overspeed trip on a steam turbine?*
Answer: The mechanical governor consists of a speed-sensing element and linkage that control the governor valve through the force acting on the speed-sensing element. The speed-sensing element consists of weights mounted on the turbine shaft. The centrifugal force on the weights is opposed by a spring as the speed of the turbine increases. The weights move and through linkage move the governor valve toward the closed position, throttling steam flow to the turbine.

The overspeed trip is separate from the main governor and prevents the turbine from exceeding its rated speed by 10%. A bolt is mounted in the turbine shaft and is opposed by an adjustable spring. It remains in one position until the shaft speed reaches a point where the centrifugal force on the bolt overcomes the spring tension. The bolt moves out and trips a latch device that releases a butterfly valve that secures the steam to the turbine.

15. *What is the function of an atmospheric relief valve?*
    Answer: An atmospheric relief valve prevents pressure buildup on the exhaust side of a reciprocating steam engine or steam turbine. It is located on the main exhaust line or directly at the top of the condenser as close to the exhaust steam inlet as possible. The atmospheric relief valve stays closed as long as there is a vacuum in the condenser. If the vacuum is lost, the atmospheric relief valve will open, allowing the reciprocating steam engine or steam turbine to exhaust into the atmosphere, thus protecting the condenser from excessive steam pressure.

16. *Define the following terms:*
    Answer:
    a. Gauge pressure: Pressure above atmospheric pressure. Read on a pressure gauge and measured in psi or psig.
    b. Absolute pressure: The sum of gauge pressure and atmospheric pressure and measured in psia.
    c. Enthalpy: A term used in steam tables rather than the term heat. Sensible heat becomes enthalpy of liquid. Latent heat becomes enthalpy of evaporation. Both are measured in Btu/lb. Total heat becomes enthalpy of steam and is the sum of sensible heat and latent heat of evaporation (starting at base of 32°F).
    d. Latent heat of evaporation: The amount of Btu necessary to change 1 lb of water at 212°F to 1 lb of steam at 212°F and is equal to 970.3 Btu.
    e. Sensible heat: Heat necessary to raise the temperature of 1 lb of water at 32°F to 1 lb of water at 212°F and is equal to 180 Btu.
    f. Specific heat: Specific heat of a substance is the quantity of heat necessary to change the temperature of 1 lb of a substance 1°F. It is measured in Btu/lb/°F and is obtained by comparison with water as a standard.

17. *A boiler carrying 200 psig has a safety valve 2½" in diameter. What is the total upward force on the safety valve disc?*
    Answer:
    $$TF = P \times A$$
    TF = total force on safety valve disc, in lb
    P = pressure on the disc, in psi
    A = area of disc, in sq in.
    $$TF = P \times A$$
    $$TF = P \times 0.7854D^2$$
    $$TF = 200 \times 0.7854 \times 2.5 \times 2.5$$
    $$TF = \textbf{982 lb}$$

18. *When is extra-heavy pipe required on bottom blowdown lines of a boiler?*
    Answer: Extra-heavy pipe must be used when the MAWP exceeds 125 psi. The blowdown line and fittings between the boiler and valves must be composed of extra-heavy bronze, brass, or malleable iron suitable for the temperature and pressures involved. They must run full size with no reducers or bushings, and each bottom blowdown line must be fitted with two 250 psi standard valves, both of which are extra-heavy.

19. *How is the proper location of a water column determined?*
Answer: The proper location of a water column can be checked by taking external or internal measurements. The water column must be located at the NOWL so that the lowest visible part of the gauge glass is 2″ to 3″ above the highest heating surface of a firetube boiler. In a watertube boiler, the lowest visible part of the gauge glass is 3″ above the lowest permissible water level as specified by the boiler manufacturer.

20. *What is the minimum size for tubing, steel pipe, or wrought iron pipe used in the installation of a steam pressure gauge?*
Answer: The pipeline that connects the steam pressure gauge to the boiler shall not be less than ¼″ diameter when using tubing. When steel or wrought iron pipe is used, it must not be less than ½″ diameter.

Test 10

1. *Describe the velocity-pressure diagram of a velocity compound impulse steam turbine.*
Answer: In an impulse steam turbine, the steam enters a first-stage nozzle where it drops in pressure and gains velocity before striking the blades on the first-stage wheel. The steam then passes through fixed blades, changing its direction, and strikes a second set of blades on the same first-stage wheel. There is no drop in pressure as the steam passes through the moving and stationary blades. The velocity drops only when steam passes through the moving blades. The steam then passes through the second-stage nozzle where again there is a drop in pressure and an increase in velocity.

2. *List the location and function of equipment used in a feedwater system.*
Answer:
- Open feedwater heater: Located at an elevation above the feedwater pump. Acts as a reservoir so there is always water available for the boiler. Water is heated here and oxygen and other noncondensable gases are vented to the atmosphere. Automatic makeup water is added here to replace water lost in system.
- Suction line: Brings water to the boiler feedwater pump. Located on the makeup water line.
- Feedwater pump: Has suction and discharge valves so it can be taken out of the system for repairs or general maintenance. Feedwater pump delivers water to the boiler at the proper quantity and pressure. Located on the feedwater line.
- Closed feedwater heater: Located on the discharge side of the feedwater pump where the water is further heated to increase plant efficiency. Usually heated with exhaust steam.
- Economizer: Located in the direct path of gases of combustion by reclaiming heat from the gases of combustion to heat boiler feedwater. It is possible to realize a savings in fuel consumption, (every 10° rise in feedwater temperature saves 1% in fuel). It also reduces thermal shock to the boiler.
- Feedwater regulator: Controls the amount of water fed to the boiler. By steady feeding it makes for a better steaming boiler. It reduces the danger of high or low water. The sensing element of the feedwater regulator is located at the

NOWL. The control valve is located in the feed water line. It is equipped with stop valves so emergency repairs can be made and a bypass line so the boiler can be fed by hand in emergency situations.

- Feedwater check valve: Located on the feedwater line to control the flow of water to the boiler and prevent a back flow if the pump stops.
- Feedwater stop valve: Located on the feedwater line between the boiler and check valve so check valve can be repaired without having to dump the boiler.

3. *List three ways to control the amount of air delivered to a furnace using a forced draft fan.*
Answer: The amount of forced draft delivered to a furnace can be controlled by a variable speed motor that controls the fan speed, using inlet or outlet dampers on the fan, and a hydraulic clutch between the fan and drive mechanism.

4. *A boiler carrying 150 psig has a safety valve 2″ in diameter. What is the total upward force on the safety valve disc?*
Answer:

$TF = P \times A$
TF = total force on safety valve disc, in lb
P = pressure on the disc, in psi
A = area of disc, in sq in.
$TF = P \times A$
$TF = P \times 0.7854D^2$
$TF = 150 \times 0.7854 \times 2 \times 2$
$TF = \mathbf{471\ lb}$

5. *Can plant efficiency be increased if a closed feedwater heater is used?*
Answer: Yes, it is possible to increase plant efficiency when heating feedwater in a closed feed water heater. A good example of this is when steam is extracted or bled from a high-pressure stage of a steam turbine and used in a closed feedwater heater. The latent heat of the steam is recovered in the feedwater instead of being lost in the condenser.

6. *A simplex reciprocating pump has a water piston diameter of 12″ and a stroke of 14″. It operates at 100 strokes/min. What is the discharge in gallons per minute if the pump is 80% efficient?*
Answer:

$$gpm = \frac{LANE}{231}$$

gpm = gallons per minute
L = length of stroke, in inches
A = area of water piston, in sq in.
N = number of strokes per minute
E = pump efficiency (sometimes expressed as % of slip)
231 = cubic inches in 1 gal.
Note: $A = 0.7854D^2$

$$gpm = \frac{14 \times 0.7854 \times 12 \times 12 \times 100 \times 0.80}{231}$$

$gpm = \mathbf{548}$

7. *A boiler operates at 100 psi steam pressure and is supplied with feedwater from a feedwater heater that has an internal pressure of 10 psi. The boiler water level is located 30′ above the pump and the heater is 45′ above the pump. What is the total head on the pump? Do not allow for friction losses.*
Answer:

$$H_t = H_d + H_f - H_s$$
$$H_t = H_d + H_f + SL$$

H_t = total head, in ft
H_d = head discharge, in ft
H_f = friction head, in ft
SL = suction lift, in ft
H_s = suction head, in ft

Note: To convert psi to foot head, divide by 0.433 psi per vertical foot.

$$H_t = \left(30 + \frac{100}{0.433}\right) + 0 - \left(45 + \frac{10}{0.433}\right)$$

$$H_t = 261 - 68.1$$
$$H_t = \mathbf{192.9′}$$

8. *How should horizontal steam lines be pitched?*
Answer: Horizontal steam lines must always be pitched in the direction of the steam flow and have traps installed at the end of the run so that condensate can be removed. If the lines were pitched back toward the boiler, the steam would pick up the condensate and cause water hammer and possible line rupture.

9. *What would the amount of expansion in a 175′ long steel steam pipe be if its temperature changed from 85°F to 355°F?*
Answer:

$Expansion = L \times C \times (T_1 - T_2) \times 12$
L = length of pipe (in ft)
C = coefficient of linear expansion, in in./in.°F (0.0000067)
T_1 = final temperature, in °F
T_2 = initial temperature, in °F
12 = inches per foot
$Expansion = 175 \times 0.0000067 \times (355 - 85) \times 12$
$Expansion = 175 \times 0.0000067 \times 270 \times 12$
$Expansion = \mathbf{3.8″}$

10. *Steam passing to a steam separator has a quality of 93%. If 5600 lb of steam/hr pass through and the separator collects 285 lb of water, what is the efficiency of the separator?*
Answer:

$$Efficiency\ of\ separator = \frac{Condensate\ removed}{Condensate\ in\ steam} \times 100$$

$$Condensate\ in\ steam = Pounds\ of\ steam \times (1 - Quality)$$

$$E_s = \frac{Condensate\ removed}{Lb\ of\ steam \times (1 - Quality)} \times 100$$

$$E_s = \frac{285}{5600 \times (1 - 0.93)} \times 100$$

$$E_s = \frac{285}{5600 \times 0.07} \times 100$$

$$E_s = \textbf{72.7\%}$$

11. *Explain how superheated steam can be desuperheated.*
    Answer: Superheated steam can be desuperheated using a desuperheating station where water is introduced into the steam line. This reduces the steam temperature for use in auxiliary steam equipment or process lines requiring saturated steam.

12. *Define the following:*
    Answer:
    a. Viscosity: The internal resistance of a fluid to flow.
    b. Flash point: Temperature at which oil when heated produces a vapor that flashes when exposed to an open flame.
    c. Fire point: Temperature at which oil burns continuously when exposed to an open flame.
    d. Pour point: The lowest temperature at which fuel oil flows as a liquid.

13. *A coal sample has the following ultimate analysis:*
    Answer:

Carbon 85%	Oxygen 5%
Hydrogen 5%	Sulfur 1%
Nitrogen 2%	Ash 2%

    Using Dulong's formula, find the heating value of the coal.

    $$Heating\ value,\ Btu/lb = 14{,}540C + 62{,}000\left(H - \frac{O}{8}\right) + 4050S$$

    $$Heating\ value, Btu/lb = 14{,}540 \times 0.85 + 62{,}000\left(0.05 - \frac{0.05}{8}\right) + (4050 \times 0.01)$$

    Heating value, Btu/lb = 12,359 + [62,000(0.05 − 0.00625)] + 40.5
    Heating value, Btu/lb = 12,359 + (62,000 × 0.04375) + 40.5
    Heating value, Btu/lb = 12,359 + 2712.5 + 40.5
    Heating value, Btu/lb = **15,112 Btu/lb**

14. *Describe an underfeed stoker and explain how it functions.*
    Answer: Underfeed stokers can be ram-fed or screw-fed. Coal is fed into a hopper and onto a ram or screw. The coal is then forced into a retort chamber where pusher blocks move the coal forward and upward, distributing the coal and leaving it over the entire grate area. Green (fresh) coal is continually fed under the fire and must be controlled to prevent the fire from burning in the retort. This would cause pusher blocks to warp.

15. *What are the disadvantages of using pulverized coal as a fuel?*
    Answer: The disadvantages are as follows:
    - Pulverized coal is explosive and requires care when it is first introduced into the furnace.
    - Pulverized coal must be kept dry to prevent caking in the mill.

- The furnace must be preheated before pulverized coal is used. Preheating is done with oil or gas.
- The high furnace temperature sometimes results in the formation of slag, which can adhere to boiler tubes. Slag removal is often accomplished using a slag trap.
- Furnace pressure must be kept below atmospheric pressure to prevent the furnace flame from entering the mill and causing an explosion in the mill.
- Flyash must be collected and removed from gases of combustion.

16. *Explain what takes place in a furnace after fuel is introduced.*
Answer: When fuel enters the furnace, it mixes with primary and secondary air to complete the combustion process. All the fuel must burn in the furnace and combustion must be completed before gases of combustion come in contact with any boiler heating surface. If this does not occur smoke and soot will result. Gases of combustion then pass through the boiler, making two, three, or more passes depending on boiler design. This is necessary so that as much heat as possible is transferred to the boiler water to produce steam. Once gases of combustion leave the boiler and enter the chimney, the heat left will be lost to the atmosphere. Larger boilers are equipped with air heaters and economizers to reclaim as much heat as possible.

17. *Discuss the routine duties and responsibilities of a boiler operator.*
Answer: The boiler operator is responsible for the safe and efficient operation of the boiler and all its auxiliary equipment. The boiler operator should always report for work 15 to 20 minutes before the shift begins and proceed as follows:
1. Check the water level of all boilers on-line by blowing down the gauge glass and water column.
2. Check steam pressure and condition of fires.
3. Check all running auxiliaries for proper temperature, pressure, and lubrication.
4. Ask operator if there were any problems on previous shift.
5. Check log for any orders.

Routine Duties:
1. Change over fuel oil strainers and clean strainer if burning oil.
2. Change over burners. Clean and then leave clean burners in the rack so they can be ready for the next shift or an emergency that requires clean burners.
3. Make routine rounds and record all readings in a log as required.
4. Be alert at all times for unusual changes in the steam load, fluctuating water level, or abnormal temperatures or pressures in the auxiliary boiler room equipment.

18. *What is the danger of high water and low water in a boiler?*
Answer: If the water level in the boiler is too high, carryover can result. This can cause water hammer and possible pipe rupture. In addition, damage to steam pumps or steam engines and steam turbines can result. Low water in the boiler can expose the heating surface to high temperatures that can cause burned out tubes and possible boiler explosion.

19. *What are the circumferential and longitudinal stresses on a boiler shell 48" in diameter, 12' long, and with a plate thickness of 1" if the steam pressure is 400 psi?*
Answer:

$$S_c = \frac{P \times D}{4T}$$

S_c = circumferential stress

P = pressure on boiler, in psi
D = diameter of boiler shell, in inches
T = thickness of boiler shell plate, in inches

$$S_c = \frac{400 \times 48}{4 \times 1}$$

S_c = **4800 psi**

$$S_l = \frac{P \times D}{2T}$$

S_l = longitudinal stress
P = pressure on boiler, in psi
D = diameter of boiler shell, in inches
T = thickness of boiler shell plate, in inches

$$S_l = \frac{400 \times 48}{2 \times 1}$$

S_l = **9600 psi**

20. *On a steam turbine, how does a mechanical governor differ from a hydraulic governor?*
Answer: Mechanical governors use a speed-sensing element and linkage to control the governor valve to regulate speed of the steam turbine. Hydraulic governors also use a speed-sensing element but with a servo-motor instead of mechanical linkage. This permits close control of steam turbine speed.

APPENDIX

Dry Saturated Steam Pressure

Abs press., psi p	Temp., °F t	Specific volume, cu ft/lb Sat. liquid v_f	Sat. vapor v_g	Enthalpy, Btu/lb Sat. liquid h_f	Evap. h_{fg}	Sat. vapor h_g	Entropy Sat. liquid s_f	Evap. s_{fg}	Sat. vapor s_g
1.0	101.74	0.01614	333.6	69.70	1036.3	1106.0	0.1326	1.8456	1.9782
2.0	126.08	0.01623	173.73	93.99	1022.2	1116.2	0.1749	1.7451	1.9200
3.0	141.48	0.01630	118.71	109.37	1013.2	1122.6	0.2008	1.6855	1.8863
4.0	152.97	0.01636	90.63	120.86	1006.4	1127.3	0.2198	1.6427	1.8625
5.0	164.24	0.01640	73.52	130.13	1001.0	1131.1	0.2347	1.6094	1.8441
6.0	170.06	0.01645	61.98	137.96	996.2	1134.2	0.2472	1.5820	1.8292
7.0	176.85	0.01649	53.64	144.76	992.1	1136.9	0.2581	1.5586	1.8167
8.0	182.86	0.01653	47.34	150.79	989.5	1139.3	0.2674	1.5383	1.8057
9.0	188.28	0.01656	42.40	156.22	985.2	1141.4	0.2759	1.5203	1.7962
10	193.21	0.01659	38.42	161.17	982.1	1143.3	0.2835	1.5041	1.7876
14.696	212.00	0.01672	26.80	180.07	970.3	1150.4	0.3120	1.4446	1.7566
15	213.03	0.01672	26.29	181.11	969.7	1150.8	0.3135	1.4415	1.7549
20	227.96	0.01683	20.089	196.16	960.1	1156.3	0.3356	1.3962	1.7319
25	240.07	0.01692	16.303	208.42	952.1	1160.6	0.3533	1.3606	1.7139
30	250.33	0.01701	13.746	218.82	945.3	1164.1	0.3680	1.3313	1.6993
35	259.28	0.01708	11.898	227.91	939.2	1167.1	0.3807	1.3063	1.6870
40	267.25	0.01715	10.498	236.03	933.7	1169.7	0.3919	1.2844	1.6763
45	274.44	0.01721	9.401	243.36	928.6	1172.0	0.4019	1.2650	1.6669
50	281.01	0.01727	8.515	250.09	924.0	1174.1	0.4110	1.2474	1.6585
55	287.07	0.01732	7.787	256.30	919.6	1175.9	0.4193	1.2316	1.6509
60	292.71	0.01738	7.175	262.09	915.5	1177.6	0.4270	1.2168	1.6438
65	297.97	0.01743	6.655	267.50	911.6	1179.1	0.4342	1.2032	1.6374
70	302.92	0.01748	6.206	272.61	907.9	1180.6	0.4409	1.1906	1.6315
75	307.60	0.01753	5.816	277.43	904.5	1181.9	0.4472	1.1787	1.6259
80	312.03	0.01757	5.472	282.02	901.1	1183.1	0.4531	1.1676	1.6207
85	316.25	0.01761	5.168	286.39	897.8	1184.2	0.4587	1.1571	1.6158
90	320.27	0.01766	4.896	290.56	894.7	1185.3	0.4641	1.1471	1.6112
95	324.12	0.01770	4.652	294.56	891.7	1186.2	0.4692	1.1376	1.6068
100	327.81	0.01774	4.432	298.40	888.8	1187.2	0.4740	1.1286	1.6026
110	334.77	0.01782	4.049	305.66	883.2	1188.9	0.4832	1.1117	1.5948
120	341.25	0.01789	3.728	312.44	877.9	1190.4	0.4916	1.0962	1.5878

continued...

Dry Saturated Steam Pressure

Abs press., psi p	Temp., °F t	Specific volume, cu ft/lb		Enthalpy, Btu/lb			Entropy		
		Sat. liquid v_f	Sat. vapor v_g	Sat. liquid h_f	Evap. h_{fg}	Sat. vapor h_g	Sat. liquid s_f	Evap. s_{fg}	Sat. vapor s_g
130	347.32	0.01796	3.455	318.81	872.9	1191.7	0.4995	1.0817	1.5812
140	353.02	0.01802	3.220	324.82	868.2	1193.0	0.5069	1.0682	1.5751
150	358.42	0.01809	3.015	330.51	863.6	1194.1	0.5138	1.0556	1.5694
160	363.53	0.01815	2.834	335.93	859.2	1195.1	0.5204	1.0436	1.5640
170	368.41	0.01822	2.675	341.09	854.9	1196.0	0.5266	1.0324	1.5590
180	373.06	0.01827	2.532	346.03	850.8	1196.9	0.5325	1.0217	1.5542
190	377.51	0.01833	2.404	350.79	846.8	1197.6	0.5381	1.0116	1.5497
200	381.79	0.01839	2.288	355.36	843.0	1198.4	0.5435	1.0018	1.5453
250	400.95	0.01865	1.8438	376.00	825.1	1201.1	0.5675	0.9588	1.5263
300	417.33	0.01890	1.5433	393.84	809.0	1202.8	0.5879	0.9225	1.5104
350	431.72	0.01913	1.3260	409.69	794.2	1203.9	0.6056	0.8910	1.4966
400	444.59	0.0193	1.1613	424.0	780.5	1204.5	0.6214	0.8630	1.4844
450	456.28	0.0195	1.0320	437.2	767.4	1204.6	0.6356	0.8378	1.4734
500	467.01	0.0197	0.9278	449.4	755.0	1204.4	0.6487	0.8147	1.4634
550	476.94	0.0199	0.8424	460.8	743.1	1203.9	0.6608	0.7934	1.4542
600	486.21	0.0201	0.7698	471.6	731.6	1203.2	0.6720	0.7734	1.4454
650	494.90	0.0203	0.7083	481.8	720.5	1202.3	0.6826	0.7548	1.4374
700	503.10	0.0205	0.6554	491.5	709.7	1201.2	0.6925	0.7371	1.4296
750	510.86	0.0207	0.6092	500.8	699.2	1200.0	0.7019	0.7204	1.4223
800	518.23	0.0209	0.5687	509.7	688.9	1198.6	0.7108	0.7045	1.4153
850	525.26	0.0210	0.5327	518.3	678.8	1197.1	0.7194	0.6891	1.4085
900	531.98	0.0212	0.5006	526.6	668.8	1195.4	0.7275	0.6744	1.4020
950	538.43	0.0214	0.4717	534.6	659.1	1193.7	0.7355	0.6602	1.3957
1000	544.61	0.0216	0.4456	542.4	649.4	1191.8	0.7430	0.6467	1.3897
1100	556.31	0.0220	0.4001	557.4	630.4	1187.8	0.7575	0.6205	1.3780
1200	567.22	0.0223	0.3619	571.7	611.7	1183.4	0.7711	0.5956	1.3667
1300	577.46	0.0227	0.3293	585.4	593.2	1178.6	0.7840	0.5719	1.3559
1400	587.10	0.0231	0.3012	598.7	574.7	1173.4	0.7963	0.5491	1.3454
1500	596.23	0.0235	0.2765	611.6	556.3	1167.9	0.8082	0.5269	1.3351
2000	635.82	0.0257	0.1878	671.7	463.4	1135.1	0.8619	0.4230	1.2849
2500	668.13	0.0287	0.1307	730.6	360.5	1091.1	0.9126	0.3197	1.2322
3000	695.36	0.0346	0.0858	802.5	217.8	1020.3	0.9731	0.1885	1.1615
3206.2	705.40	0.0503	0.0503	902.7	0	902.7	1.0580	0	1.0580

…continued

Dry Saturated Steam Temperature

Temp., °F	Abs press., psi	Specific volume, cu ft/lb			Enthalpy, Btu/lb			Entropy		
		Sat. liquid	Evap.	Sat. vapor	Sat. liquid	Evap.	Sat. vapor	Sat. liquid	Evap.	Sat. vapor
t	p	v_f	v_{fg}	v_g	h_f	h_{fg}	h_g	s_f	s_{fg}	s_g
32	0.08854	0.01602	3306	3306	0.00	1075.8	1075.8	0.0000	2.1877	2.1877
35	0.09995	0.01602	2947	2947	3.02	1074.1	1077.1	0.0061	2.1709	2.1770
40	0.12170	0.01602	2444	2444	8.05	1071.3	1079.3	0.0162	2.1435	2.1597
45	0.14752	0.01602	2036.4	2036.4	13.06	1068.4	1081.5	0.0262	2.1167	2.1429
50	0.17811	0.01603	1703.2	1703.2	18.07	1065.6	1083.7	0.0361	2.0903	2.1264
60	0.2563	0.01604	1206.6	1205.7	28.06	1059.9	1088.0	0.0555	2.0393	2.0948
70	0.3631	0.01606	867.8	867.9	38.04	1054.3	1092.3	0.0745	1.9902	2.0647
80	0.5069	0.01608	633.1	633.1	48.02	1048.6	1096.6	0.0932	1.9428	2.0360
90	0.6982	0.01610	468.0	468.0	57.99	1042.9	1100.9	0.1115	1.8972	2.0087
100	0.9492	0.01613	350.3	350.4	67.97	1037.2	1105.2	0.1295	1.8531	1.9826
110	1.2748	0.01617	265.3	265.4	77.94	1031.6	1109.5	0.1471	1.8106	1.9577
120	1.6924	0.01620	203.25	203.27	87.92	1025.8	1113.7	0.1645	1.7694	1.9339
130	2.2225	0.01625	157.32	157.34	97.90	1020.0	1117.9	0.1816	1.7296	1.9112
140	2.8886	0.01629	122.99	123.01	107.89	1014.1	1122.0	0.1984	1.6910	1.8894
150	3.718	0.01634	97.06	97.07	117.89	1008.2	1126.1	0.2149	1.6537	1.8685
160	4.741	0.01639	77.27	77.29	127.89	1002.3	1130.2	0.2311	1.6174	1.8485
170	5.992	0.01645	62.04	62.06	137.90	996.3	1134.2	0.2472	1.5822	1.8293
180	7.510	0.01651	50.21	50.23	147.92	990.2	1138.1	0.2630	1.5480	1.8109
190	9.339	0.01657	40.94	40.96	157.95	984.1	1142.0	0.2785	1.5147	1.7932
200	11.526	0.01663	33.62	33.64	167.99	977.9	1145.9	0.2938	1.4824	1.7762
210	14.123	0.01670	27.80	27.82	178.05	971.6	1149.7	0.3090	1.4508	1.7598
212	14.696	0.01672	26.78	26.80	180.07	970.3	1150.4	0.3120	1.4446	1.7566
220	17.186	0.01677	23.13	23.15	188.13	965.2	1153.4	0.3239	1.4201	1.7440
230	20.780	0.01684	19.365	19.382	198.23	958.8	1157.0	0.3387	1.3901	1.7288
240	24.969	0.01692	16.306	16.323	208.34	952.2	1160.5	0.3531	1.3609	1.7140
250	29.825	0.01700	13.804	13.821	216.48	945.5	1164.0	0.3675	1.3323	1.6998
260	35.429	0.01709	11.746	11.763	228.64	938.7	1167.3	0.3817	1.3043	1.6860
270	41.858	0.01717	10.044	10.061	238.84	931.8	1170.6	0.3958	1.2769	1.6727
280	49.203	0.01726	8.628	8.645	249.06	924.7	1173.8	0.4096	1.2501	1.6597
290	57.556	0.01735	7.444	7.461	259.31	917.5	1176.8	0.4234	1.2238	1.6472
300	67.013	0.01745	6.449	6.466	269.59	910.1	1179.7	0.4369	1.1980	1.6350
310	77.68	0.01755	5.609	5.626	279.92	902.6	1182.5	0.4504	1.1727	1.6231
320	89.66	0.01765	4.896	4.914	290.28	894.9	1185.2	0.4637	1.1478	1.6115

continued…

Dry Saturated Steam Temperature

		Specific volume, cu ft/lb			Enthalpy, Btu/lb			Entropy		
Temp., °F	Abs press., psi	Sat. liquid	Evap.	Sat. vapor	Sat. liquid	Sat. Evap.	Sat. vapor	Sat. liquid	Evap.	Sat. vapor
t	p	v_f	v_{fg}	v_g	h_f	h_{fg}	h_g	s_f	s_{fg}	s_g
330	103.06	0.01776	4.289	4.307	300.68	887.0	1187.7	0.4769	1.1233	1.6002
340	118.01	0.01787	3.770	3.788	311.13	879.0	1190.1	0.4900	1.0992	1.5891
350	134.63	0.01799	3.324	3.342	321.63	870.7	1192.3	0.5029	1.0754	1.5783
360	153.04	0.01811	2.939	2.957	332.18	862.2	1194.4	0.5158	1.0519	1.5677
370	173.37	0.01823	2.606	2.625	342.79	853.5	1196.3	0.5286	1.0287	1.5573
380	195.77	0.01836	2.317	2.335	353.45	844.6	1198.1	0.5413	1.0059	1.5471
390	220.37	0.01850	2.0651	2.0836	364.17	835.4	1199.6	0.5539	0.9832	1.5371
400	247.31	0.01864	1.8447	1.8633	374.97	826.0	1201.0	0.5664	0.9608	1.5272
410	276.75	0.01878	1.6512	1.6700	385.83	816.3	1202.1	0.5788	0.9386	1.5174
420	308.83	0.01894	1.4811	1.5000	396.77	806.3	1203.1	0.5912	0.9166	1.5078
430	343.72	0.01910	1.3308	1.3499	407.79	796.0	1203.8	0.6035	0.8947	1.4982
440	381.59	0.01926	1.1979	1.2171	418.90	785.4	1204.3	0.6158	0.8730	1.4887
450	422.6	0.0194	1.0799	1.0993	430.1	774.5	1204.6	0.6280	0.8513	1.4793
460	466.9	0.0196	0.9748	0.9944	441.4	763.2	1204.6	0.6402	0.8298	1.4700
470	514.7	0.0198	0.8811	0.9009	452.8	751.5	1204.3	0.6523	0.8083	1.4606
480	566.1	0.0200	0.7972	0.8172	464.4	739.4	1203.7	0.6645	0.7868	1.4513
490	621.4	0.0202	0.7221	0.7423	476.0	726.8	1202.8	0.6766	0.7653	1.4419
500	680.8	0.0204	0.6545	0.6749	487.8	713.9	1201.7	0.6887	0.7438	1.4325
520	812.4	0.0209	0.5385	0.5594	511.9	686.4	1198.2	0.7130	0.7006	1.4136
540	962.5	0.0215	0.4434	0.4649	536.6	656.6	1193.2	0.7374	0.6568	1.3942
560	1133.1	0.0221	0.3647	0.3868	562.2	624.2	1186.4	0.7621	0.6121	1.3742
580	1325.8	0.0228	0.2989	0.3217	588.9	588.4	1177.3	0.7872	0.5659	1.3532
600	1542.9	0.0236	0.2432	0.2668	617.0	548.5	1165.5	0.8131	0.5176	1.3307
620	1786.6	0.0247	0.1955	0.2201	646.7	503.6	1150.3	0.8398	0.4664	1.3062
640	2059.7	0.0260	0.1538	0.1798	678.6	452.0	1130.5	0.8679	0.4110	1.2789
660	2365.4	0.0278	0.1165	0.1442	714.2	390.2	1104.4	0.8987	0.3485	1.2472
680	2708.1	0.0305	0.0810	0.1115	757.3	309.9	1067.2	0.9351	0.2719	1.2071
700	3093.7	0.0369	0.0392	0.0761	823.3	172.1	995.4	0.9905	0.1484	1.1389
705.4	3206.2	0.0503	0	0.0503	902.7	0	902.7	1.0580	0	1.0580

…continued

Properties of Saturated Steam

Gauge Pressure in. Hg Vac.	Absolute Pressure psia	Temperature °F	Heat Content Sensible h_f Btu/lb	Heat Content Latent h_{fg} Btu/lb	Heat Content Total h_g Btu/lb	Specific Volume Steam V_g cu ft/lb
27.96	1	101.7	69.5	1032.9	1102.4	333.0
25.91	2	126.1	93.9	1019.7	1113.6	173.5
23.87	3	141.5	109.3	1011.3	1120.6	118.6
21.83	4	153.0	120.8	1004.9	1125.7	90.52
19.79	5	162.3	130.1	999.7	1129.8	73.42
17.75	6	170.1	137.8	995.4	1133.2	61.89
15.70	7	176.9	144.6	991.5	1136.1	53.57
13.66	8	182.9	150.7	987.9	1.138.6	47.26
11.62	9	188.3	156.2	984.7	1140.9	42.32
9.58	10	193.2	161.1	981.9	1143.0	38.37
7.54	11	197.8	165.7	979.2	1144.9	35.09
5.49	12	202.0	169.9	976.7	1146.6	32.35
3.45	13	205.9	173.9	974.3	1148.2	30.01
1.41	14	209.6	177.6	972.2	1149.8	28.00
Gauge Pressure psig						
0	14.7	212.0	180.2	970.6	1150.8	26.80
1	15.7	215.4	183.6	968.4	1152.0	25.20
2	16.7	218.5	186.8	966.4	1153.2	23.80
3	17.7	221.5	189.8	964.5	1154.3	22.50
4	18.7	224.5	192.7	962.6	1155.3	21.40
5	19.7	227.4	195.5	960.8	1156.3	20.40
6	20.7	230.0	198.1	959.2	1157.3	19.40
7	21.7	232.4	200.6	957.6	1158.2	18.60
8	22.7	234.8	203.1	956.0	1159.1	17.90
9	23.7	237.1	205.5	954.5	1160.0	17.20
10	24.7	239.4	207.9	952.9	1160.8	16.50
11	25.7	241.6	210.1	951.5	1161.6	15.90
12	26.7	243.7	212.3	950.1	1162.3	15.30
13	27.7	245.8	214.4	948.6	1163.0	14.80
14	28.7	247.9	216.4	947.3	1163.7	14.30
15	29.7	249.8	218.4	946.0	1164.4	13.90
16	30.7	251.7	220.3	944.8	1165.1	13.40
17	31.7	253.6	222.2	943.5	1165.7	13.00
18	32.7	255.4	224.0	942.4	1166.4	12.70
19	33.7	257.2	225.8	941.2	1167.0	12.30
20	34.7	258.8	227.5	940.1	1167.6	12.00
22	36.7	262.3	230.9	937.8	1168.7	11.40
24	38.7	265.3	234.2	935.8	1170.0	10.80
26	40.7	268.3	237.3	933.5	1170.8	10.30
28	42.7	271.4	240.2	931.6	1171.8	9.87
30	44.7	274.0	243.0	929.7	1172.7	9.46

continued...

Properties of Saturated Steam

| Gauge Pressure psig | Absolute Pressure psia | Temperature °F | Heat Content | | | Specific Volume Steam V_g cu ft/lb |
			Sensible h_f Btu/lb	Latent h_{fg} Btu/lb	Total h_g Btu/lb	
32	46.7	276.7	245.9	927.6	1173.5	9.08
34	48.7	279.4	248.5	925.8	1174.3	8.73
36	50.7	281.9	251.1	924.0	1175.1	8.40
38	52.7	284.4	253.7	922.1	1175.8	8.11
40	54.7	286.7	256.1	920.4	1176.5	7.83
42	56.7	289.0	258.5	918.6	1177.1	7.57
44	58.7	291.3	260.8	917.0	1177.8	7.33
46	60.7	293.5	263.0	915.4	1178.4	7.10
48	62.7	295.6	265.2	913.8	1179.0	6.89
50	64.7	297.7	267.4	912.2	1179.6	6.68
52	66.7	299.7	269.4	910.7	1180.1	6.50
54	68.7	310.7	271.5	909.2	1180.7	6.32
56	70.7	303.6	273.5	907.8	1181.3	6.16
58	72.7	305.5	275.3	906.5	1181.8	6.00
60	74.7	307.4	277.1	905.3	1182.4	5.84
62	76.7	309.2	279.0	904.0	1183.0	5.70
64	78.7	310.9	280.9	902.6	1183.5	5.56
66	80.7	312.7	282.8	901.2	1184.0	5.43
68	82.7	314.3	284.5	900.0	1184.5	5.31
70	84.7	316.0	286.2	898.8	1185.0	5.19
72	86.7	317.7	288.0	897.5	1185.5	5.08
74	88.7	319.3	289.4	896.5	1185.9	4.97
76	90.7	320.9	291.2	895.1	1186.3	4.87
78	92.7	322.4	292.9	893.9	1186.8	4.77
80	94.7	323.9	294.5	892.7	1187.2	4.67
82	96.7	325.5	296.1	891.5	1187.6	4.58
84	98.7	326.9	297.6	890.3	1187.9	4.49
86	100.7	328.4	299.1	889.2	1188.3	4.41
88	102.7	329.9	300.6	888.1	1188.7	4.33
90	104.7	331.2	302.1	887.0	1189.1	4.25
92	106.7	332.6	303.5	885.8	1189.3	4.17
94	108.7	333.9	304.9	884.8	1189.7	4.10
96	110.7	335.3	306.3	883.7	1190.0	4.03
98	112.7	336.6	307.7	882.6	1190.3	3.96
100	114.7	337.9	309.0	881.6	1190.6	3.90
102	116.7	339.2	310.3	880.6	1190.9	3.83
104	118.7	340.5	311.6	879.6	1191.2	3.77
106	120.7	341.7	313.0	878.5	1191.5	3.71
108	122.7	343.0	314.3	877.5	1191.8	3.65
110	124.7	344.2	315.5	876.5	1192.0	3.60
112	126.7	345.4	316.8	875.5	1192.3	3.54
114	128.7	346.5	318.0	874.5	1192.5	3.49

…continued…

Properties of Saturated Steam

| Gauge Pressure psig | Absolute Pressure psia | Temperature °F | Heat Content | | | Specific Volume Steam V_g cu ft/lb |
			Sensible h_f Btu/lb	Latent h_{fg} Btu/lb	Total h_g Btu/lb	
116	130.7	347.7	319.3	873.5	1192.8	3.44
118	132.7	348.9	320.5	872.5	1193.0	3.39
120	134.7	350.1	321.8	871.5	1193.3	3.34
125	139.7	352.8	324.7	869.3	1194.0	3.23
130	144.7	355.6	327.6	866.9	1194.5	3.12
135	149.7	358.3	330.6	864.5	1195.1	3.02
140	154.7	360.9	333.2	862.5	1195.7	2.93
145	159.7	363.5	335.9	860.3	1196.2	2.84
150	164.7	365.9	338.6	858.0	1196.6	2.76
155	169.7	368.3	341.1	856.0	1197.1	2.68
160	174.7	370.7	343.6	853.9	1197.5	2.61
165	179.7	372.9	346.1	851.8	1197.9	2.54
170	184.7	375.2	348.5	849.8	1198.3	2.48
175	189.7	377.5	350.9	847.9	1198.8	2.41
180	194.7	379.6	353.2	845.9	1199.1	2.35
185	199.7	381.6	355.4	844.1	1199.5	2.30
190	204.7	383.7	357.6	842.2	1199.8	2.24
195	209.7	385.7	359.9	840.2	1200.1	2.18
200	214.7	387.7	362.0	838.4	1200.4	2.14
210	224.7	391.7	366.2	834.8	1201.0	2.04
220	234.7	395.5	370.3	831.2	1201.5	1.96
230	244.7	399.1	374.2	827.8	1202.0	1.88
240	254.7	402.7	378.0	824.5	1202.5	1.81
250	264.7	406.1	381.7	821.2	1202.9	1.74
260	274.7	409.3	385.3	817.9	1203.2	1.68
270	284.7	412.5	388.8	814.8	1203.6	1.62
280	294.7	415.8	392.3	811.6	1203.9	1.57
290	304.7	418.8	395.7	808.5	1204.2	1.52
300	314.7	421.7	398.9	805.5	1204.4	1.47
310	324.7	424.7	402.1	802.6	1204.7	1.43
320	334.7	427.5	405.2	799.7	1204.9	1.39
330	344.7	430.3	408.3	796.7	1205.0	1.35
340	354.7	433.0	411.3	793.8	1205.1	1.31
350	364.7	435.7	414.3	791.0	1205.3	1.27
360	374.7	438.3	417.2	788.2	1205.4	1.24
370	384.7	440.8	420.0	785.4	1205.4	1.21
380	394.7	443.3	422.8	782.7	1205.5	1.18
390	404.7	445.7	425.6	779.9	1205.5	1.15
400	414.7	448.1	428.2	777.4	1205.6	1.12
420	434.7	452.8	433.4	772.2	1205.6	1.07
440	454.7	457.3	438.5	767.1	1205.6	1.02
460	474.7	461.7	443.4	762.1	1205.5	.98

...continued...

Properties of Saturated Steam

Gauge Pressure psig	Absolute Pressure psia	Temperature °F	Heat Content			Specific Volume Steam V_g cu ft/lb
			Sensible h_f Btu/lb	Latent h_{fg} Btu/lb	Total h_g Btu/lb	
480	494.7	465.9	448.3	757.1	1205.4	0.94
500	514.7	470.0	453.0	752.3	1205.3	0.902
520	534.7	474.0	457.6	747.5	1205.1	0.868
540	554.7	477.8	462.0	742.8	1204.8	0.835
560	574.7	481.6	466.4	738.1	1204.5	0.805
580	594.7	485.2	470.7	733.5	1204.2	0.776
600	614.7	488.8	474.8	729.1	1203.9	0.750
620	634.7	492.3	479.0	724.5	1203.5	0.726
640	654.7	495.7	483.0	720.1	1203.1	0.703
660	674.7	499.0	486.9	715.8	1202.7	0.681
680	694.7	502.2	490.7	711.5	1202.2	0.660
700	714.7	505.4	494.4	707.4	1201.8	0.641
720	734.7	508.5	498.2	703.1	1201.3	0.623
740	754.7	511.5	501.9	698.9	1200.8	0.605
760	774.7	514.5	505.5	694.7	1200.2	0.588
780	794.7	517.5	509.0	690.7	1199.7	0.572
800	814.7	520.3	512.5	686.6	1199.1	0.557

...continued

Specific Heat of Various Substances

Substance	Btu per pound per °F
Mercury	0.0333
Steel	0.117
Iron	0.129
Carbon dioxide	0.22
Air (constant pressure)	0.238
Dry flue gases (constant pressure)	0.240
Water vapor (atmospheric pressure)	0.480
Ice (0°–21°F)	0.504
Water	1.000

Properties of Superheated Steam

(v = specific volume, cu ft/lb; h = enthalpy, Btu/lb; s = entropy)

Abs press., psi (sat. temp.)		Temperature, °F									
		400	500	600	700	800	900	1000	1100	1200	1400
1 (101.74)	v	512.0	571.6	631.2	690.8	750.4	809.9	869.5	929.1	988.7	1107.8
	h	1241.7	1288.3	1335.8	1383.8	1432.8	1482.7	1533.5	1585.2	1637.7	1745.7
	s	2.1720	2.2233	2.2702	2.3137	2.3542	2.3932	2.4283	2.4625	2.4952	2.5566
5 (162.24)	v	102.26	114.22	126.16	138.10	150.03	161.95	173.87	185.79	197.71	221.6
	h	1241.2	1288.0	1335.4	1383.6	1432.7	1482.6	1533.4	1585.1	1637.7	1745.7
	s	1.9942	2.0456	2.0927	2.1361	2.1767	2.2148	2.2509	2.2851	2.3178	2.3792
10 (193.21)	v	51.04	57.05	63.03	69.01	74.98	80.95	86.92	92.88	98.84	110.77
	h	1240.6	1287.5	1335.1	1383.4	1432.5	1482.4	1533.2	1585.0	1637.6	1745.6
	s	1.9172	1.9689	2.0160	2.0596	2.1002	2.1383	2.1744	2.2086	2.2413	2.3028
14.696 (212.00)	v	34.68	38.78	42.86	46.94	51.00	55.07	59.13	63.19	67.25	75.37
	h	1239.9	1287.1	1334.8	1383.2	1432.3	1482.3	1533.1	1584.8	1637.5	1745.5
	s	1.8743	1.9261	1.9734	2.0170	2.0576	2.0958	2.1319	2.1662	2.1989	2.2603
20 (227.96)	v	25.43	28.46	31.47	34.47	37.46	40.45	43.44	46.42	49.41	55.37
	h	1239.2	1286.6	1334.4	1382.9	1432.1	1482.1	1533.0	1584.7	1637.4	1745.4
	s	1.8396	1.8918	1.9392	1.9829	2.0235	2.0618	2.0978	2.1321	2.1648	2.2263
40 (267.25)	v	12.628	14.168	15.688	17.198	18.702	20.20	21.70	23.20	24.69	27.68
	h	1236.5	1284.8	1333.1	1381.9	1431.3	1481.4	1532.4	1584.3	1637.0	1745.1
	s	1.7608	1.8140	1.8619	1.9058	1.9467	1.9850	2.0212	2.0555	2.0883	2.1498
60 (292.71)	v	8.357	9.403	10.427	11.441	12.449	13.452	14.454	15.453	16.451	18.446
	h	1233.6	1283.0	1331.8	1380.9	1430.5	1480.8	1531.9	1583.8	1636.6	1744.8
	s	1.7135	1.7678	1.8162	1.8605	1.9015	1.9400	1.9762	2.0106	2.0434	2.1049
80 (312.03)	v	6.220	7.020	7.797	8.562	9.322	10.077	10.830	11.582	12.332	13.830
	h	1230.7	1281.1	1330.5	1379.9	1429.7	1480.1	1531.3	1583.4	1636.2	1744.5
	s	1.6791	1.7346	1.7836	1.8281	1.8694	1.9079	1.9442	1.9787	2.0115	2.0731

continued...

Properties of Superheated Steam

(v = specific volume, cu ft/lb; h = entahlpy, Btu/lb; s = entropy)

Abs press., psi (sat. temp.)		Temperature, °F									
		400	500	600	700	800	900	1000	1100	1200	1400
100 (327.81)	v	4.937	5.589	6.218	6.835	7.446	8.052	8.656	9.259	9.860	11.060
	h	1227.6	1279.1	1329.1	1378.9	1428.9	1479.5	1530.8	1582.9	1635.7	1744.2
	s	1.6518	1.7085	1.7581	1.8029	1.8443	1.8829	1.9193	1.9538	1.9867	2.0484
120 (341.25)	v	4.081	4.636	5.165	5.683	6.195	6.702	7.207	7.710	8.212	9.214
	h	1224.4	1277.2	1327.7	1377.8	1428.1	1478.8	1530.2	1582.4	1635.3	1743.9
	s	1.6287	1.6869	1.7370	1.7822	1.8237	1.8625	1.8990	1.9335	1.9664	2.0281
140 (353.02)	v	3.468	3.954	4.413	4.861	5.301	5.738	6.172	6.604	7.035	7.895
	h	1221.1	1275.2	1326.4	1376.8	1427.3	1478.2	1529.7	1581.9	1634.9	1743.5
	s	1.6087	1.6683	1.7190	1.7645	1.8063	1.8451	1.8817	1.9163	1.9493	2.0110
160 (363.53)	v	3.008	3.443	3.849	4.244	4.631	5.015	5.396	5.775	6.152	6.906
	h	1217.6	1273.1	1325.0	1375.7	1426.4	1477.5	1529.1	1581.4	1634.5	1743.2
	s	1.5908	1.6519	1.7033	1.7491	1.7911	1.8301	1.8667	1.9014	1.9344	1.9962
180 (373.06)	v	2.649	3.044	3.411	3.764	4.110	4.452	4.792	5.129	5.466	6.136
	h	1214.0	1271.0	1323.5	1374.7	1425.6	1476.8	1528.6	1581.0	1634.1	1742.9
	s	1.5745	1.6373	1.6894	1.7355	1.7776	1.8167	1.8534	1.8882	1.9212	1.9831
200 (381.79)	v	2.361	2.726	3.060	3.380	3.693	4.002	4.309	4.613	4.917	5.521
	h	1210.3	1268.9	1322.1	1373.6	1424.8	1476.2	1528.0	1580.5	1633.7	1742.6
	s	1.5594	1.6240	1.6767	1.7232	1.7655	1.8048	1.8415	1.8763	1.9094	1.9713
220 (389.86)	v	2.125	2.465	2.772	3.066	3.352	3.634	3.913	4.191	4.467	5.017
	h	1206.5	1266.7	1320.7	1372.6	1424.0	1475.5	1527.5	1580.0	1633.3	1742.3
	s	1.5453	1.6117	1.6652	1.7120	1.7545	1.7939	1.8308	1.8656	1.8987	1.9607
240 (397.37)	v	1.9276	2.247	2.533	2.804	3.068	3.327	3.584	3.839	4.093	4.597
	h	1202.5	1264.5	1319.2	1371.5	1423.2	1474.8	1526.9	1579.6	1632.9	1742.0
	s	1.5319	1.6003	1.6546	1.7017	1.7444	1.7839	1.8209	1.8558	1.8889	1.9510

...continued...

Properties of Superheated Steam

(v = specific volume, cu ft/lb; h = entahlpy, Btu/lb; s = entropy)

Abs press., psi (sat. temp.)		Temperature, °F									
		400	500	600	700	800	900	1000	1100	1200	1400
260 (404.42)	v	—	2.063	2.330	2.582	2.827	3.067	3.305	3.541	3.776	4.242
	h	—	1262.3	1317.7	1370.4	1422.3	1474.2	1526.3	1579.1	1632.5	1741.7
	s	—	1.5897	1.6447	1.6922	1.7352	1.7748	1.8118	1.8467	1.8799	1.9420
280 (411.05)	v	—	1.9047	2.156	2.392	2.621	2.845	3.066	3.286	3.504	3.938
	h	—	1260.0	1316.2	1369.4	1421.5	1473.5	1525.8	1578.6	1632.1	1741.4
	s	—	1.5796	1.6354	1.6834	1.7264	1.7662	1.8033	1.8383	1.8716	1.9337
300 (417.33)	v	—	1.7675	2.005	2.227	2.442	2.652	2.859	3.065	3.269	3.674
	h	—	1257.6	1314.7	1368.3	1420.6	1472.8	1525.2	1578.1	1631.7	1741.0
	s	—	1.5701	1.6268	1.6751	1.7184	1.7582	1.7954	1.8305	1.8638	1.9260
350 (431.72)	v	—	1.4923	1.7036	1.8980	2.084	2.266	2.445	2.622	2.798	3.147
	h	—	1251.5	1310.9	1365.5	1418.5	1471.1	1523.8	1577.0	1630.7	1740.3
	s	—	1.5481	1.6070	1.6563	1.7002	1.7403	1.7777	1.8130	1.8463	1.9086
400 (444.59)	v	—	1.2851	1.4770	1.6508	1.8161	1.9767	2.134	2.290	2.445	2.751
	h	—	1245.1	1306.9	1362.7	1416.4	1469.4	1522.4	1575.8	1629.6	1739.5
	s	—	1.5281	1.5894	1.6398	1.6842	1.7247	1.7623	1.7977	1.8311	1.8936

...continued....

Properties of Superheated Steam

(v = specific volume, cu ft/lb; h = entahlpy, Btu/lb; s = entropy)

Abs press., psi (sat. temp.)		Temperature, °F											
		500	600	620	640	660	680	700	800	900	1000	1200	1400
450 (456.28)	v	1.1231	1.3005	1.3332	1.3652	1.3967	1.4278	1.4584	1.6074	1.7516	1.8928	2.170	2.443
	h	1238.4	1302.8	1314.6	1326.2	1337.5	1348.8	1359.9	1414.3	1467.7	1521.0	1628.6	1738.7
	s	1.5095	1.5735	1.5845	1.5951	1.6054	1.6153	1.6250	1.6699	1.7108	1.7486	1.8177	1.8803
500 (467.01)	v	0.9927	1.1591	1.1893	1.2188	1.2478	1.2763	1.3044	1.4405	1.5715	1.6996	1.9504	2.197
	h	1231.3	1298.6	1310.7	1322.6	1334.2	1345.7	1357.0	1412.1	1466.0	1519.6	1627.6	1737.9
	s	1.4919	1.5588	1.5701	1.5810	1.5915	1.6016	1.6115	1.6571	1.6982	1.7363	1.8056	1.8683
550 (476.94)	v	0.8852	1.0431	1.0714	1.0989	1.1259	1.1523	1.1783	1.3038	1.4241	1.5414	1.7706	1.9957
	h	1223.7	1294.3	1306.8	1318.9	1330.8	1342.5	1354.0	1409.9	1464.3	1518.2	1626.6	1737.1
	s	1.4751	1.5451	1.5568	1.5680	1.5787	1.5890	1.5991	1.6452	1.6868	1.7250	1.7946	1.8575
600 (486.21)	v	0.7947	0.9463	0.9729	0.9988	1.0241	1.0489	1.0732	1.1899	1.3013	1.4096	1.6208	1.8279
	h	1215.7	1289.9	1302.7	1315.2	1327.4	1339.3	1351.1	1407.7	1462.5	1516.7	1625.5	1736.3
	s	1.4586	1.5323	1.5443	1.5558	1.5667	1.5773	1.5875	1.6343	1.6762	1.7147	1.7846	1.8476
700 (503.10)	v	—	0.7934	0.8177	0.8411	0.8639	0.8860	0.9077	1.0108	1.1082	1.2024	1.3853	1.5641
	h	—	1280.6	1294.3	1307.5	1320.3	1332.8	1345.0	1403.2	1459.0	1513.9	1623.5	1734.8
	s	—	1.5084	1.5212	1.5333	1.5449	1.5559	1.5665	1.6147	1.6573	1.6963	1.7666	1.8299
800 (518.23)	v	—	0.6779	0.7006	0.7223	0.7433	0.7635	0.7833	0.8763	0.9633	1.0470	1.2088	1.3662
	h	—	1270.7	1285.4	1299.4	1312.9	1325.9	1338.6	1398.6	1455.4	1511.0	1621.4	1733.2
	s	—	1.4863	1.5000	1.5129	1.5250	1.5366	1.5476	1.5972	1.6407	1.6801	1.7510	1.8146
900 (531.98)	v	—	0.5873	0.6089	0.6294	0.6491	0.6680	0.6863	0.7716	0.8506	0.9262	1.0714	1.2124
	h	—	1260.1	1275.9	1290.9	1305.1	1318.8	1332.1	1393.9	1451.8	1508.1	1619.3	1731.6
	s	—	1.4653	1.4800	1.4938	1.5066	1.5187	1.5303	1.5814	1.6257	1.6656	1.7371	1.8009
1000 (544.61)	v	—	0.5140	0.5350	0.5546	0.5733	0.5912	0.6084	0.6878	0.7604	0.8294	0.9615	1.0893
	h	—	1248.8	1265.9	1281.9	1297.0	1311.4	1325.3	1389.2	1448.2	1505.1	1617.3	1730.0
	s	—	1.4450	1.4610	1.4757	1.4893	1.5021	1.5141	1.5670	1.6121	1.6525	1.7245	1.7886
1100 (556.31)	v	—	0.4532	0.4738	0.4929	0.5110	0.5281	0.5445	0.6191	0.6866	0.7503	0.8716	0.9885
	h	—	1236.7	1255.3	1272.4	1288.5	1303.7	1318.3	1384.3	1444.5	1502.2	1615.2	1728.4
	s	—	1.4251	1.4425	1.4583	1.4728	1.4862	1.4989	1.5535	1.5995	1.6405	1.7130	1.7775

...continued...

Properties of Superheated Steam

(v = specific volume, cu ft/lb; h = entahlpy, Btu/lb; s = entropy)

Abs press., psi (sat. temp.)		Temperature, °F											
		500	600	620	640	660	680	700	800	900	1000	1200	1400
1200 (567.22)	v	—	0.4016	0.4222	0.4410	0.4586	0.4752	0.4909	0.5617	0.6250	0.6843	0.7967	0.9046
	h	—	1223.5	1243.9	1262.4	1279.6	1295.7	1311.0	1379.3	1440.7	1499.2	1613.1	1726.9
	s	—	1.4052	1.4243	1.4413	1.4568	1.4710	1.4843	1.5409	1.5879	1.6293	1.7025	1.7672
1400 (587.10)	v	—	0.3174	0.3390	0.3580	0.3753	0.3912	0.4062	0.4714	0.5281	0.5805	0.6789	0.7727
	h	—	1193.0	1218.4	1240.4	1260.3	1278.5	1295.5	1369.1	1433.1	1493.2	1608.9	1723.7
	s	—	1.3639	1.3877	1.4079	1.4258	1.4419	1.4567	1.5177	1.5666	1.6093	1.6836	1.7489
1600 (604.90)	v	—	—	0.2733	0.2936	0.3112	0.3271	0.3417	0.4034	0.4553	0.5027	0.5906	0.6738
	h	—	—	1187.8	1215.2	1238.7	1259.6	1278.7	1358.4	1425.3	1487.0	1604.6	1720.5
	s	—	—	1.3489	1.3741	1.3952	1.4137	1.4303	1.4964	1.5476	1.5914	1.6669	1.7328
1800 (621.03)	v	—	—	—	0.2407	0.2597	0.2760	0.2907	0.3502	0.3986	0.4421	0.5218	0.5968
	h	—	—	—	1185.1	1214.0	1238.5	1260.3	1347.4	1417.4	1480.8	1600.4	1717.3
	s	—	—	—	1.3377	1.3638	1.3855	1.4044	1.4765	1.5301	1.5752	1.6520	1.7185
2000 (635.82)	v	—	—	—	0.1936	0.2161	0.2337	0.2489	0.3074	0.3532	0.3935	0.4668	0.5352
	h	—	—	—	1145.6	1184.9	1214.8	1240.0	1335.5	1409.2	1474.5	1596.1	1714.1
	s	—	—	—	1.2945	1.3300	1.3564	1.3783	1.4576	1.5139	1.5603	1.6384	1.7055
2500 (668.13)	v	—	—	—	—	—	0.1484	0.1686	0.2294	0.2710	0.3061	0.3678	0.4244
	h	—	—	—	—	—	1132.3	1176.8	1303.6	1387.8	1458.4	1585.3	1706.1
	s	—	—	—	—	—	1.2687	1.3073	1.4127	1.4772	1.5273	1.6088	1.6775
3000 (695.36)	v	—	—	—	—	—	—	0.0984	0.1760	0.2159	0.2476	0.3018	0.3505
	h	—	—	—	—	—	—	1060.7	1267.2	1365.0	1441.8	1574.3	1698.0
	s	—	—	—	—	—	—	1.1966	1.3690	1.4439	1.4984	1.5837	1.6540
3206.2 (705.40)	v	—	—	—	—	—	—	—	0.1583	0.1981	0.2288	0.2806	0.3267
	h	—	—	—	—	—	—	—	1250.5	1355.2	1434.7	1569.8	1694.6
	s	—	—	—	—	—	—	—	1.3508	1.4309	1.4874	1.5742	1.6452

...continued

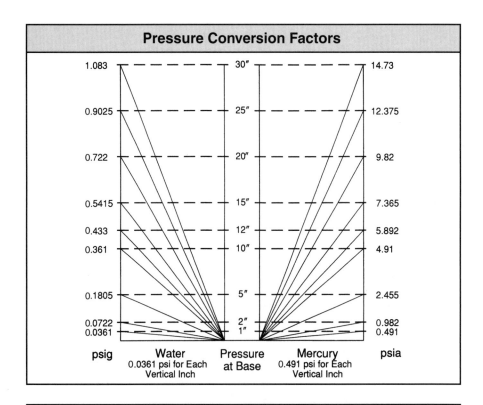

Pressure Conversion Factors

psig	Water 0.0361 psi for Each Vertical Inch	Pressure at Base	Mercury 0.491 psi for Each Vertical Inch	psia
1.083		30"		14.73
0.9025		25"		12.375
0.722		20"		9.82
0.5415		15"		7.365
0.433		12"		5.892
0.361		10"		4.91
0.1805		5"		2.455
0.0722		2"		0.982
0.0361		1"		0.491

Common ASME Boiler Classifications

Name	Description
Automatic boiler	Equipped with certain controls and limit devices per ASME code
Boiler	Closed vessel used for heating water or liquid, or for generating steam or vapor by direct application of heat
Boiler plant	One or more boilers, connecting piping, and vessels within the same premises
Hot water supply boiler	Low pressure hot water heating boiler having a volume exceeding 120 gal. or heat input exceeding 200,000 Btu/hr, or an operating temperature exceeding 200°F that provides hot water to be used externally to itself
Low pressure hot water heating boiler	Boiler in which water is heated for the purpose of supplying heat at pressures not exceeding 160 psi or temperatures not exceeding 250°F
Low pressure steam heating boiler	Boiler operated at pressures not exceeding 15 psi for steam
Power hot water boiler	Boiler used for heating water or liquid to pressure exceeding 160 psi or a temperature exceeding 250°F
Power steam boiler	Boiler in which steam or vapor is generated at pressures exceeding 15 psi
Small power boiler	Boiler with pressures exceeding 15 psi but not exceeding 100 psi and having less than 440,000 Btu/hr input

Area of Circles

Dia.	Area	Dia.	Area	Dia.	Area	Dia.	Area	Dia.	Area	Dia.	Area
1/8	0.0123	4½	15.904	16½	213.82	32	804.24	56	2463.0	80	5026.5
1/4	0.0491	5	19.635	17	226.98	33	855.30	57	2551.7	81	5153.0
3/8	0.1104	5½	23.758	17½	240.52	34	907.92	58	2642.0	82	5281.0
1/2	0.1963	6	28.274	18	254.46	35	962.11	59	2733.9	83	5410.6
5/8	0.3067	6½	33.183	18½	268.80	36	1017.8	60	2827.4	84	5541.7
3/4	0.4417	7	38.484	19	283.52	37	1075.2	61	2922.4	85	5674.5
7/8	0.6013	7½	44.178	19½	298.64	38	1134.1	62	3019.0	86	5808.8
1	0.7854	8	50.265	20	314.16	39	1194.5	63	3117.2	87	5944.6
1⅛	0.9940	8½	56.745	20½	330.06	40	1256.6	64	3216.9	88	6082.1
1¼	1.227	9	63.617	21	346.36	41	1320.2	65	3318.3	89	6221.1
1⅜	1.484	9½	70.882	21½	363.05	42	1385.4	66	3421.2	90	6361.7
1½	1.767	10	78.54	22	380.13	43	1452.2	67	3525.6	91	6503.8
1⅝	2.073	10½	86.59	22½	397.60	44	1520.5	68	3631.6	92	6647.6
1¾	2.405	11	95.03	23	415.47	45	1590.4	69	3739.2	93	6792.9
1⅞	2.761	11½	103.86	23½	433.73	46	1661.9	70	3848.4	94	6939.7
2	3.141	12	113.09	24	452.39	47	1734.9	71	3959.2	95	7088.2
2¼	3.976	12½	122.71	24½	471.43	48	1809.5	72	4071.5	96	7238.2
2½	4.908	13	132.73	25	490.87	49	1885.7	73	4185.3	97	7389.8
2¾	5.939	13½	143.13	26	530.93	50	1963.5	74	4300.8	98	7542.9
3	7.068	14	153.93	27	572.55	51	2042.8	75	4417.8	99	7697.7
3¼	8.295	14½	165.13	28	615.75	52	2123.7	76	4536.4	100	7854.0
3½	9.621	15	176.71	29	660.52	53	2206.1	77	4656.0	101	8011.8
3¾	11.044	15½	188.69	30	706.86	54	2290.2	78	4778.3	102	8171.3
4	12.566	16	201.06	31	754.76	55	2375.8	79	4901.6	103	8332.3

Pressure Equivalents

psia*	psf†	in. Hg‡	in. WC §	ft WC **	atm††
1	144	2.042	27.7	2.31	0.068
14.7	2116.3	29.92		33.95	1
0.433	62.355			1	
0.491		1	13.58	1.132	

Equivalents of PSIA

psia*	in. WC§	ft WC**	in. Hg††
1	27.71	2.31	2.041
2	55.42	4.62	4.081
3	83.14	6.93	6.122
4	110.85	9.24	8.163
5	138.56	11.55	10.20
6	166.27	13.86	12.24
7	193.99	16.17	14.28
8	221.70	18.47	16.33
9	249.41	20.78	18.37
10	277.12	23.09	20.41
11	304.84	25.40	22.45
12	332.55	27.71	24.49
13	360.26	30.02	26.53
14	387.97	32.33	28.57
14.7	407.37	33.95	29.92
15	415.68	34.64	30.61
16	443.40	36.95	32.65
17	471.11	39.26	34.69
18	498.82	41.57	36.73
19	526.53	43.88	38.77
20	554.25	46.19	40.81
21	581.96	48.50	42.85
22	609.67	50.81	44.89
23	637.38	53.12	46.94
24	665.10	55.42	48.98
25	692.81	57.73	51.02

Equivalents of In. Hg

in. Hg††	psia*	in. Hg‡	psia*
30	14.730	15	7.365
29.92	14.7	14	6.874
29	14.239	13	6.383
28	13.748	12	5.892
27	13.257	11	5.401
26	12.776	10	4.910
25	12.275	9	4.419
24	11.784	8	3.928
23	11.293	7	3.437
22	10.802	6	2.946
21	10.311	5	2.455
20	9.820	4	1.964
19	9.329	3	1.473
18	8.838	2	0.982
17	8.347	1	0.491
16	7.856	0	0

* pounds per square inch absolute
† pounds per square foot
‡ inches of mercury
§ inches of water column
** feet of water column
†† atmospheres

Linear Expansion of Metallic Pipe				
Temperature*	Expansion per 100' of Pipe†			
	Steel	Malleable Iron	Copper	Cast Iron
− 30	0.000	0.000	0.000	0.000
− 20	0.072	0.073	0.105	0.062
− 10	0.145	0.147	0.211	0.124
0	0.215	0.221	0.316	0.186
10	0.291	0.298	0.428	0.251
20	0.367	0.376	0.541	0.317
30	0.442	0.454	0.654	0.383
40	0.517	0.533	0.767	0.449
50	0.592	0.612	0.880	0.515
60	0.667	0.691	0.993	0.581
70	0.742	0.770	1.107	0.647
80	0.817	0.849	1.221	0.713
90	0.892	0.928	1.335	0.779
100	0.968	1.007	1.449	0.845
110	1.048	1.090	1.565	0.915
120	1.128	1.174	1.681	0.985
130	1.208	1.258	1.797	1.056
140	1.287	1.342	1.913	1.127
150	1.366	1.426	2.029	1.198
160	1.445	1.510	2.145	1.269
170	1.524	1.594	2.261	1.340
180	1.603	1.678	2.377	1.411
190	1.682	1.762	2.494	1.482
200	1.761	1.846	2.611	1.553
210	1.843	1.931	2.727	1.626
220	1.925	2.016	2.843	1.699
230	2.008	2.101	2.959	1.773
240	2.091	2.186	3.075	1.847
250	2.174	2.271	3.191	1.921
260	2.257	2.356	3.308	1.995
270	2.340	2.441	3.425	2.069
280	2.423	2.526	3.542	2.143
290	2.506	2.612	3.659	2.217
300	2.589	2.698	3.776	2.291
310	2.674	2.787	3.896	2.368
320	2.759	2.876	4.016	2.445
330	2.844	2.965	4.136	2.522

continued...

Linear Expansion of Metallic Pipe

Temperature*	Expansion per 100′ of Pipe†			
	Steel	Malleable Iron	Copper	Cast Iron
340	2.929	3.054	4.256	2.599
350	3.015	3.143	4.376	2.676
360	3.101	3.232	4.497	2.754
370	3.187	3.321	4.618	2.832
380	3.273	3.410	4.739	2.910
390	3.359	3.499	4.860	2.988
400	3.445	3.589	4.981	3.066

* in °F
† in in.

...continued

Heating Values and Chemical Composition of Standard Grades of Coal

Rank	Heating Value*	Chemical Composition†					
		Oxygen	Hydrogen	Carbon	Nitrogen	Sulfur	Ash
Anthracite	12,910	5.0	2.9	80.0	0.9	0.7	10.5
Semi-anthracite	13,770	5.0	3.9	80.4	1.1	1.1	8.5
Low-volatile Bituminous	14,340	5.0	4.7	81.7	1.4	1.2	6.0
Medium-volatile Bituminous	13,840	5.0	5.0	79.0	1.4	1.5	8.1
High-volatile Bituminous A	13,090	9.2	5.3	73.2	1.5	2.0	8.8
High-volatile Bituminous B	12,130	13.8	5.5	68.0	1.4	2.1	9.2
High-volatile Bituminous C	10,750	21.0	5.8	60.6	1.1	2.1	9.4
Subbituminous B	9150	29.5	6.2	52.5	1.0	1.0	9.8
Subbituminous C	8940	35.8	6.5	46.7	0.8	0.6	9.6
Lignite	6900	44.0	6.9	40.1	0.7	1.0	7.3

* in Btu/lb
† in %

Weight of Water					
Temperature in °F	Weight per Cu Ft	Weight per Gal.	Temperature in °F	Weight per Cu Ft	Weight per Gal.
32	62.418	8.344	130	61.563	8.230
35	62.422	8.345	135	61.472	8.218
39.2	62.425	8.346	140	61.381	8.206
40	62.425	8.346	145	61.291	8.193
45	62.422	8.345	150	61.201	8.181
50	62.409	8.343	155	61.096	8.167
55	62.394	8.341	160	60.991	8.153
60	62.372	8.338	165	60.843	8.134
65	62.344	8.334	170	60.783	8.126
70	62.313	8.331	175	60.665	8.110
75	62.275	8.325	180	60.548	8.094
80	62.232	8.321	185	60.430	8.078
85	62.182	8.313	190	60.314	8.063
90	62.133	8.306	195	60.198	8.047
95	62.074	8.297	200	60.081	8.032
100	62.022	8.291	205	60.980	8.018
105	61.960	8.283	210	59.820	7.997
110	61.868	8.271	212	59.760	7.989
115	61.807	8.261	250	58.750	7.854
120	61.715	8.250	300	56.970	7.616
125	61.654	8.242	400	54.250	7.252

Water Pressure to Head			
Pressure*	**Head†**	**Pressure***	**Head†**
1	2.31	100	230.90
2	4.62	110	253.98
3	6.93	120	277.07
4	9.24	130	300.16
5	11.54	140	323.25
6	13.85	150	346.34
7	16.16	160	369.43
8	18.47	170	392.52
9	20.78	180	415.61
10	23.09	200	461.78
15	34.63	250	577.24
20	46.18	300	692.69
25	57.72	350	808.13
30	69.27	400	922.58
40	92.36	500	1154.48
50	115.45	600	1385.39
60	138.54	700	1616.30
70	161.63	800	1847.20
80	184.72	900	2078.10
90	207.81	1000	2309.00

* in psi
† in ft

Steam Loss through an Orifice*						
Orifice Size[†]	Orifice Size[†]	Differential Pressure across Orifice, PSIG				
		15	30	60	100	150
#60[‡]	0.04	1	2	4	6	10
3/64″	0.0469	1	3	5	9	13
1/16″	0.0625	2	5	9	16	24
5/64″	0.0781	4	7	15	25	37
3/32″	0.0938	5	11	21	36	53
#38[‡]	0.1015	6	12	25	42	62
7/64″	0.1094	7	15	29	48	73
1/8″	0.125	9	19	38	63	95
9/64″	0.1406	12	24	48	80	120
5/32″	0.1562	15	30	59	99	148
3/16″	0.1875	21	43	85	142	213
7/32″	0.2188	29	58	116	193	290
1/4″	0.25	38	76	151	252	379
9/32″	0.2812	48	96	192	319	479
5/16″	0.3125	59	118	237	394	592
11/32″	0.3438	72	143	286	477	716
3/8″	0.375	85	170	341	568	852
7/16″	0.4375	116	232	464	773	1160
1/2″	0.5	151	303	606	1010	1515
9/16″	0.5625	192	383	767	1278	1917
5/8″	0.625	237	473	947	1578	2367
11/16″	0.6875	286	573	1145	1909	2864
3/4″	0.75	341	682	1363	2272	3408
7/8″	0.875	464	928	1856	3098	4639
1- 1/16″	1.0625	684	1368	2736	4560	6840
1- 1/8″	1.25	947	1893	3787	6311	9467
1- 5/8″	1.625	1600	3200	6400	10,666	15,999

* in lb/hr
† in in.
‡ drill size

GLOSSARY

A

absolute pressure: The sum of gauge pressure and atmospheric pressure.

accumulation test: A test used to establish the relieving capacity of boiler safety valves.

acid: A compound that contains hydrogen and is capable of reacting with a base, such as sulfuric acid or hydrochloric acid.

AC sine wave: A symmetrical waveform that contains 360 electrical degrees.

address: An identification number assigned to a specific input and output.

air atomizing burner: An inside-mixing fuel oil burner that uses compressed air from an air compressor for atomization.

airborne particulate matter: Solid or liquid matter that is suspended in flue gas and discharged to the atmosphere.

air ejector: A steam-driven device used on condensers to remove air and other noncondensable gases to maintain a high vacuum.

air flow switch: A safety device used with rotary cup burners.

air preheater: A heater used in a heat recovery system that recovers heat from gases of combustion.

alkali: A compound that contains hydroxide and is capable of reacting with an acid, such as sodium hydroxide and calcium hydroxide.

alternating current: Electron flow that reverses its direction at regular intervals of time.

ammeter: A meter used to measure current in any part of a circuit.

ampere (amp): A unit of measure used to express the quantity or number of electrons flowing through a conductor per unit of time.

analog-in (AI) point: A remotely located sensor that has a continuously varying signal that bears a known relationship to the value of the measured variable.

analog-out (AO) point: A remotely located device that receives a continuously varying signal from a computer and reacts to that signal to control equipment.

analog signal: Data that is sent about the status of pressure, humidity, airflow, voltage, or resistance.

arc blast: An explosion that occurs when the surrounding air becomes ionized and conductive.

arc flash: A short circuit through air.

area: A surface measurement expressed in square units, such as square inches or square feet.

area source facility: A facility that emits less than 10 tons per year of any single air toxic and less than 25 tons per year of any combination of air toxics.

atomization: The process of breaking up fuel into small particles to allow intimate contact with the air to improve combustion.

atomize: To break up liquid into a very fine mist.

axial clearance: The distance between the nozzle exits and leading edge of the rotor blades in an impulse steam turbine.

B

BACnet: *See* building automation and control network.

balanced draft system: A type of draft that occurs when draft pressure within a furnace is kept constant.

barometric condenser: A condenser that has steam and water mixing together, similar to an open feedwater heater.

base: *See* alkali.

bin storage system: A pulverized coal system in which coal is prepared in a separate space and stored in a coal bunker.

bleed-off: A drain line connected to a cooling water system in which water is continually removed.

bleed steam: Steam withdrawn from a stage of a steam turbine without pressure control.

blowback: *See* blowdown.

blowdown: A drop in pressure that allows a safety valve to close. Also known as blowback.

blowdown centrifugal separator: A separator that reduces boiler pressure and temperature before boiler water enters a blowdown tank.

boiler: A closed vessel containing water that is pressurized and turned into steam when heat is added.

boiler efficiency: *See* thermal efficiency.

boiler heating surface: The part of a boiler that has fire and gases of combustion on one side and water on the other side.

Boolean programming: A method of programming that uses Boolean logic operators, such as AND, OR, and NOT, to control circuit function.

Boyle's law: A gas law stating that there is a relationship between the volume, temperature, and pressure of gases and that with a constant temperature process for each change of pressure, there is an inverse (opposite) change of volume.

building automation and control network (BACnet): An open protocol standard developed and maintained by the American Society of Heating, Refrigerating and Air-Conditioning Engineers (ASHRAE) to standardize communication methods between the building automation systems of different manufacturers.

C

calorimeter: A laboratory instrument used to measure the amount of heat developed by combustion.

capacitor-run motor: A single-phase motor that has a start winding and capacitor in series at all times.

capacitor start-and-run motor: A single-phase motor that has separate start-and-run capacitors for high starting torque.

capacitor-start motor: A single-phase motor that has a capacitor placed in series with the starting windings.

carryover: Particles of water that leave a boiler with steam and enter the main steam line.

catalyzed hydrazine: An oxygen scavenger that is primarily used on high-pressure boilers to reduce or prevent return line corrosion.

caustic embrittlement: The collection of material with high alkalinity, which leads to breakdown and weakening of boiler metal.

centralized control system: A process scheme that uses a centralized computer for all information gathering, decision making, and sending of equipment control commands.

central processing unit (CPU): A computer component that interprets and controls all computer program instructions.

centrifugal feedwater pump: A feedwater pump that uses centrifugal force to develop a rise in pressure for moving a liquid.

centrifugal force: The force that tends to move an object away from a center.

centrifugal separator: *See* blowdown centrifugal separator.

Charles' first law: A gas law that states, for a constant pressure process, gas volume increases as temperature increases.

Charles' second law: A gas law that states, for a constant volume process, pressure increases as temperature increases.

chemistry: The study of the composition and chemical properties of substances.

circuit: A conductor or series of conductors through which electrons flow.

circuit breaker: An electromagnetic or thermal device that trips open to interrupt current when the current in a circuit exceeds a set point.

closed feedwater heater: A feedwater heater that is closed and pressurized to heat feedwater to a much higher temperature than an open feedwater heater.

cogeneration: The usage of a plant for the generation of electricity and process work at the same time.

colorimetric analysis: A method of determining the concentration of an element in a solution with the aid of a color reagent.

combination forced and induced draft: A mechanical draft system that uses both forced and induced draft fans on large boiler installations.

combustion: The rapid union of an element or a compound with oxygen that results in the production of heat.

combustion accessory: Boiler equipment required to store and deliver fuels to a burner safely and efficiently.

combustion modification: The alteration of a combustion process in an attempt to lower stack gas emissions.

common-bus network: A computer communication network that consists of a main local area network (LAN), or bus, to which individual controllers and workstations are connected.

communications controller: A microprocessor that controls the communication into and out of a computer device.

complete combustion: Combustion in which all fuel is burned using the minimal amount of air above the theoretical amount of air needed to burn the fuel.

compression stress: The result of force applied to an object that presses or squeezes the object.

computer-integrated control system: A system that uses a computer to control equipment in a building or production plant.

condensate return system: A collection of valves, a piping control system, and auxiliary equipment that reclaims uncontaminated condensate.

condenser: A heat exchanger that turns exhaust steam into condensate.

condensing boiler: A hot-water heating boiler designed to recover heat that is normally discharged up a stack.

condensing steam turbine: A steam turbine that allows condensate to be reclaimed for use in the system.

conduction: A method of heat transfer in which molecules come in direct contact with each other and energy is passed from one molecule to another.

conductor: Material that permits free movement of electrons, such as copper or aluminum.

connecting rod: A rod that connects the crosshead pin at the piston rod to the crank pin at the flywheel in a steam engine.

constant voltage transformer: A transformer in which voltage supplied to the controller varies more than the normally acceptable 10% and that compensates for voltage variations at the primary side to maintain a consistent voltage on the secondary side.

contactor: A control device that uses a small control current to energize or de-energize a load connected to it.

continuous blowdown: A procedure used in a heat recovery system to control boiler water total dissolved solids with minimal heat loss.

convection: A method of heat transfer in which heat is transferred by currents through a fluid.

convection superheater: A superheater that receives its heat from the flow of gases of combustion in a boiler.

cooling pond: A body of water used to provide cooling.

cooling tower: A large auxiliary device where air moves upward to mix with falling water, resulting in cooling of the water.

corresponding pressure: The pressure of steam at saturation temperature.

CPU: *See* central processing unit.

crank pin: A pin connected to one end of a connecting rod in a steam engine.

critical pressure: The point at which water and saturated steam are indistinguishable; the critical pressure is 3206 psi and its corresponding temperature is 705°F.

crosshead guide bar: A guide bar attached to the frame of a reciprocating steam engine to keep the crosshead moving in a straight line.

crosshead pin: A pin that connects to one end of a connecting rod in a steam engine.

current: The flow of electrons through a circuit that is measured in amps.

cutoff governor: A device used to control the volume of steam on a reciprocating steam engine when full steam pressure is admitted to the steam chest.

cycle: One complete positive and negative alternation of a wave form.

cycle of concentration: The concentration of dissolved solids in boiler water relative to the makeup water.

cyclone separator: A separator with a number of cylinders (cyclones) set side-by-side along the length of a boiler drum and a baffle arrangement used to direct steam into a cyclone that separates water droplets from the steam using centrifugal force and by changing direction.

cylinder: Casting closed at the ends by cylinder heads that are removable.

D

DDC system: *See* direct digital control system.

dead center: The point at which a piston reaches the end of its stroke and is ready to move back in the opposite direction.

deaerator: A feedwater heater operating under pressure that is used to separate oxygen and other gases from steam before releasing the gases to the atmosphere through a vent.

diaphragm gland clearance: The distance between the shaft and labyrinth packing on the diaphragm in an impulse steam engine.

digital-in (DI) point: A remotely located sensor that represents a distinct predetermined value.

digital-out (DO) point: A remotely located device that receives a distinct predetermined value from a computer and reacts to the signal to control a staged procedure.

digital signal: Data that has a distinct predetermined status as one of two states: ON or OFF.

direct-acting interface module: A special I/O interface module that is capable of processing low-level and fast-input signals.

direct current: Electron flow in one direction only in a circuit.

direct digital control (DDC) system: A solid-state system in which a building automation system controller is wired directly to controlled devices to turn them ON or OFF.

direct-fired unit system: A pulverized coal system that stores coal in an overhead coal bunker.

discharge head: The vertical distance between the centerline of a pump and the level of the water above the pump.

distributed control system: A control system that uses controllers located at the equipment throughout a plant for information gathering, the decision-making process, and the processing of commands.

draft: The difference in pressure between two points that causes air or gases to flow.

drain cock: The stationary part of a reciprocating steam engine that is located at each end of the cylinder and removes condensate from the cylinder during warm-up.

dry pipe separator: A separator used to control carryover, consisting of a closed pipe perforated at the top with drain holes on the bottom that remove moisture from steam.

E

economizer: A boiler accessory used to reclaim heat from the gases of combustion to preheat feedwater.

electric boiler: A very clean special-purpose boiler that does not require fuel storage or ash removal, but has a high operating cost.

EMC system: *See* energy management control system.

energy: The ability to do work.

energy management control (EMC) system: A solid-state control system that usually supplements other stand-alone control systems such as pneumatic controls.

enthalpy: The total heat of steam.

equivalent evaporation: The amount of steam that a boiler would produce if converted back to standard conditions at 212°F.

excess air: Air supplied to the burner that exceeds theoretical amount of air needed to burn the fuel.

exhaust lap: The amount the exhaust edge of a slide valve overlaps the steam port on the exhaust side when the slide valve is in midposition on a reciprocating steam engine.

exhaust port: The opening located in the steam chest of a reciprocating steam engine.

external treatment: The treatment of boiler water before the water enters a boiler to remove scale-forming salts and gases.

extraction steam: Steam withdrawn from a steam turbine that is under pressure control.

F

factor of evaporation: A correction factor used to compensate for the difference between evaporation at 212°F and evaporation at the corresponding temperature in a boiler.

feedwater heater: An auxiliary component on a boiler that is used to increase plant efficiency by preheating the feedwater with exhaust steam or heat from the gases of combustion leaving the boiler.

feedwater injector: An auxiliary or secondary means of feeding water to a boiler.

feedwater pump: A pump used to deliver the correct amount of water to a boiler at the proper temperature and pressure.

feedwater regulator: A control device used to maintain a normal operating water level (NOWL) in a boiler, which reduces the danger of high or low water.

feedwater system: A basic boiler room system that provides feedwater at the correct temperature and pressure to a boiler.

feedwater valve: A valve located on the feedwater line to control feedwater flow to a boiler.

field panel controller: A controller that contains all hardware, firmware (programmed software), and applications software to provide continuous control of connected components.

fire point: The temperature at which fuel oil gives off vapor that will burn continuously.

flash point: The temperature at which fuel oil gives off vapor that will flash when exposed to an open flame.

flash tank: The tank in a heat recovery system that is used in conjunction with continuous blowdown to recover flash steam from water being removed from the steam and water drum.

flex-tube watertube boiler: A bent-tube boiler that has shaped tubes surrounded by gases of combustion.

float feedwater regulator: A feedwater regulator consisting of a float chamber, float, and mercury switch or microswitch.

float thermostatic steam trap: A steam trap that has a ball float and a thermostatic element.

fluid: Any substance that flows, such as a liquid or a gas, and takes the shape of the container it is in, regardless of the container shape.

flywheel: A device that stores the kinetic energy of rotation in a steam engine and causes the piston to continue to rotate the main shaft through top (head) dead center or bottom (crank) dead center.

foaming: A formation of layers of steam bubbles trapped below the water surface in a steam and water drum.

force: A pushing or pulling motion, commonly measured in pounds.

forced circulation boiler: A special-purpose boiler that uses pumps to increase water circulation and was developed because natural circulation was not sufficient in some boilers.

forced draft: Mechanical draft produced when the fan or blower located in front of a boiler setting forces (pushes) air into a furnace.

foundation: The stationary part of a reciprocating steam engine on which other parts are mounted.

frame: The stationary part of a reciprocating steam engine that is commonly constructed of cast iron and anchored to the foundation to prevent movement during operation.

frequency: The number of cycles per second of alternating current that is expressed in hertz (Hz).

frequency meter: A meter used to indicate the electrical frequency in a system.

friction head: Head equal to a loss caused by the friction of water on piping, valves, and fittings.

fuel oil correction: An allowance for the expansion of fuel oil when heated, which affects the volume of fuel oil.

fuel oil pressure gauge: A pressure gauge that indicates when a strainer is dirty.

fuel oil pump: A pump that delivers fuel oil from storage tanks aboveground or underground to burners at a controlled pressure.

fuel oil strainer: A fuel oil accessory that removes foreign matter in a fuel oil system before the fuel oil reaches the burner assembly.

function block programming: A method of programming that uses a library of preprogrammed routines (blocks) that can be linked together in order to meet job-specific requirements.

fuse: An electrical device that contains a fusible metal strip inserted in series with a circuit in which the metal strip melts and interrupts current flow when the current exceeds a certain level.

G

gauge pressure: Pressure recorded on a steam pressure gauge.

general gas law: A gas law stating that the product of the pressure and volume divided by the temperature before change is equal to the product of the pressure and volume divided by the temperature after change.

grade: The size, heating value, and ash content of coal.

grafcet programming: A method of programming that uses a symbolic graphic language to represent the control program as steps or stages in a process.

H

heat: Kinetic energy caused by molecules in motion within a substance.

heat balance: Accounting for all the heat units in the fuel used or wasted in a boiler.

heater: A thermal overload device primarily used in starters that is connected in series in the motor circuit and opens the heating coil circuit in the relay when a motor is overloaded.

heating surface: *See* boiler heating surface.

heating value: The amount of Btu per pound or gallon of fuel.

high/low/OFF control: A combustion control strategy that senses steam pressure and sends a control signal to start and stop the burner to maintain the steam pressure.

high-temperature, high-pressure hot water boiler: A special-purpose boiler used where high-temperature, high-pressure water is circulated to various buildings.

horizontal return tubular (HRT) boiler: A firetube boiler that consists of a drum fitted with tubes suspended over a furnace.

HRT boiler: *See* horizontal return tubular boiler.

hydrometer: An instrument used to measure specific gravity.

hydrostatic test: A water pressure test made on a boiler after it has had a low-water condition or repairs on the steam or water side.

I

impeller: The rotating element in a centrifugal pump through which water passes and is the means by which energy is imparted to the water.

impulse steam trap: A steam trap that consists of a valve with an extended, hollow stem that has a control disc attached at the top.

inclined plane: A simple machine formed by two surfaces at an acute angle to each other.

incomplete combustion: Combustion that occurs when all the fuel is not burned, resulting in the formation of soot and smoke.

indicator card: A reading taken from a reciprocating steam engine that shows what occurs inside the cylinder from the time the piston is on top (head) dead center until it returns to top (head) dead center.

indicator cock: *See* pressure cock.

induced draft: Mechanical draft produced when the fan or blower located between a boiler and chimney removes (pulls) gases of combustion from the boiler and discharges them into the chimney.

industrial waste boiler: A special-purpose boiler designed to dispose of and utilize industrial waste and/or by-products for the production of steam.

input module: A device that receives a signal from an input field device classified as a sensor and transmits that signal to a computer.

input/output (I/O): The way in which a controller receives and transmits data to and from field devices.

integral economizer: A feedwater economizer that is part of a boiler and replaces some of the boiler heating surface.

intelligent interface module: A special I/O interface module that incorporates onboard microprocessors to add intelligence to the interface.

internal treatment: The treatment of water after it has entered a steam and water drum.

inverted bucket steam trap: A steam trap in which steam enters the bottom of the trap and flows into an inverted bucket.

I/O: *See* input/output.

isolation transformer: A transformer that eliminates electromagnetic interference with other equipment.

J

jet condenser: A condenser that mixes steam and water together, similar to an open feedwater heater.

K

kinetic energy: Energy due to a body in motion.

L

labyrinth packing: A seal used to prevent the leakage of steam from a steam turbine.

ladder programming: A method of programming that uses ladder diagrams and electrical contact symbols to detail the electrical relationship between controls and loads.

latent heat: Heat added to a substance that changes its state without a change of temperature.

latent heat of evaporation: The heat necessary to change the state of a substance from a liquid to a gas at a constant temperature.

latent heat of fusion: The heat necessary to change the state of a substance from a solid to a liquid at a constant temperature.

law of moments: A law that states for a lever to be in equilibrium, the sum of clockwise movements must equal the sum of counterclockwise movements.

lever: A bar that pivots on a fixed body (fulcrum) with both resistance and effort (force) applied.

linear clearance: A measurement of the distance between the top of a piston and cylinder head when the piston is at top (head) or bottom (crank) dead center in a reciprocating steam engine.

line programming: A method of programming that uses a sequence of computer commands in lines to initiate an operation.

lockout: The process of removing the source of electrical power and installing a lock that prevents the power from being turned on.

low water fuel cutoff: A safety device located slightly below the normal operating water level (NOWL) that shuts off fuel going to a burner if a low-water condition occurs.

M

magnetism: Invisible lines of force (magnetic field) that exist between the north and south poles of any permanent magnet or electromagnet.

main steam stop valve: A valve on the main steam line leaving a boiler that is used to cut the boiler in on-line or take the boiler off-line.

major source facility: A facility that emits 10 tons or more per year of any single air toxic or 25 tons or more per year of any combination of air toxics.

manufactured gas: A gaseous fuel consisting of methane, ethylene, and small amounts of hydrogen, nitrogen, carbon monoxide, carbon dioxide, oxygen, and heavy hydrocarbons.

MAWP: *See* maximum allowable working pressure.

maximum allowable working pressure (MAWP): Maximum boiler pressure determined by the design and construction of a boiler in accordance with the ASME code.

mechanical draft cooling tower: A cooling tower in which air is forced or induced through it.

megohmmeter: A meter used to measure resistance in a circuit.

membrane boiler: A boiler that has water directed through formed metal membranes connected to chambers that serve the same purpose as upper and lower drums.

memory: Computer storage capacity.

mini programmer: A small handheld device for programming computer-integrated controls.

mixed-bed demineralizer: A demineralizer that purifies makeup water as it passes through a single vessel equipped with a cation and an anion solution.

modulating pressure control: An automatic device that provides local control of the firing rate proportional to the steam demand.

moment: Product of an effort times its distance from the fulcrum.

motor starter: A contractor with overload protection included.

N

National Fire Protection Association (NFPA): A national organization that provides guidance in assessing the hazards of the products of combustion.

National Electrical Code® (NEC®): A standard of practices for the installation of electrical products published by the National Fire Protection Association (NFPA).

natural circulation boiler: A boiler that uses the movement of heated and cooled water (natural circulation) for circulation.

natural draft cooling tower: A cooling tower that cools water with air provided by natural draft.

natural gas: A gaseous fuel that consists of methane, ethylene, and small amounts of hydrogen, nitrogen, carbon monoxide, carbon dioxide, oxygen, and heavy hydrocarbons.

NEC®: *See* National Electric Code®.

New Source Performance Standards (NSFS): Environmental standards for ambient pollutant emissions from new sources of emissions.

NFPA: *See* National Fire Protection Association.

noncondensing steam turbine: A steam turbine that exhausts at atmospheric pressure or above.

nonvolatile memory: Computer memory that is not lost if power is interrupted.

number of expansions: The ratio of exhaust steam volume to admission steam volume.

O

Occupational Health and Safety Administration (OSHA): A federal agency that requires all employers to provide a safe environment for their employees.

ohm: The unit of measure used to express resistance to electron flow in a conductor.

ohmmeter: A meter used to measure the amount of resistance in a circuit or part of a circuit.

Ohm's law: A law stating that current in a DC circuit is proportional to the voltage and inversely proportional to the resistance.

ON/OFF control: A combustion control strategy that is used to start and stop a burner without any modulation of the flame.

open feedwater heater: A feedwater heater that is open to the atmosphere and in which steam is added to water to raise the water temperature.

OSHA: *See* Occupational Health and Safety Administration.

output module: A device that receives a signal from a computer and transmits that signal to field devices.

P

packing gland: The movable part of a stuffing box in a reciprocating steam engine that compresses the packing to prevent steam from leaking past the piston rod and valve stem.

parallel circuit: A circuit that has two or more paths for current to flow.

perfect combustion: Combustion achieved when all fuel is burned using only the theoretical amount of air.

personal protective equipment (PPE): Clothing and/or equipment worn by a technician to reduce the possibility of injury in a work area.

PID: *See* proportional-integral-derivative action.

piston rod: A device that transfers the reciprocating motion of a piston to the crosshead in a steam engine.

piston valve: A cylindrical inside-admission outside-exhaust valve with supply steam admitted in the middle and exhausted outside the valve on a reciprocating steam engine.

PLC: *See* programmable logic control system.

point: Analog input, analog output, digital (binary) input, or digital output that has characteristics determined by a particular manufacturer.

polarity: The positive (+) or negative (–) electrical state of an object.

pollutant: Matter that contaminates air, soil, or water.

poppet valve: A valve on a high-speed reciprocating steam engine that is fitted with a spring on top to give a positive closing action and has its own eccentric, which opens the valve through a cam.

popping pressure tolerance: The difference between the set pressure and the popping pressure of a safety valve, given in percent or psi.

potential energy: Energy associated with position, not motion.

pour point: The lowest temperature at which fuel oil flows as a liquid.

power: The rate at which work is performed.

power supply: A computer component that provides DC voltage for the operation of processors, memory devices, and input/output modules.

PPE: *See* personal protective equipment.

pressure: Force acting on a unit of area that is equal to force divided by area.

pressure atomizing burner: A fuel oil burner that atomizes fuel oil without using steam or air.

pressure cock: A device located on the sides of a cylinder at each end that is used to take indicator card readings by a reciprocating steam indicator.

pressure gauge: A gauge used to indicate suction and discharge pressures of fuel and water lines, air pressure, gas pressure, and pressures on steam engines and steam turbines.

primary air: Air supplied to a burner that controls the rate of combustion, which determines the amount of fuel that can be burned.

primary pollutant: A pollutant emitted directly from identifiable sources.

processor: A computer device that contains integrated circuits for performing mathematical operations, data interpretation, and diagnostic routines.

program: A sequence of coded instructions designed to initiate a desired function.

programmable logic control (PLC) system: A solid-state control system that is programmed to automatically control an industrial process or machine.

programming device: A human interaction device that is used for controlling the actions of a computer.

programming language: A standardized communication method used to develop a control program for a controller in a digital control system.

proportional-integral-derivative (PID) action: A control scheme in which control action occurs by comparing a sensed condition to a setpoint for determining the output.

proportional-integral-derivative (PID) processor: A processor that performs independent control tasks such as closed-loop control utilizing PID actions.

proximate analysis: A mechanical process used to determine the moisture, volatile matter, ash, fixed carbon, and sulfur content in a coal specimen.

public utility boiler: A special-purpose boiler designed for high steam output and built on a large scale.

pulverized coal: Coal that has been ground into a fine dust by a pulverizer to the consistency of talcum powder and is highly explosive.

pulverizer: A machine used to grind and pulverize coal so that it is suitable for combustion.

Q

qualified person: A person who is trained in, and has specific knowledge of, the construction and operation of electrical equipment or a specific task and is trained to recognize and avoid electrical hazards that might be present with respect to the equipment or specific task.

R

radial clearance: The distance between the top of the revolving blades and casing in an impulse steam turbine.

radiant superheater: A superheater that receives its heat from the radiant heat zone in a boiler.

radiation: A method of heat transfer in which heat is transferred without a material carrier.

rank: The degree of hardness of coal.

receiver separator: A large piece of equipment that stores steam in large volumes.

reciprocating feedwater pump: A positive displacement pump where a piston stroke moves a specific amount of water.

rectification: The process of converting alternating current (AC) to direct current (DC).

relay: An electromechanical switching device that controls one electrical circuit by opening and closing contacts in another circuit.

resistance: The opposition of a material to the flow of electrons in a circuit, measured in ohms.

rim speed: The distance covered by the surface of a rotating element in a given unit of time.

ring network: A computer communication network that consists of workstations and controllers linked in series to each other.

rotary cup burner: A fuel oil burner that consists of an atomizing or a spinning cup, a blower to supply air for combustion (primary air), an air nozzle to mix air and fuel oil, a means of driving the spinning cup and blower, and a means of delivering fuel oil to the spinning cup.

S

safety valve: A boiler valve that releases pressure to protect the boiler from exceeding its maximum allowable working pressure (MAWP).

saturated steam: Vapor at a temperature that corresponds with its pressure.

scale: Buildup on boiler surfaces caused when calcium carbonate or magnesium carbonate is deposited on hot metal surfaces, forming a hard, brittle substance.

scotch marine boiler: A firetube boiler equipped with an internal furnace that is completely surrounded by water, which increases the boiler heating surface and in turn increases boiler efficiency.

secondary air: Air supplied to a burner that controls combustion efficiency by controlling how completely fuel is burned.

secondary pollutant: A pollutant formed by the interaction between two or more primary pollutants.

sensible heat: Heat that changes the temperature of a substance, but not its state.

sensor: A device that senses or measures the condition of a variable, such as temperature, relative humidity, pressure, flow, level, electrical units, and the position of various mechanical devices.

series circuit: A circuit that has one path for current to flow through all devices in a circuit and that has electrical devices and/or controls in-line with each other.

series-parallel circuit: A combination of series and parallel circuits.

shaded-pole motor: A single-phase motor that operates with a two-phase magnetic field to develop the torque necessary for starting.

shaft bearing: A bearing that supports a reciprocating steam engine shaft.

shear stress: The result of force applied to an object that shears or cuts through an object.

shrink: A process in which the water level in a boiler momentarily drops with a decrease in steam demand.

simple D-type slide valve: An outside-admission inside-exhaust valve for a reciprocating steam engine that has steam admitted over the outside of the valve and steam exhausted through the inside of the valve.

simple machine: Tools used to convert energy to useful work in boiler and auxiliary equipment.

SIP: *See* state implementation plan.

soot blower: A device used to remove soot deposits from around tubes and to permit better heat transfer in a boiler.

special I/O interface module: An input/output interface that provides a link between PLCs and field devices by using special types of signals.

specific gravity of fuel oil: The ratio of the weight of any volume of fuel oil at 60°F to the weight of an equal volume of water at 60°F.

specific heat: The amount of heat required to raise the temperature of 1 lb of a substance 1°F.

split-phase motor: A single-phase motor that has starting windings with a higher resistance than the running windings.

spray pond: A cooling pond in which water is sprayed in order to increase the surface area of water in contact with the air by breaking up the water into a fine spray.

squirrel-cage motor: A three-phase induction motor with a rotor that has conductor bars embedded below the surface of the rotor.

star network: A computer communication network that consists of a multiport host computer with each port connected to the programming port of a controller.

state implementation plan (SIP): A plan that gives the states the responsibility for developing their own programs to reduce air pollution.

stay: A device used to reinforce flat surfaces subjected to boiler pressure.

steam: Water in a vapor state.

steam admission valve: A valve located in the steam chest of a reciprocating steam engine.

steam atomizing burner: A fuel oil burner that uses steam at a pressure approximately 20 psi higher than the fuel oil pressure to atomize fuel oil.

steam chest: The stationary part of a reciprocating steam engine that is commonly cast integral with the cylinder and insulated to reduce radiant heat losses.

steam lap: The amount of the steam edge of a slide valve that overlaps the steam port on the steam admission side when the valve is in midposition.

steam lead: The amount the slide valve has opened the steam ports on the steam admission side when a piston is on top (head) or bottom (crank) dead center.

steam port: An opening at the head end or crank end of the cylinder of a single cylinder reciprocating steam engine.

steam pressure gauge: A gauge used on a boiler to indicate the steam pressure in the boiler.

steam strainer separator: A separator that removes impurities in addition to condensate before steam enters a steam turbine.

steam supply line: An insulated line that provides steam to a reciprocating steam engine.

steam trap: An automatic device that increases the overall efficiency of a plant by removing air and water from steam lines without loss of steam.

stoker: A mechanical feeding device for feeding coal to a burner.

stress: The result of forces acting on an object.

stroke: The distance traveled by a piston from top dead center to bottom dead center on vertical reciprocating steam engines, or from head end center to crank end center on horizontal reciprocating steam engines.

stuffing box: A part on a reciprocating steam engine fitted with packing to prevent steam leaking past the piston rod and valve stem.

suction head: The vertical distance from the centerline of a pump inlet to the surface of the water above the pump.

suction lift: The vertical distance from the level of the water below a pump to the centerline of the pump inlet.

superheated steam: Steam at a temperature higher than its corresponding pressure.

superheater: A nest of tubes used to raise the temperature of steam but not its pressure.

surface condenser: A condenser in which steam and water are not in intimate contact with each other.

surface tension: A condition caused by impurities floating on top of water in a steam and water drum.

swell: A process in which the water level in a boiler momentarily rises with an increase in steam demand.

synchronous motor: A three-phase induction motor with solid iron conductor bars and coils of conductor wound in the rotor slots.

synchroscope: A meter used to show the phase relationship between two voltages.

system architecture: Computer equipment, the strategies used to control the equipment and operations in a plant, and the connections between them that form a complete control system.

T

tagout: The process of placing a danger tag on the source of electrical power, which indicates that the equipment may not be operated until the danger tag is removed.

TDS: *See* total dissolved solids.

temperature: The measurement of the degree or intensity of heat.

tensile stress: The result of force applied to an object that stretches the object.

terminal equipment controller: A small stand-alone controller that has a limited number of input and output point connections.

theoretical amount of air: The amount of air used to achieve perfect combustion in a laboratory setting.

therm: The energy equivalent of 100,000 Btu.

thermal efficiency: 1. In boilers, the ratio of heat output to heat input. Also known as boiler efficiency. **2.** In steam turbines, the ratio of heat energy used in the turbine to heat energy available in the steam.

thermistor: An electrical device that changes electrical resistance with a change in temperature.

thermodynamic steam trap: A steam trap that has a single movable disc that raises to allow the discharge of air and cool condensate.

thermoexpansion feedwater regulator: A feedwater regulator with a thermostat that expands and contracts when a steam space increases or decreases with changes in water level.

thermohydraulic feedwater regulator: A feedwater regulator consisting of a regulating valve, bellows, a generator, and stop valves.

thermostatic steam trap: A steam trap that has a bellows filled with a fluid that boils at steam temperature.

three-element feedwater regulator: A feedwater regulator that controls the amount of feedwater fed to a boiler by monitoring the steam flow, feedwater flow, and water level.

throttling governor: A device used to control reciprocating steam engine speed by controlling (throttling) steam flow before it enters the valve chest, with steam admitted at less than full steam pressure.

torque: The force on a body that causes it to turn.

total dissolved solids (TDS): The amount of sludge that is in suspension in boiler water.

turbo-generator: A steam turbine connected to a generator.

two-element feedwater regulator: A feedwater regulator that controls the water level in a boiler with a superheater by monitoring the steam flow through the superheater and by monitoring the boiler water level.

two-vessel demineralizer: A demineralizer that purifies makeup water as it passes through one vessel with a cation solution and one vessel with an anion solution.

U

ultimate analysis: A chemical process used to determine the nitrogen, oxygen, carbon, ash, sulfur, and hydrogen content in a coal specimen.

unit economizer: A feedwater economizer that handles the gases of combustion and feedwater for one boiler.

V

vacuum condensate tank: A tank equipped with a vacuum pump that creates a vacuum on a return line.

vacuum gauge: A gauge that measures pressure below atmospheric pressure in inches of mercury.

velocity compounding: The change in steam velocity as it passes through a multistage turbine.

vertical lift: *See* suction lift.

viscosity: Internal resistance of fuel oil to flow.

volatile memory: Computer memory that is lost if power is interrupted.

volt: A unit of measure used to express the electrical pressure differential between two points in a conductor.

voltmeter: A meter used to measure the difference of voltage between two points in a circuit.

volumetric clearance: The measurement in volume of space between the top of the piston and the cylinder head, including volume of the steam port when the piston is at top (head) dead center or bottom (crank) dead center of a reciprocating steam engine.

W

waste heat boiler: A boiler that eliminates industrial waste and uses heat to produce steam for plant process or plant heating.

water column: A fitting used to slow down the turbulence of boiler water so that an accurate reading of the water level in a gauge glass can be obtained.

water rate: The amount of steam or water required per unit of power output.

watertube boiler: A boiler in which water passes through tubes that are surrounded by gases of combustion.

watt: A basic unit of electrical power.

wheel and axle: A simple machine formed by an outer circular rim (wheel) fastened to an inner rim (axle).

work: The movement of an object by a constant force to a specific distance.

wound-rotor motor: A three-phase induction motor that has coils of conductor wound in the rotor slots, replacing the bars of a squirrel-cage motor.

INDEX

*Page numbers RSE-1 to RSE-30 refer to the chapter in the learner resources.
*Page numbers in italic refer to figures.